W9-CNC-804

Britannica Yearbook of Science and the Future

Encyclopædia Britannica, Inc.
William Benton, Publisher

Chicago Toronto London
Geneva Sydney Tokyo Manila

1972 Britannica Yearbook of Science and the Future

Library of Congress Catalogue Number: 69-12349 Standard Book Number: 0-85229-165-5

THE UNIVERSITY OF CHICAGO
The Britannica Yearbook of Science and the Future is published with the editorial advice of the faculties of the University of Chicago

Editorial Advisory Board

Wm Benton
The Power and Perils of Science

Francis Bacon, living at the beginning of the 17th century, was one of the first great enthusiasts of science in our modern sense of the term. Before his time, Bacon thought, science had been concerned only with studying the ways of nature for its own sake. But in his view, science should be studied for its potential benefits to man. He therefore explained and extolled scientific method. He pointed the way to man's control of nature. He tried to convince his fellowmen that through understanding nature they could learn to live "like kings."

In the three and a half centuries since Bacon's death man has been ingenious in solving the riddles of nature. He has learned to exploit external nature as never before—and thereby enormously improved his life. Almost all of us live better now than kings did in Bacon's time. And it is the attitude toward science that Bacon helped develop that has given us this new wealth and power.

Recently, however, we have come to the bitter realization that *control* of nature is not enough. Mere power over external nature is not necessarily to man's benefit. Indeed, just the opposite may be the case. Science, now, casts such ambiguous portents over the human adventure that the future may hold more hazard than hope.

This insight is not wholly new. "It is the business of the future to be dangerous," said Alfred North Whitehead. But he could not imagine the enormity of the perils that we are just beginning dimly to see. Science has delivered to a few national leaders a power that hitherto was credited only to Bronze Age gods, one mistake in the use of which—through greed, pride, or ambition—could indeed wipe civilization from the face of the earth. But beyond that science has given even the least of us enormous and little understood powers for both good and ill. A simple act like spraying the roses with an insecticide can have consequences we cannot foretell.

Man's response has not been to put an end to science, but to give it a new urgency. Science now arouses not only the imagination of man but also his conscience. People today are sensing the fact that they are dealing with essential values that they do not want corrupted, either by science or within it. Many of the articles in our *1972 Britannica Yearbook of Science and the Future* reflect this new view. Margaret Mead, for example—whose fame as a social reformer has perhaps eclipsed her renown as an anthropologist—says that she learned early in her career that it was not only research but also "a climate of opinion ready to let it be used" that was important. Thus, she says, she purposely has divided her time between science and the climate in which it might flourish beneficently.

The impact on science of the social imperative can be seen in the newspaper headlines—for example, the vigorous current effort toward arms control, and the decision of the U.S. Congress in 1971 not to finance an SST. We see it from a different, and deeper, perspective in this Yearbook—in the kind of laboratories that man sets up (the article "A Freedom to Excel") and in the places he hopes to establish them ("Colonizing the Moon"). We see it in the way that man relates to himself ("Instant Intimacy") and to the integrity of his art ("How Science Detects Art Fakes"). We see it in such an article as "The UN: Its Science Mission," which describes how man has shaped his supranational institutions the better to use the gifts of science. The article is written by Dr. Walter Kotschnig, my long-time friend and my associate in the State Department when I served as Assistant Secretary of State.

We see it also in the attempts to define the implications of man's technological triumphs. One of the many disturbing facts discussed in this Yearbook—you'll find it in the article titled "Buried in Affluence"—is that the average family of four in America discards many tons of trash and garbage every year. Much of that trash turns out to be indestructible—plastic containers, for example—because technology made it so. But as you will discover in the article "Materials from the Test Tube," technology also is being directed to find ways to induce decay in the seemingly indestructible—for example, by attaching ultraviolet-sensitive chemical groups to the polymer molecules that bind the plastic, so that the containers will be perishable when exposed to sunlight. Thus the connective tissue of technology—the relations of the various scientific disciplines to one another as well as their relation to social needs—can be traced in this volume, from the articles on "Materials" and "The Energy Crisis" through those on "Washing Our Dirty Water" and "Buried in Affluence." Man is adapting to the linked opportunities of science. He is facing up to the disillusioning discovery that science is neutral, not necessarily beneficent—that it will kill as readily as it will heal.

This generation has learned from the past: "Experience," as Arnold Toynbee has said, "is another word for history." But there is still much to learn. The basic problem, perhaps, is to realize that control over nature is essentially perilous, just as control over men is essentially immoral. Nature is our partner on Spaceship Earth, and we must treat her like a friend. If we treat her like a slave, if we try to dominate her to our will, she will rise against us in revolt. Thus man, who has used science to bring about understanding of the natural world, must in the future use it increasingly to bring about order in his own.

Contents

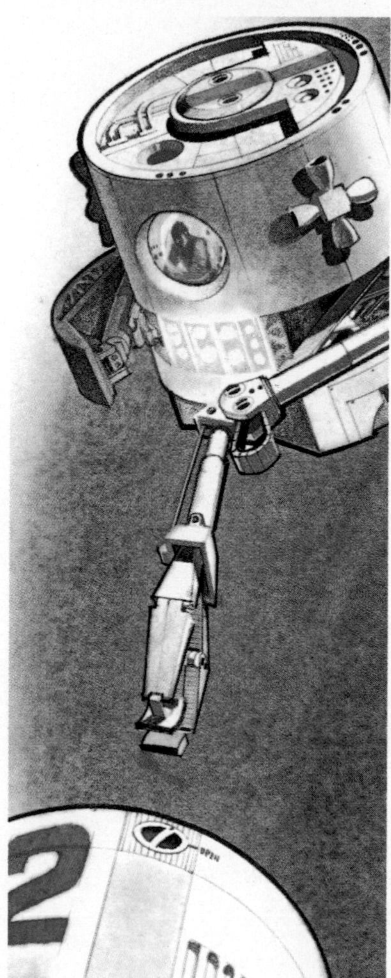

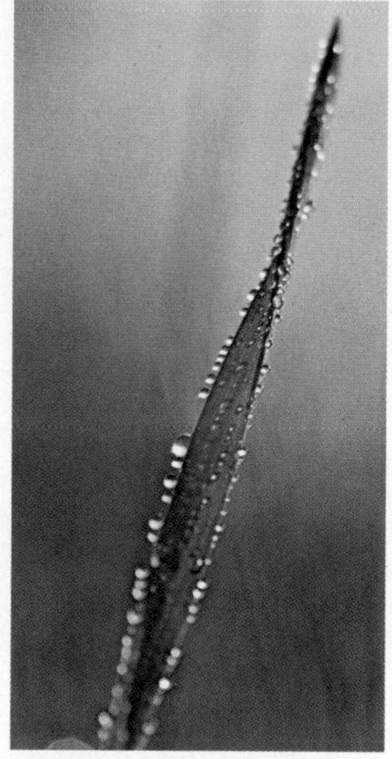

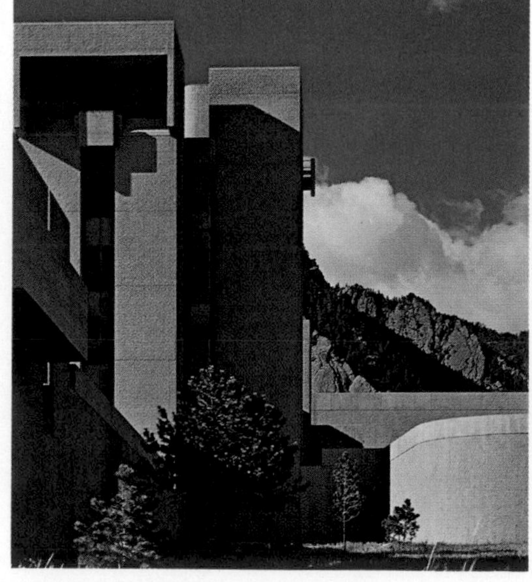

Contributors to the Science Year in Review

William J. Bailey *Chemistry: Chemical Structure.* Research Professor of Chemistry, University of Maryland, College Park.
Hyman Bass *Mathematics.* Professor of Mathematics, Columbia University, New York, N.Y.
F. S. Beckman *Computers.* Manager, Special Studies Computer Sciences Department, Thomas J. Watson Research Center, IBM Corporation, Yorktown Heights, N.Y.
Ajay K. Bose *Chemistry: Chemical Synthesis.* Professor of Chemistry, Stevens Institute of Technology, Hoboken, N.J.
Michael J. Brennan *Medicine: Malignant Diseases.* President, The Michigan Cancer Foundation and Professor of Medicine, Wayne State University, Detroit, Mich.
George M. Briggs *Foods and Nutrition.* Professor of Nutrition, Department of Nutritional Sciences, University of California, Berkeley.
D. Allan Bromley *Physics: Nuclear Physics.* Professor and Chairman, Department of Physics, and Director, A. W. Wright Nuclear Structure Laboratory, Yale University, New Haven, Conn.
Walter Clark *Photography.* Formerly head of the Applied Photography Division, Research Laboratory, Eastman Kodak Co., Rochester, N.Y.
Morris F. Collen *Medicine: Biomedical Engineering.* Director, Medical Methods Research, Permanente Medical Group and Kaiser Foundation Research Institute, Oakland, Calif.
Carl W. Condit *Architecture and Building Engineering.* Professor of Art and Urban Affairs, Northwestern University, Evanston, Ill.
John M. Dennison *Earth Sciences: Geology and Geochemistry.* Professor and Chairman, Department of Geology, University of North Carolina, Chapel Hill.
F. C. Durant III *Astronautics and Space Exploration: Earth Satellites.* Assistant Director (Astronautics), National Air and Space Museum, Smithsonian Institution, Washington, D.C.
Robert G. Eagon *Microbiology.* Professor of Microbiology, University of Georgia, Athens.
Gerald Feinberg *Physics: High-Energy Physics.* Professor of Physics, Columbia University, New York, N.Y.
Avram Goldstein *Medicine: Drug Addiction.* Professor of Pharmacology, Stanford University, Stanford, Calif.
Jesse L. Greenstein *Astronomy.* Professor of Astrophysics and Executive Officer for Astronomy, California Institute of Technology, Pasadena.
Philip F. Gustafson *Molecular Biology: Biophysics.* Associate Director, Radiological Physics Division, Argonne National Laboratory, Argonne, Ill.
Robert M. Hamilton *Earth Sciences: Geophysics.* Geophysicist, National Center for Earthquake Research, U.S. Geological Survey, Menlo Park, Calif.
Leo A. Heindl *Earth Sciences: Hydrology.* Executive Secretary, U.S. National Committee for the International Hydrological Decade, National Academy of Sciences—National Research Council, Washington, D.C.
Robert L. Hill *Molecular Biology: Biochemistry.* Professor of Biochemistry, Duke University, Durham, N.C.
George W. Irving, Jr. *Agriculture.* Administrator, Agricultural Research Service, U.S. Department of Agriculture, Washington, D.C.
Daniel F. Johnson *Behavioral Sciences: Psychology.* Associate Professor of Psychology, Virginia Polytechnic Institute and State University, Blacksburg, Va.
Richard S. Johnston *Astronautics and Space Exploration: Manned Space Exploration.* Deputy Director, Medical Research and Operations,

NASA Manned Spacecraft Center, Houston, Tex.
Lou Joseph *Dentistry.* Assistant Director, Bureau of Public Information, American Dental Association, Chicago, Ill.

C. Henry Kempe *Medicine: Child Abuse.* Professor and Chairman, Department of Pediatrics, University of Colorado Medical Center, Denver.

Henry Lansford *Atmospheric Sciences.* Information Officer, National Center for Atmospheric Research, Boulder, Colo.

Joshua Lederberg *Molecular Biology: Genetics.* Professor of Genetics, Stanford University School of Medicine, Stanford, Calif.

John G. Lepp *Zoology.* President, Marion County Technical Institute, Marion, Ohio.

Howard J. Lewis *Science, General.* Director, Office of Information, National Academy of Sciences—National Academy of Engineering—National Research Council, Washington, D.C.

J(ohn) David Lockard *Education, Science.* Professor of Science Teaching, Associate Professor of Botany, and Director of the Science Teaching Center and The International Clearinghouse on Science and Mathematics Curricular Developments, University of Maryland, College Park.

William D. McElroy *The Science Year in Review: A Perspective.* Director, National Science Foundation, Washington, D.C.

Norman Metzger *Chemistry: Applied Chemistry.* Assistant Managing Editor, News Service, American Chemical Society, Washington, D.C.

James K. Page, Jr. *Environmental Sciences.* Co-editor, *Law and the Environment,* and Member of the Board of Editors, *Smithsonian* magazine, Washington, D.C.

Willard J. Pierson, Jr. *Marine Sciences.* Professor of Oceanography, School of Engineering and Science, New York University, New York, N.Y.

Froelich Rainey *Archaeology.* Professor of Anthropology and Director of The University Museum, University of Pennsylvania, Philadelphia.

Peter H. Raven *Botany.* Associate Professor of Biological Sciences, Stanford University, Stanford, Calif.

Byron T. Scott *Medicine: General Medicine.* Executive Editor, *Medical Opinion* magazine, New York, N.Y.

Mitchell R. Sharpe *Astronautics and Space Exploration: Space Probes.* Science writer, author of *Satellites and Probes, the Development of Unmanned Space Flight,* Huntsville, Ala.

Frank A. Smith *Transportation.* Vice President—Research, Transportation Association of America, Washington, D.C.

J. F. Smithcors *Veterinary Medicine.* Associate Editor, American Veterinary Publications, Santa Barbara, Calif.

William E. Spicer *Physics: Solid-State Physics.* Professor of Electrical Engineering and Materials Science, Stanford University, Stanford, Calif.

Charles Süsskind *Electronics.* Professor of Engineering Science, University of California, Berkeley.

Sol Tax *Behavioral Sciences: Anthropology.* Professor of Anthropology, University of Chicago.

Nicholas J. Turro *Chemistry: Chemical Dynamics.* Professor of Chemistry, Columbia University, New York, N. Y.

Irwin Welber *Communications.* Associate Executive Director, Toll Transmission, Bell Laboratories, North Andover, Mass.

James A. West *Fuel and Power.* Chief, Branch of Interfuels and Special Studies, Division of Fossil Fuels, U.S. Bureau of Mines, Washington, D.C.

Colonizing the Moon

by Mitchell R. Sharpe

Man has walked on the moon, explored it briefly, and returned to earth. The challenge ahead is to establish moon bases for further scientific research and to exploit lunar resources.

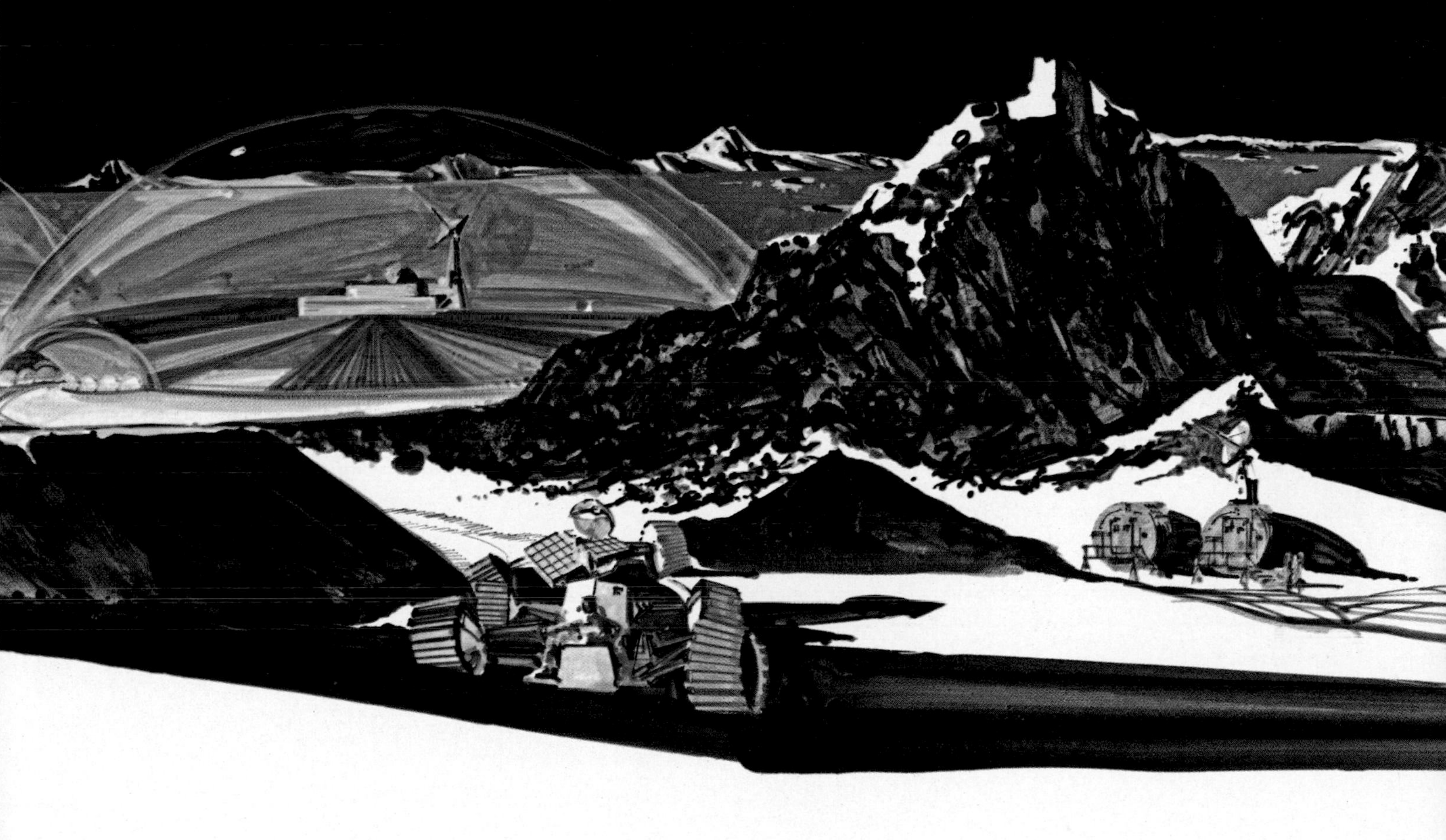

Late in 1972, two U.S. astronauts are scheduled to lift off from the moon, rendezvous with a spacecraft orbiting it, and then return to the earth. They will be the last of 12 Apollo crewmen who will have walked on the moon since the first men landed in the Sea of Tranquillity on July 20, 1969. When men will set foot on the moon again is uncertain. There may be a lapse of more than two decades before Americans return, although in the interim Soviet cosmonauts might make the journey.

Ultimately, however, it is likely that Americans and others will return to the moon and establish a permanent colony on it. The environmental hardships facing these pioneer lunar colonists will be far more inhospitable than those encountered by the Viking colonists of eastern North America in the 11th century and the later English settlers of Roanoke Island in the 16th and Plymouth in the 17th centuries. Yet their chances of survival conceivably might be greater because of the knowledge of astronautics that will have been achieved during the first few decades of manned spaceflight.

Why colonize the moon?

Cogent scientific and utilitarian reasons can be presented for colonizing the moon. The very hostility of its environment offers attractions for earth-bound scientists. Astronomers are intrigued with the possibility of establishing both optical and radio telescopes on the moon, while geologists and cosmologists see the moon as a laboratory for delving into some of the secrets of the age and origin of the solar system.

For astronomers, the moon has a number of advantages:

—There is no atmospheric absorption or interference across the entire electromagnetic spectrum (the continuous range of frequencies, from the lowest to the highest wavelengths, of all radiations).

—Because the moon rotates so slowly on its axes, the length of the lunar night affords long periods of continuous observation, ranging from 27 days in the polar regions to 14 days near the equator.

—The drift of the star field across the lunar sky is much slower than it is across the earth's sky—0.5° per hour as opposed to 15° per hour. This also is a result of the moon's slower rate of rotation.

—There is no stray or scattered man-made light.

—Seismic disturbances are relatively few, based upon data returned by instruments left on the moon by the Apollo 12 astronauts; also, aside from the landings of space vehicles, there are no man-made seismic disturbances.

—The 1/6 (of the earth's) gravity makes manned operations easier than the zero gravity of an orbital observatory; it also allows larger reflectors to be used with less gravitational distortion than on the earth.

Considering all the advantages, astronomers J. A. Hynek and W. T. Powers of Northwestern University point out: "In a rough way, then, we can see that a 25.4-centimeter [10-inch] . . . telescope on the moon, if used to full advantage, can in fact surpass the performance of the

MITCHELL R. SHARPE, a science writer, is a contributor to the Britannica Book of the Year, Compton's Encyclopaedia, *and the author of several books and articles about rockets and space exploration.*

Palomar telescope [200-inch] as it is usually used on earth, even in wavelengths suitable for earth-based astronomy. In the near-infrared [region of the spectrum], its relative capabilities are greater still. . . . A count of galaxies in the near-infrared, extending to objects only 1½ magnitudes fainter (a factor of 4), will double the size of the observable universe and quite probably will enable the determination of the answer to the crucial cosmological question of whether space has positive, negative, or no curvature."

In the field of radio astronomy the potential for new knowledge is equally as great. The far side of the moon is especially attractive as a site for radio astronomical instruments because of the shielding provided by the moon itself. This vast area also may offer the geography, being heavily pitted with craters, for building large radio telescopes similar to the 1,000-foot-diameter Arecibo Ionospheric Observatory in Puerto Rico. Instruments several times as large could be made on the moon. Natural craters could be lined with aluminum-coated mylar plastic, and the structural members for supporting the antenna equipment would be correspondingly lighter and easier to fabricate because of the

Initial lunar dwellings will be underground to protect inhabitants from rugged temperature changes, cosmic radiation, and meteors. One technique suggested is drilling a bore hole and expanding it with high explosives. The resulting cavity is then sealed with a plastic balloon, making it airtight (left). Floors are then put in (center), and the dwellings are furnished with the necessities for life and work (right).

reduced gravity. However, because of the relative shallowness of large lunar craters, they might require considerable excavation. Such radio telescopes permit the observation of galactic and extragalactic radio sources at very long wavelengths in the hectometer and kilometer ranges.

Probably the most disappointed scientists at the decision to cancel 3 of the originally planned 10 Apollo lunar landing missions were the geologists. The photographs, observations, and samples of soil and rock brought back by the astronauts of Apollos 11, 12, and 14, merely whetted their curiosity. The soil, for example, seems to be 3,600,000,000 –3,800,000,000 years old in the Sea of Tranquillity and 3,100,000,000 –3,400,000,000 years old in the Ocean of Storms. Rocks from both areas appear to have an age of about 4,500,000,000 years. Geological observations and samples from the remaining three Apollo missions will add more knowledge and will be enhanced by the observations made by a geologist-astronaut.

But the Apollo program was not designed to do more than literally scratch the surface of the moon. Extensive geological investigation will have to await the semipermanent and permanent lunar colonies. Only then can the more puzzling questions be answered. These include such matters as whether the moon originated as a hot or cold body; whether or not it has a small metallic core, as possibly indicated by its small magnetic field; and why it reverberated like a gong when the ascent stage of the Apollo 12 lunar module crashed onto it.

The earth's weather is largely the result of the energy the atmosphere receives from the sun, and lunar meteorologists would have a particularly good opportunity to study the sun's role in this relationship. Measuring and recording the solar constant and earth's albedo (fraction of incident light reflected) from the moon can be done with excellent accuracy at any moment in time. These quantities can lead to a continuous computation of the heat balance between the earth's surface and its atmosphere, which is a main factor in weather.

In the early phases of lunar exploration, men will use highly specialized vehicles (opposite page). Functionally designed, they bear little resemblance to earth vehicles where necessities such as windshields and fenders would be considered superfluous extras. Double Ackerman steering system permits very tight turning radii. Low center of gravity makes the vehicles extremely hard to turn over on the rough terrain. Since drivers wear spacesuits, no elaborate and expensive closed-cabin life-support systems are required. The vehicles are especially designed to limit top speeds to those consistent with the rough lunar surface. Large-diameter wheels permit the vehicles to cross relatively wide crevices and craters.

A lunar meteorological observatory working in conjunction with weather satellites in orbit around the earth might become a valuable extension of the present World Weather Watch sponsored by the World Meteorological Organization (WMO) of the UN. Indeed, such an observatory, staffed with experts from various nations, would certainly help WMO realize the three goals of its program: development of a capability for making dependable weather predictions up to two weeks, exploration of the degree to which weather can be modified on a large scale, and the furtherance of international cooperation.

The moon's economic potential

The commercial potentialities implicit in a lunar colony must not be overlooked. Ultimately, if exploration and colonization of space follows its historical pattern on the earth, then manufacturing will eventually exist on the moon. In regard to this, Elmer P. Wheaton, vice-president of Lockheed Missiles & Space Co., observed: "The surface area of the

The first observatories on the moon will be underneath the surface. Optically flat plates (below) will reflect images into an optical system shown on the opposite page. Early lunar astronomical observatories will be built into the walls of craters for protection against the environment. An airlock will permit the astronomers to enter and leave the observatory.

moon totals more than 90% of the land area of North, Central, and South America combined. In size, then, the moon just about matches the New World of 500 years ago."

As with the men who first reached the Western Hemisphere, we today have no real way of knowing the economic potential of the moon. Indeed, one of the reasons for the continued exploration and colonization of the moon is to determine the degree to which it can be exploited commercially. Yet even without the hindsight of 500 years, some areas of technology seem to be candidates for the moon. In an attempt to identify such potentialities, in 1964 Neil P. Ruzic, editor and publisher of *Industrial Research,* made a survey of 1,742 members of the American Vacuum Society. The question he put to them was, "Which of the following devices or materials do you believe could be *produced* easier or better if the production were to be done in a Moon factory?" The list included the following processes or products (with percentages of

replies in favor): vacuum-cast alloys (70%), vacuum welds (56.8%), electron-optical systems (28.4%), optical components (17.5%), pharmaceuticals and biologicals (13.3%), industrial chemicals (6.7%), petrochemicals (4.1%), plastics (3.3%), and miscellaneous (5.8%).

The most probable and obvious market for products manufactured on the moon is, of course, the lunar colony. Except for certain items, the cost of exporting manufactured items from the moon to the earth would be extremely high. But there is also another potential for lunar manufacturing. As educator Lloyd V. Berkner pointed out: "Manufacturing on the Moon may not be established for at least the next several decades. But the lunar environment may offer the opportunity to develop experimentally certain manufacturing processes which, after being worked out, then can be established better on the Earth."

In his *The Case for Going to the Moon*, Ruzic mentions a number of products that might be made more efficiently and economically on the

moon or in orbiting spacecraft than on the earth. These include precision ball bearings; pharmaceuticals; industrial diamonds; semiconductor materials; crystals with fewer dislocations; cathode-ray and electron tubes; metallic foams; metallic "whiskers" of extreme strength; microelectronic computer circuits; ultra-pure metals; precision glass mirrors for astronomical instruments; vacuum-deposited aluminized plastic; ferrite magnets; liquid propellants for rockets; precision-coated optics; and freeze-dehydrated, high-protein food. While many of these have obvious markets in a permanent lunar colony, others, especially industrial diamonds, might be economically competitive with earth-produced goods in spite of the transportation costs involved.

In order to decide what can be manufactured better on the moon than on the earth, the lunar environment must be analyzed for those features that are conducive to specialized production techniques. Thus, industrial processes which depend upon vacuum, extreme cold, or sterility are obvious candidates. The vacuum of the moon, which is about 0.000000019 pounds per square inch, is particularly attractive because it is much more nearly total than the best that can be produced on earth. Processes employing a vacuum thus can exploit this aspect of the moon's natural environment. Vacuum operations also can take place anywhere on the moon rather than being confined to the small chambers necessary on the earth, and production therefore can be on a much larger scale.

Man returns to the moon

When man will return to the moon after the last astronauts of Project Apollo lift off from it in 1972 is much more difficult to predict than how he will do it. Plans for the manned exploration of the moon by the Soviet Union are unknown and likely to remain so until the first cosmonauts land there; future plans of the United States are nonexistent. In the mid-1960s NASA's proposals for extended lunar exploration were ambitious, as exemplified by the Apollo Extension System (AES). Suggested in 1965, this plan was based on a series of studies made by several aerospace companies. In brief, it proposed that, after the initial Apollo program, Saturn V and Saturn IB launch vehicles would be used to boost modified Apollo command, service, and lunar modules to the moon for extensive geological exploration. The time spent on the surface by the astronauts would have been increased from 36 hours to 14 days. The radius of exploration from the lunar module would have expanded from slightly more than a half mile to five miles. And instead of bringing back to earth only 75 to 100 pounds of lunar rocks and soil, the astronauts of the AES would have brought back 250 pounds.

Likewise, estimates of when various events in lunar colonization would occur were, at one time, optimistic. In 1966, William G. Purdy of the Martin Marietta Corp. predicted that between 1985 and 1995 there would be a manned lunar observatory in operation. Similarly, Krafft Ehricke of the North American Rockwell Corp. foresaw the establish-

Among the features of the permanent lunar colony will be gigantic cryostats to produce extremely low temperatures for a variety of manufacturing and scientific research purposes. A cryostat for use on the moon was designed and patented by Neil P. Ruzic, publisher of Industrial Research *magazine. Shown under construction (opposite page), the cryostat consists of a bowl that insulates a condensor from all sources of heat from space and the moon itself. The cryostat should be able to produce temperatures as low as 5° K (—448° F) on the moon.*

ment of a deep-space communications network and radio astronomical telescope on the moon between 1985 and 1988.

Nevertheless, even before the first moon landing of Apollo 11, it became apparent that such predictions were overly optimistic. Confronted by pressing needs to improve the quality of life in such areas as education, health care, and housing, it became obvious that, from the point of view of public support and funding, such timetables were impossible. As the economic recession of 1970 and 1971 deepened, it was obvious that such schedules were unattainable and that any attempt to estimate when Americans would return to the moon was useless.

How colonization probably will begin

While it is pointless to speculate specifically *when* man will begin his lunar colonization, it is enlightening to study *how* hc would do it if and when the time arrives. Colonizing of the moon would be accomplished in three logical and technically attainable phases:

—Gradually increasing periods of exploration utilizing the integrated space transportation system (*see* pages 28–33) and multimanned shelters on the surface;

Water and metals will be produced by heating lunar soil and rocks in large solar furnaces. Such furnaces will use a parabolic reflector and are expected to produce temperatures as high as 4,000° F.

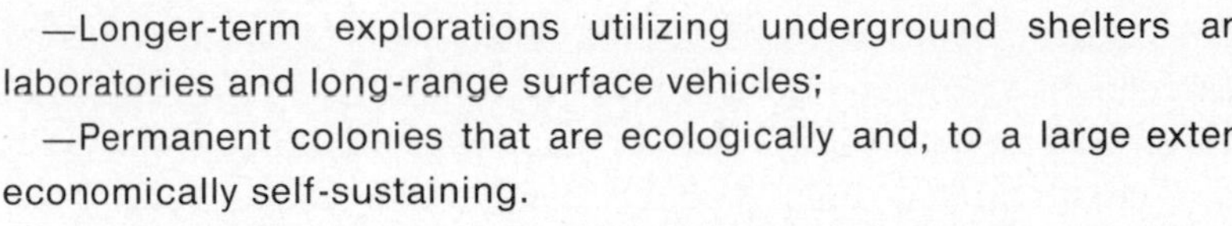

—Longer-term explorations utilizing underground shelters and laboratories and long-range surface vehicles;

—Permanent colonies that are ecologically and, to a large extent, economically self-sustaining.

In the initial phases of the colonization of the moon, men would make extensive explorations of its surface on both the near and far side from a station in lunar orbit. The lunar-orbiting station would be in a polar orbit to facilitate observation of all of the moon and to permit the best landing site to be selected for the base. After a site had been selected, the explorers and their cargo would descend from the moon-orbiting satellite in a space tug. The first crew of three men would stay on the moon's surface for about three months, living in the crew compartment of the space tug and operating from it.

While much of the exploring these first men do would be of a scientific nature, they always would be oriented toward observations that later would be useful for more long-term manned stations. Thus, a primary consideration in this phase would be the search for sources of water, such as ice. Kenneth Watson, Bruce C. Murray, and Harrison S. Brown, geologists at the California Institute of Technology, have suggested that the permanently dark craters near the polar regions may be likely sites. Soviet scientists V. S. Safronov and Y. L. Ruskol disagree, but believe that ice deposits are possible in the continental areas, where there have been no lava overflows. Ice beneath the surface, they think, could well exist because it would be protected by the insulating properties of the soil above it. The explorers also would search for such natural features as caves, crevices, and over-hangs that could be converted into rudimentary shelters at a later date.

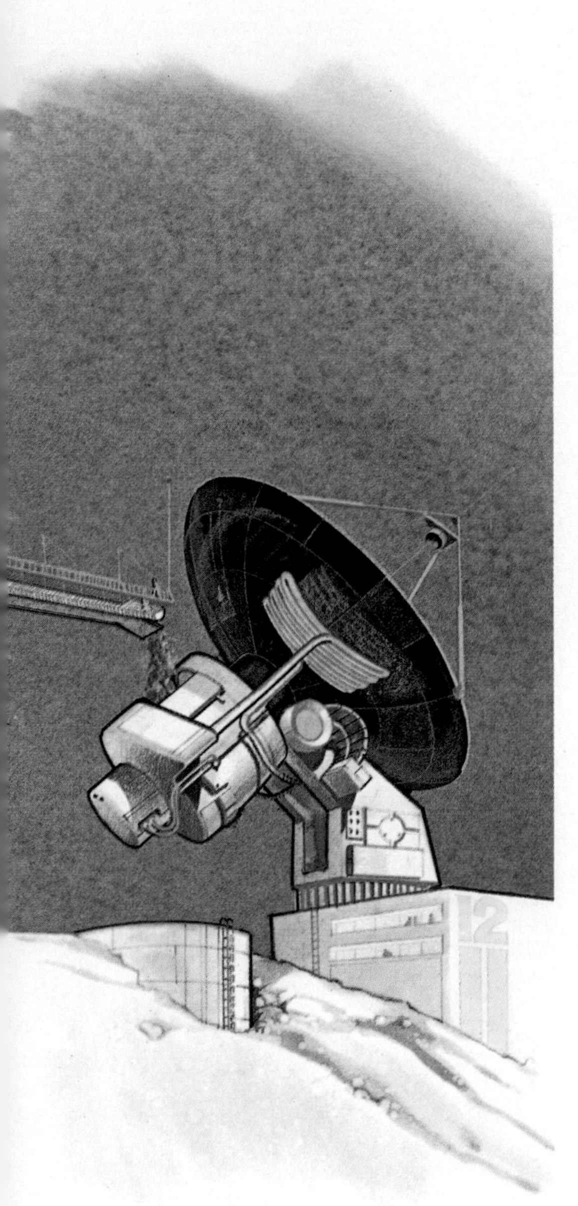

These men would be scientists with little or none of the training associated with astronauts today. They would be geologists and geochemists, civil engineers, and astronomers; and they would be flown to their site by the space tug's pilot crew. This same crew would assist the scientists in their explorations by driving an advanced model of the lunar rover vehicle that is scheduled for use by Apollos 15, 16, and 17. However, it would have a far greater range than the Apollo lunar rover's 40 miles. Powered by solar cells, it would have a total lifetime of 15 days; but it probably would still have a top speed of only 10 miles per hour (as does the current lunar rover) for the sake of safety. On the other hand, the lunar drivers would have an advantage not available on the earth. Because of the moon's low gravity, should a 200-pound vehicle tip over, the driver and passengers could easily turn it upright.

The lunar rover also could carry an inflatable shelter for two men. It would add to the flexibility of the rover and its mission by permitting extended exploration of geological features of special interest. The two men simply would remove the shelter from its container, inflate it with oxygen, and move in.

A shelter of this type already has been developed by the Goodyear Aerospace Corp. Called Stem (Stay Time Extension Module), the structure is a cylinder 13 feet long and 7 feet in diameter. It can accommodate two men for eight days, maintaining a constant temperature of 75° F even though the exterior temperature ranges from 250° F to as low as −240° F. The shelter is made of high-strength, stainless steel filaments in a composite with flexible materials.

Larger crews that would live on the surface for a longer time would follow these early pioneers. A six-man crew, for example, could stay six months equipped with a larger lunar rover having a lifetime of at least 30 days. In similarly increasing increments, a highly specialized team of scientists and engineers totaling as many as 18 could spend a year and a half on the moon, performing detailed investigations and planning facilities that would permit even larger numbers at later dates. Such a group could utilize a fleet of lunar rovers with a useful lifetime

Life-sustaining oxygen for the lunar colony will be manufactured from soil and rocks and by utilizing algae in the process of photosynthesis. Tending such plants will employ a good portion of the permanent lunar colonists.

Expandable structures that can easily be transported and erected will provide shelter for the early lunar explorers and for scientific outposts in remote regions.

of 120 days. During this period, it is possible that many of the technical problems of lunar colonization would be solved. In addition, much useful data might be obtained concerning the social and psychological dynamics of relatively large groups of men and women in the isolation and potentially hazardous environment of the moon.

Beginnings of technology on the moon

After this period of intensive surface exploration, during which men on the moon would be to some degree dependent upon the lunar base orbiting overhead and the lifeline tying it to the earth, there would be a phase in which man literally would dig in to stay—and to survive. It would be of indeterminate length but marked by a prevailing attitude of permanency on the part of the people participating. Like the earlier phase, it would be characterized by a basically scientific and technological orientation, and it is likely to be international in composition. The emphasis would still be on searching for lunar materials essential for man's independent existence on the moon. And at this time, there probably would be the modest beginnings of technology on the moon, with a prototype manufacturing process for the conversion of lunar rocks and soil into life-sustaining oxygen and water. In short, it would mark the beginning of a period in which man on the moon would begin proving he can become independent of earth for at least two of those essentials for life: oxygen and water.

During this phase it is likely that men would try to establish dwelling places beneath the moon's surface. Moving beneath the lunar soil would help solve two problems in living on the hostile moon. It would simplify the problem of maintaining a comfortable temperature in a prefabricated shelter, and it would provide shielding against the intense radiation from the sun and galactic sources. While the surface temperature may vary through a range of 500° F during the course of a lunar day and night, the temperature of the soil a few inches beneath the surface remains much more stable. The dusty topsoil has good insulating properties; at a depth of only four inches or so the temperature apparently never gets below 30° F.

One of the suggested means of constructing an underground shelter is to drill a hole to the desired depth and then detonate explosives to create a cavity. The cavity is then filled by inserting a plastic balloon and inflating it. A group of such cavities could be interconnected by tunnels to provide a kind of protocity for men on the moon. The balloon technique could also be used to seal the interior surfaces of naturally occurring caves or other features suitable for habitation.

During this phase of colonization, man would live in these subsurface structures and perform some of his work in them. But most of his activities would be on the surface. Typical of them might be the manufacture of water, hydrogen, and oxygen. In order to release water from hydrous (water-bearing) rocks, lunar engineers would harness the energy of the sun, which is readily available for two-week periods. The water would be produced as steam and then condensed in a solar furnace, which would have the appearance of a contemporary radar antenna or reflecting telescope. Sunlight gathered by a large parabolic reflector would be focused by a secondary reflector into a chamber containing the lunar raw material. The resulting steam then would be conducted to a chamber on the shaded side of the furnace, where it would condense into water because of the low temperature in the shadow.

Even fairly simple furnaces of this design could produce temperatures as high as 4,000° F, and precision-built models could create temperatures as high as 8,500° F. With the latter model it would be possible to produce some metals as by-products, as well as water. However, this furnace and its products must be considered as an experiment rather than a full-fledged production facility for supplying oxygen and water. The experience gained with it would be applied to more advanced equipment that would be needed for the permanent colonies that would follow.

A permanent colony on the moon

After extensive exploration of the moon from a series of increasingly self-sustaining bases, man would establish a permanent colony. For the protection it offers against radiation and the simplification of

The huge craters on the far side of the moon will make ideal locations for the construction of extremely large radio telescopes. They can be several times the diameter of the largest one on the earth, the 1,970-foot reflector at Zelenchuk in the Soviet Union.

For people born and raised on the moon an elaborate program of physical conditioning will be necessary before they can make the trip to earth. Centrifuges and weights on the body will accustom the lunarian to the six times greater weight he will experience on earth. Training will be required in simple tasks such as using knives and forks and more complicated ones such as walking, which will be far different from the loping gait that is characteristic of locomotion on the moon.

temperature regulation, the permanent colony would be largely beneath the lunar surface. However, certain features of it would be on the surface. The colony would be covered by two transparent plastic domes, one within the other for safety purposes. The domes would be made of strong plastic in sections supported by mullions and anchored firmly by tension lines buried deep within the lunar soil. Stretched over the outermost dome would be a meteoroid and cosmic-ray shield consisting of a flat sheet of the same material.

A dome would be used not for esthetic purposes but for practical engineering ones. As a section of a sphere, a dome is a very efficient structural shape for withstanding the high internal gas pressure that will be present because of the moon's vacuum. A full sphere would be even better, but it would be impractical because of the relatively small utilizable space inside. The dome, then, offers an acceptable ratio of pressurized-to-usable volume.

The domes would be extremely strong—far stronger than similar structures on earth. On earth such structures are designed so that the internal air pressure almost balances the opposing atmospheric pressure of 14.7 pounds per square inch. Thus, the material can be fairly thin, as it need only be strong enough to withstand the small pressure difference between the interior and exterior. On the moon, however, such is not the case. To construct a structure capable of containing an internal pressure of 14.7 pounds per square inch (the normal sea level atmospheric pressure on the earth) against an external vacuum would be practically impossible. In view of the engineering problems involved with such a differential, the atmospheric pressure within the dome probably would be considerably less than 14.7 pounds per square inch.

The domes would serve two essential functions: they would provide a means of entering and leaving the underground colonies, and they would provide a sealed surface area in which oxygen could be manufactured by photosynthesis. Each dome would also have an airlock through which men and vehicles could pass to the outside. Within each dome, there would be ample area in which to grow a variety of green plants that would consume carbon dioxide and produce oxygen through photosynthesis. While oxygen would be the primary product, the plants could also produce additional dividends for the lunar colonist in the form of edible foods. The extent to which the lunar soil, with no nitrogen supplied by decomposing organic matter, could be used as a base for growing plants fed by fluid nutrients is speculative but not to be ruled out.

The major problem to be resolved in lunar agriculture would be to find plants that could adapt to the moon's 14-day alternations of light and dark rather than the familiar 12-hour pattern on the earth. It would be helpful to be able to breed edible plants that could mature in 14 days.

The most important task for the lunar colony would be the manufacture and recycling of oxygen and water—and there is water on the moon. Rocks brought back by Apollo 11 and 12 astronauts contained water in minute amounts. Tests of the water released from these rocks

showed that it had a lower percentage of deuterium than does terrestrial water. (Deuterium is an isotope of hydrogen having twice the mass of ordinary hydrogen.) Thus, scientists are sure that the water in the lunar rocks assayed was not contamination from terrestrial sources. There may be other sources of water on the moon as well. Although the probability of surface water or ice can be ruled out because of the high temperatures and vacuum, there could be underground deposits of ice since the lunar soil is a very efficient insulator and thus acts as a natural refrigerator. Harold Urey, a Nobel laureate, believes that underground water could be the reason why certain parts of the moon are more dense than the body as a whole.

Possible places for finding underground ice include the curious domes near the crater Copernicus. According to some geologists, these domes may be buried lunar glaciers. Similarly, the wrinkle ridges that are found in many places on the moon, such as those in the Sea of Serenity, may contain mineral deposits in which moisture escaping from lower depths may be trapped. Ice found trapped underground in any of these locations might not be safe for use as drinking water without elaborate filtration. Thus, it would be better that it be electrolyzed into hydrogen and oxygen. These two gases then could be stored or recombined to make pure water.

In anticipating the manufacture of water and rocket propellants on the moon, scientists at the Manned Spacecraft Center in Houston, Tex., have developed a process for producing oxygen and water from

ilmenite, a mineral containing large amounts of iron and titanium oxides that was found in the rocks and soil returned by the Apollo 11 and 12 astronauts. In this process, a mirror focuses the sun's rays into a transparent container of lunar soil, raising its temperature to 2,372° F. Hydrogen is then introduced into the container, changing the oxygen in the ilmenite to steam, which passes through electrolytic cells and separates into oxygen and hydrogen. For the process to work, an initial supply of hydrogen must be brought from earth, but a portion of the gas subsequently produced then can be recycled. It also may prove possible to manufacture the initially needed hydrogen from lunar materials. Hydrogen has been released from lunar rocks returned to earth by heating them to 550° F.

Calculations show that this process could produce a pound of water from every 100 pounds of lunar soil processed. The system could be made even more productive by magnetically concentrating the iron oxide. Using this concentrate, 14 pounds of water could be recovered from each 100 pounds of lunar ore.

The major industry on the moon, then, would be mining or processing lunar soil and rocks for water and oxygen. Assuming that the process described above is used, metallic iron would accumulate as a by-product. Undoubtedly, the lunar colonists would use this metal as a component in their structural materials.

Future uses of a lunar colony

The permanent lunar colony also would provide an ideal spaceport for future launchings of manned and unmanned spacecraft to the planets, and, ultimately, to worlds beyond the solar system. This prospect is made even more attractive because the propulsion necessary to escape from the moon's gravity is much lower than that needed to escape the earth. A speed of 25,090 miles per hour is needed to escape the earth's gravity, compared to only 5,270 miles per hour for the moon. Furthermore, the lunar colony would make an excellent holding and quarantine area for astronauts returning from other planets before they are allowed to return to the earth.

Although colonization of the moon presents complex problems, they are not beyond the intellectual or technological capabilities of man. Eventually, perhaps within the next few decades, he will return to the earth's only natural satellite and begin to live on it. In exploring and exploiting the moon, he will accrue knowledge and technology that will allow him to stay there for long periods of time and help prepare him for establishing colonies on more distant worlds in space.

The first long-term, but not permanent, men on the moon will live in multistory dwellings made from the spacecraft in which they landed. Such structures will be stocked with water, food, and oxygen to last for long periods of time, and they can be periodically resupplied.

FOR ADDITIONAL READING:

Maisak, Lawrence, *Survival on the Moon* (Macmillan, 1966).

Ruzic, Neil P., *The Case for Going to the Moon* (Putnam, 1965).

Ruzic, Neil P., *Where the Winds Sleep* (Doubleday, 1970).

Sharpe, Mitchell R., *Living in Space: The Astronaut and His Environment* (Doubleday, 1969).

Space Task Group Report to the President, *The Post-Apollo Space Program: Directions for the Future* (U.S. Government Printing Office, 1969).

Thomas, Davis (ed.), *Moon: Man's Greatest Adventure* (Abrams, 1970).

AUDIOVISUAL MATERIALS FROM ENCYCLOPÆDIA BRITANNICA EDUCATIONAL CORPORATION:

Films: *The Moon; A Trip to the Moon.*

Filmstrip: *Exploring the Moon.*

8mm Film Loop: *Trajectory of a Moon Probe.*

An integrated space transportation system

Before colonization of the moon can become a reality, there must be an economical system of transportation between it and the earth. The Saturn V rocket that boosts the Apollo astronauts to the moon is too costly for the traffic requirements of even early phases of colonization. Saturn V is a one-shot vehicle; having served its purpose of launching the Apollo spacecraft onto the proper trajectory for the moon, its job is done. The first two of its spent stages plunge back into the atmosphere and burn; the third stage either crashes on the moon or continues past it into an eternal orbit around the sun.

Obviously, such a rocket is too expensive for the necessary exploration of the moon that would precede permanent colonization. Indeed, using the Saturn for such a purpose would make no more sense than driving a new car into a salvage yard after the first trip to the office.

There are various ways in which a more economical system could be designed. However, the prospect is remote for development of a single, all-purpose space vehicle that would take off from earth, go into earth orbit or lunar trajectory, go into orbit around the moon or land on its

Key to an economical space transportation system for extended exploration and colonization of the moon is the space shuttle vehicle. The orbiter craft (below) is launched into space by a reusable booster that returns to the earth and lands like a conventional airplane. The orbiter can also return, and it can deliver a variety of payloads into earth orbit.

surface—and still have enough propellant to make the return trip to the earth. The inefficiency and prohibitive costs of such a vehicle can be easily imagined when one is reminded that the Apollo 11 mission cost $375 million. The unreusable Saturn V vehicle itself cost some $185 million, while it and the Apollo spacecraft burned approximately 5,993,200 pounds of liquid propellants at the rather modest cost of only $165,534 to send 495 pounds of men to the moon and return them to earth with 65 pounds of rocks and film.

The system envisioned by the National Aeronautics and Space Administration (NASA) and the aerospace industry consists of several specialized vehicles, some of which remain permanently in space, touching neither earth nor moon. The economy of the system lies in the design concept of reusability. Reusable space vehicles can be used between earth and earth orbit, earth orbit and lunar orbit, and lunar orbit and the moon. Some idea of the cost savings to be expected from one type of reusable vehicle is apparent in an estimate by George E. Mueller, former associate administrator for manned space flight. In discussing a reusable space vehicle for the earth-to-earth-orbit and return trip, he said: "It would lower the cost of flying into orbit and returning to earth from the present [1970] $1,000 per pound (for the

The long cylindrical orbit-to-orbit shuttle will become the space ferry of the future, moving men and materials from earth orbit to lunar orbit. Unlike the space tug, which it is ferrying in this view, or the space shuttle, it will be designed to remain in space permanently. It will never land on either the moon or earth.

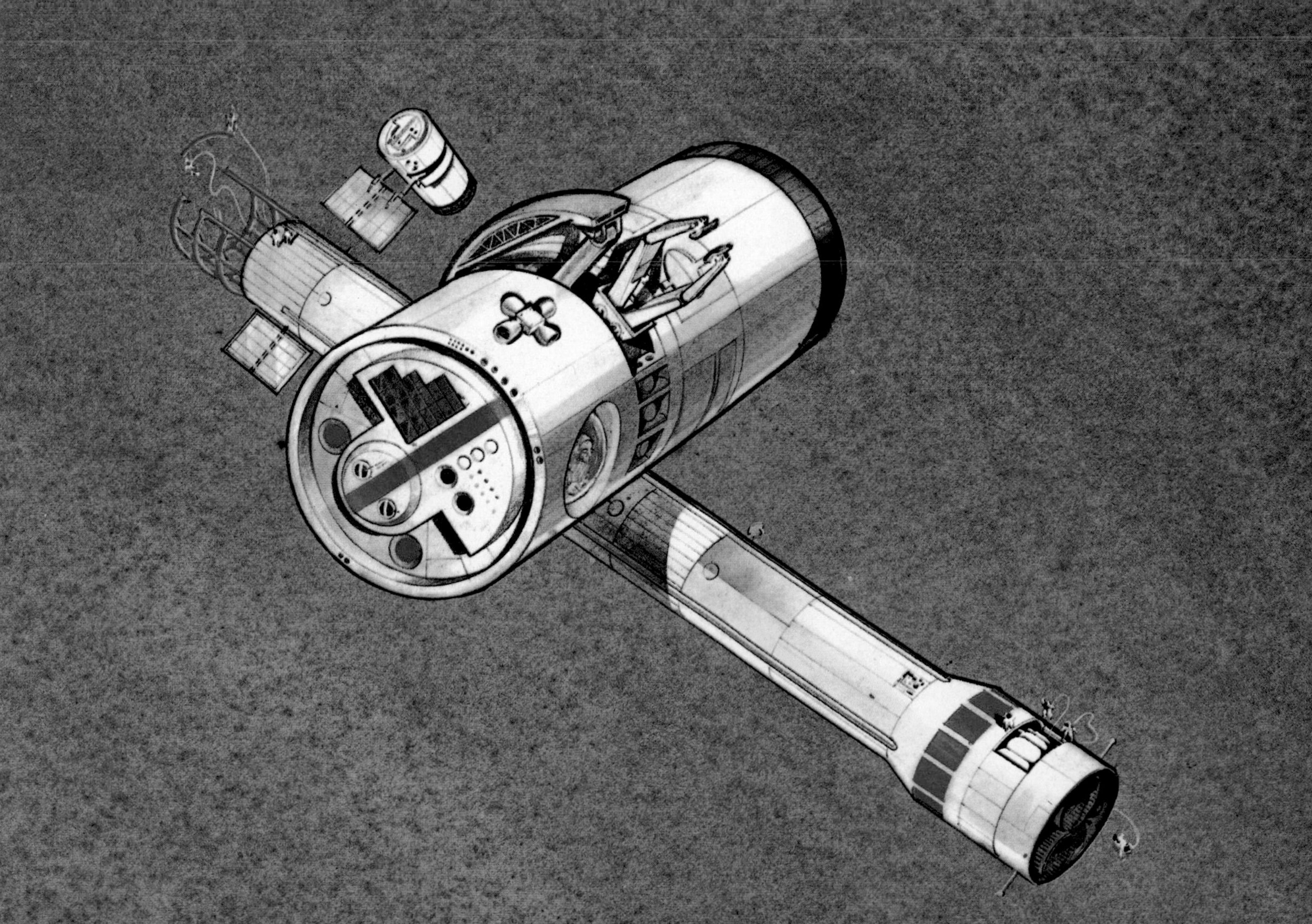

one-way upward flight) to between $20 and $50 per pound for the round trip." In considering such a new vehicle, however, the cost of its development must also be taken into account, an estimated $5 million to $6 million according to LeRoy E. Day, manager of NASA's space shuttle task group. Taking inflation into account, however, a figure of $8 million as a minimum seems more realistic for the five vehicles envisioned. Day also believes that additional shuttles beyond this number could be built for about $200 million apiece, slightly more than the current Saturn V.

Another characteristic of the system is *commonality,* or design of the minimum number of vehicles and engines for the maximum number of missions or uses. For example, a single propulsion module could be fitted with standardized manned modules or cargo modules to permit its use as a space taxi or as a space cargo tug in either earth orbit or lunar orbit.

The space shuttle. The first vehicle in the integrated system to be developed undoubtedly will be the space shuttle. In December 1968, NASA's Science and Technology Advisory Committee for Manned Space Flight met in La Jolla, Calif., to study the applications of space flight to scientific and technological objectives in the decade 1975–85. Out of the deliberations of this committee and a series of technical feasibility studies sponsored by NASA among aerospace companies came a recommendation for the space shuttle that was sent to Pres. Richard M. Nixon on September 15, 1969. On that date, Vice-Pres. Spiro T. Agnew, chairman of the Space Task Group, submitted to Nixon the group's report, *The Post-Apollo Space Program: Directions for the Future.* Specifically, it recommended: "A reusable chemically fueled shuttle operating between the surface of the Earth and low-Earth orbit in an airline-type mode."

As currently conceived, the space shuttle would look more like a jumbo jet airliner than the Saturn V or Titan IIIC rocket. To be more exact, it would look like one large jet airplane mounted piggy-back on a much larger one. The two-stage vehicle would consist of a booster and an orbiter, and it would rise vertically when launched.

The shuttle would weigh approximately 5 million pounds at launch. The 12 rocket engines of the booster and the orbiter would use liquid oxygen and liquid hydrogen as propellants; each engine would furnish 550,000 pounds of thrust, or more than twice that produced by the J-2 engine of the Saturn V's second and third stages. Shuttle engines, unlike those of the Saturn V, would be designed so that they can be throttled. As a result, accelerations during ascent would reach no more than three *G*'s (three times the force exerted by gravity on a body at rest). Thus, passengers would be subjected to less than half the acceleration of the Apollo astronauts. During lift-off and the first few minutes of flight, the orbiter's two crewmen would have little to do. The two crewmen of the booster section would be in control.

At an altitude of about 40 miles, the two craft would separate. The booster would make a 120° roll and dive back into the atmosphere to maneuver and land on a conventional airfield runway, provided it is at least 10,000 feet long. The orbiter would ignite its two engines and continue into space for about 60 additional miles, after which it would go into orbit around the earth. Once its mission was completed, it would briefly fire its rocket engines, brake out of orbit, and return to the atmosphere. Like the booster, it would land on a conventional runway.

With a total payload capability of 50,000 pounds, the orbiter would haul both passengers (probably 12) and cargo. Its cargo compartment would be approximately 60 feet long and 15 feet in diameter. Thus, the compartment would be large enough to carry a satellite and its high-energy launch stage, making the orbiter a potential first stage of a two-stage launcher for satellites and deep-space probes as well as a cargo craft for lunar colonies. It would carry enough food, oxygen, and water to remain in orbit for a week.

In regard to the clothes they wear, travelers and crewmen of the shuttle would resemble contemporary aircraft passengers rather than astronauts bundled in their bulky spacesuits. The orbiter and booster would be designed for a shirt sleeve environment. However, for safety reasons, passengers and crew of the shuttle and other vehicles of the space transportation system may have a special, lightweight spacesuit for emergency use. The prototype of such a suit already has been developed. Weighing only 10.2 pounds, it can be folded and carried in a briefcase. It could be worn during potentially hazardous phases of spaceflight, such as launch, atmospheric reentry, and vehicular transfer. Because of its flexibility, it can be put on in about a minute.

The space shuttle would be economical mainly because of its re-usability. The vehicle would be designed for making as many as 100 missions, though more may be possible. Also, the time required to ready it for another mission would be only two weeks. By working around the clock, however, it should be possible to reduce the time to 3½ days. Maxwell Hunter, shuttle engineer for the Lockheed Aircraft Corp., believes that a 3½ day turn-around would permit more missions and eventually drop the price per pound of payload via shuttle to as low as $10.

The space tug. A second vehicle recommended by the Space Task Group is the space tug. Like the shuttle, it would be reusable and would feature commonality. The space tug would consist of two parts: a propulsion module, which, as currently envisioned, would be approx-imately 22 feet in diameter and 25 feet long; and, attached to the propulsion module, a manned compartment with space for three to six people. In some cases, this manned compartment could be provided with remotely operated mechanical arms like those used for handling radioactive materials in laboratories. Thus equipped, the space tug could assist in the building of space stations, repairing of satellites, and loading and unloading of the space shuttle orbiter. If equipped with legs, like the Apollo lunar module, the tug could also be used to ferry personnel and cargo between a space station in orbit around the moon and the moon itself. Employed in this use, it could ferry as much as 25,000 pounds of payload down to the moon's surface and return with 7,000 pounds (without refueling on the surface).

Orbit-to-orbit shuttle. The third vehicle of the integrated transportation system would be the orbit-to-orbit shuttle. As the name implies, it would remain in space, never touching down on the moon or earth but travelling from vehicles in orbit around one body to vehicles in orbit around the other. Ideally, this shuttle should be nuclear-propelled for greater efficiency and economy; however, early models probably would be chemically propelled in keeping with the concept of commonality.

Assuming the integrated transportation system thus described, transportation of men and women as well as cargo to the moon still would be a relatively expensive operation because delivery of personnel and materials to the moon is about 15 times the cost of their delivery from the surface of the earth to a vehicle in orbit around the

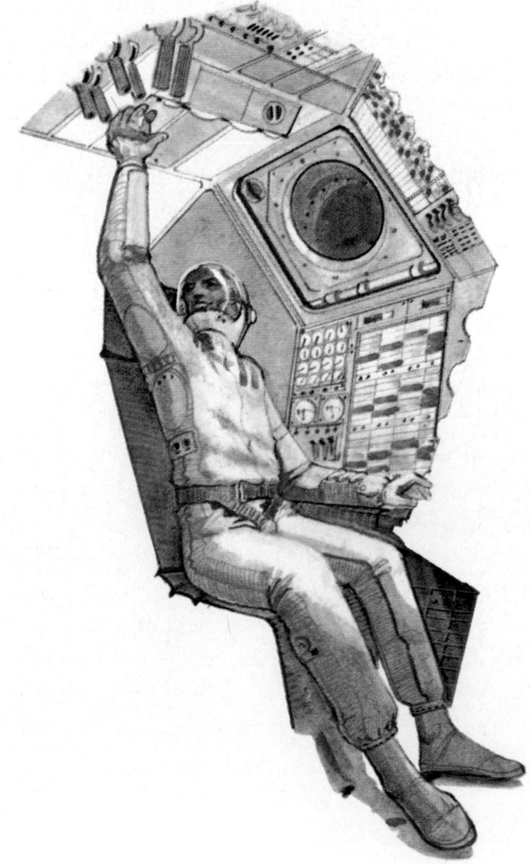

Space traveler in the shuttle will be able to wear lightweight clothing in comparison to the bulky space suits of present-day astronauts. The space tug (opposite page) will become the workhorse of future space programs, including the colonization of the moon. In addition to performing a variety of tasks in earth or lunar orbit, the tug will also have legs to permit its landing on the moon.

earth. With the shuttle, orbit-to-orbit vehicle, and space tug, the estimated cost is approximately $1,500 a pound. With increasing use of the integrated transportation system over the next three decades, studies by NASA show that the cost could drop to $75 a pound by the year 2000. Cost could be considerably reduced overall by the manufacture of liquid hydrogen and liquid oxygen on the moon for use in the lunar vehicular components of the system.

Significantly for the future colonization of the moon in view of NASA's prospect for annual budget cuts during the early 1970s, the European Launcher Development Organization (ELDO) in early 1970 sponsored a competitive study between two European aerospace consortia on design and capabilities of a transportation system. Equally significantly, NASA participated fully in those studies, supplying technical data on the proposed space shuttle vehicle. ELDO members also took part in a presentation on space tug technology held at NASA's Manned Spacecraft Center, in Houston, Tex., on October 21, 1970. As ELDO continues to demonstrate its growing competence in this particular aspect of the integrated space transportation system, a solid foundation is being laid for an international approach to the colonization of the moon.

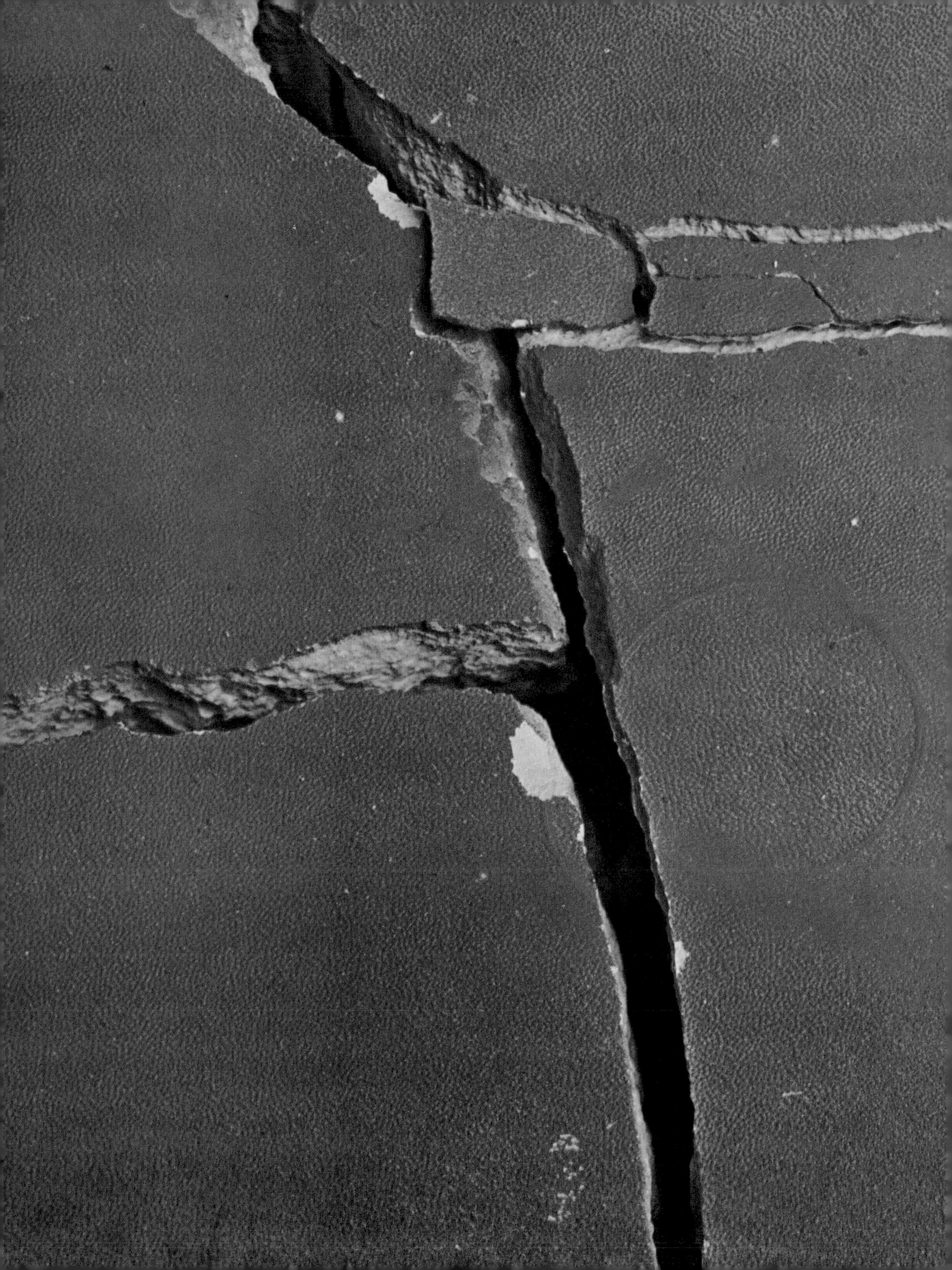

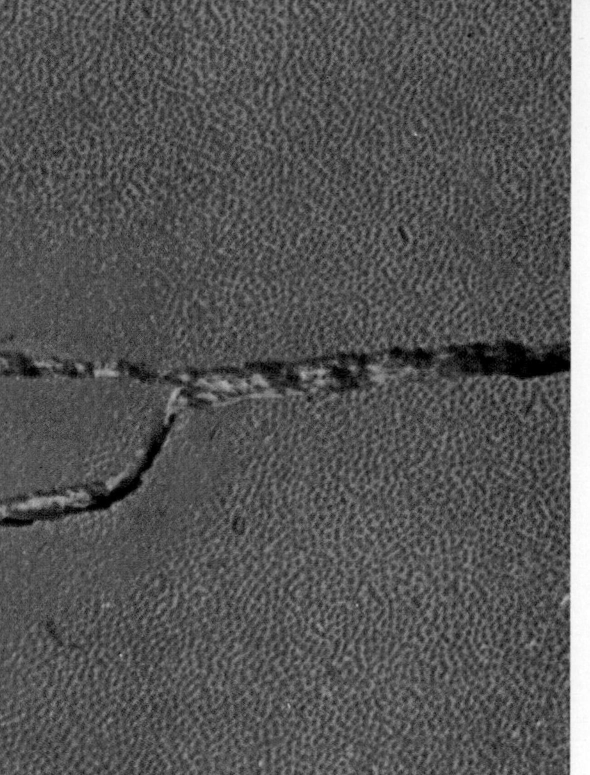

The Trembling Earth

by David Perlman

Earthquakes are among the most feared of all natural disasters. How do Californians who live in "earthquake country" react to the danger? What are scientists doing to unlock the secrets of these destructive events?

The massive escarpment of California's San Gabriel mountains seemed comfortably stable above the prosperous communities of the San Fernando Valley until precisely 41.1 seconds past 6 A.M. on the warm winter morning of February 9, 1971. At that instant, and without warning, an earthquake thudded through a long-inactive fault that surfaced where valley and mountains met. The mountains leaped upward and sideways from three to five feet; the valley shook, cracked, and buckled. Ten million Californians felt the rolling jolt.

More than 60 people died in the ruins of hospitals, homes, and crashing highway bridges. Nearly 1,000 were injured; 80,000 fled while a reservoir was drained to save a weakened dam; 3,400 homes and buildings were destroyed or damaged; costs soared to $550 million. Once again, here was evidence that man's most modern settlements, his proudest communities, are at the mercy of forces deep within a planet that is still hot, still in primitive motion, still not completely understood.

Cracks in the earth's crust

California's earth trembles, shakes, and occasionally wrenches asunder in literally thousands of earthquakes every year. This is because the state is built above a broad fault system whose most dramatic feature is the San Andreas, a constantly grinding fracture in the earth's crust that extends for more than 650 miles along the western edge of the continental land mass from the Pacific Coast off Cape Mendocino down into the Gulf of California.

An observer in the air can clearly trace many miles of the San Andreas fault through north-south rift valleys that slice the folds of hills, through chains of linked lakes, and across the sharply-offset beds of old streams. But millions of people live directly on the zone of the San Andreas, and where their dwellings stand it is hard to pick out the hidden fault trace. Subdivisions now straddle the 1906 trace just south of

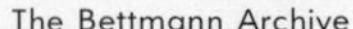

The Bettmann Archive

This old engraving dramatically depicts the death and destruction caused by a major earthquake that struck Lisbon, Portugal, in 1755.

San Francisco. Housing tracts "float" on artificial fill covering the unstable muds of nearby San Francisco Bay. Schools, hospitals, and even a reservoir are immediately atop the moving track of the Hayward and Calaveras faults on the eastern side of the bay.

The leaky offshore oil fields of Santa Barbara Channel probe into a heavily seismic seabed related to the San Andreas system. Near the San Joaquin Valley city of Bakersfield, a quake on the subsidiary White Wolf fault killed 14 people in July 1952. The San Gabriel, the San Jacinto, and the Banning faults all branch away from the San Andreas between Los Angeles and San Bernardino. This is earthquake country with a vengeance.

The "ring of fire"

Nor is California's San Andreas system an isolated one in the world. It is part of what geologists have come to call the "ring of fire"—a vast, nearly circular band of the earth that loops around the Pacific Ocean from Chile to Alaska; from the arc of the Aleutians to Japan, the Philippines, and New Zealand. Here the seas are slashed with abyssal trenches, the coasts of continents and islands are studded with erupting volcanoes, and earthquakes tremble daily. Swinging eastward away from the "ring of fire" in the southwest Pacific, the Trans-Alpine earthquake region cuts a swath through northern India, Turkey, and Italy.

Even the eastern half of the United States is not immune to violent quakes. In 1811–12 thinly populated southern Missouri (near New Madrid) was rocked by what may have been the most shattering earthquake ever to strike North America. Actually, there were three shocks: Dec. 16, 1811; Jan. 23, 1812; and Feb. 7, 1812. They were felt as far away as Canada in the north and the Gulf Coast in the south. The prin-

DAVID PERLMAN, science editor for the San Francisco Chronicle, *has been president of the National Association of Science Writers and is a member of the Council for the Advancement of Science Writing.*

Courtesy, The Society of California Pioneers

cipal shock shook down chimneys in Cincinnati, Ohio, 400 miles away. Charleston, S.C., suffered severe damage from an earthquake in 1886, and tremors have hit there again from 1960 to 1968.

Altogether, according to estimates by geologists, several million quakes strike the earth's crust each year; 700 of these are strong enough to cause at least some damage, and a few rank as catastrophes; 80% occur around the "ring of fire."

A pose of boastful insouciance

As in the San Fernando Valley, earthquakes may cause destruction by shaking or by buckling or rupturing the ground beneath buildings. They may cause unconsolidated or waterlogged soils to magnify ground motion, or to liquefy almost like jelly. They may touch off landslides on steep hillsides. They may churn up huge seismic sea waves, known as "tsunamis," that speed across oceans to drown coastal communities thousands of miles away.

Californians, of course, are used to earthquakes, and despite the terror when a strong one hits, the appropriate California pose is one of boastful insouciance. Many San Franciscans preserve intact for visitors the cracked walls that their houses suffered during a damaging temblor in 1957. The moment of fear, when walls buckle, glassware shatters, and lamps sway, lasts only for a few seconds, or perhaps a minute. As Edward Stainbrook, a University of Southern California psychiatrist says, "When the ground on which a man lives begins to move, it tends to make children of us all."

But fright soon subsides. People already are rebuilding in the San Fernando Valley—as they did in San Francisco in 1906—and the earthquake engineers who try to push more rigid safety standards often find themselves frustrated by political and economic resistance.

Because of its proximity to the San Andreas fault system, San Francisco has been subjected to many large and small earth tremors. Pictured here are scenes of the city following the earthquakes of 1906 (top) and 1868 (above).

(Below) J. Eyerman from Black Star; (others) Bert Van Bork

Illustrating the severity of the San Fernando Valley quake of February 1971 are these pictures of the weakened structure of the Van Norman Dam (top, left), which necessitated the evacuation of about 60,000 residents who lived below it until the water level could be lowered, crumbled bridges along a freeway (above), and a toppled circuit breaker at a power station (left).

(Below) Curt Gunther from Camera 5; (others) Bert Van Bork

Also destroyed by the San Fernando quake were buildings at the Veterans Hospital (top, left), where at least 44 persons were killed, and the Olive View Medical Center (left), where three people died. The crumbled wall and cracked street shown above are other examples of damage caused by the quake.

A severe earthquake that struck Turkey's Kutahya Province in March 1970 caused widespread destruction (top and opposite page, top) and took more than 1,000 lives. Above is a refugee camp set up after the quake to house some of the 90,000 homeless victims, such as the woman shown at the bottom of the opposite page.

Needed: better safety standards

"A major earthquake in California today (the one in the San Fernando Valley was only moderately severe) would cost over $30 billion in damage and result in an inestimable loss of life," cautions Karl V. Steinbrugge, a noted structural engineer who directs earthquake insurance activities for the Pacific Fire Rating Bureau and who recently headed a White House Task Force on Earthquake Hazard Reduction. California seismologists agree unanimously that a great quake along the San Andreas fault is virtually inevitable before the end of this century. According to Louis C. Pakiser, Jr., of the U.S. Geological Survey's National Center for Earthquake Research in Menlo Park, Calif., there will be at least 10 damaging earthquakes on the San Andreas within the next 25 years, and "one is likely to be a really bad one." Robert E. Wallace, also of the Geological Survey, has measured displacement rates along sections of the curving San Andreas and calculates that quakes as strong as the 1906 San Francisco temblor recur every 100 years or so; and that severe ones may strike every 15 years.

Despite these scientific findings, California officials estimate that 250,000 children in the state now attend schools too hazardous to meet earthquake safety standards set by a law passed in 1933. In San Francisco, as recently as 1970, voters turned down a $40-million bond issue to make 62 of the city's 120 public school buildings earthquake-resistant. Forty-six schools, four hospitals, and a football stadium on the east shore of San Francisco Bay are athwart the Hayward fault or within 1,000 feet of it.

Earthquake insurance is available to all Californians at an average cost of less than $60 a year for a $30,000 home. Yet almost no one carries this insurance, according to Steinbrugge, because deductibility clauses do not cover the first 5% of damage, and few people feel any sense of urgency. Steinbrugge himself concedes, cautiously, that he would not build his own home in many of the heavily-populated subdivisions close to active faults. But most Californians do not think the same way. "Economic forces are often stronger than tectonic forces," says Don Tocher of the Earthquake Mechanisms Laboratory at the U.S. Department of Commerce in San Francisco.

Building codes in Los Angeles and San Francisco are among the most stringent in the world and serve as models for other quake-prone areas. But their aim is primarily to minimize the danger of death-dealing collapse, not to insure against lesser damage. Well-funded builders of large office buildings and apartment houses may retain earthquake engineering specialists to make soil analyses; determine proper siting; and see that the most modern techniques go into foundations, steel frames, and reinforced concrete. Safety specialists maintain, however, that speculative developers cut corners and barely meet the building codes. As a result of the San Fernando quake, California officials are urging federal enforcement of earthquake safety standards and higher criteria to assure structural integrity of vital buildings, such as hospitals, power plants, and fire departments.

Bert Van Bork

Until 1958, Los Angeles banned buildings taller than 13 stories, but during the San Fernando quake the new high-rises in the downtown area suffered no structural damage at all, although they swayed and shook with unexpected severity. Because of a recent Los Angeles law requiring quake-recording instruments to be installed in all large buildings, the structural effects of the San Fernando temblor were measured by at least 200 strong-motion seismographs. Information on movement, vibration, and resistance will prove invaluable for earthquake safety engineers.

Measuring a quake, from jiggle to jolt

Large earthquakes send strong shock waves coursing through the ground; they travel along the surface and down through the inner layers of the earth, and their vibrations can be detected all over the world by sensitive seismographs tuned to measure and locate the shocks.

The awesome power of an earthquake is measured by its magnitude and intensity. Magnitude describes a quake's ground motion as recorded on seismographs and is universally rated on the familiar scale devised in 1935 by Charles F. Richter of the California Institute of Technology. On the Richter scale, an earthquake rated 2 is the smallest that people might normally feel. Theoretically, there is no upper limit to the scale, but the strongest quakes so far recorded have had magni-

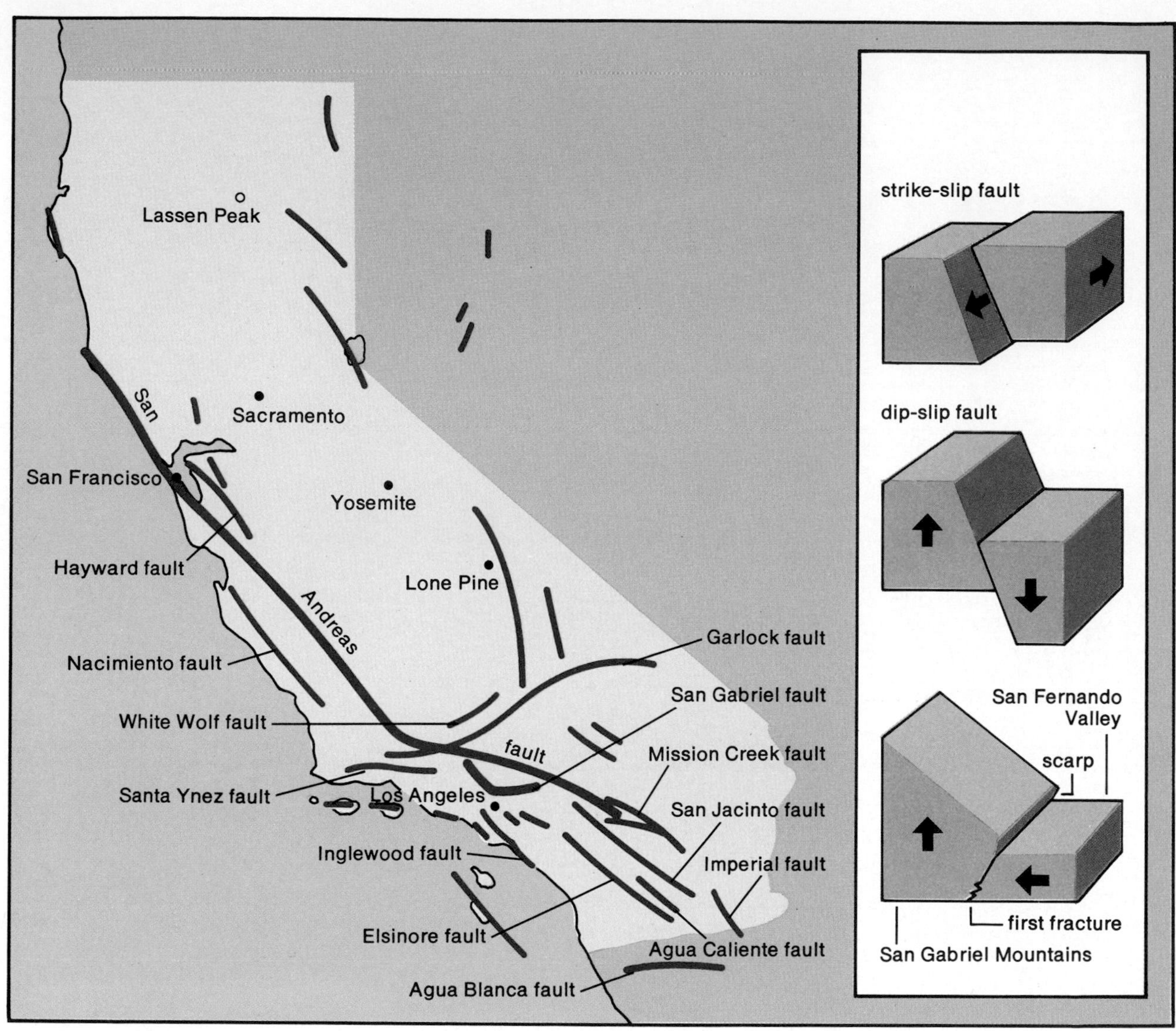

The diagram next to the map, which shows the location of active faults in California, illustrates three ways in which a fault can move: in the strike-slip fault, movement is horizontal; in the dip-slip fault, movement is vertical; and along many faults, such as the one in the area of the San Fernando Valley, movement can be both vertical and horizontal.

tudes of about 8.9. The San Fernando quake rated 6.6 on the Richter scale; the San Francisco disaster in 1906 was 8.3.

Richter's scale is logarithmic, which means that each increase of one number on the scale implies a ground motion—a seismic wave through the earth—10 times larger than the preceding number. Ground motion, in turn, is related to the energy released in a quake, and each step up on the Richter scale implies an energy release of about 30 times greater than the step before. Seismologists calculate that a "great earthquake" of magnitude 8 would release as much energy as 10,000 World War II atom bombs.

Earthquake *magnitude* and *intensity* are vastly different concepts. Intensity measures a quake's effect in a given area, regardless of its magnitude. It is most often described by the Modified Mercalli scale, named for a 19th-century Italian geologist. The scale ranges from Roman numeral I ("not felt except by a very few, favorably situated")

DESTRUCTIVE EARTHQUAKES IN CALIFORNIA HISTORY

Year	Area	Richter Magnitude	Deaths	Damage
1812	San Juan Capistrano	7-8	50+	missions badly hit
1857	Fort Tejon	8+	1	no record
1868	Hayward	7(?)	30	$350,000
1872	Owens Valley	8+	27	$250,000
1899	San Jacinto	7(?)	6	no record
1906	San Francisco	8.3	700 to 1,000	$1 billion
1915	Imperial Valley	6-7	9	$1 million
1925	Santa Barbara	6.3	13	$8 million
1933	Long Beach	6.3	115	$40 million
1940	Imperial Valley (El Centro)	7.1	9	$6 million
1952	Tehachapi (Kern County)	7.7	14	$60 million
1954	Eureka	6.5	1	$2.1 million
1971	San Fernando Valley	6.6	65	$550 million

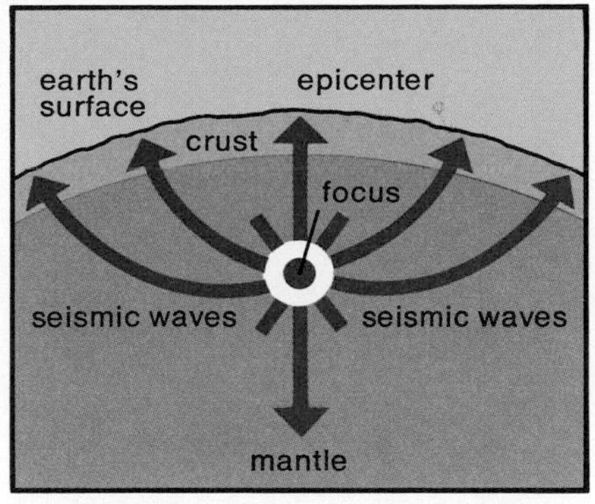

The diagram above indicates the manner in which seismic waves radiate outward from the focus of an earthquake. The location of a quake is usually described by the geographic position of its epicenter (the point on the surface of the earth immediately above the focus) and by its focal depth (the distance from the epicenter to the focus).

to XII ("damage total, lines of sight distorted, objects thrown into the air"). The 1906 San Francisco earthquake's maximum intensity was XI ("few, if any, masonry structures remain standing, bridges destroyed, broad fissures in ground, underground pipelines completely out of service, rails bent greatly").

The land moves north

Long before rails were bent or cities destroyed, earthquakes along the San Andreas were deforming the land. Some geologists, measuring the abrupt offset of rock formations and canyons on either side of the great fault system, believe that the oldest traces of earthquakes can be dated as far back as 150 million years; evidence is strong that seismic activity was occurring 60 million years ago. In historic times single earthquakes have shifted the ground horizontally as much as 30 feet in one lurch; Pakiser and Wallace at the Geological Survey estimate

Steve and Dolores McCutcheon

Pictured above is an example of the destruction caused by the Alaskan earthquake of March 1964, one of the strongest ever in North America. The creep meter shown at the top of the opposite page is a device used to measure crustal movement along a fault. Such movement can cause cracks in the earth, like those located in the desert at Ocotillo Wells, east of San Diego (opposite page, bottom).

that portions of the San Andreas are now moving two inches or more a year.

Although some earthquakes originate many tens or even hundreds of miles beneath the surface, the focal depths of the San Andreas system are rather shallow—no more than 10 miles deep. In the San Fernando Valley, the depth of origin of the recent quake was measured by calculations from many seismographs at about 10 kilometers, which is a little more than six miles deep.

Earthquakes can shift opposite sides of a fault in varying directions; horizontally, vertically, or a twisting combination of the two. Most quakes on the San Andreas move horizontally; they involve what seismologists call right-lateral strike-slip, which simply means that an observer on the east side of the fault would see the ground on the west side shifting to his right during a quake. Relief maps and aerial photos of California's valleys show this strike-slip motion clearly: stream beds have been offset as much as 1,200 feet by quake upon quake over thousands of years. Fences and roads, too, often display more recent offsets.

Like a juggernaut, this motion of the fractured earth is steadily driving the western side of the San Andreas system farther and farther to the northwest. Because the fault runs east of Los Angeles and passes westward of San Francisco, the two cities are implacably drifting toward each other. In 10 million years or so they will be neighbors. Meanwhile, in the south, the Gulf of California marks a great fault-torn rip in the land that began perhaps four million years ago and is still pulling Baja California away from Mexico.

Like most faults, however, movement rates along the San Andreas vary strikingly from place to place. There is almost no movement of the land along the fault's northern traces near San Francisco. Here the fault is said to be "locked." East of San Francisco, however, the

Hayward fault is in constant motion; and to the south, the San Andreas creeps and slips for 120 miles between the towns of Los Gatos and Parkfield. In this area the concrete foundations of a winery near Hollister are cracking all the time because of "fault creep," and seismologists have studded the surroundings of the winery with creep meters, strain gauges, and minutely sensitive seismographs. A little farther south, near Parkfield, Wallace has found two sections of a farm fence that are moving steadily apart at a rate of more than two inches a year. Still farther south, from Bakersfield to Los Angeles, the San Andreas curves sharply eastward and is locked again.

"Locking" may have ominous implications for the motionless sections of the San Andreas. To determine just how ominous, scientists at Menlo Park have been using powerful presses to squeeze, twist, and deform faulted rock samples at pressures higher than 250,000 pounds per square inch and temperatures of more than 1,000 degrees. From these experiments they have coined the term "stick slip," which means that certain types of granular rock will resist an enormous build-up of strain along their fracture lines until they succumb in one large lurch, while smoother rocks will slide more easily, or "creep," under far less strain.

Scientists like Wallace and Pakiser believe that the locked areas of the San Andreas now are undergoing a constant buildup of strain in the deep crustal material below the surface, and that eventually this strain will cause the fault blocks to snap past each other in a catastrophic release of energy. The creeping areas of the San Andreas system, by contrast, display innumerable tiny tremors each year; it is here that the subterranean strain may be releasing energy more steadily, with less danger of malevolent build-up.

Why the earth quakes

Understanding the crustal strains that lead to earthquakes has taken geology into the realm of geophysics. In the past decade research in this field has led to spectacular new findings and remarkable new concepts about the entire dynamic earth.

Oceanographic vessels from Columbia University's Lamont Geological Observatory and the University of California's Scripps Institution of Oceanography at San Diego have traced an extraordinary network of mid-ocean ridges that cleave the Atlantic and the Pacific. It is believed that out of these ridges, the plastic rock of the earth's sub-crustal mantle oozes upward by convection and spreads across the sea floor, forming great plates of new crust. Current theory holds that these thick "tectonic plates" are in constant motion. The Atlantic plate, which is coupled to the North American continent, is slowly pushing the entire continent westward. The Pacific plate, on the other hand, is driving generally eastward, and where it strikes the edge of the continent in California, the two plates meet in the great "transform fault" that is the San Andreas. Thus, as the westward side of the San Andreas

continued on page 47

Courtesy, Dr. Robert Nason, NOAA Earthquake Mechanism Laboratory

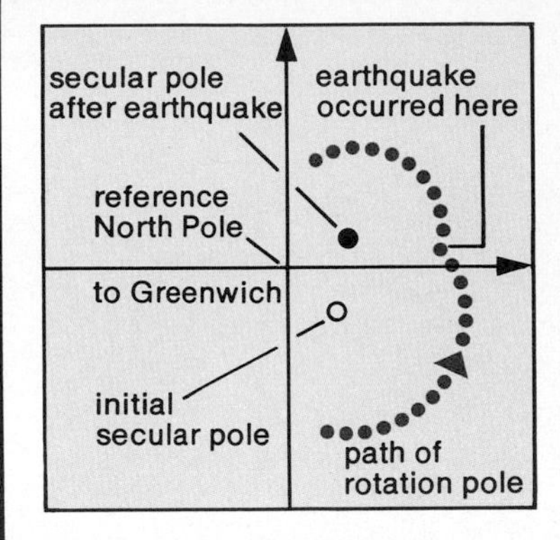

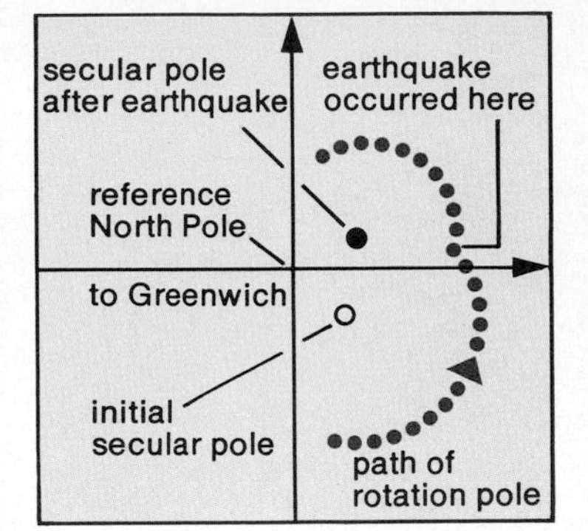

Mr. Chandler's mysterious wobble

Like all planets, the earth moves in many ways. It sweeps around the sun each year; it spins like a top to mark each day and night; and it wobbles drunkenly on its axis in a most curious fashion.

In 1891 an American astronomer, Seth Carlo Chandler, studied tiny variations in latitude that had been measured from time to time at observatories around the world. The shifts in apparent latitude were due to the earth's eccentric wobbling motion, and Chandler found that they recurred in a 14-month cycle. His calculations showed that if someone could stand a pen upright at one of the poles, and have it trace the polar motion, the pen would draw an irregular counterclockwise circle about 40 feet in diameter every 14 months.

The "Chandler Wobble" became a puzzling phenomenon for geophysicists. They wondered why the elasticity of the earth and its other rotational forces did not gradually "damp out" the wobble so that the spin axis could become smooth again. Something must excite the wobble periodically, earth scientists maintained, some force large enough to unbalance the globe's axis.

As early as 1906, John Milne of England, the inventor of the modern seismograph, proposed that earthquakes might have something to do with Mr. Chandler's wobble. Then, in 1968, two geophysicists at Canada's University of Western Ontario offered a new theory. The Chandler motion, they suggested, shows a peak about every six or seven years, and the record of large earthquakes around the world correlates closely with detectable spurts in the wobble.

The Canadians have left their colleagues puzzled, however, by two apparently conflicting ideas. Major quakes, they suggested, displace segments of the earth's crust along thousands of kilometers, and these massive changes would be enough to excite the Chandler wobble, giving the earth just the kick it needs to keep the wobble on schedule. At the same time, however, they also noted that many of the abrupt changes in the "wobble path" of the earth's poles seemed to precede major earthquakes by five to ten days, apparently suggesting that the wobble itself was helping to excite the quakes.

Early in 1971, Charles A. Whitten of the U.S. National Ocean Survey suggested that the year might be a big one for earthquakes because Mr. Chandler's wobble was at its seven-year peak. Soon after Whitten predicted that a series of earthquakes might be coming, several major ones struck: in Italy, off South America, and near New Guinea.

Whether the wobble will prove an aid in predicting quakes, or whether the quake record will prove that temblors excite the wobble, remains to be seen. Mr. Chandler's mysterious motion, at any rate, has served to whet the scientific appetites of those who hope one day to understand all the forces of the dynamic earth, from its dense metallic core to its unstable crust.

continued from page 45
presses against the east side, all that massive strain energy is converted to a grinding action in the upper layers of crustal rock.

Similar opposing movements of great continental and oceanic plates create other fault systems around the Pacific's "ring of fire." In the areas of deep-sea trenches the oceanic plates are believed to thrust themselves beneath the continents in a slow, gigantic process that lifts entire mountain chains, crumples coasts, and injects volcanoes with fresh charges of molten magma from the deep interior.

The new geophysical concepts about sea-floor spreading, continental drift, and tectonics (literally, the study of the earth's deforming forces) are giving scientists new and promising theoretical leads in their efforts to predict and even control earthquakes. While some seismologists, including Richter, are dubious about the near-term possibilities for prediction and control, others like Pakiser believe that if research funds are forthcoming, reasonably accurate earthquake forecasts might be possible before 1980.

Distinguished scientific panels, including one appointed by the U.S. National Academy of Sciences, call the new discoveries in tectonics "a major scientific breakthrough with important implications for all aspects of earth sciences." The Academy recently recommended a federal seismology program totalling $500 million over the next 10 years. According to the Academy, the investment would be returned manyfold, such as improving practical applications of seismology in prospecting for oil and minerals, reducing hazards in seismic areas of the nation, and making possible an earthquake prediction capability that could prove as economically useful as weather forecasting.

Can earthquakes be predicted . . .

There is solid evidence that earthquake prediction is indeed a reasonable goal. Scientists at Stanford University, for example, have learned how to measure changes in the magnetic properties of rocks that precede earth movements by a few hours. They believe that as crustal strains magnify just before a fault gives way and the rocks are distorted, magnetometers can measure such premonitory changes in the magnetic fields around stressed faults.

In Japan, seismologists have been using tiltmeters and laser beams to measure minute tilting, shifting, and subsidence of the earth in active fault zones. Coupled with observations that swarms of tiny earthquakes frequently occur as "foreshocks" to a strong temblor, Japanese scientists were able to issue valid predictions of quakes at Matsushiro in 1966. T. Rikitake of the University of Tokyo reported that the forecasts were scientifically successful and "helpful for local governments." But, he added wryly, "those engaged in the sightseeing and hotel business were not really pleased." In Japan's Kanto area, south of Tokyo Bay, seismologists are trying to determine whether accumulated strain in the earth's crust alters the density of rock and thereby changes the travel time of seismic waves touched off by experimental

The diagram at the top of the opposite page illustrates the effect of an earthquake on the pole path, which moves in a counterclockwise direction before and after the quake. When a large earthquake occurs, the center of rotation (secular pole) is shifted. However, because there is a nearly equal and opposite vector contribution to the wobble, the position of the rotation pole is hardly altered; only the path it describes in time is changed. The graph beneath the diagram shows the correlation between the earth's polar wobble and major earthquakes. The dotted line indicates the mean daily shift of the polar wobble from 1900 to 1970; the white area indicates the energy released by recorded earthquakes during the same period. Note the frequency with which spurts in the wobble correspond to the occurrence of earthquakes.

Courtesy, NOAA
and California Institute of Technology

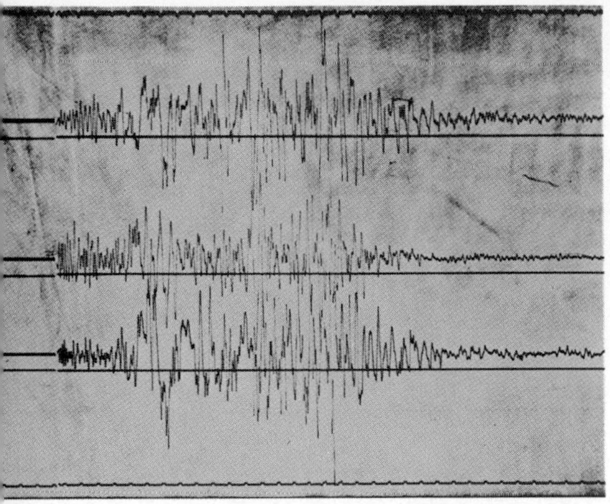

The strong-motion accelerograph record shown above indicates how violently the ground moved during the San Fernando earthquake, which lasted 12 seconds. Because the accelerometer that made this record was situated near the quake's epicenter, it was the strongest ground motion ever recorded. The recording below was made during the main shock of the same quake by a seismoscope located at the crest of the lower Van Norman Dam. Such a recording reveals how a structure responds to the forces exerted on it by movements of the earth.

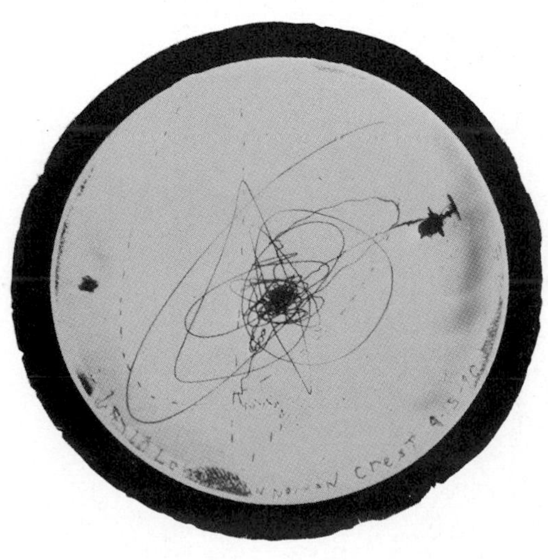

explosions. The Japanese are following up a technique pioneered at Stanford Research Institute, and they hope the measurement of explosive shock waves will enable them to monitor crustal strain as it builds up.

In California the San Andreas is crisscrossed with lines surveyed by geodimeters and geodolites, sensitive devices that use light waves and mirrors to detect tiny displacements of land across fault lines. Tiltmeters precisely register uplift and subsidence. More than 100 seismographs, measuring the ground motion of small and large quakes, are arrayed throughout the state and transmit their continuous signals by telemetry to a central computer at the Geological Survey's Menlo Park center. According to Pakiser: "If we could intensify all these arrays of instruments, and combine the measurements of creep, tilt, seismic activity, and magnetic changes, we could move into high gear and develop scientifically useful quake predictions within five years. Such predictions eventually could provide the basis for practical earthquake warnings. And that's remarkable, because a few years ago if anyone talked about forecasting earthquakes, people thought he was trying to peddle clairvoyance."

. . . And can they be controlled?

Beyond prediction lies the even more exotic goal of controlling earthquakes. Since 1966 seismologists have noted that underground nuclear explosions in Nevada touched off tremors along nearby fault lines. The one-megaton Benham blast of 1968 caused a swarm of more than 1,000 small quakes within an eight-mile radius of ground zero the day after the explosion. More than 100 aftershocks a day still were being recorded 24 days later. Some scientists have even proposed setting off small nuclear blasts within locked faults to release them before strain builds up to the danger point.

Perhaps more realistic, however, is an earthquake-control experiment now under way at an oil field in Rangely, Colo. In 1966 geologists determined that a sequence of increasingly alarming earthquakes around Denver were being caused by injection of liquid wastes from a U.S. Army poison-gas plant into a 12,000-foot well. The liquid was lubricating an old locked fault and causing it to slip in a series of jolting steps. After the Army stopped pumping wastes into the well, the quakes began to subside. The study of earthquakes that had occurred near large dams strengthened the lubrication concept, and in 1969 a team of Menlo Park scientists pinpointed this phenomenon as the cause of an earthquake that claimed 200 lives near the new Koyna dam in India.

These findings have led to the dramatic Rangely experiment, where Chevron Oil Co. has turned over four empty oil wells to Barry Raleigh and his Geological Survey colleagues to serve as a field laboratory. Chevron engineers had been using the wells to inject up to 120,000 gallons of water a day into the oil field to increase subterranean pressures so that more oil could be recovered from other producing wells nearby. While the water was being injected, an array of 16 seismographs emplaced by the Geological Survey team detected up to 60 earthquakes a week, most of them tiny, but some registering as much as 3.5 on the Richter scale. In November 1970, Raleigh and his team began removing the water from the wells. By early 1971 the earthquake count dropped spectacularly; during some weeks there were none at all. The experiment was scheduled to continue into 1972, with alternating cycles of water injection and withdrawal, to see if quake frequency would rise and fall the way Raleigh predicted. "We're by no means certain, but the results so far look very promising," Raleigh said.

If experiments like the one at Rangely do prove successful, Raleigh envisions clusters of fluid-injection wells drilled every 12 to 15 miles along dangerously locked faults, so that strains can be released gradually and predictably. Lubricating the faults might touch off moderate man-made quakes with Richter magnitudes up to 5 every few years in relatively safe areas of a fault zone, according to Raleigh. But this, he believes, would be better than allowing tectonic strains to build up for a century until they triggered a single catastrophic 8-magnitude quake.

Thus, the basic science of geophysicists, the modern instruments of seismologists, the skills of safety-conscious engineers, and the ingenuity of research teams working on prediction and control are bringing the day closer when the trembling earth will no longer arouse superstitious dread or place men's lives and works in jeopardy.

See in ENCYCLOPÆDIA BRITANNICA (1971): EARTHQUAKE; FAULT; SEISMOLOGY.

Courtesy, Dr. Robert Nason,
NOAA Earthquake Mechanism Laboratory

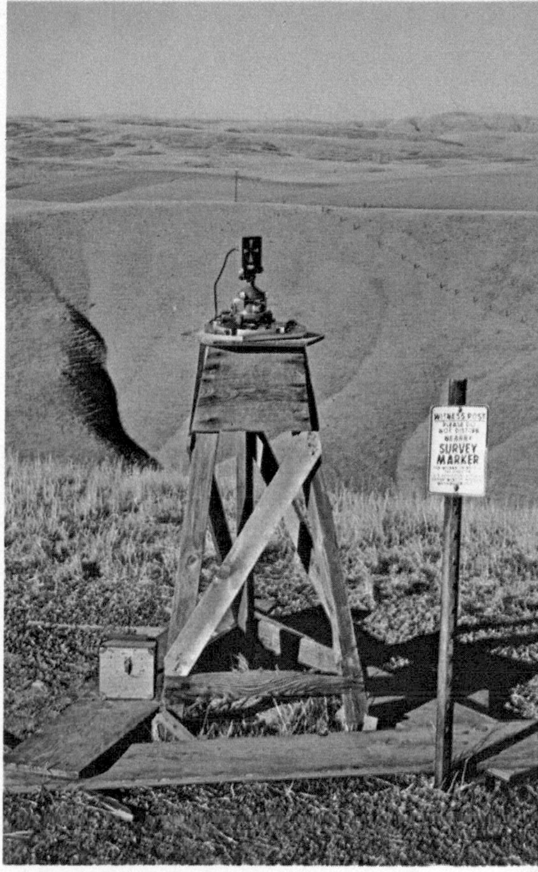

This surveying instrument, which is located near Cholame, Calif., is one of many used to monitor changes along the San Andreas fault. By taking periodic sightings on markers located on the other side of the fault line, surveyors can determine if there has been any slippage along the fault.

FOR ADDITIONAL READING:

Environmental Science Services Administration, *Earthquakes* (U.S. Government Printing Office, 1969).

Heintze, Carl, *The Circle of Fire* (Meredith Press, 1968).

Iacopi, Robert, *Earthquake Country* (Lane Book Co., 1964).

In the Interest of Earthquake Safety, findings of a White House Task Force (Institute of Governmental Studies, University of California, Berkeley, 1971).

Pakiser, L. C., *et al.*, "Earthquake Prediction and Control," *Science* (Dec. 19, 1969, pp. 1467-74).

Seismology: Responsibilities and Requirements of a Growing Science (National Academy of Sciences, 1969).

AUDIOVISUAL MATERIALS FROM ENCYCLOPÆDIA
BRITANNICA EDUCATIONAL CORPORATION:

Films: *Earthquakes: Lesson of a Disaster; San Andreas Fault.*

Film Loops: *Earth Structures*; *Overthrust Faults*; *Tectonic Movements.*

Asphalt Cemetery of the Ice Age

by Theodore Downs and George J. Miller

In downtown Los Angeles, a popular tourist attraction is providing scientists with information about the animal and plant life of southern California 40,000 years ago.

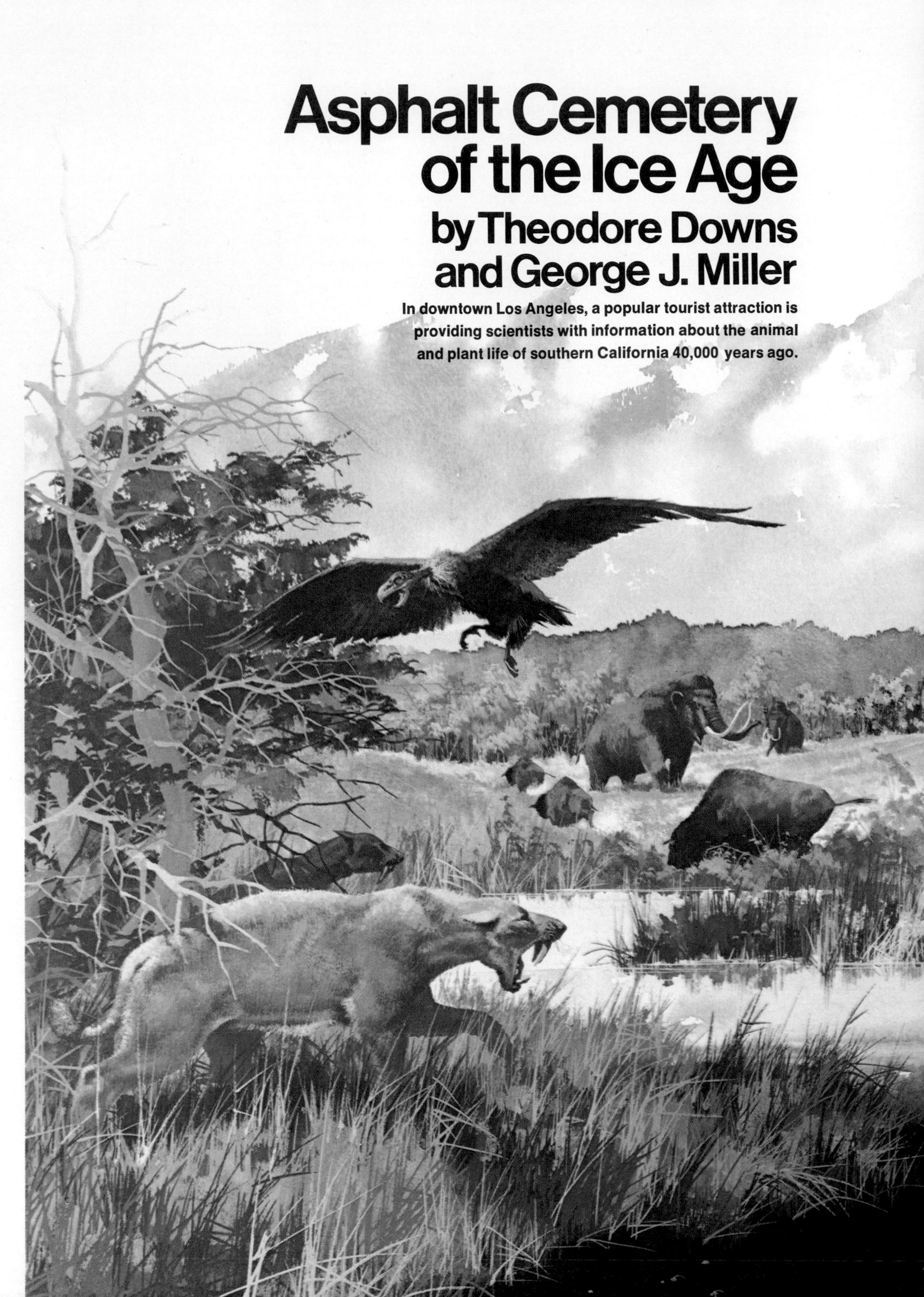

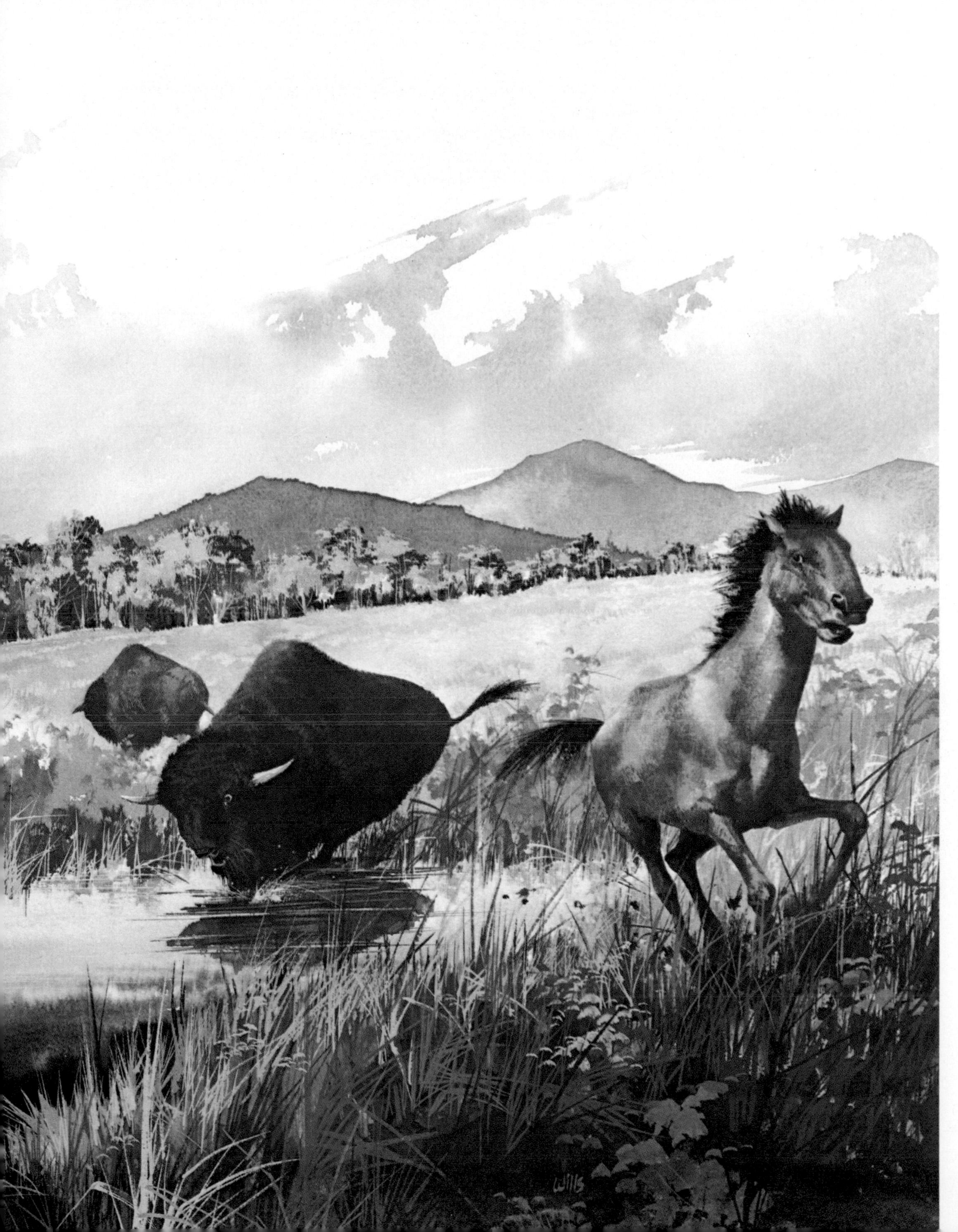

The diagram on the opposite page shows the glacial and interglacial stages of the Pleistocene Epoch, which began about two million years ago and ended 11,000 years ago. The Rancho La Brea flora and fauna are believed to date from late in the Upper Pleistocene, just before the great Ice Age glaciers began their final recession.

Rancho La Brea in Los Angeles is an unlikely location for an active paleontological dig. Located in 23-acre Hancock Park, it is overshadowed by the banks, office buildings, and department stores that line Wilshire Boulevard's bustling Miracle Mile. One corner of the park is occupied by the Los Angeles County Art Museum. In the "Pleistocene meadow," families, young couples, and senior citizens bask in the occasionally smog-obscured southern California sun. But in addition to being a popular tourist attraction, Rancho La Brea, a satellite of the Los Angeles County Museum of Natural History, is also the site of a major attempt to reconstruct what Ice Age life was like from 5,000 to 40,000 years ago.

To the half million visitors who come to Rancho La Brea every year, it may not be obvious immediately that scientific history is in the making. Clues are fragmentary and seemingly unrelated. A slight odor of petroleum hangs in the air, and walkers along the paths that cut through the meadow must be on the watch for little puddles of black tar that ooze through the sandy soil. In an artificial lake adjoining the meadow, oily bubbles rise and break with a pop at the surface. At the edge of the lake, a life-size replica of a huge female mammoth appears to be struggling to pull itself free, while its mate and calf look helplessly on.

A few yards from a neat wooden building that looks very much like a Swiss chalet is a pit some 10 feet deep and 28 feet square. The chalet is a privately donated temporary laboratory, and within the pit scientists and an enthusiastic crew of volunteers gather around a large gray block, carefully chipping away the asphalt coating from a collection of ancient bones. They are bringing the scientific excavation of Rancho La Brea back to life nearly 60 years after many paleontologists said that "the La Brea tar pits," the richest treasure house of Ice Age fossils in the world, had little more to offer.

As the scientists in charge of the new Rancho La Brea dig, we believe that the earlier excavations literally only scratched the surface of the scientific information that still lies encrusted with asphalt in the depths of pit 91. We hope that, as we delve deeper than anyone else has ever done before, we will be able to show how plant and animal life responded to the advance and retreat of the great ice sheet of late Pleistocene time. We are particularly interested in the Wisconsin stage, 500 to 700 centuries ago, which witnessed the coming and going of the many extinct life forms now buried in the asphalt cemetery of Rancho La Brea.

THEODORE DOWNS is chief curator of the Earth Sciences Division at the Natural History Museum of Los Angeles County. GEORGE J. MILLER is the Rancho La Brea project coordinator for the Los Angeles County Museum of Natural History Foundation.

A new look into pit 91

Much of what the workers in pit 91 are looking for is obvious. Peering over their shoulders, one might see, packed tightly into the asphaltic matrix, an animal skull with two enormous tusks jutting down from its upper jaw. Another fossil might appear to be the head of a dog, or perhaps a wolf. Also visible might be the hipbone of an extinct mammal that must have been as large as an elephant.

As valuable as such large fossils are, we will never be able to recreate the ecology of the late Pleistocene period until we can relate them to the nearly microscopic life forms that lived and died at the same time as the larger mammals, fowls, insects, trees, and plants that left their marks in the pit. We also would like to know what role, if any, man played in that paleontological scenario.

Supported by the Los Angeles County Museum of Natural History Foundation, a grant from the National Science Foundation, and public donations, the new excavation at pit 91 was begun on "Asphalt Friday," June 13, 1969. Before we finish the job, we will have drawn on the talents of chemists, geologists, geophysicists, pathologists, computer programmers, and other scientists who are making paleontology a more precise and sophisticated science than it has ever been.

With their help, it should be possible to achieve a better understanding of how and why so many animal species became extinct in the extremely short span—geologically speaking—of a few thousand years. In addition, they may add to the growing evidence that man roamed California at the same time those extinct animals were entombed in asphalt, perhaps 10,000 or 20,000 years earlier than anthropologists had once believed.

Some La Brea history

Although the new knowledge we hope to uncover will be built on the findings of a series of earlier excavations, carried out sporadically from 1901 to 1915, the asphalt deposits that still bubble up from far below the earth's surface were objects of interest long before that.

In 1769 the explorer Gaspar de Portolá visited the area and observed "extensive swamps of bitumen, which is called chapapote." In his journal, Portolá wondered whether or not "this substance which flows melted from beneath the earth, could occasion so many earthquakes." Thirty-three years later, another traveler commented with some awe on the sight of small animals and birds sinking into the "pitch," especially in hot weather. He theorized then, as many still do, that the victims mistook the asphalt pools for water holes and became trapped inextricably before they realized their error. In fact, even today it is not uncommon for rabbits and birds to become mired in the tar seeps at Rancho La Brea.

By 1828 the pits had been included in a large Mexican land grant of about 4,500 acres that became known as Rancho La Brea, and the practical residents of the small town of "The Angels," or Los Angeles, began to use the soft asphalt to tar the roofs of their adobe huts. In the 1870s Rancho La Brea was purchased by Major Henry Hancock, a lawyer who planned to sell the asphalt in San Francisco for fuel and street paving material. Until then, most of the bones found in the pits were believed to be from local cattle. But when Major Hancock, in 1875, presented a sabertooth canine and other peculiar-looking bones to William Denton of the Boston Society of Natural History, the latter immediately recognized their scientific significance.

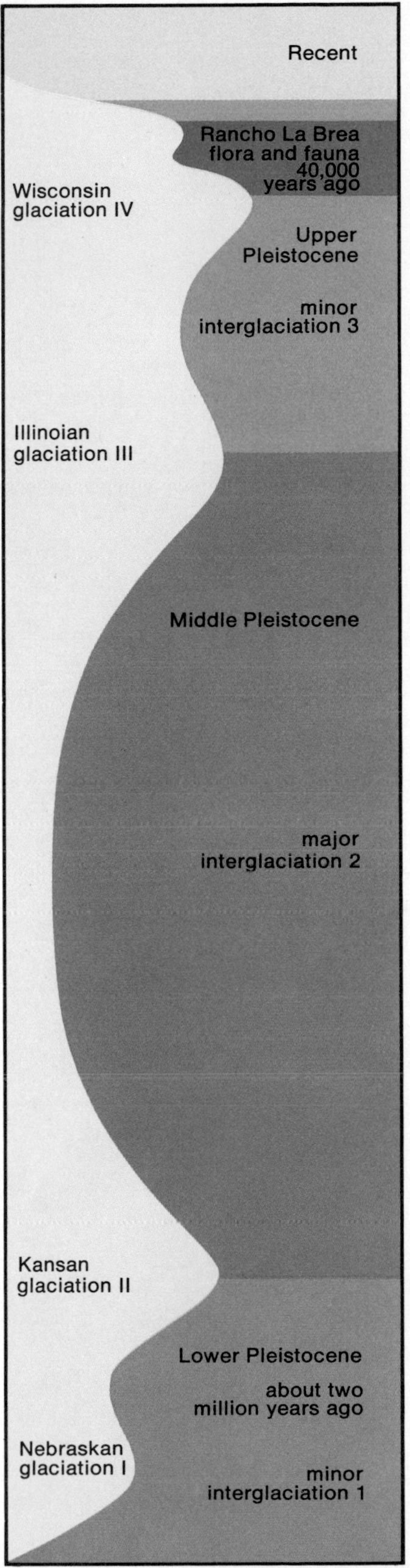

A skull of a California saber-toothed "cat," Smilodon californicus, *is revealed in a fossil concentration at Rancho La Brea. Except for some discoloration caused by the asphalt, bones found in the tar pits are remarkably well preserved.*

In 1901 the first serious excavation of the pits was initiated by William Orcutt, a geologist from the Union Oil Company who was interested chiefly in the area's oil potential. He had enough foresight, however, to forward some of the interesting bones he unearthed (a saber-toothed "cat" skull, the jaw of a large wolf, and bones from a ground sloth) to John C. Merriam, a paleontologist at the University of California. Although Merriam was immediately impressed with the site, it was not until 1912 that he was able to obtain sufficient funds for a major excavation, which resulted in the removal of more than 40 tons of bones from the pits.

A year later, when Allan Hancock, the major's son, gave exclusive excavation rights to the Los Angeles County Museum, the museum's L. E. Wyman began the first controlled study of the La Brea pits. He divided the excavation area into a three-foot grid system and identified each bone as to grid number and its depth below the surface. But Wyman's controls were difficult to maintain, since the walls of the excavated area often collapsed and the pits were inundated with water in the rainy season. In spite of these difficulties, however, Wyman's group did open 100 pits, some of them only shallow trenches. The deepest pit went down 27½ feet, compared with our own goal of 50 feet. They unearthed great quantities of sabertooth, wolf, bison, and sloth material, along with a large number of camel and horse fossils.

Because of the activities of unauthorized bone hunters and outright poachers at the unguarded pits between 1901 and 1915, it is impossible to say just how many bones have been recovered. We estimate that well over one million specimens, representing more than 200 plant and animal species, have been removed, and we are adding to that total every day. We find, as Wyman did, that even a few cubic yards of asphalt are packed with hundreds of bone specimens.

One of the outstanding features of the Rancho La Brea excavations —past and present—is the remarkable preservation of many of the fossils. Except for a brown to blackish discoloration of bones and teeth caused by oil and asphalt, they are, in all important aspects, indistinguishable from contemporary teeth and bone. Some mammalian skulls have been recovered in which the threadlike network of the inner ear is still intact. Wood found at Rancho La Brea also has a very fresh appearance, while cones from ancient pine and cypress trees retain much of their original detail.

Harvey Singer © EB Inc.

Volunteer workers remove fossils from the asphaltic matrix at the Rancho La Brea dig. The excavation would not have been possible without the dedicated effort of some 2,000 volunteers serving under the direction of a small permanent staff.

The La Brea mammals

Most prevalent among the larger fossils are the carnivores, especially the dire wolf (*Canus dirus*) and the California saber-toothed "cat" (*Smilodon californicus*). Herbivorous mammoths, mastodons, and giant ground sloths also are well represented.

The dire wolf ranged from east to west in North America during the late Pleistocene. It was a little smaller than today's northern timber wolf and had strong jaws and teeth. Because its remains are found in large numbers, scientists suspect that the dire wolves traveled in packs that preyed on animals mired in the asphalt, then became trapped themselves. The saber-toothed "cat" was about as large as an African lion, had powerful front limbs, and possibly used its two dagger-like canine teeth to stab its prey to death.

Largest denizen of the La Brea pits is the mammoth, an elephant-like mammal that stood 13 feet high at the shoulder and had curved tusks that sometimes attained a length of 15 feet. Also found at Rancho La Brea were the smaller mastodon and the lumbering giant ground sloth. The latter were plant eaters with enormous front paws and kangaroo-like tails. They were possibly prey for the dire wolf and saber-toothed "cat."

Other animals taken from the asphalt pits help characterize the breadth of life in the Pleistocene. They include the California lion (*Panthera atrox*), the largest cat that has ever lived, and horses, which became extinct in the Western Hemisphere long before the Spanish reintroduced them to North America in the 16th century. Camels, bison, antelopes, rodents, rabbits, tapirs, peccaries (wild pigs), and deer are also among the many reconstructed animals at the Los Angeles County Museum of Natural History. Many of these animals are now extinct, though as yet we do not know why.

Interestingly enough the pits contained many more fossils of the dire wolf than of its surviving relatives, the coyote and gray wolf. The saber-toothed "cats" far outnumber the lion-like cats and pumas. This preponderance of extinct forms is a strong indication of the antiquity of

Representative mammals and birds from Rancho La Brea illustrate the great variety of animal life that flourished in southern California near the end of the Pleistocene.

the pits. Accordingly, we hope that the present dig will enable us to determine much more accurately when and why the various forms were prevalent, how they became entombed, when they disappeared, and what factors may have caused their extinction. Judging by what we already know, the extinction of many mammalian species began about 14,000 years ago and hit a peak about 8,000–10,000 years ago.

Birdlife at Rancho La Brea

An impressive analysis of the birdlife of Rancho La Brea from Pleistocene days to the present was prepared in 1962 by Hildegarde Howard,

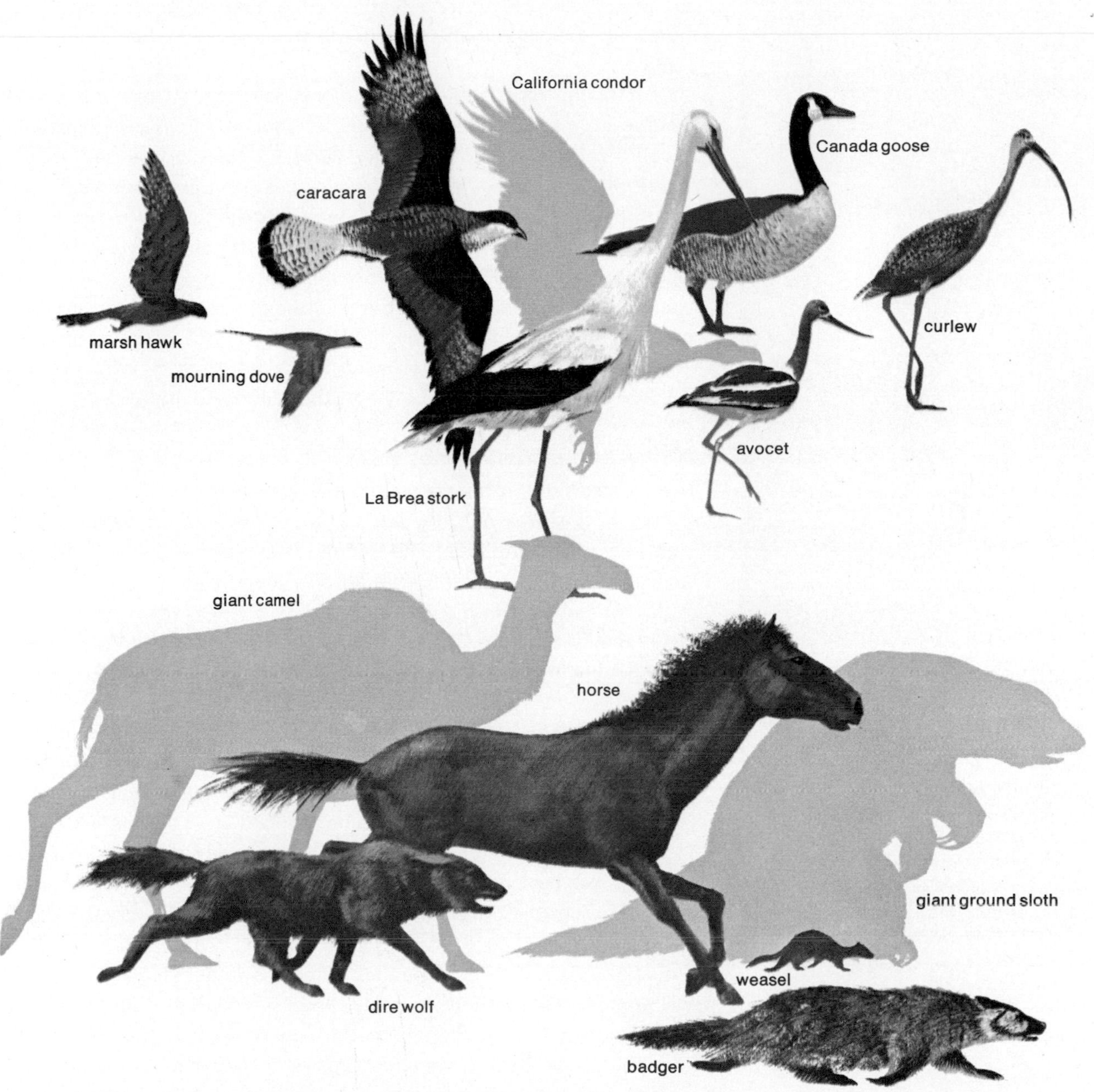

former chief curator of the Science Division at the Los Angeles County Museum. Of the 125 Ice Age birds represented in the asphalt deposits at La Brea, she notes, 20 are extinct. Howard supports the now generally accepted hypothesis that the pits were active at different times and thus might be excellent indicators of shifts in climate, environment, and the makeup of the animal populations. For example, certain water birds were found in great numbers in some pits but were entirely absent in others. In some of the pits, dated as representing the most recent past, brush- and tree-dwelling birds increased, perhaps mirroring similar changes in the plant life of the area. David I. Axelrod of the Uni-

Courtesy, Los Angeles County Museum of Natural History

This photograph of pit 91 was taken in 1915, just before the pit was covered over. This marked the end of the first controlled excavation at Rancho La Brea, carried out between 1913 and 1915 under the direction of L. E. Wyman of the Los Angeles County Museum.

versity of California at Davis has concluded that at least two basic floral communities were preserved at La Brea, one suggesting a slightly more humid climate than exists in the area today and the other a drier climate.

Most of the birds represented in the Rancho La Brea fauna were predators or scavengers, including condors, vultures, eagles, hawks, falcons, and owls. Many of these, such as the sharp-beaked vulture (*Teratornis merriami*), a huge bird with a 12-foot wingspread, undoubtedly fed on the remains of dead animals already mired in the asphalt pits. Among the nonpredators, the most abundant is the extinct California turkey (*Parapavo californicus*).

Judging from plant fossils found in the pits, oak, juniper, cypress, and pine trees provided resting sites for many of the larger bird species. Howard theorizes that low-growing brush would also have provided excellent cover for smaller bush- and ground-dwelling birds.

Where the asphalt comes from

The causes of the formation of the asphalt pits at Rancho La Brea are at best conjectural, but they are surely the result of the peculiar geologic features of the area. The Rancho is part of an alluvial plain that fans out from the rugged Santa Monica Mountains, two miles to the north. Beneath a more or less porous upper layer, which is about 40 to 140 feet deep in the La Brea area, are marine shales, sandstone, and oil sands that extend backward in time to the Pliocene (about 3 million to 12 million years ago).

The underlying strata slope downward as a southwesterly directed limb of an anticline. After deposition in a marine environment, the beds were folded and faulted, then eroded during Pleistocene time. The Pleistocene La Brea sediments lie horizontally over the marine strata. This structural situation is ideal for the formation and accumulation of oil, as was perfectly clear to the petroleum prospectors of 1914 who opened up the old Salt Lake oil field just to the northeast of Rancho La Brea.

Asphalt apparently accumulates when the underlying oil works its way along the older Pliocene strata, then seeps up through a network of natural pipes or chimneys to the surface. When the lighter petroleum fractions evaporate into the atmosphere—as they still do—the residual asphalt may enter natural pools and streams, or permeate the buried sediments and the fossils they contain.

It is easy to speculate that animals mistook the asphalt traps for water holes and became mired in them. Or, perhaps, they blundered into them when the pits were covered with a thin layer of dust, and predators and scavengers pursued them and were caught themselves. However, since we recently found the rocky traces of a Pleistocene stream bed meandering through pit 91, we suspect that the picture may be much more complicated. Weathering marks and abrasions on some of the bones from the earlier excavations suggest that they could have been washed down to Rancho La Brea from as far away as the Santa

Harvey Singer © EB Inc.

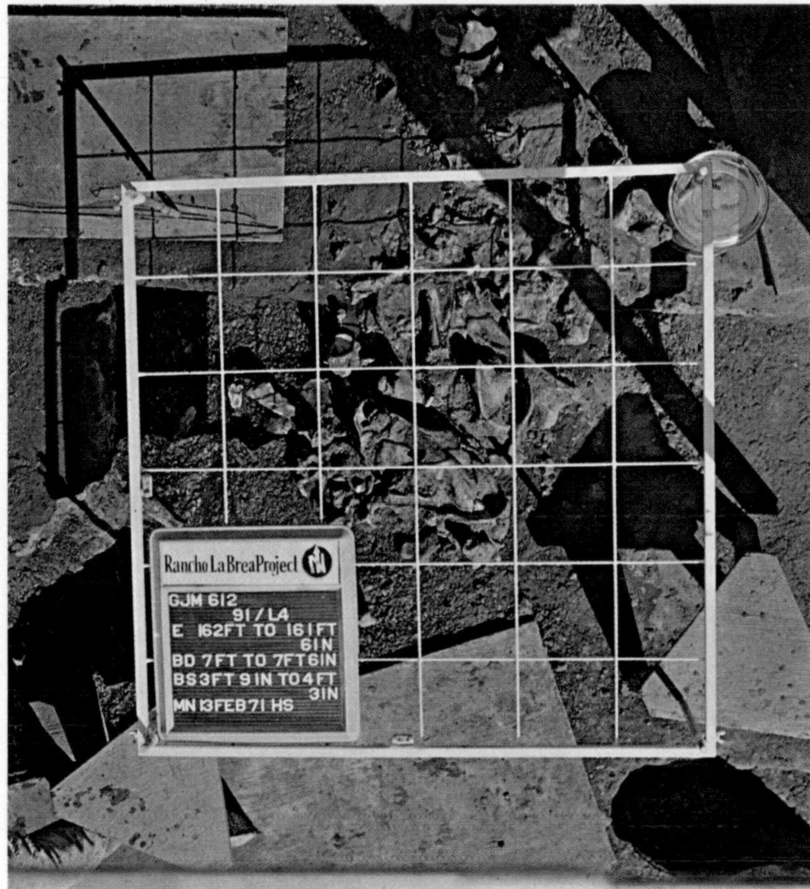

Pit 91 is shown here as it appeared in 1971, following new excavation. The grid system, shown in place over an exposed bone concentration, correlates with that used by Wyman but permits the position of finds to be recorded in greater detail.

Monica Mountains, their surfaces becoming battered and eroded as they tumbled through the rushing water.

With the present dig at Rancho La Brea, we hope to establish a meaningful time frame for a paleontologic and geologic history that began at least 40,000 years ago and extends into the relatively recent past. Radiocarbon and other modern archaeological dating techniques have indicated that wooden artifacts in pits 61–67 are about 4,500 years old. Mammoths found in pit 9 may date as far back as 40,000 years ago. A single excavation in pit 3 revealed layered deposits of saber-toothed "cat" bones ranging in age from 12,650 years near the top of the pit to 21,400 years in the lowest deposits of the same pit.

La Brea Woman and Ice Age man

Another major objective of the present excavations is to determine whether a link exists between the flora and fauna of Rancho La Brea and man himself. There is a growing body of evidence that man appeared in North America much earlier than the 5,000 to 10,000 years ago previously believed by most anthropologists. Some scientists now argue that man could easily have followed Ice Age mammals across the Bering land bridge from their Asian homes, perhaps as much as 25,000 years ago. This argument was fortified recently when scientists

This generalized cross section of the geologic structures underlying Rancho La Brea illustrates the faulting and folding of the older Pliocene strata. Oil works its way to the surface where the lighter fractions evaporate, and the residual asphalt permeates the fossiliferous sediments.

at the University of California at Los Angeles determined the age of a male human's skeletal remains, found by Los Angeles sewer workers during the 1930s, as about 24,000 years before the present. The bones were found only a few miles from Rancho La Brea, but it is difficult to say whether there is any real link between "Los Angeles Man" and "La Brea Woman."

La Brea Woman was found in 1914, when Merriam's group removed portions of a female human skeleton in pit 10. Recently, her age was determined by a modification of the radiocarbon dating technique to be 9,000±80 years. She appears to have been five feet tall, with characteristic Indian features, and about 25–35 years old when she died.

How she got into the pit will probably always remain a mystery, but the frustrating fact that no other human remains have been found at Rancho La Brea suggests that she was probably thrown into the pit, alive or dead, rather than having wandered in accidentally. We are hopeful—but not optimistic—that we will find other human remains. It seems reasonable that if a man or woman did stumble into the pit, one of the victim's friends would have come to his or her assistance.

Regardless of the reasons for her appearance in pit 10, the age determination of La Brea Woman is an excellent example of the sophisticated techniques that must be developed to answer the many questions that still enshroud this particularly challenging excavation. The conventional radiocarbon dating technique is based on the observation that all living or once living things contain traces of carbon-14, a radioactive isotope that decays at a constant, measurable rate. The problem has been that both the bones and the asphalt have their own radiocarbon age, the asphalt having been formed a million or more years before La Brea Woman lived. Rainer Berger and Richard Reynolds of the University of California at Los Angeles and the Natural History Museum were primarily responsible for overcoming this difficulty.

In 1968–69 they and their colleagues used a variety of chemical techniques to separate the asphalt from the bones, and then measured the age of a specific protein found in bones, called collagen. When its chemical building blocks are separated and burned, the radiocarbon contained in the resulting carbon dioxide gas can be measured with a radiation counter.

Although the 9,000 ± 80 year age of La Brea Woman does not conflict seriously with earlier estimates of man's appearance in North America, peculiar markings on the leg bones of some saber-toothed "cats" and other Ice Age animals suggest that man may have roamed Rancho La Brea at a much earlier date. These 15,000-year-old fossils, removed from pit 4, are marked with parallel grooves, usually about half an inch apart and about a quarter of an inch deep, cut into the bone. We suspect they were made by man. On the other hand, it has been argued that the grooves could have been caused by one bone rubbing against another in the pit, or perhaps by the teeth of some animal. However, the nonrandom nature of the cuts, and our own attempts to duplicate the conditions for accidental marking, convince us that these explanations are questionable.

Using flakes of chalcedony (a form of quartz), similar to the tools used by primitive man, one of the authors (Miller) sawed the chips back and forth across other bones, producing cuts very similar to those found on the sabertooth fossils. We also have given fresh bones to animals at several zoos, and in no case did their teeth leave marks similar to those found on the 15,000-year-old bones.

As yet we have no explanation for the bone markings if they were man-made, but it is possible that they may have had some ceremonial significance. Or perhaps man worked bone near the pit sites, and we are finding his rejects. However the marks were produced, most of the experts who have seen them state either that they were probably man-made or that they have seen nothing exactly like them in the past.

New techniques for an old age

In attempting to learn as much as we can about the interplay of plant and animal life, climate and environment during a 40,000-year period, we must record the location of every fossil, from tiny pollen to the bones of a mammoth, more precisely than has been done in the past. Accordingly, we have modified the grid system developed by Wyman in 1913, in which each fossil was located as to its position in a three-foot grid and the depth at which it was found. To correlate our own findings with the earlier digs, we are using a three-foot grid system that will also record the precise position of each fossil within a grid. The grids are subdivided into six-inch sections and are lined up with the magnetic north for horizontal control. Our vertical controls are measured relative to sea level as well as to depth below the present surface. The position of each specimen is located by measuring the distance of each end of the specimen north and west of the southeast corner of the grid. Elevation is also recorded.

Courtesy, George J. Miller

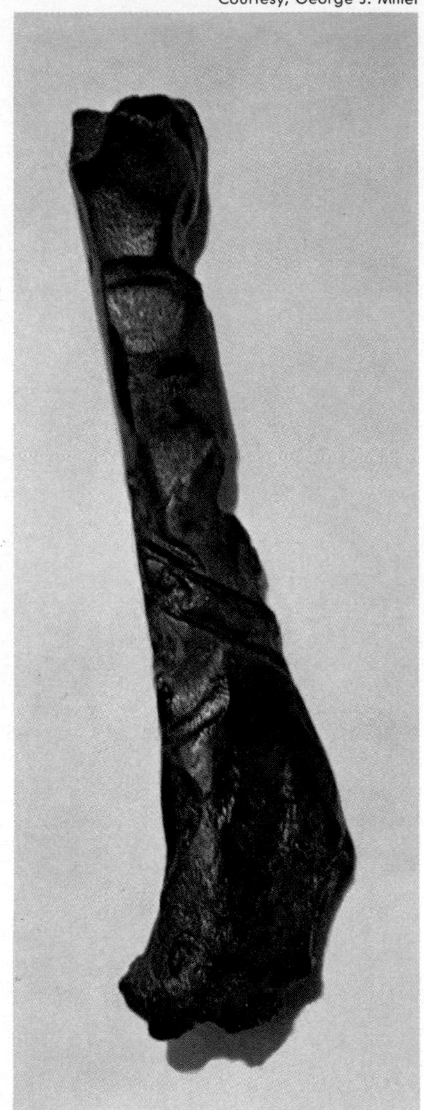

In earlier excavations, the microfossils in the asphaltic matrix surrounding the larger fossils were usually lost when the matrix was thrown away. Today, however, all matrix is saved and washed in a vapor degreaser of the kind that industrial concerns use to remove oil from machine parts. The matrix is boiled with an organic solvent in the degreaser in order to separate the tars, silts, muds, and other fine material from the microfossil concentrate. From five gallons of matrix we usually get about a quart of concentrate. This is then painstakingly examined by volunteer workers and separated into groups—tiny teeth, insect parts, bones, seeds, snails, and other microfossils. Often the fossils recovered in this manner are no bigger than the period at the end of this sentence.

Although we operate the La Brea dig with a small permanent staff, the brunt of the effort is borne by our army of 2,000 volunteer workers. Working in small contingents, they contribute about 3,000 man-hours a month, both in pit 91 and in our temporary laboratory facility. The volunteers include a large number of housewives, many students and their teachers from local high schools and colleges, artists, musicians, physicians, airline stewardesses, nurses, and at least one professional wrestler. Their ages range from 16 to 88, and they are universally enthusiastic about their work, which in many cases can be extremely tedious and arduous.

Some of the tedium connected with the separation of fragile specimens and microfossils was reduced recently when we found that we could easily recover them from the asphalt matrix by placing them in ultrasonic cleaning devices, similar to those used to clean jewelry, surgical tools, and dentures. By passing high-frequency sound waves through jars of solvent containing the matrix-encrusted microfossils, we can now recover as much as 95% of the fragile specimens in three minutes. Previously it had taken a volunteer, working with a microscope and a tiny needle, as much as two and a half weeks to separate less than 10% of the available fossils. Moreover, there was always the danger that a mere slip of the hand could seriously damage the tiny specimens.

We have also been able to clean larger fossils with an industrial agitator ordinarily used to deburr metal parts. The tublike agitator contains millions of small plastic pellets. Matrix-covered fossils can be placed in the agitator and then picked out, completely clean, a second or two later.

Another comfort not available to earlier diggers at Rancho La Brea is an air conditioner, especially welcome when the temperature in pit 91 soars well past 100° F. Unlike Wyman's hard-pressed crew, we also have the assurance that the walls of pit 91 will not come tumbling down around us. Thanks to modern shoring techniques developed by the construction industry for use in the high-rise buildings along the Miracle Mile, we hope to be able to sink our excavation straight down, perhaps to a depth of 50 feet. That would be nearly twice the depth reached in 1915.

Lesson for mankind

As the new information is collected, it will be correlated with the old in a computer program. This should present a unique opportunity to focus a "microscope" on a 40,000-year era that witnessed the extinction of so many animal and plant species. Since Rancho La Brea does not represent one "catastrophic" event or mass burial, but rather a series of catastrophies spread over thousands of years, the quantity and variety of information to be revealed is almost infinite.

Finally, when we have learned more about the degree and rate of environmental change in that 40,000-year period, it may be instructive to measure the effects of those slowly accruing changes against the effects of the changes man has imposed on his environment in the last 200 years. Perhaps there is a lesson in the asphalt pits of Rancho La Brea that mankind cannot afford to ignore.

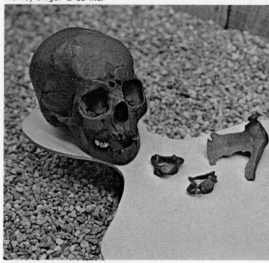

La Brea Woman, above, removed from pit 10 in 1914, represents the only human remains found at Rancho La Brea to date. Possible further evidence of man's presence in the area is the existence of regular cuts on some animal bones, such as that of Smilodon *shown on the opposite page.*

FOR ADDITIONAL READING:

Downs, Theodore, *Fossil Vertebrates of Southern California* (University of California Press, 1968).

Heric, T. M., "Rancho La Brea: Its History and Its Fossils," *Journal of the West* (April 1969, pp. 109–120).

Howard, Hildegarde, *Fossil Birds*, no. 17 in the "Los Angeles County Museum Science Series" (Los Angeles County Museum, 1962).

Hopkins, David M. (ed.), *The Bering Land Bridge* (Stanford University Press, 1967).

Miller, George J., *Man and Smilodon: A Preliminary Report on Their Possible Coexistence at Rancho La Brea*, no. 163 in the series "Los Angeles County Museum Contributions in Science" (Los Angeles County Museum of Natural History, 1969).

Sibley, Gretchen, *La Brea Story* (Los Angeles County Museum of Natural History with W. Ritchie Press, 1968).

Stock, Chester, *Rancho La Brea, a Record of Pleistocene Life in California*, no. 20 in the "Los Angeles County Museum Science Series" (Los Angeles County Museum, 1956).

Wright, Herbert E., and Frey, David G. (eds.), *The Quaternary of the United States* (Princeton University Press, 1965).

AUDIOVISUAL MATERIALS FROM ENCYCLOPÆDIA BRITANNICA EDUCATIONAL CORPORATION:

Filmstrip: *Fossils* (series).

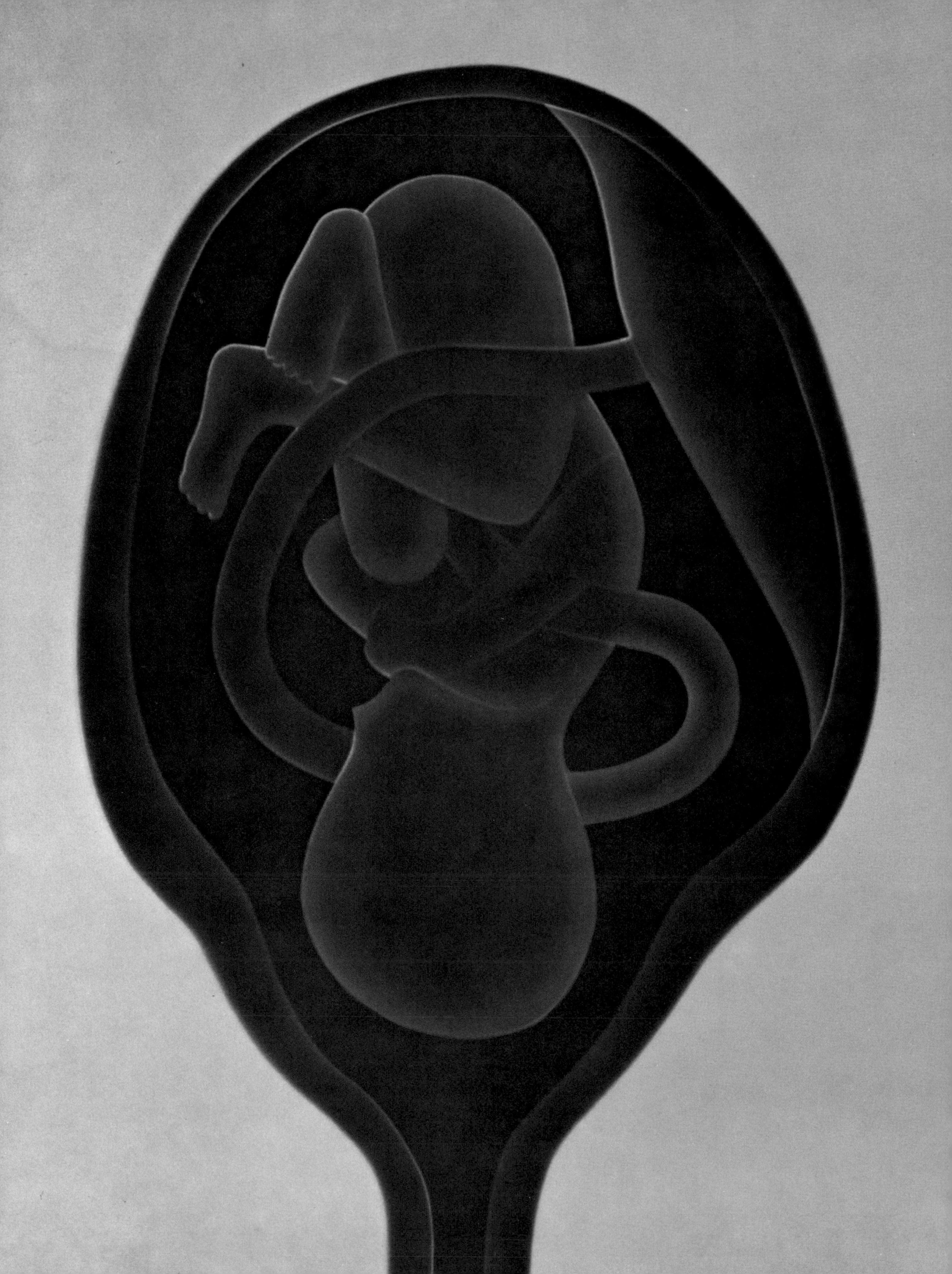

Treating the Unborn and Newborn

by Frederick C. Battaglia

In the last decade there have been dramatic improvements in medical care for infants during the fetal period and at birth. The results are heartening: fewer deaths or abnormalities and better general health for the newborn.

Normal human development requires a delicate interplay of genetic and environmental factors. Although we have acquired much information about the role of genes in regulating development, much less is known about the effect of environment on man's growth during the first 20 years of his life. This is especially true of fetal development, which, because it takes place within the uterus, is not readily accessible for study.

Traditionally, parents and physicians have had to accept the natural results of pregnancy. For a small fraction of the 3.5 million babies born in the United States each year, life before and after birth is perilous indeed. For these infants a variety of problems may interfere with their health and normal development.

In the past, medical care of the mother during pregnancy was directed at preventing her death from gross obstetrical catastrophe, such as severe infection or hemorrhage. Care of the infant was directed at providing a sheltered environment that reduced its chances of acquiring an overwhelming bacterial infection, with additional attention to feeding and general supportive care. Relatively little effort was directed at specific therapeutic measures for the unborn child or at reducing morbidity (the incidence of disease or disorder) in the premature or low-birth-weight infant.

During recent years, however, obstetricians and pediatricians have worked to fill these gaps in medical care. Their efforts have given rise to a new field of medicine that is known by a relatively new term—perinatal medicine (*peri*, around; *natal*, birth). Specialists in this field already have developed obstetrical techniques for aiding the fetus, for evaluating newborn infants, and for detecting those conditions that produce damage in the central nervous system (brain and spinal cord).

Courtesy, The Yale-New Haven Hospital, Yale University

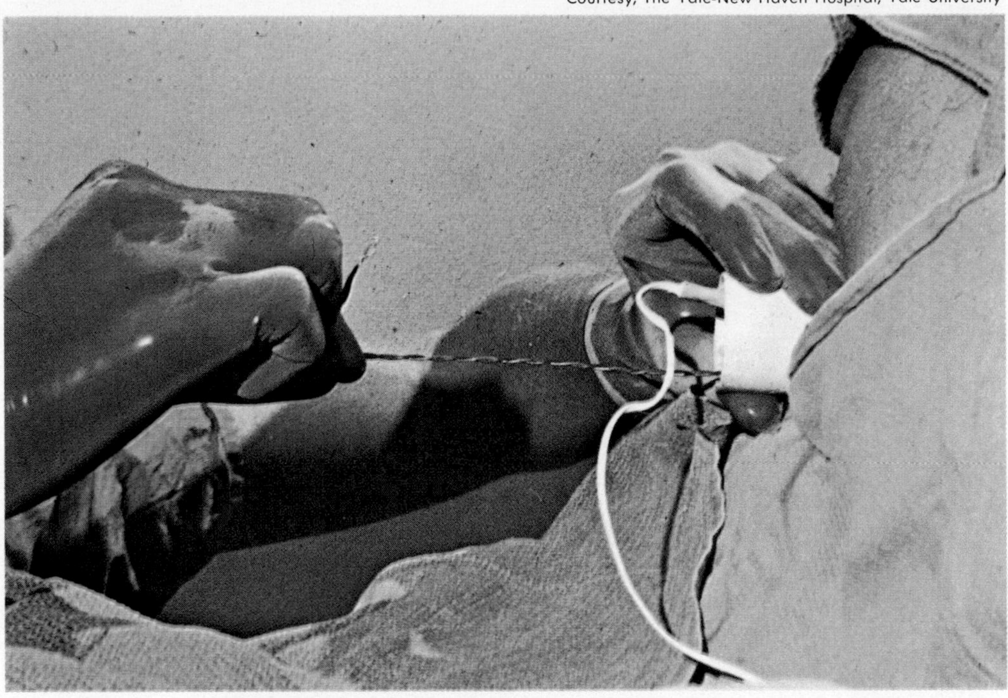

Robert Child

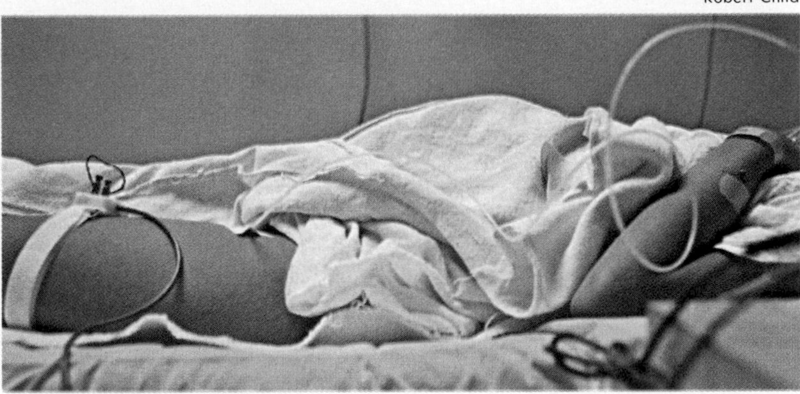

Sophisticated electronic equipment used in fetal monitoring to assess fetal distress during labor and delivery is a recent phenomenon in the medical world. Electrodes are attached directly to the fetus (top) by placing an endoscope in the vaginal canal of the mother. A leg plate (bottom), tied around the patient's thigh, connects the electrode to the monitoring unit.

The perinatal period

At birth, life in the uterus is ended, and along with it the baby's complete dependence on the mother for respiration, nutrition, and excretion. Before birth, all these functions are carried out through the placenta, that admirable arrangement of membranes of mother and child that brings their circulations into sufficient proximity for these vital functions to be carried out on behalf of the child. The importance attached to the event of birth can be seen from the fact that we date our ages from this point in time. When we say "a six-month-old infant," we mean that six months have elapsed since its birth. We make no allowance for the length of time that the infant had been living inside the uterus before birth or for the fact that, biologically, the time from the second half of pregnancy to about four years of age is actually a single phase of continuous development and functional maturation of many organ systems.

FREDERICK C. BATTAGLIA is director of the Division of Perinatal Medicine and professor of pediatrics as well as of obstetrics and gynecology at the University of Colorado School of Medicine.

Courtesy, The Yale-New Haven Hospital, Yale University

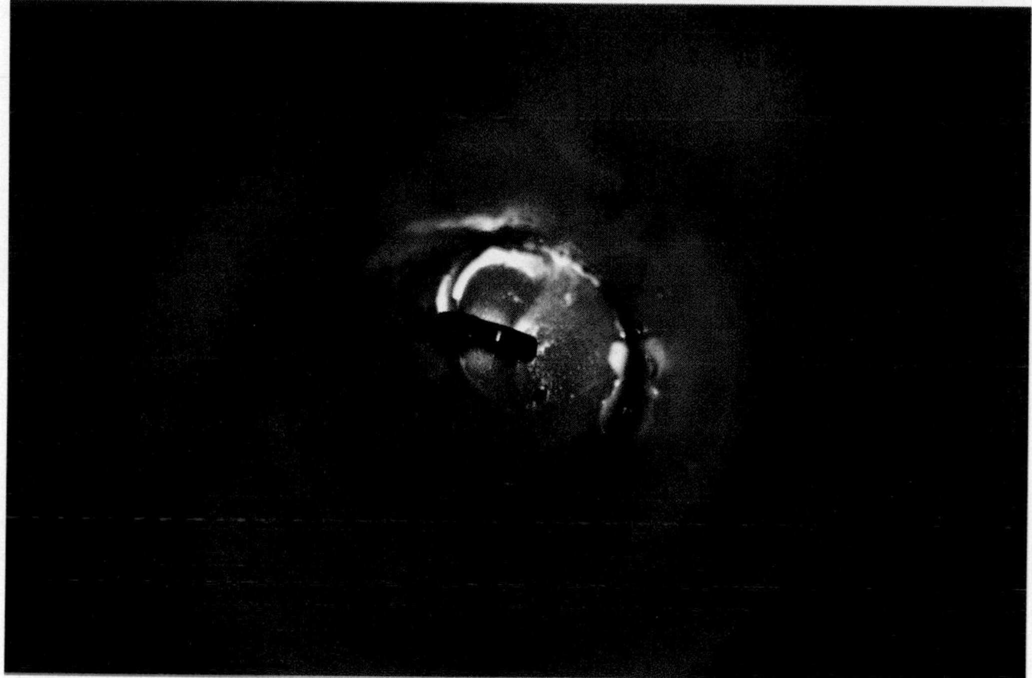

Seen through an endoscope is an electrode attached to the unborn baby's head to monitor signs of fetal distress.

Broadly speaking, the perinatal period includes the first year of life, beginning with conception and usually ending three months after birth. In practice, however, it is generally understood to mean the time from the beginning of the second half of pregnancy to the end of the neonatal (newborn) period, which is that time when the child attains the maturity (but not necessarily the chronological age) of the normal two-month-old infant. It is the perinatal period that is becoming a common area of interest for scientists and physicians. Reduction of infant mortality, prevention of abnormalities, and promotion of better development certainly must begin with greater knowledge of the biology of this period and the factors that may alter this biology for better or worse.

The fetal environment

The primary problem in studying the fetus has been that of getting at it without damaging mother or child and without interrupting the pregnancy. Now, however, it is possible, at least in animals, to deliver the fetus from the uterine cavity, carry out fetal surgery, return the fetus to the uterus, and have the pregnancy proceed to normal delivery of a healthy baby animal.

The fetus develops in a totally different environment from that in which it must live after birth. Many characteristics of the fetus that physicians used to regard as "immature" now are recognized as highly specific adaptations to its environment. For example, the fetus synthesizes a form of hemoglobin, the oxygen-carrying pigment in the blood, that is particularly suited to the very low concentration of oxygen in the fetal blood. Although it is just as complex as adult

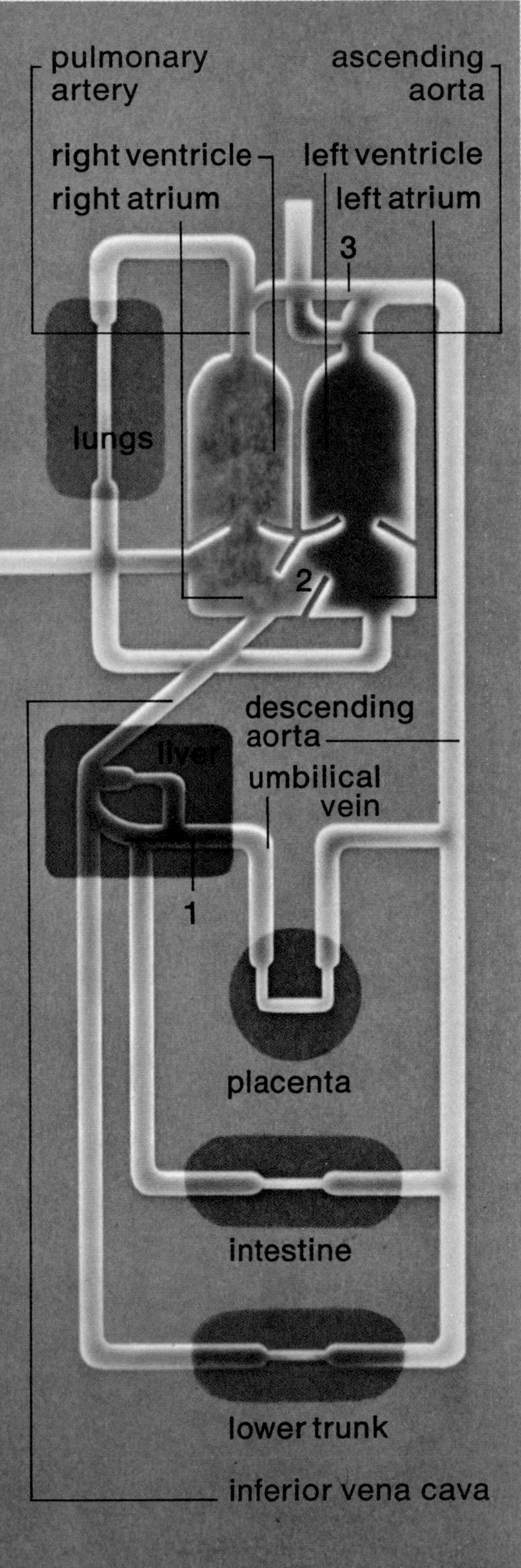

hemoglobin, the fetal form—together with other characteristics of fetal blood—gives this blood a much higher affinity for oxygen than adult blood.

Animal studies have demonstrated that the environment within the womb is also remarkably stable. Measurements of oxygen pressure, acidity, blood sugar, and concentrations of other substances dissolved in the blood fluctuate only slightly from day to day. This contrasts rather sharply with the situation after birth, when periodic feeding of the baby causes large swings in the amount of nutrients in the blood.

Stability of the fetal environment depends almost entirely upon the health and welfare of the mother. While the embryo of a bird or reptile is nourished by a fixed and continually diminishing supply of yolk, a mammalian fetus is supplied by a continuing, dynamic exchange of substances across the placenta. Much research in fetal physiology has been aimed at defining how efficient an organ of exchange the placenta really is. This has led to a growing recognition of the importance of the flow of blood to the uterus and its contents in regulating fetal growth.

An example of factors affecting this blood flow is found in the role played by the estrogens, which are a group of female sex hormones. The blood vessels within the uterus are exposed to a high concentration of these hormones because they are manufactured by the trophoblast, which is the tissue of the fetal part of the placenta. One role of the estrogens may be to keep the uterine blood vessels so that the enormously increased flow of blood required during pregnancy can pass through these dilated channels. Certainly a promising area of study is the search for ways to increase uterine blood flow significantly over long periods of time in those pregnancies where fetal growth retardation is suspected.

Fetal oxygenation and growth

The most elementary need of the fetus, just as of the adult, is for oxygen. Fetal distress, which occurs rather suddenly during a pregnancy, often is attributed to fetal hypoxia (lack of oxygen). Because of the importance of oxygen, recent experiments have explored the factors regulating its supply to the fetus. At high altitudes, for example, oxygen pressure is lower than at sea level. When pregnant sheep unaccustomed to high altitudes were taken to different heights in the Colorado Rockies, there was a definite reduction of oxygen in fetal blood, particularly at elevations greater than 10,000 feet. This raised the question of the maximum safe altitude for acute exposure of the human mother during pregnancy.

Other studies have shown that the fetal oxygen supply can be improved by having the mother breathe high concentrations of oxygen. They also have demonstrated the importance of the mother's posture in late pregnancy, when the uterus is quite large and heavy. When the mother lies on her back, the weight of the pregnant uterus can press upon major blood vessels, thereby indirectly affecting the fetus.

It is an old clinical observation that the size of the infant and the size of the placenta are quite in keeping with each other. Small babies usually have small placentas; large babies have large placentas; twins, triplets, or higher multiple pregnancies have considerably larger total placental tissue. There are, of course, exceptions, particularly when some of the placental tissue is diseased. But there is no exception to the rule that no large infant is born with a small placenta. Given a small organ supply, fetal weight has an upper limit that it cannot exceed.

Recent studies of placental hormones have led to the concept of a fetal-placental unit that may have some bearing on placental growth. According to this concept, the adrenal gland of the fetus and the trophoblastic tissue of the placenta form a unit for the synthesis of several estrogens, particularly estradiol. Neither the adrenal gland nor the placental tissue alone has all the enzymes required to produce estradiol; each makes its own contribution to a chain of synthesis that leads to the final product. One of the interesting points observed in fetal and maternal endocrinology is the fairly consistent correlation between the general well being and development of the fetus and the excretion in the mother's urine of estriol, which is an end product in the metabolism of estradiol and other estrogens.

Prenatal care

"Fetal distress" is a term commonly used by obstetricians and pediatricians to indicate that they have reason to believe a fetus is in trouble. The baby's heart rate changes, it moves about a great deal, it may pass meconium (the debris that is the only content of the fetal bowel) into the surrounding amniotic fluid (the fluid in which the fetus floats). Because these rather nonspecific signs often have been associated with infants born dead or badly damaged, they justifiably have become regarded as distress signals. "Distress" may be a matter of stress or strain, requiring relief in the form of some kind of improvement in the condition of the fetus in a specific or general way. Or it may be in the form of immediate delivery, often by cesarean section.

One of the more useful indicators of the fetal condition is the urinary excretion of estriol by the mother. Since estriol is one end product of estrogen metabolism that is dependent upon the function of both placental and fetal adrenal tissue, it is not surprising that its repeated measurement in pregnancy has been helpful in detecting chronic fetal distress, which arises most often in high-risk pregnancies. The measurement of estriol provides very helpful information for the physician who must decide whether to allow a particular fetus to continue its development in the uterus or to deliver it prematurely and raise the baby in the physiologically less-than-ideal environment of the intensive-care nursery.

Another obstetrical technique is the use of ultrasonic devices, which produce sounds above the audible range, to determine the location of the placenta, a factor that is particularly important in the detection of

The circulatory system of the fetus (opposite page) has three distinguishing structures: (1) The ductus venosus in the liver permits the mother's oxygen-rich blood, which has flowed from the placenta through the umbilical vein, to be conveyed into the principal vein (inferior vena cava) leading to the right atrium of the heart. (2) The foramen ovale, an opening between the right atrium and the left atrium, permits this same rich blood to flow to the left side of the heart so that it can be pumped into the ascending aorta to the head and upper parts of the body. The venous blood, which remained on the right side of the heart, is pumped into the pulmonary artery as if it were being carried to the lungs to be purified. (3) The ductus arteriosus, however, a special artery branching off the pulmonary artery before it reaches the lungs, shunts the venous blood to the descending aorta, which carries it to the lower parts of the body and to the placenta to be purified.

Courtesy, Dr. E. I. Kohorn, Yale University

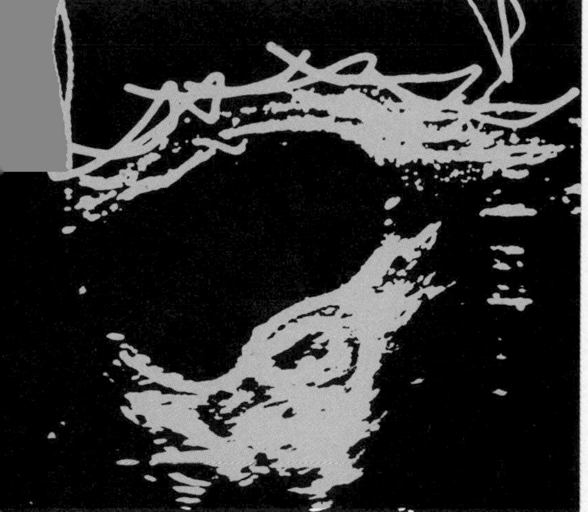

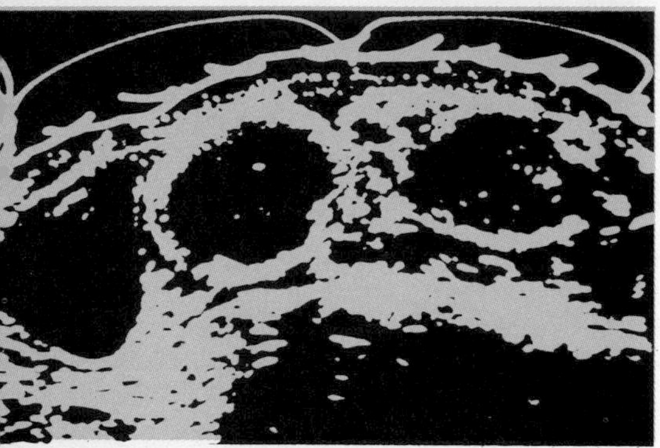

A longitudinal ultrasound scan shows fetal development (top) at 8 weeks and (center) at 34 weeks gestation. (Bottom) Both fetal heads of twins appear on this transverse ultrasound scan at 22 weeks gestation.

placenta previa (a placenta located across the cervix, the opening of the uterus). Ultrasound can also detect multiple births and measure fetal size, particularly that of the head, the largest part to pass through the birth canal. It is a very attractive technique because apparently it can be used repeatedly in a pregnancy with safety and without the risk of producing congenital malformations. Indeed, if its usefulness and safety continue to be demonstrated, it may replace many of the uses of X-ray techniques in obstetrics.

A test that has had dramatic impact on obstetrical management is diagnostic amniocentesis, the insertion of a needle through the abdominal wall of the mother, through the uterine wall, and into the amniotic sac, to withdraw a sample of the amniotic fluid surrounding the fetus. This technique has been most helpful in the management of Rh-sensitized pregnancies resulting from a father who is Rh-positive and a mother who is Rh-negative. If the fetus inherits the Rh-positive blood factor from the father, its blood cells may be destroyed by antibodies produced by the Rh-negative mother. By using amniocentesis to measure the concentration of bilirubin in the amniotic fluid—bilirubin is one of the products formed by the breakdown of hemoglobin in red blood cells—it is possible to determine more precisely how severe the disease is in a particular baby and whether the condition warrants an intrauterine blood transfusion to the fetus.

The technique of intrauterine transfusion to a fetus suffering from Rh disease is one example in medicine where surgical entry into the uterus and manipulation of the fetus has been successful therapy for the infant. Many infants have survived severe hemolytic disease (the liberation of hemoglobin from red blood cells) because of one or more intrauterine transfusions, and the success of this technique in the management of Rh-sensitized pregnancies has spurred other aggressive approaches to the therapy of the fetus. So far, however, these attempts have not met with equivalent success.

There are, of course, instances when samples obtained by amniocentesis for diagnostic purposes may lead to the decision to perform an abortion, but this is hardly therapy for the fetus. In fact, what has happened with Rh disease illustrates some of the complexities involved in predicting the consequences to society when the prevention or interruption of pregnancy is used as a solution for certain diseases (see *1970 Britannica Yearbook of Science and the Future,* HUMANICS AND GENETIC ENGINEERING, pages 82–85). The relatively recent discovery of a vaccine that, when administered to unsensitized Rh-negative mothers after each birth, prevents the disease in future children is another factor that has changed the approach to family counseling for this specific problem. The research that led to these preventive measures was prompted in part by the fact that Rh sensitization was a relatively common obstetrical problem associated with a high fetal and newborn infant mortality. Similar research may find cures for other diseases for which either abortion or the prevention of pregnancy are now the only solutions available.

Robert Child

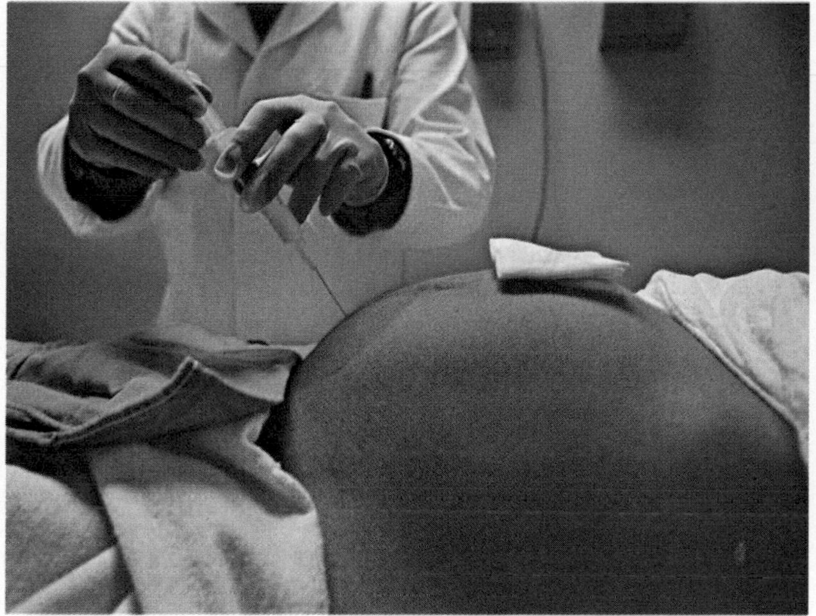

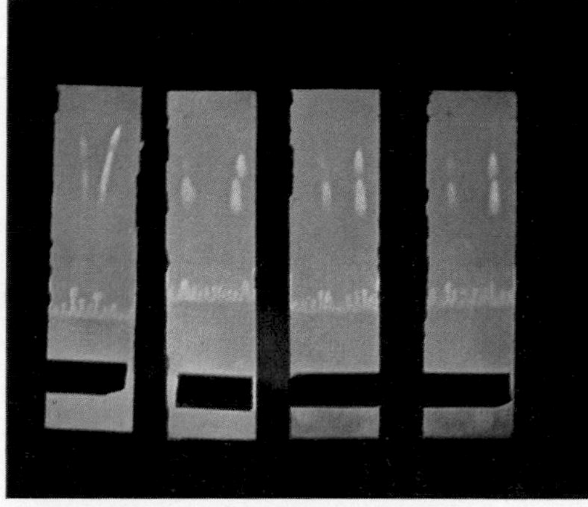

(Left) To obtain a sample of the amniotic fluid surrounding the fetus, a needle is inserted into the amniotic sac through the abdominal and uterine walls of the mother. Under ultraviolet light (above), changes in the phospholipides, or fats, in the amniotic fluid reveal the degree of lung maturity in the unborn child.

Labor and delivery

The role of the fetus in initiating labor has been demonstrated by experiments in which either the adrenal or the pituitary gland of a sheep fetus was removed. In either case, pregnancy continued far beyond the expected date, giving evidence that both fetal glands are required for the normal onset of labor. Such studies are important because the most common obstetrical complication associated with maternal disease is premature onset of labor, resulting in the delivery of an infant who is required to begin life outside the uterus at too early an age.

The careful synchronization of events at birth is an obvious hallmark of normal development. This has been illustrated dramatically by studies on kangaroos. While a kangaroo infant is nursing in its mother's pouch, another egg can be fertilized and will go through the early stages of development. As long as the first infant still is nursing, however, the fetus will not progress beyond this point. Instead, it will remain quiescent until some months later, when the first infant leaves the pouch. The fetus then resumes its development and ends up as a normal infant, indistinguishable from the one with no interruption in growth. This observation shows that an arrest in growth and development at one stage need not necessarily imply a loss of the potential for normal development at a later stage.

While the occurrence of chronic fetal distress at some time during the 40 weeks of pregnancy is probably the most common complication of fetal development, there are more sudden problems that develop in the comparatively short time span of labor and delivery. Physicians now have techniques for continuous monitoring of the course of labor by recording pressures within the amniotic sac. In

HEAD COMPRESSION

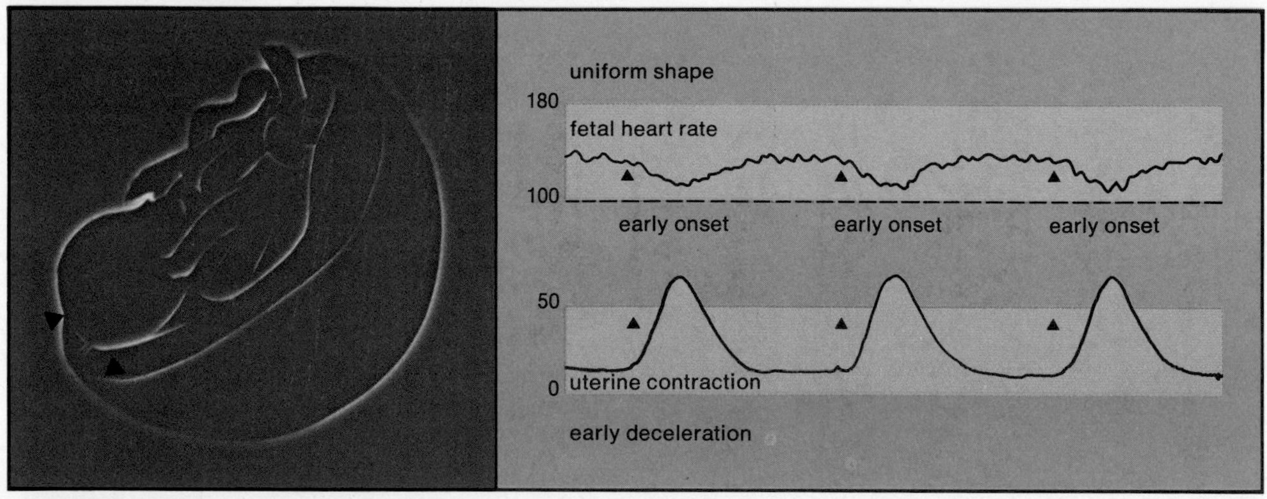

UTEROPLACENTAL INSUFFICIENCY

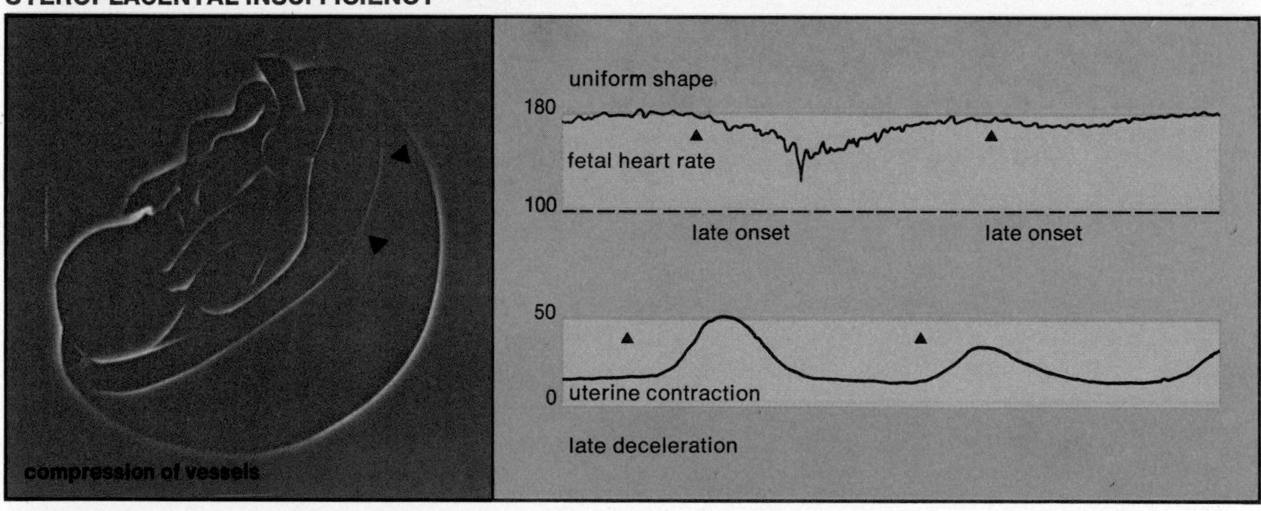

UMBILICAL CORD COMPRESSION

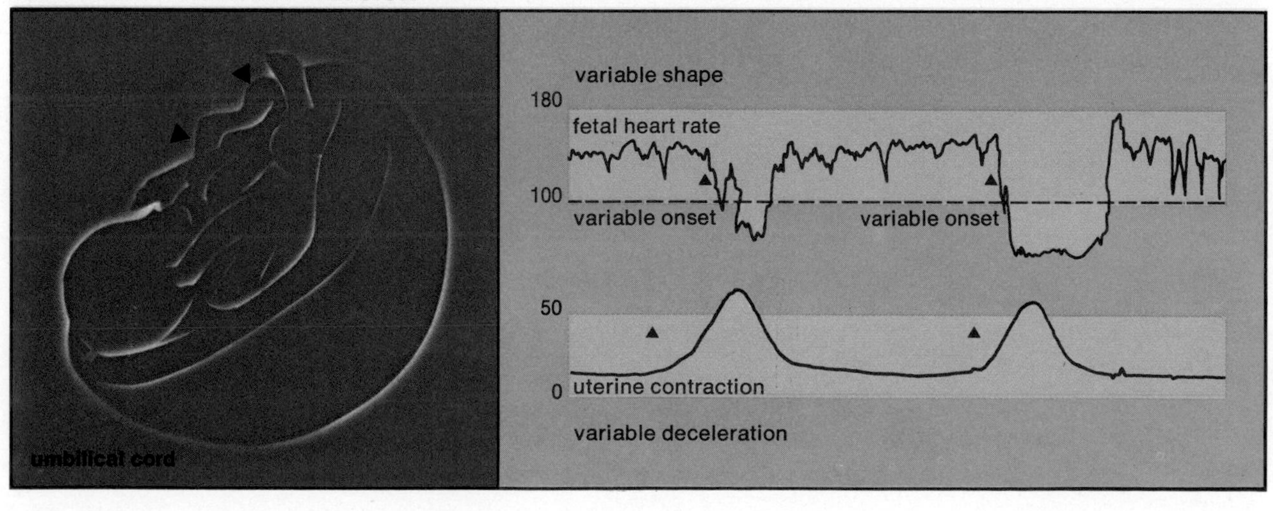

addition, from electrocardiographic leads placed on the fetus and the mother, the pattern of contraction of the uterus can be compared with changes in the fetal heart rate. More recently, it has been shown that fetal blood samples can be obtained safely by the equivalent of pin-pricks of the fetal scalp during labor and delivery. At the present time, a great deal of research in obstetrics is aimed at evaluating the real usefulness of these different techniques. Enough clinical work has been done, however, to show that such techniques will certainly have a place in the management of high-risk pregnancies, particularly those developing complications at the time of labor and delivery.

Newborn medicine

The impact of technology on patient care has been felt even more dramatically in newborn medicine. Numerous studies, for example, have shown the importance of constantly monitoring and supporting a normal body temperature in newborn infants. Physicians now have reliable equipment for maintaining normal body temperature in the infant, not only in incubators in the nursery, but in the delivery room as well. In addition, there is now equipment for the detection of apneic spells (suspension of respiration) before the infant is damaged by them, for measuring blood pressure and heart rate in critically ill infants, and for the safe and accurate infusion of very small amounts of solutions over long periods of time into tiny infants.

In neurophysiology, progress also has been made in recognition of patterns of motor activity, patterns of sleep and wakefulness, and responses of the infant to specific stimuli. These techniques will have considerable influence on the clinical care of newborn infants. They will provide much earlier recognition of abnormal behavior, which in turn will prompt other diagnostic tests earlier in the course of an illness. They also will sharpen the physician's ability to predict the later development of a child in the light of his neonatal course.

Neonatal hematology (the study of newborn blood and its disorders) is another area in which clinical research has suggested a relationship between changes in the baby's circulation just before birth and problems that it may encounter after birth. Quite rightly, most attention in this field has been devoted to the problem of blood group incompatibilities (such as Rh disease), where the mother's antibodies, which are hostile to the fetal red blood cells, cross the placenta and coat the fetal red cells, shortening their life span considerably and overloading the blood with the products of their disintegration. There are, however, other hematologic problems associated with events that occur before delivery. For example, certain instances of thrombocytopenia (the shortage of blood platelets, which play a large part in the clotting process and whose absence leads to a serious bleeding problem) are caused by antibodies from the mother that are hostile to the fetal platelets.

Two major concepts have altered our approach to newborn care. The first is the recognition that infants can grow at abnormal rates in

The relationship between the heart rate of the fetus and the pattern of the mother's uterine contractions, as detected by continuous monitoring techniques, can tell an obstetrician when sudden problems develop during labor and delivery. When the fetal heart rate shows a consistent onset of deceleration early in each contraction (opposite page, top), the uterus is presumed to be compressing the head of the fetus. When the flow of blood to the fetus through the uterus/placenta complex is insufficient during a contraction (center), the fetal heart rate begins to decrease well after the contraction has begun. If the umbilical cord is being blocked (bottom), the onset of the heart-rate deceleration does not reflect an association with the onset of the uterine contraction.

Robert Child

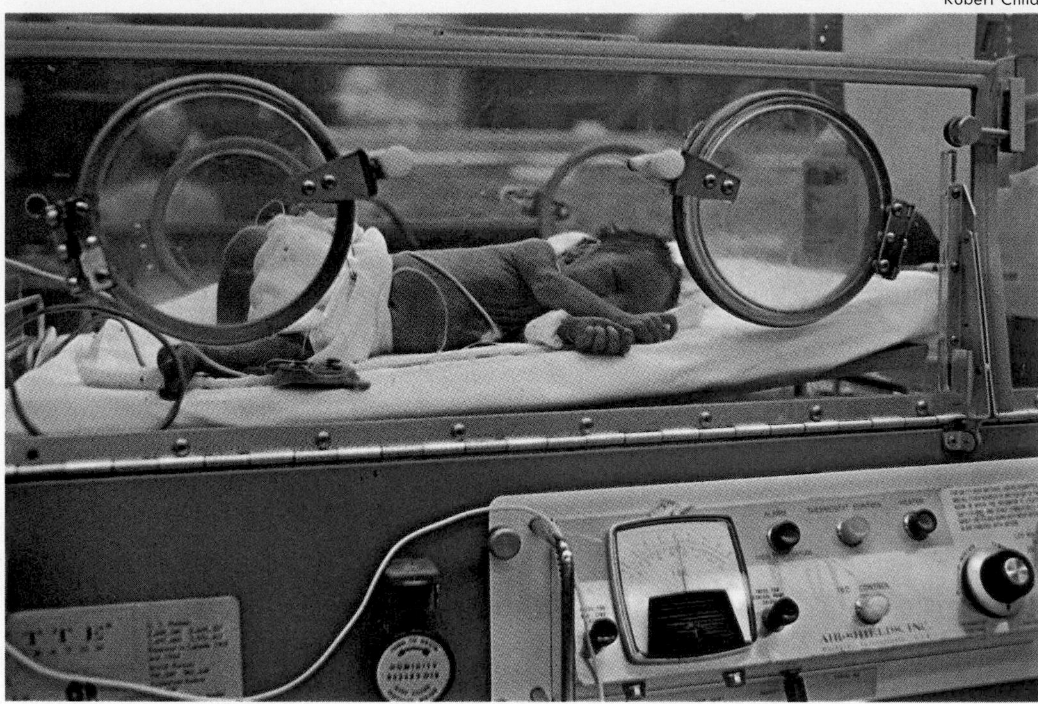

A newborn baby suffering from a respiratory disorder is being monitored by an electronic system designed to prevent mortality or brain damage in premature infants with respiratory difficulties.

the uterus; hence, not all infants born with a low body weight are premature. This recognition of intrauterine growth retardation and of some specific problems associated with its management has helped to provide better guidelines for the care of the growth-retarded infant, and to clarify problems of the true preterm or premature infant. Moreover, integration of clinical care during an infant's perinatal development also has done much to draw obstetricians and pediatricians more closely together.

What intrauterine growth retardation implies for future development still must be determined. A classic dilemma in obstetrics has been whether a particular fetus should be allowed to continue development within the uterus or should be delivered prematurely to develop within the nursery environment. The question of which environment, intrauterine or extrauterine, can ensure better development, given a set genetic endowment, cannot be answered intelligently without much information about intrauterine conditions in an individual patient. The answer also requires a knowledge of fetal physiology and biochemistry in order to interpret properly how the environmental changes detected in a pregnancy will affect the infant's development. Although we are still a long way from this dual goal in perinatal medicine, the frequency of the problem of intrauterine growth retardation has acted as a major stimulant to research in this area.

The second conceptual change in newborn care has come in our growing awareness of abnormal patterns of behavior, even in the immediate newborn period. Many clinical studies, stimulated by recognition of intrauterine growth retardation, have shown that infants generally tend to behave more in keeping with their age than with their size. This has made us more aware of abnormal patterns of behavior. For example, a number of clinical studies showed that some infants develop extremely low blood sugars (hypoglycemia) and that this is associated with such immediate signs as jittery behavior or even frank convulsions and with long-term problems in development. Thus physicians reluctantly came to accept the idea that hypoglycemia could be harmful to newborn infants.

The lesson of hypoglycemia prompted the search for other causes of abnormal behavior in infants. Many such causes now have been found, including drug intoxication brought about by maternal ingestion of drugs harmful to the infant and the deficiency of essential elements in the blood, such as calcium or magnesium. The importance of determining the causes of neurologic dysfunction in the immediate newborn period is that most of the causes recognized in the last few years have been easily treatable, either by the administration of glucose, calcium, magnesium, or other compounds, or by exchange transfusion or other techniques that remove the damaging substances from the newborn infant's circulation.

Thus, nowhere in medicine is the impact of environment on growth and development more clearly evident than in perinatal medicine. As a result, basic research and the rapid improvement in bioengineering technology are providing the means for more aggressive intervention during pregnancy. Such intervention, however, is not for the sake of sacrificing the infant. Rather, it is for the time-honored and old-fashioned concept in medicine of healing the infant *in utero* by providing an environment more conducive to optimal growth and development without risk to the mother.

FOR ADDITIONAL READING:

Clarke, C. A., "The Prevention of Rhesus Babies," *Scientific American* (November 1968, pp. 46–52).

Ingelman-Sundberg, A., and Wirsen, C., *A Child Is Born: The Drama of Life Before Birth;* photos: Lennart Nilsson (Delacorte Press, 1966).

Windle, W. F., "Brain Damage by Asphyxia at Birth," *Scientific American* (October 1969, pp. 76–84).

AUDIOVISUAL MATERIALS FROM ENCYCLOPÆDIA BRITANNICA EDUCATIONAL CORPORATION:

Film: *Biography of the Unborn.*

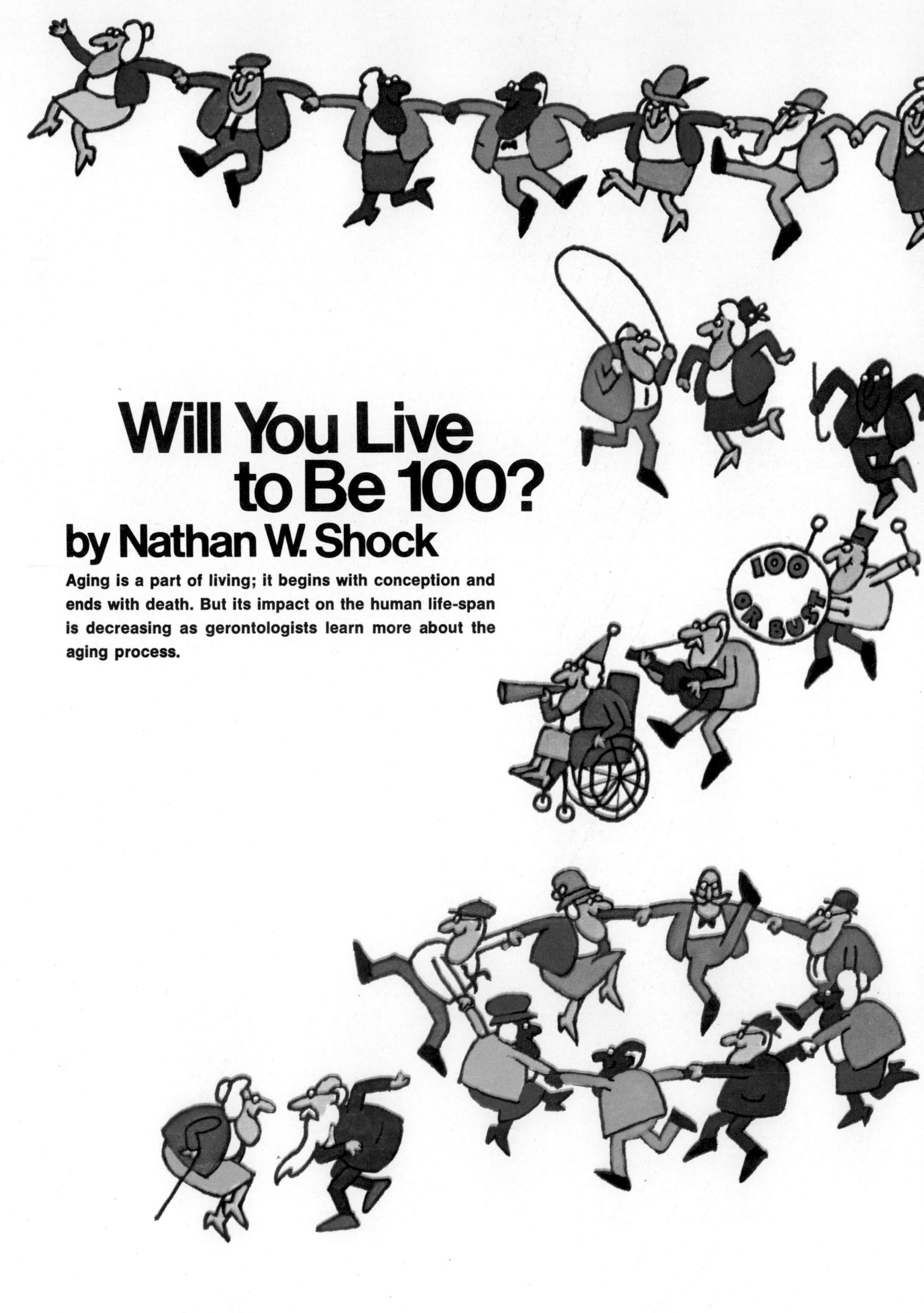

Will You Live to Be 100?

by Nathan W. Shock

Aging is a part of living; it begins with conception and ends with death. But its impact on the human life-span is decreasing as gerontologists learn more about the aging process.

Happy Birthday
100 YRS

Man either accepts the mortality of his body as the limit of his existence or he envisions a life after death. In either case, his aspiration is to prolong his identity as far into the future as possible. Few people, however, are interested in simply prolonging life; they want a healthy and vigorous long life.

The knowledge of what changes occur with advancing age and the underlying causes of these changes may provide insight into ways of altering or retarding aging processes. Although the gerontologists who study these changes can now identify certain conditions that tend to shorten life-span, including exposure to radiation, obesity, and cigarette smoking, they still cannot identify those factors that tend to lengthen life.

The span of life

The present-day life-span of almost 70 years, which is the average age at death of all persons dying in a given year, is a relatively recent development in the history of man. Life expectancy was between 35 and 40 years prior to 1900, but at about that time a significant increase occurred. Most of this increase stemmed from a reduction in infant mortality rates and from medical advances in the prevention and cure of infectious diseases. In fact, between 1900 and 1965, the years of life remaining after attainment of age 60 increased by only about four years. Further medical advances and general improvement of living conditions will probably add five or perhaps even 10 years to the average human life-span by the end of the 20th century.

Life-spans vary greatly among different species of animals. The adult May fly, for example, flits about in one of the shorter spans—only one day. Blowflies live 30 days, fruit flies 90; rats live two years, beagles 12, horses 20, man 69, and some turtles 100. These species differences suggest that the length of life is genetically determined. But individuals within a species also vary with respect to genetic instructions about life-span. In other words, it seems that at fertilization a total life program is set for each individual. But because the environment determines the expression of this genetic program, few individuals live the fullness of their programmed life-span.

The sexes also age at different rates. Females, on the average, outlive males, not only in the human species but throughout all of the animal kingdom. It is difficult to ascribe this discrepancy to greater life stresses, as has been claimed. Apparently, femaleness imparts some survival value in terms of longevity. What it is, we do not know.

The interest of man in his own longevity is understandable. He dwells, hopefully perhaps, on the reports of supercentenarians that appear from time to time. Most famous among the long-lived is probably Old Parr, whose tomb in Westminster Abbey credits him with an age of 152 years. But Old Parr had no birth certificate and was, beyond reasonable doubt, an imposter with respect to his age.

Many centenarians, some claiming to be over 130, have been reported from the Georgian area of the U.S.S.R. Most of these claimants

NATHAN W. SHOCK is associated with the Baltimore (Md.) City Hospitals, where he is Chief of the Gerontology Research Center for the U.S. National Institute of Child Health and Human Development.

live in isolated mountainous areas where birth certificates are unavailable. Few are as old as they say they are. In some cases the life-span of father, son, and grandson of the same name appear to be joined in one record.

Walter Williams, who died in 1959 in Texas, claimed to be 117 years old and the last survivor of the Confederate army. Errors in his army records, however, had overstated his age by 10 years. In Britain, which has had compulsory birth certification since 1836, authenticated birth records of persons living 111 years have been reported. In the U.S., Social Security payments are now being made to 4,574 individuals reported to be more than 100 years old; the oldest client on the rolls claims to have been born in 1842. The progressive increase in mortality rate between ages 30 and 80 is so large that even this number of centenarians is surprising; it strongly suggests that the rate of increase in the change of dying slows down in extreme old age.

Factors affecting life-spans

Life usually is terminated by disease. The primary causes of death among older people are heart disease, cancer, and stroke. It has been calculated from mortality statistics, however, that elimination of both heart disease and cancer as causes of death would add only three to five years to the average life-span; deaths would simply be redistrib-

uted at somewhat later ages among a wide variety of other diseases. Because vulnerability to stress increases with increasing age, many diseases not necessarily fatal to younger persons prove to be more than the declining capabilities of the older person can meet.

It is generally agreed that life-span can be significantly increased by reducing food consumption. In rats food restriction begun early in life, increased the life-span by 50% (from two to three years). Although the rats were retarded in growth, they were vigorous and had increased resistance to diseases. The same life-prolonging effect of reduced feeding is also found in a wide range of other species. While the mechanism causing this effect is still not understood, it appears that the low level of food intake somehow demands less of the biochemical machinery in terms of age-generating effort.

Although there is no assurance that reduced food intake would extend the life-span of humans, life insurance statistics indicate that overweight persons tend to have shorter life-spans than normal weight persons. Death rates for almost all diseases except tuberculosis are higher for the overweight than for those with normal weight. Among other factors influencing the span of life are traumatic experiences of all kinds: exposure to temperatures unusual for the species, crowding beyond the usual conditions, and exposure to ionizing radiation. These influences, which can be seen to operate rather clearly in other animals, may prove to have parallels for the human species.

Changes caused by age

The snowy head, wrinkled and drooping skin, stooped shoulders, and shuffling gait are the stereotype of extreme old age. These signs steal upon us slowly. The average performance of many organs of the body shows a gradual decrease beginning at about age 30: the heart pumps less blood, the lungs take in less air, the blood vessels and joints stiffen, the skin loses its elasticity, the endocrine glands slacken their production of hormones, muscle strength declines, and hearing and vision diminish.

Senescence also brings behavioral changes. The range and scope of activities often diminish; responses are slower; memory for recent events may be impaired. Difficulties may be encountered in responding when many items of information are presented simultaneously, as, for example, extracting the essential information from a welter of signs on a superhighway. Older persons may also become more rigid in their behavior; cautiousness increases and stereotypes in thought and action may ensue. Many of these changes, however, can be avoided by conscious efforts to maintain mental alertness through continued learning. Although the learning of new material may take longer, there is no evidence that "you can't teach an old dog new tricks." Learning in the aged also benefits from the wealth of experience accumulated over a lifetime.

When elderly people are subjected to stresses, they require more recovery time than do the young. The loss of reserve capacity is most dramatically shown by slower recovery after physical exercise, which increases respiration rate, heart rate, and blood pressure. Each organ of the body is made up of many functioning cells, the gradual loss of which can reduce reserves. Evidence for cell loss appears only in those tissues, such as muscle, brain, nerves, and kidney, where cells in the adult have lost their ability to divide and thus replace any losses. Cell loss is not the whole story, however, because tissues that retain their ability to divide in adults still show age changes. Nevertheless, because cell loss does occur in certain key tissues that are critical for survival, programs for extending life-span must focus primarily on methods for reducing the incidence of cell death.

Genetic theories of aging

Biological theories of aging fall into two major categories: genetic and post-genetic, or environmental, theories. Genetic theories assume that increasing age causes some event within the genetic information of a cell that is deleterious and causes cell death. Post-genetic theories assume that with the passage of time changes occur in key molecules that are not renewed or repaired, or that deleterious substances accumulate in cells and interfere with normal chemical reactions. Gerontologists agree that genetics plays a role in determining life-span. Species differences in longevity and the fact that selective breeding within a given species can produce family lines with significantly different life-spans offer evidence of this fact. But just how genetic factors operate and the extent to which environmental factors alter the underlying genetic program of aging are not yet clear.

Experimental inbreeding studies indicate that over successive generations, individual animals in a population become more and more alike in terms of the genes they carry. It is interesting to note that the lifespan of the inbred stock is always shorter than that of the original "wild" stock containing many more different genes. This effect has given rise to the hypothesis that the broader the spectrum of genes an organism possesses, the better prepared it is to meet the wide variety of stresses in living.

Recent advances in molecular biology have shown that the information required for the development of a whole organism, or more directly for the formation of proper and appropriate enzymes in specialized cells, lies in the DNA (deoxyribonucleic acid) molecule. In the cell, molecules of proteins, especially enzymes, that are essential for the chemical working of the cell are formed according to instructions stored in DNA. With the passage of time, it is believed by some that the DNA molecule may become damaged or may deteriorate so that the information required to form essential proteins either becomes garbled or is no longer present. Another idea is that certain histones (a type of protein) are more tightly bound to the DNA molecule in old cells than in young ones. This suggests that aging in cells might occur because the tightly bound histones cover parts of the DNA molecule, thereby causing a cell to age and die because it lacks access to essential information.

If, on the other hand, errors occurred during the transfer of information from DNA to protein synthesis, abnormal proteins or enzymes might be formed that would sabotage the metabolic processes of the

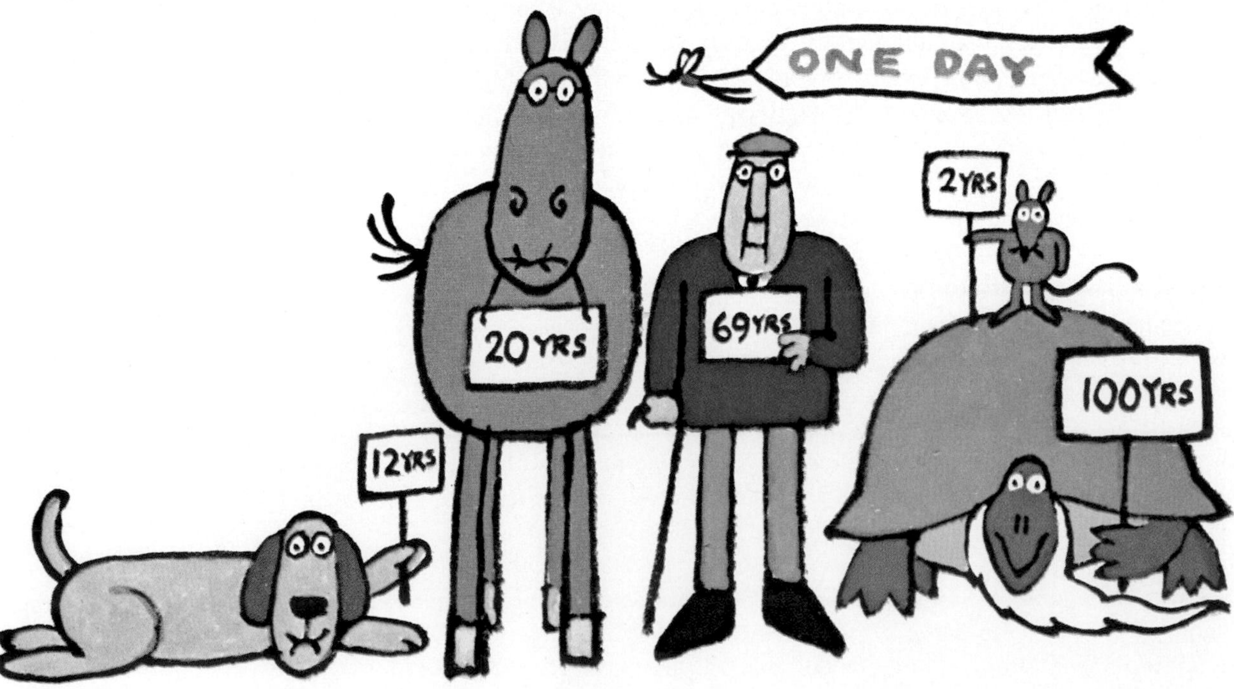

cell. Over the lifetime of the cell these abnormal molecules might accumulate and eventually outnumber the useful molecules; the cell would then die.

According to another theory, the somatic mutation theory, the general body cells develop spontaneous mutations (inheritable changes of characteristics) in the same way as do germ (reproductive) cells. Once a mutation has been formed, based on an alteration in DNA, subsequent cell divisions perpetuate it. As more and more cells develop mutations, the time comes when an appreciable fraction of the cells in an organ are mutated. Since most mutations are deleterious, the cells carrying mutations are either unable to perform their functions at all or may do so less efficiently. As the proportion of mutated cells increases, the function of the organ will be impaired and death of the animal may ensue.

A number of arguments have been raised against the somatic mutation theory. The first is that the impact would fall primarily on dividing cells whereas aging has its greatest effects in nondividing cells. Another objection has been that the mutation rate appears to be too low to permit the formation of a large enough number of aberrant cells to account for known age decrements. Furthermore, when chemical substances known to greatly increase the rate of mutation have been administered to experimental animals, there has been no evidence of a shortened life-span.

Another theory of aging assumes that immune reactions—normally directed against disease-causing organisms—become directed against the cells of the individual's own body and ultimately result in death. This autoimmunity theory is based on the presumption that at least some body cells undergo changes with age. The immune system of the body overreacts to these atypical cells and destroys them.

Errors in the immune system itself may also result in the loss of the ability to recognize "self." This enhances the destruction of cells even though they may not have been significantly altered. According to this idea, aging is assumed to result from reactions between cells rather than from reactions that occur primarily within individual cells.

Post-genetic theories of aging

The theory that animals and cells, like machines, simply wear out has a long history. This theory, although simple and conceptually attractive, fails to recognize the importance of self-repair mechanisms, which are present in animals but not in machines.

A theory that aging results from an accumulation of deleterious waste products either within the blood or within the cell represents a corollary to the wear-and-tear theory. A variation of the waste-product theory stems from the observation that highly insoluble particles accumulate in some cells as a function of age. These particles, often referred to as "age pigment," may occupy as much as 30% of the space inside an old cell.

continued on page 86

Prolonging your life

Even before the days of Ponce de Leon's search for the "Fountain of Youth," man dreamed of ways to retain his vigor and lengthen his life-span. After the discovery of the action of hormones in the late 1800s, there was a renewed interest in rejuvenation. At about that time a noted physiologist startled the medical world by claiming rejuvenating effects in himself following the injection of monkey testicular extracts. Subsequently, a rash of attempts to transplant glands from other animals into humans spread over Europe. The chief proponent of this technique was Serge Voronoff of France, who transplanted glands from monkeys into humans. For a time, this treatment was popular on the Continent, especially among the rich.

A variation of this theme was the Steinach operation, in which the spermatic ducts were tied off. This method was based on the assumption that by preventing the loss of sperm the sex glands could be stimulated to produce hormones that would be absorbed into the blood stream and produce rejuvenating effects. After the first burst of enthusiasm, the operation was found to produce no long-lasting effects and was consequently abandoned.

At the present time, two methods of rejuvenation or restoration have received much publicity: one is a cellular therapy practiced by Paul Niehans of Switzerland; the other is the procaine, or H-3, therapy espoused by Anna Aslan of Romania. Many famous persons claim to have been benefited by cellular therapy, which involves the injection of suspensions of cells prepared from the organs of embryonic sheep. On theoretical grounds, the treatment is not without hazard, and no controlled laboratory experiments have demonstrated a lengthening of life-span or improvement in vigor of animals subjected to it. Cellular therapy is not generally accepted by the medical profession in Europe and it is not officially permitted in the United States.

The other current therapy for senescence is the injection of procaine, the compound routinely used as a local anesthetic in dentistry. This therapeutic approach gained worldwide attention in the early 1960s. Although a number of physicians in the U.S. have prescribed the drug,

the Food and Drug Administration has now banned its use for this purpose on the grounds that its effectiveness has not been demonstrated.

It is doubtful whether any single substance will ever be found to reverse the accumulated changes of a lifetime. There are, however, several ways of reducing the probabilities of developing disabilities in old age. With advancing age there should be a gradual reduction in the daily caloric intake. Avoid obesity by reducing fats but continue sufficient protein intake. Maintain at least a moderate level of physical exercise—although there is no evidence that athletes who engage in heavy, systematic exercises have greater-than-average life-spans. Avoid smoking; it has been calculated that smoking two packages of cigarettes a day may reduce the average life-span by 12 years.

Future improvements in social conditions and advances in the prevention or cure of such illnesses as heart disease, arteriosclerosis, cancer, and diabetes will undoubtedly result in a further increase in the average life-span to at least 80–85 years. Although cancer, atherosclerosis, and other degenerative diseases may still develop in certain individuals, they will occur at a more advanced age, and some of them might very well be postponed right out of the human lifetime. While current research in these diseases is extremely important, it is not apt to produce a sudden jump in human longevity in the immediate future.

Factors Influencing Average Life-Span (Average Life Expectancy at Birth for Males = 68 years)

	Years
Living in country rather than city	+ 5
Married rather than single, widowed, or divorced	+ 5
Female rather than male	+ 6
4 grandparents living to age 80 years	+ 4
2 grandparents living to age 80 years	+ 2
Father lived to 80 years	+ 2.2
Mother lived to 80 years	+ 1.5
25% overweight	− 3.6
35% overweight	− 4.3
45% overweight	− 6.6
55% overweight	−11.4
67% overweight	−15.1
Smoking 1 pkg. cigarettes each day	− 7
Smoking 2 pkg. cigarettes each day	−12

continued from page 83

On the other hand, the cross-linking theory of aging assumes that, with the passage of time, molecules important in the structure or function of cells undergo structural changes that impair their usefulness. The chemical change is the formation of cross-links either between different parts of the same molecule or between molecules. When these cross-links are formed, the molecule can no longer carry out the proper chemical reactions, or its physical properties may be so altered that the tissue of which it is a part loses its characteristics.

Collagen, the primary protein constituent of connective tissue, represents the supporting framework for many tissues and organs of the body. In young animals, collagen fibers impart elastic properties to organs and tissues. With increasing age elasticity is lost primarily because cross-links form between collagen fibers or within the collagen molecules themselves.

Exposure to radiation may also disrupt cellular function by the formation of free radicals, which are highly reactive and readily combine with other molecules in the cell. Free radicals may thus induce the formation of cross-linkages in intra-cellular proteins. If the formation of free radicals plays a role in aging, it might be possible to slow the rate of damage by means of chemical agents called antioxidants. These agents resemble those used commercially to prevent the spoilage of food and other materials.

Because it is chemically possible for cross-linkages to form either within or between molecules of DNA or proteins, all of which are located within the cell, some investigators have proposed this as a generalized mechanism for aging. Thus far, however, chemical evidence for the formation of cross-links in living tissue has been obtained only from collagen, which lies outside of cells. Although the cross-linking

theory is an attractive one, as yet no experimental evidence has definitely proved that such cross-linkages do occur in molecules inside of cells. Until such evidence is obtained, this theory can only be regarded as speculation.

Potentials for the future

Whereas most of the biological theories about aging focus attention on changes at the cellular level, age changes are actually most apparent in the overall performance of an animal. Thus, the decrement in overall performance of physical work is greater than age changes that can be detected in the enzymatic activities of the muscle cells that perform the work. It is possible, therefore, that aging as we see it in the total animal is more a reflection of a breakdown in the coordinating mechanisms than a failure of the individual parts. Perhaps the future for gerontology may lie in investigations of the control mechanisms that produce and integrate complex performances.

Some authorities have stated that by the year 2000, it may be possible to add 20–40 years to the middle part of life. Their predictions are based on the development of successful methods to transplant organs, such as the heart and kidneys, which will remain healthy, the design of artificial organs, which can be implanted into the body to replace defective ones, or the administration of drugs that will slow the aging process. Even if long-term survival of transplanted organs can be achieved or if satisfactory artificial organs can be designed, they can be applied only to a small number of people. It is also unlikely that effective drugs can be found until more is known about the basic mechanisms of aging.

Although significant extension of life-span is not impossible, it will require an entirely different approach. This approach involves the ability to control the genetic program at a cellular level or to discover methods of altering the time sequence for the expression of such a program. Already molecular biologists have begun to isolate mechanisms for the repair of DNA and to alter the timing of biological clocks in some lower animal forms. As British gerontologist Alex Comfort has stated, "Biological gerontology may well be to the medicine of the 1970s what chemotherapy was in the 40s and 50s and immunology [was] in the 60s."

FOR ADDITIONAL READING:

Birren, J. E., *The Psychology of Aging* (Prentice-Hall, 1964).

Comfort, A., *The Process of Ageing* (New American Library, 1964).

DeRopp, R. S., *Man Against Aging* (St. Martin's Press, 1960).

McGrady, P. M., Jr., *The Youth Doctors* (Coward-McCann, 1968).

Prehoda, R. W., *Extended Youth: The Promise of Gerontology* (G. P. Putnam's Sons, 1968).

Shock, N. W., "The Physiology of Aging," *Scientific American* (January 1962, pp. 100–110).

Strehler, B. L., *Time, Cells, and Aging* (Academic Press, 1962).

How Science Detects Art Fakes

by Perry T. Rathbone

Museum curators and private collectors increasingly rely on modern scientific techniques to detect fraudulent art and to learn more about art objects.

Some art forgers have been successful because they have been talented artists in their own right. Such is the case with the well-known Italian forger Alceo Dossena (1878–1937), who created numerous fakes in archaic Greek, medieval, and Renaissance styles. The terracotta relief "Virgin and Child" (opposite page), done in early Renaissance style, is an example of Dossena's work.

Recently, a private collector brought to Boston's Museum of Fine Arts an unknown and unattributed 15th-century Florentine portrait of a young lady. To the practiced eye of the collector, the panel painting looked authentic enough, but there were certain puzzling aspects about it that troubled him. The collector wanted the museum's research laboratory to subject the painting to a careful scientific examination.

Preliminary dating tests conducted by the museum laboratory revealed that the panel on which the work was painted was of the appropriate period. The results of an X-ray examination, however, aroused some suspicion. Beneath the top layer of paint on the face of the painting were traces of a secondary layer. When these two layers were matched, the crackle—surface cracks that appear on the face of old paintings—did not correspond. This meant that the original surface had been partially removed and then overpainted.

But when was the overpainting done? In the 15th century or in recent times? To find out, the paint surface was chemically analyzed using a microbeam probe. Needle samples of the pigment were removed in four places on the painting. These core samples then were bombarded by electrons to generate measurable fluorescent X rays that were recorded on a spectrometer. The resulting pattern of lines identified the chemical substances in the pigments.

The lady in question was much younger than she looked. Among the pigments found was titanium white, a substance unknown until 1920. The lady was a fake!

The collector was making use of the well-established procedures most museums employ to consider carefully any proposed gifts or purchases before accepting them. At Boston's Museum of Fine Arts,

The Victoria and Albert Museum

The Courtauld Institute of Art, University of London

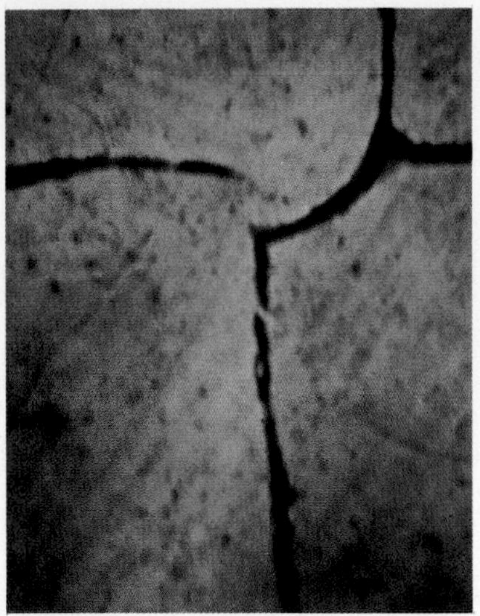

The painting "Religious Procession" (right) is the work of an unknown modern forger done in the style of Pieter Brueghel (c. 1525–69). The forgery was discovered when the painting's crackle—the surface cracks that naturally appear in old paintings as a result of expansion and contraction—were found to have been simulated (close-up, left). It was also discovered that the green pigment used in the painting was chromic oxide, a substance not in use before the mid-19th century.

for example, every piece presented to the collection committee has a supporting document that tells what it is, where it is from, its importance, and why it belongs in the collection. The object's validity is assessed carefully not only by the well-trained eye of the curator but also by a battery of scientific instruments that use techniques ranging from simple optics to nuclear bombardment.

The scientific examination is designed to extract as much information as possible about the original structure and formation of the object for a variety of reasons: (1) to give the curator, scholar, or connoisseur more data about the historical, cultural, and creative context in which the object was made; (2) to show how the object was formed or molded; (3) to conserve the object by arresting or limiting the natural process of decay; (4) to guide the restorer in reconstituting the object to compensate for damage wrought by decay or by such man-made hazards as breakage, amateur restoration, and environmental pollution; and (5) to detect a fake or imitation.

Art forgery—an ancient tradition

Science and art must go hand in hand today. Because collecting art treasures is so much in vogue, there has been a decrease in supply as the number of bidders has increased. In most cases this demand can be met by customary legal methods—art auctions, reputable galleries, private collections, etc.—but, unfortunately, it also creates a ready market for counterfeits and forgeries. It is in the detection of such frauds that science plays an important role in the art world.

Forgery is certainly as old as art. Ancient Egyptian papyri reveal the trade secrets of art copyists who practiced their craft thousands of years ago. When the Romans grew prosperous, they faked as well as copied Greek sculpture. With the rise of easel painting during the Renaissance, what began as imitation in time became forgery. One documentation of this exists in the form of a diary by Hans Hieronymus

PERRY T. RATHBONE is director of the Museum of Fine Arts in Boston, Mass., and has served as president of the Association of Art Museum Directors.

Imhoff. In 1633, Imhoff, the grandson of a friend and patron of the German master Albrecht Dürer (1471–1528), sought to repair his tattered fortune by selling the remains of the family art collection, the value of which he had enhanced by attributions to great masters and by spurious signatures. "The things we sold," Imhoff wrote in his diary, "were small watercolors . . . . It may well be doubted of many among them whether they were actually painted by Dürer."

The middle 19th century brought the age of museums and great collectors to the United States and with it a new class of forgers, who reached their height in the 20th century with the Dutch painter Han van Meegeren (1889–1947), the Italian sculptor Alceo Dossena (1878–1937), and the contemporary European copyist Elmyr de Hory (1906–), who can produce a "Picasso" or a "Matisse" on order. Most recently, it has been the victims rather than the forgers who are best known. But they are not alone. Almost every great museum has former "treasures" that are no longer on exhibition.

Courtesy, The Museum of Fine Arts, Boston

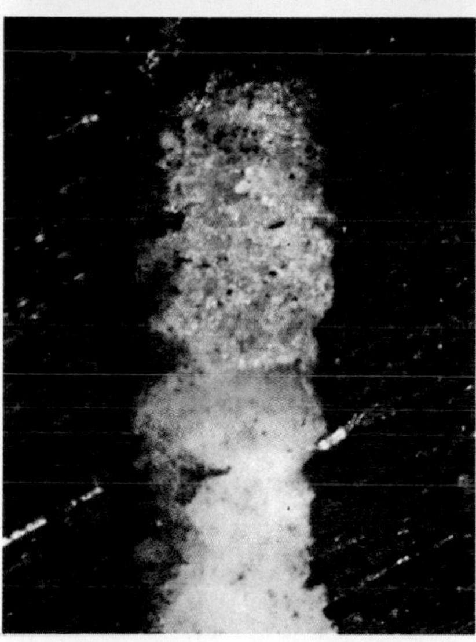

A purported 15th-century Florentine portrait (top) was discovered by Boston's Museum of Fine Arts to be a fake when a microbeam probe revealed the presence of titanium white (bottom), a pigment unknown until 1920.

The educated man's eye and instinct

Powerful and effective as it is in the service of art, science is still a subordinate helpmate. The prime reference remains the educated man's eye and instinct, which are not reflected in even the most sophisticated instrument of science. Two examples illustrate this point.

Once, the administrators of a charitable organization brought an old painting to the Boston Museum to find out if it had any value. Most such objects that are submitted to the curatorial offices of a museum can be spotted immediately as copies, fakes, or worthless attic bric-a-brac. Occasionally, however, some great masterpieces are discovered in this way. The curator who examined this particular painting recognized it as a 15th-century Italian work, "Portrait of a Doge" by Gentile Bellini. Subsequently, his identification was verified by tests in the laboratory.

More recently, a private collector of Oriental art was particularly taken with the work of the 18th-century Japanese artist Soga Shohaku. His work is rare even in Japan, where the National Museum in Tokyo has only four examples. But the collector had assembled five paintings that purportedly came from the hand of Shohaku. To test his judgment, the collector invited the Boston Museum's Curator of Asiatic Art to see them. With speed matched by sureness, the curator accepted all but two of the paintings. "Why are you so sure?" the collector asked in bewilderment. The curator explained that two of the museum's 19th-century collectors had both shared an enthusiasm for Shohaku, and the museum's collection, by far the largest extant, contains a solid body of almost 40 works by this master from all his periods. The curator had carefully studied the work for an exhibition and had gained a complete picture of Shohaku's development—his changes in style, subject matter, and technique. Combining art history, aesthetic judgment, and the sureness of intuition, the curator could unhesitatingly pinpoint Shohaku's work.

National Gallery, London

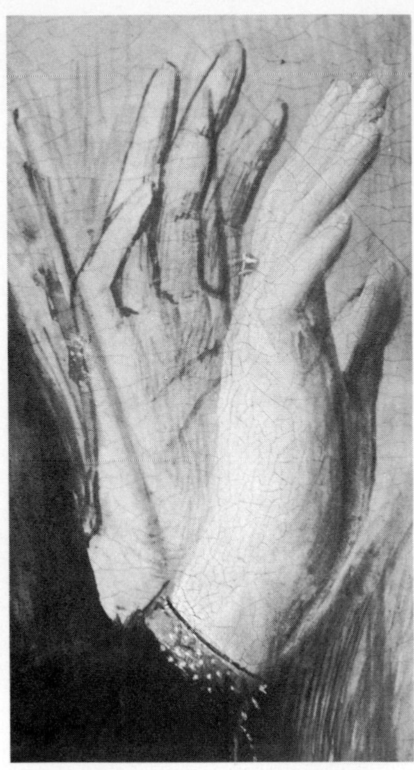

An infrared photograph of a detail from the painting "The Marriage of Giovanni [?] Arnolfini and Giovanna Cenami [?]" by Jan van Eyck (active 1422–41) reveals how van Eyck changed his mind about the placement of the hand between his preliminary sketch and final painting. The scientific techniques used by museum laboratories frequently may provide such information about the creative development of a work of art.

Art's need for science

The human eye is fallible as well as intuitive, of course, but fallibility can be checked by scientific tests. Individually, such tests may not be conclusive because they can answer only one question at a time. Taken in the aggregate, however, they move questions and answers from suspicion to proof.

Consider a Greek vase that *was not* accepted by the Boston Museum. To the curator's eye the vase, circa 450 B.C., seemed perfect—perhaps too perfect. Covered with a jet-black glaze on which was painted a cheerful domestic scene of two female figures bathing, the vase was immaculate inside and out. But laboratory tests revealed a very different story. The figures were not in their original form; examination under ultraviolet light revealed that they had been repainted. The vase actually was badly fragmented and had been pieced together. Infrared and X-ray tests indicated that the cracks had been concealed by overpainting. The overpainting itself was a combination of modern paint and varnish and not the ancient glaze of iron oxide, as revealed by spectroscopy, a Geiger counter, and X-ray diffraction tests. Although it allegedly had been taken from an excavation in the Nile Delta, microscopic examination indicated that the vase actually originated in a pottery kiln in the Peloponnesus. The vase was a beautiful object, but the curator, his suspicions confirmed about its too obvious perfection, rejected it. It was not up to the standards of the museum.

Art investigation remains a double equation—the scholar on one side; the scientist on the other. William Young, head of the research laboratory at the Boston Museum, puts it this way: "While visual observation cannot be superseded, it often can be profitably supplemented with more specialized investigation. Scholarship cannot be overemphasized. A stylistic judgment can be a rewarding one, but it has certain shortcomings when not supported by scientific fact."

It is only in recent years that science has been conjoined to art. The earliest museum laboratory is probably that of the former Imperial Museum in Berlin, which was founded in 1888. Many more were established in the 1920s and 1930s; now most of the world's great museums have such laboratories or access to them.

A Sienese triptych (below), purportedly of the 15th century, was created by a forger active in the 1920s and 1930s. An X ray of the triptych (bottom) reveals how the forger used modern nails to mend the center panel and filled in the wormholes in the old wood before repainting it.

The Courtauld Institute of Art, University of London

Optical tests

In the museum laboratory the first step in the scientific examination of a work of art is a series of optical tests. First, the object is examined microscopically in order to gain a full understanding of its surface phenomena. Microscopy can answer a wide variety of questions. For example, is the patina on a Chinese bronze a natural one or has it been induced? Are wormholes in the polychromed walnut of a saintly 15th-century figure the result of animal life? Was a bronze figure cast by the lost wax process, the piece-mold method, or was it hammered? What kind of engraving tool was used to make the incised design on a piece of ancient Greek jewelry? In an oil painting microscopy will reveal the condition of the paint, its composition and glazes, the structure of the crackle—the cracks that pattern the surface of a painting as it is disturbed by the drying of the paint and by the expansion and contraction of the material on which it is laid, be it canvas, linen, or board.

The next step is a careful investigation of the object under light. A side light or raking light will differentiate between original and new surface textures on a variety of objects. On a canvas such light may throw into relief the brush techniques of an artist or reveal shadowed outlines of an underlying painting. Materials used to alter the surface of an object—painting, vase, or bronze—will show up as a contrast between old and new under the yellow monochromatic light of a sodium vapor lamp.

Next, the object is subjected to various kinds of invisible light rays. In general, the shorter the wavelength, the greater the penetration. Ultraviolet light will cause some substances to glow or fluoresce in visible colors, thereby revealing any repairs, patches, or touch-ups that

Courtesy, Edward V. Sayre, Brookhaven National Laboratory

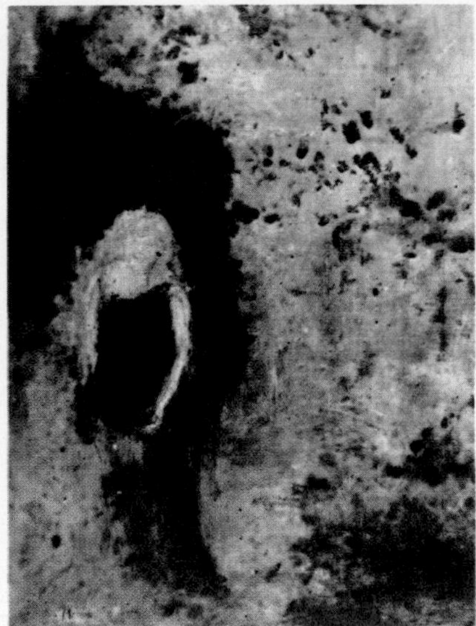

The series above reveals how a Marian Blakelock painting was altered to appear to have been done by her father, the U.S. painter Ralph Albert Blakelock (1847–1919). The painting is shown (left) as it appears under normal light, (center) in a conventional X-ray radiograph, and (right) in a neutron activation autoradiograph. Only the autoradiograph shows the partially removed, overpainted Marian Blakelock signature.

may have been added to the original. It will bring out faded inscriptions or semiobliterated signatures on a painting. The color of the glow sometimes will reveal data about the materials used in an object. Ancient Egyptian wood will fluoresce with an amber color; modern wood will fluoresce purple. Infrared light will go beyond the topmost surface of a painting and penetrate the brown varnish that may have been applied to the surface to conceal alterations. It will show up chemically different substances that may indicate restoration in textiles, ceramics, marbles, or wooden sculptures. X rays probe even deeper into a painting to reveal whether a canvas is old or new, or whether a newer painting has been emplaced over an older one. X rays also will show the structure of wood and bronze objects and how they have been fabricated. For example, X rays may reveal that an "old" piece of wood sculpture is fastened together with modern nails.

Finding fingerprints

After completion of the optical tests, the art scientist then examines other characteristics of an object. Here he may be concerned with the elements and compounds that make up the object. These chemical substances, which can be identified by a growing battery of instruments now at the scientist's command, provide a revealing "fingerprint" that can tell the scientist a great deal about the object.

The fingerprints in a work of art, for example, can be matched with known samples to give the object's technical history, which, in turn, will reveal where it came from, how old it is, how it was made or fashioned, and whether it is what it purports to be—that is, whether it is genuine, a copy, or a forgery. Consider a Greek vase made of clay. Clay has different properties and characteristics that vary from site to site because of geological history—the way the earth was formed. An unknown sample of clay may be matched with a known sample, and the origin of the vase pinpointed to a specific area in Greece. Similarly,

Japanese ceramics can be traced accurately to their exact kiln site.

Age also reveals itself to the scientist. The Romans, for example, successfully produced a high zinc alloy which they used mainly for coinage but rarely for the statuary they copied from Greek originals. Thus, an "ancient" bronze which comes to light today and contains a high zinc content is more likely to be of 19th-century rather than Roman origin. Bronze Buddhas from Thailand are copied from generation to generation and defy stylistic dating. But their zinc content, negligible before the 14th century and as high as 30% in the 19th century, helps establish their chronology. Today, dating analysis is achieved most often by such methods as thermoluminescence and the carbon-14 and lead-210 tests, which rely on the presence of radioactivity.

The carbon-14 test is used to determine the age of organic materials —in simple terms, anything living or once living; in scientific terms, any compound that contains carbon. Although it is imprecise, the carbon-14 test is adequate enough to detect a forgery and to aid the scholar in establishing reference points since numerous materials used by the artist are organic—linseed oil, canvas, paper, resinous varnishes, wood or textile dyes, even food. Food residues found in ancient Egyptian storage vessels, for example, can provide a useful dating reference.

The lead-210 test is a radioactive dating test for inorganic material

The Metropolitan Museum of Art, Bequest of Collis P. Huntington, 1925

The painting "Portrait of a Bearded Man" (above) by the Dutch artist Govaert Flinck (1615–60) was painted over an earlier portrait done by the artist of a young girl, as an X-ray radiograph (left) reveals.

Courtesy, The Museum of Fine Arts, Boston

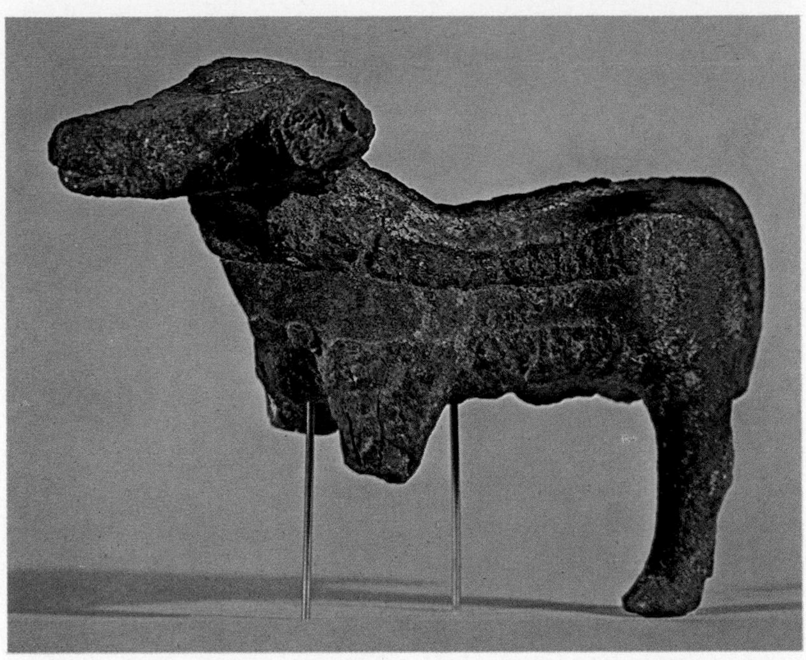

*An Anatolian bull (above), thought to be of the pre-Hittite period (*c. *2100 B.C.), is shown after cleaning by electrolysis (opposite page, left), and after being subjected to a laser beam (laser target area shown greatly magnified, opposite page, right). The laser drilled a hole invisible to the naked eye, one-fifth the diameter of a human hair, volatilizing the area it struck. The volatilizing reaction was then measured by spectrographic means, which revealed the exact chemical composition of the bull.*

that involves dating the time when the lead ore used in lead white paint was smelted. The test is primarily of use in distinguishing between works of art produced before the 19th century and those produced since the late 19th century. With pottery, the most promising dating technique is the thermoluminescence method developed by Edward T. Hall and the Research Laboratory for Archaeology and the History of Art at Oxford University. (See *1971 Britannica Yearbook of Science and the Future*, Feature Article: DATING THE PAST, pages 386–399.)

A clever forger can know about and take advantage of such tests, of course. It is perfectly possible, for instance, for a knowledgeable forger to find a piece of timber dating from the 13th century, fashion it into a stylistic duplicate of a madonna and child of the period, and thereby pass the dating test. Similarly, a forger could take a known 17th-century painting of little value, remove the paint from the canvas, and substitute his own version of a prized master like Vermeer. Indeed, this is what the Dutch painter Han van Meegeren did in 1937 with "Christ and the Disciples at Emmaus," deceiving all eminent art authorities of the day. So painstaking was his method of forgery that he even copied the chemical makeup of the pigments that Vermeer would have used. A supplier upset his scheme, however, by giving him the wrong ingredients for lapis lazuli, ingredients that were adulterated with the modern pigment cobalt blue, which was not known until the 19th century. Microchemical analysis subsequently revealed the presence of cobalt blue in two of van Meegeren's later forgeries and provided an important link in the chain of circumstances that eventually led to his apprehension and conviction as a forger. Van Meegeren's paints, in this instance, had been successfully identified and "fingerprinted" by science.

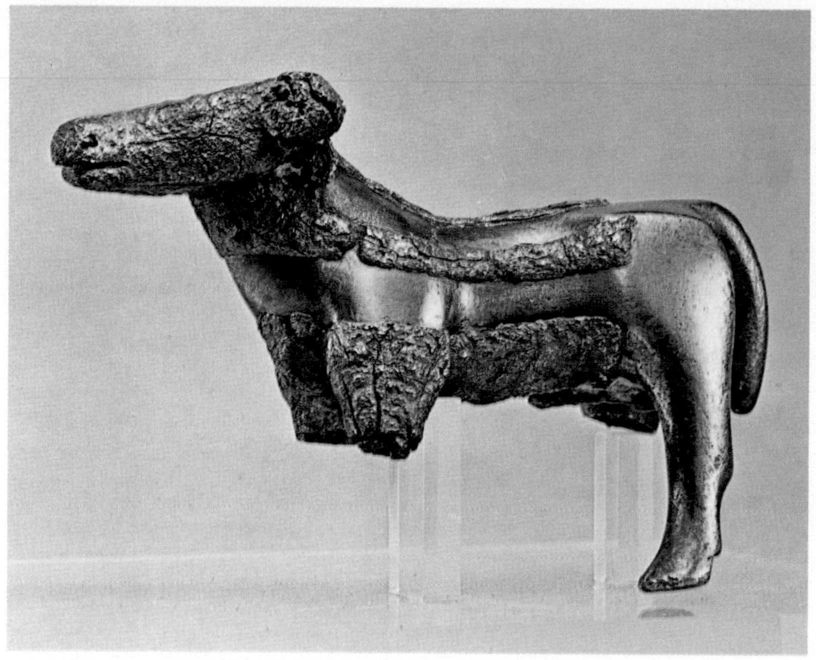

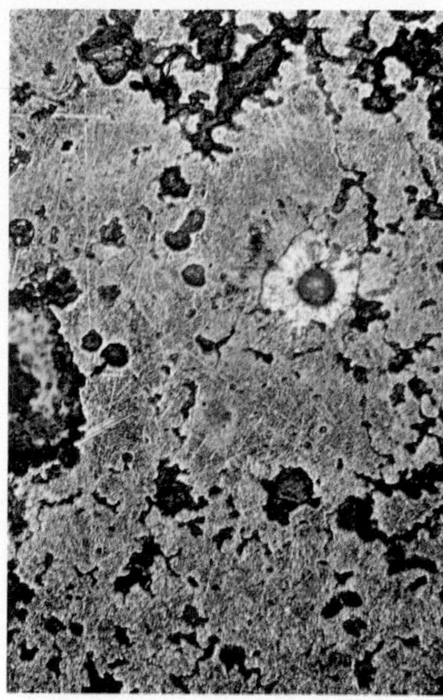

A 'bureau of standards'

To assist in this fingerprinting, the art scientist and museum laboratory set up a bureau of standards. That is to say, they set up standards provided by known and proven objects against which any newcomers can be measured. For instance, the laboratory at the Boston Museum has assembled such standards as spectrographic analyses of:

• 600 samples of stone from identified quarries around the Aegean Sea;

• 3,000 samples of ancient bronzes in Egypt, Asia Minor, Asia, and Europe;

• ceramics from 50 kiln sites in China, Japan, and Korea;

• all of the 150 or so American woods and about 90 European and African woods;

• 600 samples of paint pigments from all over the world, including those that Rembrandt, Giotto, or Poussin would have used.

Other research centers have similar standards. Brookhaven National Laboratory in Upton, N.Y., for example, has analyzed more than 400 specimens of ancient glass. Salmi Agusti, director of the Physicochemical Laboratory of the Museo e Gallerie Nazionali di Capodimonte in Naples, Italy, has recorded 240 samples of pigments used in Pompeiian mural painting. Researchers at the Henry Francis du Pont Winterthur Museum in Winterthur, Del., have analyzed 100 silver objects of British origin and 100 silver objects of American origin classified at 50-year intervals between 1700 and 1900.

New analytical techniques

In recent years physical analysis has been made much easier and faster with the advent of such specialized techniques as X-ray diffraction, emission spectrography, nuclear spectrography, and infrared and ultraviolet spectrophotometry. These analytical methods make it possible to determine the chemical composition of an object from a mere fragment—the pigment in a painting, the glaze on a ceramic, a sample of a bronze, or a piece of a sculpture in wood. In labeling the chemical elements, the scientist is also helping to label the object.

A further refinement on spectroscopy has been the electron microbeam probe, which was developed and applied to art by the Massachusetts Institute of Technology in collaboration with the research laboratory at the Boston Museum. The probe utilizes a minute needle core sample from the object. The core sample is bombarded with electrons to generate fluorescent X rays that can be measured by a Geiger counter to give the chemical composition. Another method of chemical analysis is to direct a laser beam at a specific target area on a painting or object. The laser drills a hole 15 microns wide, volatilizing the area it strikes. The reaction is measured by spectrographic means. The hole is not visible to the naked eye; it is one-fifth the diameter of a human hair.

An Italian 15th-century Majolica bowl is shown (below) as seen under normal light and (bottom) as seen under ultraviolet light. The lighter areas indicate areas of repair. "Christ and the Disciples at Emmaus" (opposite page) was painted by the Dutch forger Han van Meegeren (1889–1947) in the style of Jan Vermeer (1632–75). Van Meegeren was so successful in imitating Vermeer's style that the painting was not only accepted by all the eminent art authorities of the day but even hailed as a masterpiece.

Courtesy, Conservation Center of the Institute of Fine Arts, New York University

One problem with these methods is their destructive nature. Curators and collectors are not anxious to have even minute samples of their precious objects removed and destroyed in the arc of a spectrograph or the needle for a microbeam probe. In an extreme case, the textile collector would have to part with six square inches of a tapestry or weaving before the carbon-14 test could be served with an adequate sample. In this respect, 1970 was a breakthrough year with the first application at the Winterthur Museum and the Boston Museum of a nondispersive X-ray spectrometer, a device originally intended for nuclear spectroscopy. This instrument analyzes the elements present in an object without in any way physically altering the specimen being analyzed. It can pinpoint 80 elements even if an element is present only in trace amounts. The system not only has the advantage of being totally nondestructive, it also can be manipulated easily to scan the entire surface of a painting or an object. Because it produces immediate readouts by visual means on an oscilloscope screen with a built-in memory, elements can be compared. Readouts also are produced by printed means on a graph (x-y recorder), on a typewriter, or on a punched paper tape or magnetic tape. The information therefore can be stored and utilized for future analyses, thus saving enormously on laboratory time. The results are so promising that the U.S. Internal Revenue Service reportedly has purchased a nondispersive X-ray spectrometer to test the validity of tax deductions for works of art donated to museums.

Experiments currently are being carried out by the National Gallery of Art in Washington, D.C., the Mellon Institute in Pittsburgh, Pa., and the U.S. Atomic Energy Commission on yet another analytical method—neutron activation analysis. This method is capable of detecting trace elements in paint pigments so small that they are in the parts-per-million scale. These trace elements form impurities within the pigment,

Museum Boymans-van Beuningen

particularly in white lead. It is thought that, because of improved technology, modern white lead may contain lesser amounts of these trace elements than old white lead. It thus may be possible to establish a pattern of these impurities that is characteristic of certain regions, periods, and artists. If so, art scientists will be able to develop yet another set of standards for fingerprinting works of art.

The scientist, however, is not always a hero in the art community. Curators and collectors are not necessarily overjoyed to have their suspicions confirmed and hopes dashed when the scientist finds that a painting, vase, bronze, or precious jewel is not what it is purported to be. But science has become an essential and invaluable part of the art world. Although stylistic judgment remains of paramount importance, curators and collectors increasingly are relying on scientific research to provide knowledge about art, point the way to new methods of conservation, and, finally, to provide a reliable way to determine the authenticity of a work of art and to frustrate the forger.

FOR ADDITIONAL READING:

Application of Science in Examination of Works of Art (Museum of Fine Arts, Boston, Volume I, 1959; Volume II, 1967).

Coremans, P. B., *Van Meegeren's Faked Vermeers and DeHooghs* (Amsterdam: J. M. Meulenhoff, 1949).

Esterow, Milton, *The Art Stealers* (Macmillan, 1966).

Godley, John, *The Master Forger* (London: Home and Van Thal, 1951).

Kurz, Otto, *Fakes* (Dover Publications, 1967).

Margaret Mead: Anthropologist and Social Reformer

by Malcolm C. Webb

"We are squarely up against the dilemma of whether out of fear and desperation we will seek to prop up a crumbling old pattern or too hastily run up a new one, intent only that the new shall be a bulwark against the destruction of the old, . . . or whether we can believe that we can build a new world suited to men's needs. "In these words from *New Lives for Old*, anthropologist Margaret Mead expresses her long-held view that the present age, despite its mounting crises and constant turmoil, offers man his first real opportunity to escape the deadening hand of tradition and to shape the future to his needs.

At a time when gloom and doubt are the prevailing fashion, such opinions may seem bold to the point of rashness. There is nothing pretentious about their author, however. Dr. Mead could easily be taken for a moderately prosperous suburban housewife or schoolteacher, and she has been known to pass as her own secretary when wishing to avoid tedious visitors. Rather heavy-set and short (5 feet 2 inches), with short, wavy gray hair that still shows traces of light brown, Mead favors simple print dresses or severely restrained formal outfits, as the occasion demands. The one dramatic touch is provided by her well-known forked staff or "thumb stick," which she has carried everywhere since injuring her ankle some years ago—and which she flourishes on occasion when trying to gain the floor at scientific meetings.

Conversation with Mead reveals an unexpected homeliness of speech and a notably calm, outgoing, and easy manner—which must, nevertheless, mask a fantastic competence and organizational ability, since she could not possibly manage her diverse activities otherwise. One's attention, however, is most strongly drawn to her plain but lively face, which displays the good humor and kindliness of a grandmother (her sympathy for students and younger colleagues is proverbial), combined with the intelligence, alertness, and self-assurance of an internationally renowned scholar. (*See* "Mead on Mead," pages 108–110.)

Apotheosis of a scholar

Now at the peak of a distinguished career, Mead has achieved a degree of eminence rare in the history of science—for almost a generation she has symbolized and represented an entire scientific discipline to the general public. If a nonanthropological audience is asked to name one anthropologist, the name suggested will almost surely be that of Margaret Mead. In my experience this is so on every level, from high school students to academicians in other fields. Moreover, this public familiarity is understandable, since few other anthropologists—or professionals in any field—have devoted so much time and energy to presenting their discipline both to the layman and to specialists in related areas, while at the same time ably satisfying the exacting demands of their own work.

The aura of romanticism that still surrounds anthropology in the public mind is largely due to Mead's work. So is the current vogue of the central anthropological concept of culture, the patterned way of life of any people, which has in large measure been shaped by their

MALCOLM C. WEBB, associate professor of anthropology at Louisiana State University, is an authority on the evolution of early civilizations, especially in Mexico and Central America.

past history and which, by conditioning their response to new situations, will in turn help to shape their future. At the same time, the popularity of these ideas has assured Mead of wide attention. When to the public's willingness to hear one adds Mead's confident willingness to speak on a great variety of issues, as well as her considerable talents in doing so, it is scarcely surprising that by the end of World War II her reputation had ceased to be confined to the scholarly world and had become that of a public figure, a person of affairs, even a "personality." This is a truly remarkable position for one holding such formal positions as curator at the American Museum of Natural History in New York City and part-time Columbia University professor.

Such renown is not without its problems, however. Many of Mead's suggestions for improving our society have met strong criticism. Her recent assertions that traditional attitudes toward marihuana are inconsistent, and that our current draconian prohibition laws are both ineffective and corrupting, aroused predictable ire among conservative politicians. In her *Redbook* magazine column of July 1966 she argued that long-term "parental" marriage, with its financial and child-rearing responsibilities, should be preceded by "individual" or "student" marriage that would carry no support obligations and could be terminated at will. The concept upset many readers, including both the young people toward whom the plan was directed and a panel of family life experts.

Nor is disagreement with Mead confined to the traditionalists. During a heated plenary session on war held at the 1967 meeting of the American Anthropological Association, Mead avoided polemics against the Vietnamese war, urging instead the development of truly multinational institutions that would cut across existing territorial boundaries. Not surprisingly, many in the audience saw this as taking refuge in utopian generalities in the face of a burning moral issue.

Status within the anthropological tribe

If Mead's place on the national scene is slightly ambiguous, her position among her colleagues is equally interesting. With very few exceptions, fellow anthropologists consider her fieldwork to be of a very high order. Mead has made more than 12 field expeditions, beginning with Samoa in 1925 and including the Admiralty Islands (five times), an American Indian community, New Guinea (four peoples), and Bali (twice). Plans for yet another expedition to the Admiralty Islands were announced in the spring of 1971. In a discipline that continues to stress firsthand observation of social behavior, and that has always doubted the validity of generalizations about mankind unsupported by observations taken from a wide range of societies, this work earns instant respect. So does Mead's well-known ingenuity, industry, and constantly renewed freshness of approach in recording her material.

Mead's often expressed wish is to observe *everything*. Her employment of new data-recording techniques that often require specialized training and involve considerable effort while in the field have won wide acclaim—more acclaim than emulators, perhaps, since few of her colleagues possess her energy and determination. One would expect that Mead's orientation toward the study of "personality and culture" (the influence of cultural norms in shaping the individual personality and the role of individuals in cultural innovation) would cause her to be an anthropological pioneer in the use of psychological testing and the study of child-rearing techniques, and indeed this is so. Less to be expected are her employment of both still and motion-picture photography and her observation that body posture and relative space carry cultural messages, long before these became fashionable. Her work in these areas is still considered worthy of emulation.

Predictably, admiration becomes less unanimous when interpretation begins. Anthropology's emphasis on unitary interpretations of entire societies places an extraordinary burden of clarity, sobriety, and impartiality on the observer, who may be the sole witness of the system he describes. Lacking clear-cut methods of verification, the discipline has always been bedeviled by disputed interpretations. The conventional way of meeting this problem has been to provide copious and repetitious specifications as to events, times, places, actors, and presumed relationships, generally using native terms and categories. This explains the impression of extreme dryness that notoriously overwhelms students who have just passed beyond a superficial knowledge of the subject.

Mead consciously rejected this approach from the first. Her descriptions of non-Western peoples—*Coming of Age in Samoa* (1928), *Growing Up in New Guinea* (1930), *Sex and Temperament in Three Primitive Societies* (1935), *Balinese Character: A Photographic Analysis* (1942), *New Lives for Old* (1956)—are almost novelistic, both in their emphasis on a few basic themes and in their subjective and "literary" descriptions of character, circumstance, and motivation. Unlike many of her peers, Mead can be read for pleasure, and her civilized style, along with the intrinsic interest of the material, has helped ensure a steady supply of readers. Anthropologists repelled by jargon, as well as laymen, have been attracted not only to the books mentioned above but also to such topical or theoretical works as *And Keep Your Powder Dry*, a wartime study of American character (1942); *Male and Female* (1949); *Soviet Attitudes Toward Authority* (1951); *Continuities in Cultural Evolution* (1964); and *Culture and Commitment* (1970).

Admirers of Mead's style tend to feel that the gains in clarity and empathy are well worth the costs in precision and scientific documentation. Such an approach does increase the difficulty of providing a concise summary of her substantive contributions—a problem compounded by the extremely wide nature of her interests and the fact that her work is scattered through the literature of many fields. In all, she has written over 20 books, edited seven more, and contributed hundreds of scholarly and popular articles. On the whole, however, it can be said that her research has focused on the degree to which personality traits characteristic of many individuals are due to social conditioning, as opposed to being innate human nature, and upon how such conditioning might be altered to produce more cultural innovators without, at the same time, giving rise to excessively deviant or unstable individuals.

Coming of Age in Samoa describes a society in which the adolescent stress that seems so inevitable to us is largely absent, apparently because of a corresponding lack of strong sexual prohibitions or other potential sources of value conflict. (It is worth noting that, despite the importance of sex in her books—and the lurid covers of some paperback editions—Mead's treatment of this topic has always been quite chaste or even flat.) Among the previously little known Manus of the Admiralty Islands, Mead discovered a seemingly strong incongruity between childhood and adult behavior (reported in *Growing Up in New Guinea*). This led her to doubt the effectiveness of education as a technique for social reform, while the relative ease with which this tribe later adopted Western customs (the theme of *New Lives for Old*) suggested to her that societies with a well-developed "youth culture" are especially open to innovation. The importance she places on this latter point is indicated by her frequent visits to the Manus, as well as by her continual reworking of the hypothesis during the last two decades. Similarly, Mead found that three neighboring New Guinea peoples apparently assign what we would regard as typical male and female traits to the two sexes in ways quite different from

one another or from Western society, thus establishing the basis for her insistence that such personality traits are largely, though not entirely, conditioned by cultural expectations. These findings obviously have far-reaching implications.

Still, a great deal besides beauty may lie in the eye of the beholder. Mead described the peaceful adolescence of the Samoans, but other students of the island have given differing accounts. Did Mead see or only imagine the tranquillity she reports? Another doubtful case is the presumed variation in sexual temperament among Mead's three New Guinea tribes. The occurrence of such wide differences within a small region is a fantastic coincidence, to say the least. It is not surprising, therefore, that some critics have intimated that in this instance also Mead saw what she wished to see. Marvin Harris, author of a recent major history of anthropology and Mead's colleague at Columbia, has even suggested that many of her conclusions are totally unverified and, perhaps, unverifiable.

Many anthropologists apparently feel that, although her fieldwork is impressive and her role in bringing anthropological insights into other fields (and the reverse) has been useful, Mead's contributions to anthropological theory have been rather minor. The extent of these views is somewhat problematic since, except for Harris and a study made some years ago by David Mandelbaum of the University of California at Berkeley, criticisms of Mead by her colleagues consist chiefly of casual comments and attitudes. Nevertheless, they do exist and have tended to reawaken doubts concerning the validity of her reporting.

There is also a tendency to regard Mead the public figure as inherently antithetical to Mead the anthropologist. One senses a feeling of disappointment that so much of her time and the greater part of her interest have been devoted to reformist activity. This is considered not only a diversion of energy but also as implying either a lack of seriousness in her theorizing or a quality of irrelevance in her reformist efforts. Feeling as strongly as she obviously does, can she be fair to the data? To understand this uneasiness, we must examine briefly the concept of culture as it is used by anthropologists and as it bears on Mead's work.

The constraining circle of tradition . . .

For anthropologists, the term "culture" denotes the entire inherited tradition of a people: technological knowledge, customs, kinship, wider social organization, law, religion, art, world view—in a word, all learned behavior patterns. A cultural tradition continues through time because the younger generation has (or, until recently, had) no other complete model of humanness except that presented by their elders. No one can be objective about his culture because it shapes the way he sees the world (including his standards of objectivity). Anthropologists therefore feel that it is no more possible to stand outside the encircling culture to appraise it than it is for a fish to escape the water.

There is less consensus about the extent to which cultures are systems, but two rather opposed views have tended to predominate. The first is that individual societies are built out of cultural "shreds and patches," as Robert Lowie of the University of California put it, by their unique histories. They are, therefore, unpredictable, and the little we can understand of any complex society provides scant guidance for corrective action with regard to it. The other approach regards societies as organic, with each part integrated into the whole in a determined way. While anthropologists who hold this view reject crude determinisms, they generally feel that environmental possibilities, in conjunction with wide categories of primary subsistence types (hunting, hoe agriculture, industry, etc.), limit other aspects of culture. Change occurs in response to new conditions, but it is of a self-directing sort, as in biological evolution. Combining these two views, one may conclude that if a custom is predictable enough to be understood and basic enough to be worth changing, it will be extremely resistant to modification, while if it is susceptible to alteration, it is probably peripheral. This "cultural determinism" offends some non-anthropologists, but many in the field consider it the key contribution of the discipline.

But if this is so, what of efforts toward social reform? It seems doubtful that even so perceptive a person as Margaret Mead could alter significantly the two million-year history of culture. Or, if the desire for a particular reform is itself an outgrowth of our culture, is it not frivolous to think of ourselves as consciously shaping that culture? So when Mead, for example, calls upon anthropologists to invent social institutions that will eliminate war, her colleagues understandably

continued on page 111

Mead on Mead

When I entered anthropology in 1923, the main emphasis was on the importance of the few surviving primitive cultures in testing hypotheses that otherwise would be hopelessly culture-bound, like the origin of our type of head shake in a child's turning away from the breast. By the late 1930s it became clear that anthropological fieldwork conducted in living laboratories could generate new hypotheses as well as testing and criticizing old ones. At the end of World War II we were faced with such a drastic cultural change and such an urgent need to prevent disaster that a third function for anthropologists, culture building, could be added to our two earlier roles. I also consider that members of the anthropological profession, among the sadly segmented and divided human sciences, have the potential competency to criticize ourselves, since they comprise the professional group that has the training to be more culture conscious than those human scientists who work within our own tradition.

Beginning with my first research, *Coming of Age in Samoa,* I thought that whatever I found should be made available to those who might make the most use of it, and it was for that reason that I eschewed technical jargon and my book became, quite without my expectation or intent, a best seller. This happened again, 21 years later, when I was editing what I had conceived of as a specialized and technical handbook for specialists—*Cultural Patterns and Technical Change*. But because it had to be written for several groups of specialists, each lay to the other, it again had to be written in English, not jargon, and turned into a paperback sold on newsstands.

Before World War II, I thought of the function of anthropological reports for the layman as primarily providing greater self-awareness of our own culture, and it was not until the war that I became interested in specific institutions in our own and other modern cultures—and how their structure and functions were or were not appropriate to the ends they purported to serve. I do not feel myself to be a "reformer" but rather someone who works for the transformation or reconstruction of existing forms, which will not introduce the disastrous breaks introduced by violent revolution—transformations that are within the character and the capabilities of those who must make them if society is to survive.

Thus I did not set out on a crusade on behalf of marihuana. I was asked and consented to testify for a Senate committee on the use of prescription drugs, and my reference to marihuana in my original written testimony simply remarked, in a subclause, that legalizing marihuana would do something toward correcting the dangerous amount of drug abuse, but would not be sufficient. It was the questioning by members of the committee and the emphasis given by the press—because this testimony came at the exact moment that public opinion was shifting—that provided the publicity. I believe also that I made a technical error in using the term "legalize." But if I had used the phrase "repeal the laws against," which relieves the voter of any complicity in acts in which he does not believe, when he, responsibly choosing between two evils, abuse or corruption, votes for repeal, I would not have learned as much as I have about the extent to which those who use their country's laws to reform their neighbors think of legalization as sanctification.

In my participation in our own society, I learned a great deal from the late Lawrence K. Frank, who felt that research, training to use the results of research, and a climate of opinion ready to let it be used were all important. Since 1941, my work has been divided between research, including research on modern cultures, and especially upon methods of group process, conference techniques, interaction between public policy and individual beliefs, and teaching, writing, and lecturing. I teach some 100 students a year, sometimes many more, including training for the use of modern methods of fieldwork, and my writing, lecturing, and organization work are specifically directed toward establishing the climate of opinion within which action can occur.

I have never thought I could write for an anonymous audience, or say anything intelligible to the country as a whole, unless I moved around it, tried out ideas and ways of phrasing things for all kinds of audiences, and wrote, and received comment, for magazines and newspapers directed to large popular audiences. The proceeds of these activities have made it possible for me to promote the kind of research for which there are few funds and few sympathetic ears. In this I have been fortunate.

I have always thought that it was possible to make choices in such a way that the outcome would be felicitous, "Get the distaff ready and God will send the flax." This maxim was regarded as too feminine when I wanted to use it as a title for *And Keep Your Powder Dry*. It might be more acceptable in these days of renewed interest in women's rights.

My mother was a feminist and my grandmother was also a professional woman, so I was spared the conflicts of the first generation of professional women. I have persistently advocated greater freedom for both men and women, to follow either domestic or public careers, and I am conscious today that every woman who is liberated, who is no longer dependent upon the economic support of some male, inevitably also frees that male from responsibilities which in most cases have been as corrosive as those of his wife.

When I went back to the field in 1953, after 13 years of working on the problems connected with the war, I knew that we were facing a situation about which we knew almost nothing, the nature of change within one generation. The people I had studied in 1928 were purported to have made a miraculous leap into the modern world. I found them, and so reported in *New Lives for Old*, to have made a true transformation, so I was able to begin to formulate the significance of change that is self-initiated and not imposed, change that has a model, change that is across the board, leaving no detail of the former life-style unscrutinized, and change that includes all three generations, leaving no grandparents to pull the others back. It is this kind of relevance to the tasks that confront us that we can find in studies of small societies—especially small societies that we knew before. Such studies also throw into high relief how different our situation is, for they had models and we, moving into an era in which no one has ever grown up before, will have to devise totally new models.

I am presently devoting my efforts to the Scientists' Institute for Public Information, in which groups of scientists form voluntary groups to organize information on problems of the environment; to the World Society for Ekistics, which is concerned with worldwide problems of human settlements; to Glyphs, Inc., an effort to develop uniform signs around the world for the simpler needs of all travelers; to the Working Group of the World Council of Churches on the relationship between church and society; to the U.S. Task Force of the National Council of Churches and Union Seminary on a science-based, future-oriented society; on completing the Hall of the Peoples of the Pacific at the American Museum of Natural History; teaching at Columbia University; and preparing to go back to New Guinea, for return visits to three of the peoples I have studied in the past. In the fall of 1970 I did a dialogue on race with James Baldwin, which was published in May 1971. I am working on an autobiography that will answer more specifically why I find it not impossible to understand the young. The answer, *because I am an anthropologist used to listening and learning the life-styles of people far more different than our own children*, doesn't seem to satisfy my questioners, because, alas, they know too many anthropologists to whom the young are either allies in their own disgruntlement or threats to an established way of life, and who in any case don't consider our own society to be any of their business.

I believe the Vietnam war has turned from a more or less predictable post-world war activity into a disaster for the people of Vietnam and for ourselves. But there will be more Vietnams unless anthropologists are able to convince the layman that war is not innate, and unless human scientists are able to devise 20th-century institutions for 20th-century problems. The times are out of joint, and I, for one, consider it a privilege, not "a cursed spite," that ever I was born to help "set them right."

—Margaret Mead
Curator Emeritus of Ethnology
The American Museum of Natural History

continued from page 107

find it difficult to take her quite seriously. On the face of it, she is either paying mere lip service to the concept of culture or she is contradicting herself.

. . . And how to escape it

Mead's answer to these charges is both interesting and (I think) hopeful. For her, as for Sir Edward B. Tylor, the great 19th-century English anthropologist who first defined culture, mankind is seen as shifting from a stage of unconscious change, such as characterizes primitive peoples, to a stage of conscious change. Although Mead is well aware that, strictly speaking, there is no such thing as a stable culture—if by that we mean one that is undergoing no change at all—she has long regarded cultures as falling into two broad groups: the primitive and traditional, in which change takes place very slowly; and the advanced, or those strongly influenced by advanced societies, in which change is rapid. The kinds of change characteristic of these two types of culture differ not only in rate but in kind.

In a primitive or traditional culture (Mead's Postfigurative type), all children receive the same instruction from their parents, and every new generation experiences much the same sort of cultural inculcation as those that have gone before. On the other hand, the history of industrial civilization (Cofigurative cultures) has been characterized by an increasing rate of change. More and more, every child is markedly different from every other child, and the enculturation of each generation varies widely from that of its predecessor. By the time a couple has had a child, the culture has changed so much that they *cannot* pass on what they received themselves.

This has both bad and good results. On the one hand it means that many members of the society are rootless and desirous of simplistic solutions. On the other, it means that society is in a constant state

of flux and receptivity. Societies where there is a marked discontinuity in life experience between generations are more open to social innovation. It is of course the most rapidly changing societies, such as our own, that display this quality to a supreme degree (the emerging Prefigurative culture). Another advantage of our more evolved society is that we possess a much larger pool of potential innovations, the result of our greater knowledge, our more advanced technology, and our wider range of cultural contacts and consequent awareness of how other peoples have solved their problems. Further, our knowledge includes an increasing understanding of how societies operate and how individuals relate to them. To Mead it follows that precisely because we have this knowledge we have the opportunity to modify our society toward desired ends.

Institutions and practices resembling those that Mead proposes for our own society are, in fact, found among peoples well known to anthropologists. Thus, the prevention of war through nonlocalized institutions of social control brings to mind the Iroquois League, in which war was inhibited by the existence of clans, each of which drew a portion of its membership from each component tribe; since clansman could not fight clansman, disputes between tribes were settled peacefully. One thinks also of the Tiv of Nigeria, a people who had no day-to-day organization above the extended family compound but who could, in emergencies, create vast alliances along far-flung kin lines. Trial marriage has clear analogues in the adolescent sexual freedom that once characterized many Pacific and North American Indian groups, while universal service for both sexes in war and peace was provided by the age-grades, through which all youths passed, of the Zulu and various other East African peoples. Finally, the difficulty, if not the folly, of prohibiting the milder mind-altering drugs will be the more firmly impressed upon one familiar with the immense array of materials—betel nut, coca leaf, datura, kava root, kola nut, hashish, tobacco infusions, alcohol in numberless forms, to name only a few—that man has used to relieve the tedium of existence. All these examples (and many more) provide instances of more or less successful solutions to problems similar to ours.

The final question is: Just which portions of the innovation pool will be selected and for what reasons? For Mead the answer is clear. We must strive to create a culture in which all individuals will be as free as possible—free to develop in such a way that their unique personalities will have full expression. This will lead not only to greater individual happiness but also to the most perfectly functioning, rapidly advancing society possible. To achieve this it will be necessary to change our whole society—not merely the education of the young but the entire system and the attitudes that go with it. Mead has never doubted the reality of cultural constraints, but she believes that if—through observation of many peoples—we learn how these constraints operate, we can make them work for, not against, us. Again like Tylor, Mead is not ashamed to refer to anthropology as a reformer's science.

Why anthropology?

Far from being a contradiction, Mead's sense of mission is the key to her entire approach to anthropology. Her desire to discover the way in which mankind might be both excellent and happy is her reason for being in the field at all. As she herself has stated, the family into which she was born 69 years ago—with an economist father, a sociologist mother, and a psychologist grandmother, and with a concern for members of other ethnic groups—formed in her the belief that greater knowledge was the most effective route to greater human freedom and happiness. Interested in reform politics and the arts while an undergraduate, she turned to anthropology as the one field that combines all these themes.

I think it is clear, however, that for Mead the time for study and debate is long past. As she has said, "Anthropology is made for man; not man for anthropology." Today the very survival of our society—perhaps even of mankind as a whole—is increasingly doubtful, but if we at least try to use our knowledge to control our culture, we may yet succeed. Although the task will be difficult, it is uniquely human. As Mead put it so aptly many years ago in her book *Male and Female*:

> Every step away from a tangled situation, in which moves and counter-moves have been made over centuries, is a painful step, itself inevitably imperfect. Here is a vicious circle in which it is not possible to assign either a beginning or an end. . . . Those who would break the circle are themselves a product of it . . . may only be strong enough to challenge it, not able actually to break it. Yet once identified, once analyzed, it should be possible to create a climate of opinion in which others, a little less the product of the dark past . . . may in turn take the next step. Only by recognizing that each change in human society must be made by those who carry in every cell of their bodies the very reason why the change is necessary can we school our hearts to the patience to build truly and well, recognizing that it is not only the price, but also the glory of our humanity that civilization must be built by human beings.

FOR ADDITIONAL READING:

Dempsey, David, "The Mead and Her Message," *The New York Times Magazine* (April 26, 1970).

Harris, Marvin, *The Rise of Anthropological Theory* (Thomas Y. Crowell, 1968).

Hays, H. R., *From Ape to Angel: An Informal History of Social Anthropology* (Knopf, 1958).

Mead, Margaret, *Coming of Age in Samoa* (William Morrow, 1928).

Mead, Margaret, *Male and Female: A Study of the Sexes in a Changing World* (William Morrow, 1949).

Mead, Margaret, *New Lives for Old: Cultural Transformation-Manus, 1928–1953* (William Morrow, 1956).

Mead, Margaret, *Culture and Commitment: A Study of the Generation Gap* (Natural History Press/Doubleday and Company, 1970).

Sargeant, Winthrop, "It's All Anthropology," *The New Yorker* (Dec. 30, 1961).

Dan Morrill

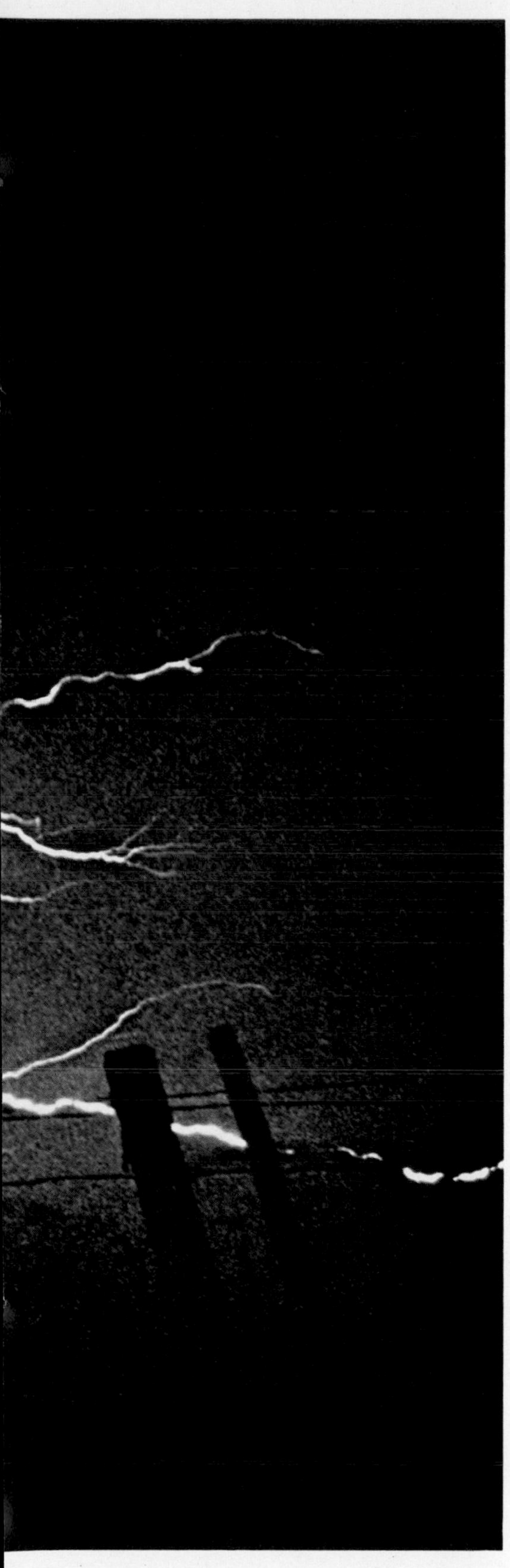

Killer Storms

by Louis Battan

Tornadoes and hurricanes strike many parts of the world each year, leaving death and destruction in their wake. Scientists are studying these violent storms in the hope they can learn to control them.

Ivan Massar from Black Star

Nature struck one of its most violent blows at a favorite target, the lower Mississippi Valley, on February 21, 1971. During a period of several hours, more than 25 tornadoes formed over Louisiana, Tennessee, and Mississippi. There were three main tracks of destruction, and some storms struck several towns as they moved along. About 5 P.M. the small town of Inverness, Miss., was especially hard hit. In only a few seconds more than three-quarters of the buildings were demolished, and 18 people were killed. Overall, the three-state death toll was at least 120, and more than 1,400 homes were destroyed or damaged.

Weather records of the U.S. National Weather Service abound with similar examples of the devastating force of a tornado. In a matter of minutes, this concentrated vortex of high-speed winds can wipe out a whole community, cause buildings virtually to disintegrate, and scatter the debris like shrapnel.

Tornadoes are not the only storms that can cause great destruction. On August 14, 1969, a U.S. Navy reconnaissance plane observed over the Caribbean the first signs of the formation of a hurricane, which was later given the name Camille. At the time it was centered south of Cuba. When it swept across the Mississippi coastline on August 17–18, it was a storm of mammoth proportions. Surface winds may have exceeded 200 miles an hour, and near its center the wind-induced waves were higher than 24 feet. Water from the Gulf of Mexico flooded coastal lowlands, while the large amount of rainfall from the storm added to the destruction.

As Hurricane Camille passed northeast over the United States, it weakened but did not die easily. Early on the morning of August 20, the storm produced torrential rains over the Virginias. In the valleys, water levels were tree-top deep, and the flooding caused more than 100 drownings. The overall toll attributed to Camille was staggering: almost 300 dead and damage amounting to $1.5 billion.

Tornadoes and hurricanes are the two most violent and destructive types of storms. But they have their origins in more common and less extreme forms of unsettled weather. What causes such killer storms to emerge from these milder conditions, and what, if anything, can be done to stop them?

Origin of tornadoes

Although tornadoes occasionally are reported in Western Europe, Australia, India, and Japan, it is only in the United States and the Soviet Union that these intense storms are common. For the 10-year period ending in 1964, the number of tornadoes reported in the United States averaged almost 700 a year, but many sightings were never verified officially.

The seasonal and geographical distribution of tornadoes depends largely on the temperature, moisture, and wind patterns in the atmosphere. The storms commonly form when a relatively shallow layer of warm, moist air is overlain by a deep layer of dry air that is moving rapidly from a southwesterly direction. If such a body of air is forced upward by a weather disturbance, such as a cold front, the air mass becomes very unstable. In such an environment violent thunderstorms commonly occur, and some of them produce tornadoes.

These conditions occur most often in the spring, as tropical air from the Gulf of Mexico moves over the southern parts of the central United States. As the season progresses, the warm, moist air penetrates progressively farther northward, reaching into Canada by summer. This northern extension of the tropical air sometimes is accompanied by southwesterly currents of fast-moving dry air aloft. The result is a northward progression of the region of maximum tornado occurrence.

LOUIS BATTAN is an associate director of the Institute of Atmospheric Physics and professor of atmospheric sciences at the University of Arizona. He is also a member of the Editorial Advisory Board for the Britannica Yearbook of Science and the Future.

Composition of a tornado

A tornado most often has the form of a funnel, a column, or a rope extending from the base of a thunderstorm to the ground. The overall diameter of the funnel is usually quite small, almost always less than a mile, and more often less than a quarter of a mile. In general, an individual funnel is on the ground for periods of only a few minutes or less, but there have been some notable exceptions. For example, a huge tornado reportedly remained on the ground for more than three hours as it cut a 220-mile path through Missouri, Illinois, and Indiana in 1925.

The visible part of the funnel consists of water droplets resembling those we experience in fog. Air moving into the funnel expands rapidly because of the low atmospheric pressure there. This causes a temperature reduction, an increase in relative humidity, condensation of water vapor on tiny particles in the air, and the formation of the visible water droplets.

The small size, short duration, and violent nature of tornadoes make it difficult to obtain detailed information about their structure. Because few direct measurements have been made inside the funnels, most of what is known about tornadoes has been obtained from a study of photographs of the funnel cloud and patterns of damage on the ground. Despite this paucity of data, however, enough is known about tornadoes to attribute their destructiveness to several distinctive features: small diameters, low pressures at the center, and the strong winds.

The few reliable measurements of the maximum wind speeds and pressures at the center of tornadoes indicate that the central pressure may be as low as 80% of the average atmospheric pressure at sea level. Thus, when a funnel moves over a building, the pressure on the inside of the structure can be much higher than the pressure on the outside. This is particularly true if the building's windows and doors are closed, thereby preventing the high-pressure air from rushing out. These pressure differences can develop enormous forces, more than enough to blow off the roof and possibly blow out the walls of even a well-constructed building. The crumbling remains can then be picked up by the winds and scattered.

Recently, Tetsuya T. Fujita of the University of Chicago proposed

A computer prints out its interpretation of what a typhoon in the Pacific looks like (opposite). An actual weather-satellite photograph of the storm was scanned electronically and its various shades of gray converted into the letters and symbols seen on the printout. A Doppler radar scope (below) measures the movement of rain and snow particles by noting changes in the frequency of the reflected radar waves. The rightmost pulse on the broad green line indicates the height, size, and number of the particles and their speed up or down.

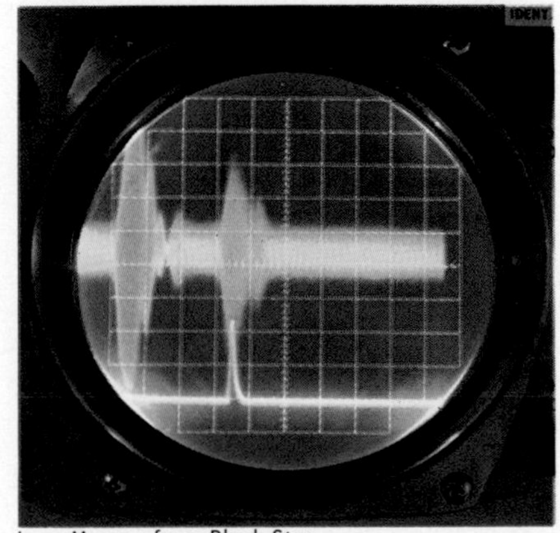

Ivan Massar from Black Star

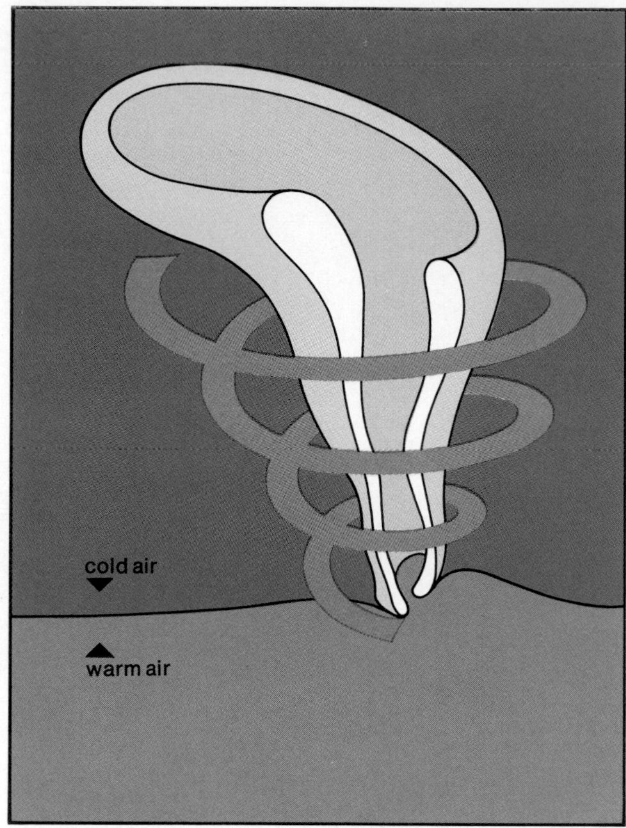

Two of the leading theories advanced to account for the formation of a tornado are the thermal and mechanical. According to the thermal theory (above, left), a tornado results from the imbalance created when cool air overrides warm. Rapid upward convection from the warm air compensates for the imbalance; this updraft becomes a rotary flow and forms the tornado vortex. The mechanical theory (above, right) proposes that the vortex is formed when slowly rotating air currents are compressed by external forces. Whatever the cause, the result is a storm with a potential for great destructiveness (opposite page).

that within a tornado vortex there are small regions of very low pressure which he has called "suction spots." Perhaps 1/20 the diameter of the tornado core, the suction spots are carried along by the tornado's strong winds and are thought to be responsible for much of its destruction as well as for the strange nature of some tornado damage. It is not rare for one house to be demolished while another 10 feet away is left unscathed.

Because few anemometers can withstand the strong forces involved, few direct measurements exist of the maximum wind speeds in tornadoes. However, wind speeds have been estimated by studying storm damage as well as by studying the cycloidal-shaped patterns in open fields. These patterns are caused by the suction spots, which are carried in a counterclockwise direction around the tornado funnels as they move across the fields. On the basis of these patterns, Fujita and his colleagues have estimated maximum wind speeds of nearly 200 miles per hour. This value is substantially below earlier estimates, which in some instances were two or three times higher. Fujita did estimate that the winds might gust briefly to speeds of 300 miles per hour.

Tornado detection, warning, and control

At the present time, we do not have a suitable means for detecting and tracking tornadoes. The most useful instrument is radar, but exist-

Courtesy, NOAA

ing techniques of identification by radar work only part of the time. On the other hand, conventional radars do an excellent job of tracking funnel-producing thunderstorms. Once such a storm has been identified by visual means, it can be followed, and warnings can be issued to people in its path.

Some tornadoes generate electromagnetic energy over a wide range of frequencies and therefore produce considerable "radio noise." As a result, they can cause a television screen to glow if the tube brightness had previously been set at the appropriate threshold. Because of this effect, it has been proposed that an ordinary television set can be an effective tornado detector. The limited available evidence indicates, however, that not all tornadoes produce an electromagnetic "signature" allowing simple television detection. Therefore, the procedure of tornado identification by means of television can be relied on only some of the time. The absence of a distinctive signal cannot be taken as proof that there are no tornadoes in the vicinity.

By means of modern electronic technology, it should be possible to develop a reliable tornado detection and tracking technique. One promising avenue of attack is to employ a suitably designed Doppler radar. This device would not only observe the location and intensity of thunderstorms, as does conventional radar, but also would be able to determine the direction in which they are moving and the velocity of the debris carried by the high winds.

Curtis Kitchen from Black Star

Debris litters the ground in the wake of a tornado that struck Belvidere, Ill., in April 1967. The storm was one of a series in northern Illinois that killed more than 50 people and caused property damage of more than $20 million within only a few hours.

When the National Weather Service finds atmospheric conditions favorable for tornado formation, it issues a "tornado watch." It is a prediction that tornadoes are expected in a designated area. Citizens are urged to be alert to the possibility of a violent storm and to keep informed of developments by means of radio or television. When a tornado has been sighted, a "tornado warning" is released, giving the location of the funnel or funnels.

Tornadoes may move as fast as 70 miles per hour, but usually their speeds are less than half that. For persons unable to get away or into an underground storm shelter, it is generally recommended that they go into a basement or against an interior wall, preferably under some cover away from windows. It is wise to leave some windows and doors open to equalize the pressure inside and outside of the house.

Ultimately, it may be possible to prevent the formation of tornadoes or to detect them early and dissipate them. This will be a difficult task, however, because of their small sizes, short durations, and the difficulty of pinpointing where they will occur. A few scientists have offered speculations on how tornadoes might be modified, but at this time successful control of tornadoes must be put into the category of a long-range goal.

Hurricanes, typhoons, and cyclones

Violent tropical storms occur in many parts of the world. When they have winds exceeding 73 miles per hour and form over the western Atlantic or the Gulf of Mexico, they are called hurricanes. When the same type of storm occurs in the Pacific and heads toward Japan, it is called a typhoon. When one moves over the Indian Ocean into the Asian subcontinent, it is called a cyclone. Since different names are given to the same type of storm, it is sometimes easy to be confused. Moreover, meteorologists use the term "cyclone" to designate a closed circulation around any low-pressure area. In India and Pakistan, however, the same term is used to represent the intense storm we call a hurricane. In mid-November 1970, such a storm swept into East Pakistan and produced one of the worst natural disasters in the history of man. Estimates of deaths from the storm run as high as 300,000.

Hurricane winds usually are lighter than those in a tornado, but a hurricane is much larger and lasts much longer. As a result, it can do more damage and kill more people than a tornado. But a hurricane's size and long lifetime makes it easier to find, follow, and predict. Thus, it can be dealt with more effectively by means of modern science and technology.

Origin of a hurricane

Hurricanes form during the same season, late summer and early autumn, throughout the world. The reason for this appears to be that the higher the temperature of the ocean water, the greater the transfer of energy into the atmosphere and the more likely the development of a hurricane. In the tropical North Atlantic, hurricanes occur most often during the period from August to October, but they sometimes occur as early as June and as late as November. Hurricanes usually weaken rapidly when they move over land or over colder ocean water because they then have become separated from their major source of energy, the warm ocean water.

Although hurricanes develop over warm, tropical oceans, their formation sometimes can be related to events occurring over continental areas thousands of miles away. There is evidence, based mostly on data from weather satellites, that hurricanes occurring over the tropical Atlantic and affecting the United States have their origins in low-pressure centers originating over Africa. As these regions of low pressure move westward in the trade-wind regions, they ingest warm, moist air. This leads to the formation of clouds and rain, and supplies energy for the intensification of the low-pressure center. In a small fraction of the cases, the atmospheric pressure drops markedly and wind speeds increase to hurricane force. In the Northern Hemisphere, air moving into the storm goes in a counterclockwise direction around the center; in the Southern Hemisphere, the wind blows clockwise around the eye. Air enters the storm at low levels, spirals upward, and moves out of the hurricane at high levels.

The process of hurricane formation may take several days as the cloud systems become organized and the wind speeds increase. When they reach hurricane speeds, the storm is named officially so that it can be identified easily. It has become common practice to use women's names, with the first storm having a name starting with A, the second with B, and so forth. The names for Atlantic hurricanes expected in 1972 have already been selected: Abby, Brenda, Candy, and Dolly will be the first four storms.

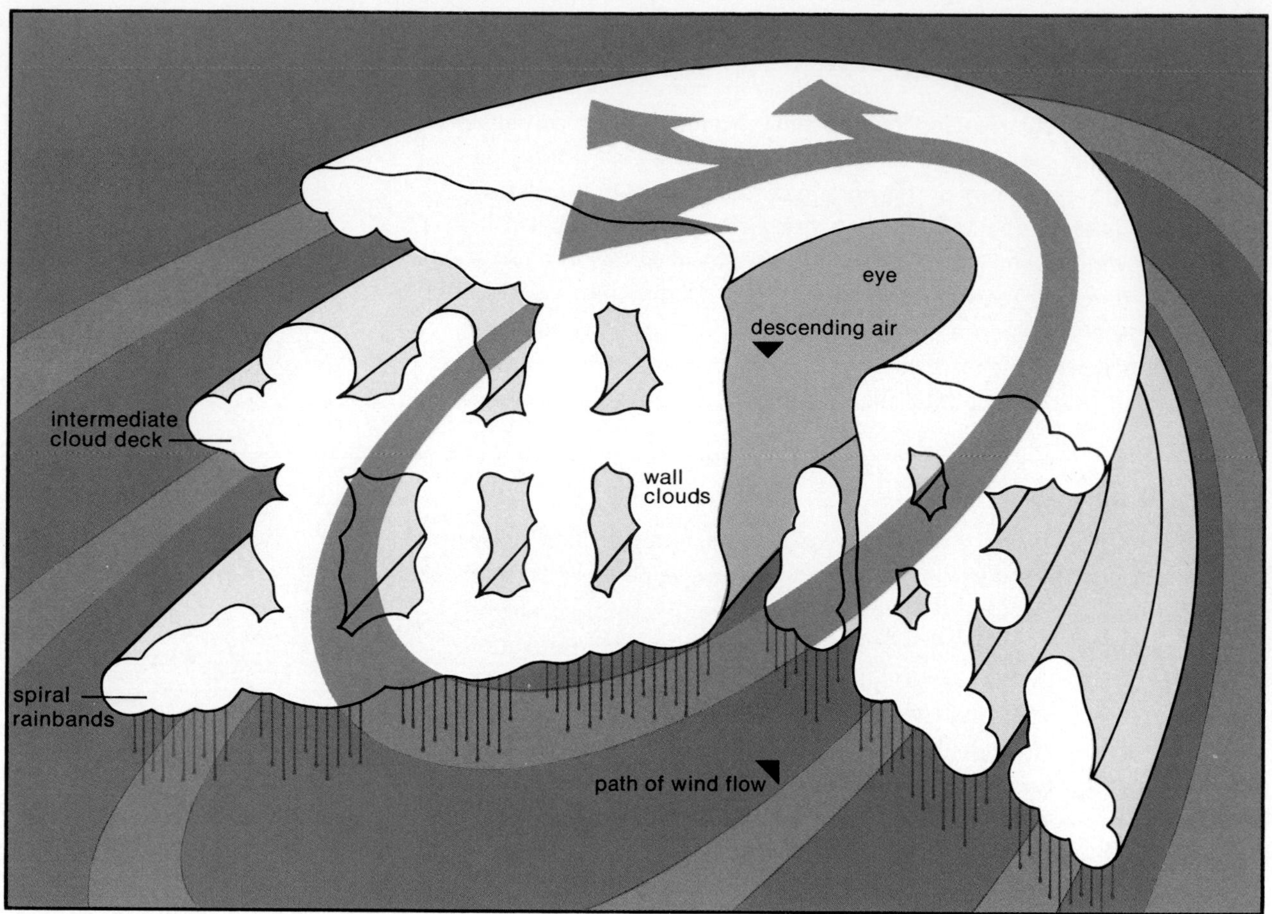

This cross section of a hurricane shows the relatively calm eye bounded by towering "wall clouds," within which are spiral bands of rain and the highest wind velocities of the storm.

The storm and the eye

A distinctive feature of a hurricane is its "eye," inside of which winds are light and skies are often only partly cloudy. This relatively calm weather is associated with generally sinking air, inhibiting cloud growth. The diameter of a hurricane eye averages about 15 miles, but the size varies from storm to storm and from time to time in the same storm. The eye is bounded by so-called "wall clouds," which are towering thunderstorm clouds that extend to great altitudes.

There is a marked wind-speed increase just beyond the edge of the eye, with maximum speeds associated with the wall cloud region. These velocities often exceed 100 and sometimes may go over 200 miles per hour. As the distance from the center of the storm increases beyond the radius of maximum velocity, wind speeds decrease fairly gradually until they fall to perhaps 20 miles per hour at distances about 100 miles from the vortex center.

As a hurricane passes—particularly if it moves at a slow speed—it can deposit enormous quantities of rain. For example, in November 1909 a storm moving over Silver Hill, Jamaica, yielded about eight feet of rain in four days. Taylor, Texas, was inundated with almost two feet of precipitation on September 9–10, 1921.

The strong winds in hurricanes generate waves of tremendous

Courtesy, NASA

The rotary wind currents of a cyclone are revealed in this photograph of a Pacific storm 1,200 miles north of Hawaii, taken by the Apollo 9 spacecraft.

heights. An average storm over the Atlantic Ocean produces waves 35 to 40 feet high, and in a great hurricane the wave heights may exceed 45 feet. A typical tropical storm travels about 12 miles per hour, while the waves go perhaps three to four times faster. As a result, they move ahead of the storm, and, as they do so, decrease in height.

When a hurricane approaches a coast line, several factors, including the comparative shallowness caused by the underwater continental shelf, can lead to a pronounced rise of sea level. In some cases, waves approximately 20 to 30 feet high sweep over the land. Evacuation to higher ground is the only escape. If there is inadequate warning, a warning to leave is not heeded, or the people have no way of escaping, widespread disasters are inevitable. The cyclone that devastated East Pakistan in 1970 was detected by satellites and radar, but there were no measurements of its wind velocities. Nevertheless, the Pakistani weather service predicted the arrival of a storm of great danger. For various reasons, however, large numbers of people failed to heed the prediction. Because they did not move inland or to higher ground, they became victims of a hurricane wave that was some 20 feet above mean sea level.

Clearly, then, hurricanes can do much harm, but they also do a great deal of good. If they did not pass periodically over the southeastern

United States, there would be a crucial reduction in the average annual rainfall. The climate of the region would become more arid and, as a result, many plant species would disappear, cities would experience water shortages, and other activities requiring water would suffer.

Detecting and tracking hurricanes

Because various weather satellites map the cloud pattern over the entire earth at least once each day, it is now virtually impossible for hurricanes anywhere in the world to go undetected for more than 24 hours. Some areas are viewed much more frequently than once a day. For example, the Applications Technology Satellite 3, which was launched into geosynchronous orbit at an altitude of about 22,300 miles over Brazil, can view the Atlantic Ocean at intervals of 20 minutes or less on demand. (In a geosynchronous orbit, a satellite's speed is the same as the earth's rotational speed. Hence, the satellite appears to be "stationary" in the sky.)

A ship founders, caught in the fury of 1967's Hurricane Beulah. The storm raged through the Caribbean Sea and the Gulf of Mexico, inflicting severe damage to many islands and to coastal areas of Mexico and Texas.

Shel Hershorn from Black Star

When hurricanes come within about 200 miles of land, radar can detect the large rain and snow particles in the clouds and track the storm minute by minute. At the present time, however, neither satellites nor radar can measure with adequate precision the wind speeds inside the storms. To do this, specially instrumented reconnaissance airplanes are needed. In the United States, such aircraft are operated by the Navy and the National Oceanic and Atmospheric Administration. When a satellite detects a hurricane within flight range, these airplanes go out to sample it, locate its eye, and measure its winds.

The National Weather Service issues a "hurricane watch" whenever its forecasters believe a hurricane may be threatening a populated area. The public is then urged to listen for further advisory bulletins. When hurricane conditions are expected within 24 hours, a "hurricane warning" is issued by weather officials.

Controlling hurricanes

Society would benefit greatly if the maximum wind speeds in a hurricane could be reduced. In general, weaker winds would mean smaller ocean waves, less flooding, less damage, and fewer fatalities. Scientific interest in modifying or controlling hurricanes dates back many years. The first attempt to do so was carried out on October 13, 1947, when Irving Langmuir and his colleagues at the General Electric Research Laboratory, Schenectady, N.Y., dispersed about 200 pounds of crushed dry ice into a hurricane off the southeast coast of the United States. After it was seeded, the storm followed an erratic course and moved inland. Langmuir was inclined to believe that the storm's behavior was caused by the seeding.

For a variety of reasons, little more was done in studying the modification of hurricanes until about 1962, when the U.S. Weather Bureau (now the National Weather Service) and the Navy formed Project Stormfury. Over the years, this project, under the leadership of Robert H. Simpson, Joanne Simpson, and Cecil Gentry, studied the properties

Ivan Massar from Black Star

An antenna that resembles a bedspring is part of an Automatic Picture Transmission (APT) receiver that is used to gather weather pictures from U.S. meteorological satellites.

of hurricanes and conducted a number of modification experiments. They evolved a simple hypothesis which proposed that by seeding the wall clouds on the edge of the eye with substances that form ice crystals, the pressure pattern in a hurricane could be changed and the maximum wind speed decreased.

Instead of seeding hurricanes with dry ice as Langmuir had done, Project Stormfury used special devices dropped into the storm from a high-flying airplane. The devices, which were designed by Navy scientists under the guidance of Pierre St. Amand, contained a mixture of silver iodide and a pyrotechnic substance. When the mixture was ignited with a fuse, it burned and released huge numbers of tiny particles of silver iodide. The particles caused ice crystals to form in liquid-cloud regions in which temperatures were below about −5° C. When this took place, heat was released, resulting in a small but important amount of warming of the air. It has been theorized that such warming can alter the dynamics of a hurricane.

Two hurricanes were seeded in the early 1960s, but opportunities were few after that because of a decision to seed only storms out at sea, far away from inhabited areas. In August 1969, an opportunity to test Project Stormfury ideas occurred when Hurricane Debbie was sighted 750 miles east-northeast of Puerto Rico moving toward the northwest. On August 18, when the storm was in a mature stage with maximum winds of 113 miles per hour at an altitude of 12,000 feet, it was seeded five times at two-hour intervals. Five hours after the seeding, maximum measured wind speeds were 78 miles per hour, or 31% less than at the start of the seeding. By August 20 the storm had reintensified, and another seeding test was carried out. At the start of

Ivan Massar from Black Star

An experimental radar station helps scientists learn how rain is created. The small antenna at the left focuses a radar beam that measures the height, sizes, and vertical speeds of snow and rain particles. The jet plane on the opposite page is used for cloud-seeding experiments in Project Stormfury.

this part of the experiment, maximum winds were 114 miles per hour; six hours later they had slowed to 97 miles per hour.

Some support for the view that seeding can cause a decrease in wind speed comes from a simplified mathematical model of a hurricane developed by Stanley L. Rosenthal at the National Hurricane Research Laboratory in Miami, Florida. In a simulated seeding test, he assumed that the hurricane was seeded in a particular way, and calculations were made of the consequences. The computer revealed that in some circumstances the maximum wind speeds near the surface should decrease.

By 1971, the evidence on hand was encouraging to scientists concerned with the development of procedures for reducing the intensity of hurricanes. Nevertheless, considering how little was known about the mechanisms controlling hurricane dynamics, the possibility still existed that the wind reductions in Hurricane Debbie could have occurred by chance, and that, in some instances, seeding might lead to greater storm damage. Also, little was known about the effects of seeding, if any, on the path followed by a hurricane. Nevertheless, a few scientists proposed that hurricanes threatening populated areas be seeded no matter what the risk. The preponderance of informed scientific opinion, however, favored the performance of more experiments before attempting to control hurricanes that were about to strike populated areas.

Effects of storm control

As they learn to modify violent storms, scientists will start to influence in important ways the lives of a great many people. Inevitably, many would benefit if a hurricane were weakened or deviated from its "natural" path; others would be harmed. Changes in the character of the weather would change the climate and lead to profound alterations of human, animal, and plant ecology. Who should decide whether or not a hurricane threatening the southeastern United States should be modified? If it is, and inland farmers suffer a costly drought, who would pay their bills? If a hurricane headed for Florida is diverted and strikes the Carolinas instead, who would help the victims?

Questions such as these were academic in the past. However, as we learn to control violent storms, they will have to be answered—not only by atmospheric scientists, but also by sociologists, ecologists, lawyers, politicians, and the general public. It is essential that governments, by means of appropriate legislation, adopt procedures for making decisions having such potentially widespread social effects as those that will arise when the control of violent storms becomes a reality.

See also A Gateway to the Future: CHALLENGING THE RESTLESS ATMOSPHERE.

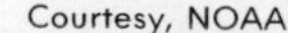
Courtesy, NOAA

FOR ADDITIONAL READING:

Battan, L. J., *The Nature of Violent Storms* (Doubleday, 1961).

Battan, L. J., *Harvesting the Clouds* (Doubleday, 1969).

Dunn, G. E., and Miller, B. I., *Atlantic Hurricanes.* (Louisiana State University Press, 1960).

Gentry, R. C., "Hurricane Debbie Modification Experiments, August 1969," *Science* (April 24, 1970, pp. 473–475).

Kessler, E. "Tornadoes," *American Meteorological Society Bulletin*, (October 1970).

Sewell, W. R. Derrick (ed.), *Human Dimensions of Weather Modification* (University of Chicago Press, 1966).

Taubenfeld, H. J., *Controlling the Weather: A Study of Law and Regulatory Processes.* (Dunellen, 1970).

AUDIOVISUAL MATERIALS FROM ENCYCLOPÆDIA BRITANNICA EDUCATIONAL CORPORATION:

Film Loop: *Story of a Storm.*

The Energy Crisis

by Harry Perry

Electricity shortages have caused widespread brownouts along the U.S. east coast. Dwindling fuel supplies and increased demand make this a continuing problem; more generating capacity and new energy sources are needed.

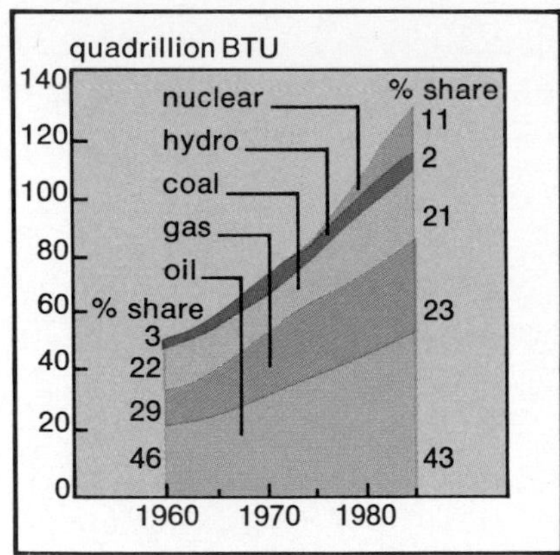

The above graph indicates the percentage contribution of each fuel source to the total energy supply of the U.S. over a period of 25 years. The 1971–85 percentages are based on projected figures. The graph below shows how rapidly annual energy consumption in the U.S. began to increase in the early 1960s.

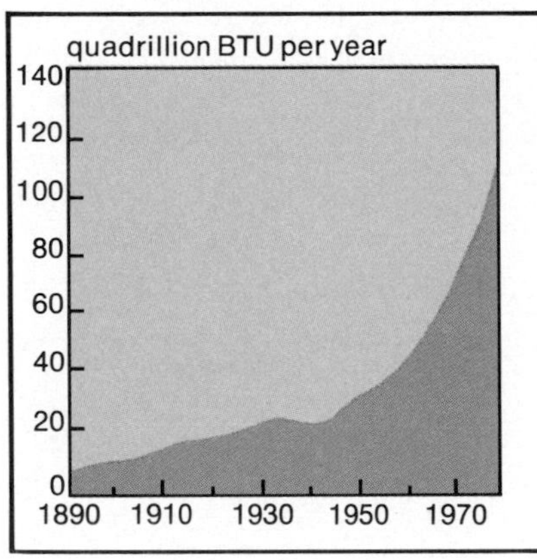

HARRY PERRY, a chemical engineer and a senior specialist in the U.S. Library of Congress in the area of fuels technology, was research adviser to the assistant secretary of the interior from 1967 to 1970.

It is mid-summer with the temperature in the 90s, but the use of air conditioners has been prohibited. Subway trains have been slowed to a snail's pace, and television viewing is restricted to an hour or two each day. Except for street lights, the downtown area is in almost total darkness because advertising signs and the lights in store windows have been turned off.

Six months later, when temperatures are hovering near zero, children are sent home from school because there isn't enough fuel to keep the building warm. The local gas company advises its customers to turn down their thermostats at home to conserve the rapidly dwindling supply of heating gas. Even coal companies have curtailed deliveries because their stockpiles are being depleted.

Exaggerations, perhaps, but only slight ones. Brownouts have been more and more frequent in New York City. During a recent winter Cleveland found itself without enough gas.

Why should a country that is as technically advanced as the United States suddenly experience a shortage of fuel and of electric-generating capacity? Other highly industrialized countries face a similar problem, despite the fact that the world's fuel reserves seem adequate to supply man's needs for a long time. What is the reason for this critical energy shortage? What can be done about it?

The problem of supply and demand

In the United States, the insistence on using energy in its most convenient form has caused an increasing percentage of the nation's energy resources to be consumed in generating electricity. Between 1950 and 1969 the amount of electricity used in the United States increased from 15 to 24% of the total energy consumed. This trend will continue because of the nation's dependence on a constantly proliferating variety and number of electrically powered aids to modern living. By the year 2000, nearly half of the fuels consumed in the United States will be used to generate electricity.

Despite the need for more fuels to meet increasing energy demands, during 1968—for the first time since records were kept—explorers for natural gas in the United States found about 5.6 trillion cubic feet less than was consumed that year. In 1969 this gap became even greater, as consumption exceeded discoveries by 12.2 trillion cubic feet. By the middle of 1969 gas-distributing companies in some major cities of the country were having difficulties in obtaining supplies and no longer were seeking new customers. Transmission companies, unable to contract for enough gas to keep their pipelines filled, announced that shortages would occur unless there were new discoveries. During the 1969–70 winter heating season, local shortages did hit the Cleveland area, making it necessary to interrupt industrial deliveries during peak periods of demand.

The gas shortage of 1969–70 was followed by one involving electrical generating capacity during the summer of 1970. This shortage, which caused "brownouts" (the curtailment of service and voltage

Larry Dale Gordon

Water rushes through a spillway of the Guri Dam on the Caroní River in Venezuela. The Guri generates hydroelectric power used to help release the rich mineral resources of Venezuela's Guiana region. The availability of hydroelectric power for such purposes, however, is limited by geographical factors.

reductions) along the east coast during the hottest parts of the summer, resulted in large part from a delay in the construction of nuclear-electric plants that were originally ordered during the 1966–67 period. By 1971 it appeared that it would take nearly twice as long as had been anticipated for some of these plants to become operational—up to seven years. Some of them, however, were scheduled to begin generating power in 1971 and 1972, resolving temporarily the problem of meeting peak demands for most utility systems. Other factors favoring a dependable supply of electricity were the increase in high-voltage transmission lines and the greater number of interties between utility systems to permit better exchange of electricity.

After the brownouts, when the public was asked to reduce its use of electricity to the lowest practical level, there were complaints that utilities should no longer promote the use of electricity through advertising. Instead, it was suggested that they should spend their advertising budget in search of ways to minimize the pollution attending power plant operations.

Other environmental considerations also were affecting the supply of electricity. Because fossil-fueled power plants cause air pollution and because nuclear-electric generation pollutes lakes and streams with its discharges of hot water, conservationists have sought to ban the construction of such plants until they could be made pollution-free. In several cases they have been successful in delaying plant construction, but while such action has protected the environment, it also has contributed to the growing power shortage.

Causes of fuel shortages

Although the electricity shortage during the summer of 1970 stemmed largely from limited generating capacity, the major problem in the future threatens to be the supply of fuels. Stockpiles of coal have been

Courtesy, Union Oil Company of California

Steam from hot reservoirs far below the earth's surface thunders into the air during a depletion test of the first geothermal generator in the United States. Called The Geysers, this facility generates electricity by directing the force of the steam against turbine blades. By 1975 its production is hoped to be about 600,000 kilowatts, enough to supply the needs of a city of 600,000. Like hydroelectric power, however, geothermal steam is also a limited energy source.

depleted by conventional power plants because coal-mine operators reduced their output in anticipation of increased competition from nuclear generating plants. Aggravating the situation even more have been the tight supply of gas and the sharp increase in residual oil prices. Public concern about the environment also could contribute to the problem. For example, sulfur standards imposed by many large municipalities could prohibit use of certain fuels. Some delay might be necessary in applying these standards, however, if there are to be no major brownouts or blackouts in the future.

Sulfur standards also have been important factors in creating a shortage of fuel oil of the quality desired. Most residual oils burned by electric utilities in the United States are imported from the Caribbean and have sulfur contents ranging from 2.5 to 3%. New regulations in many cities place an upper limit of 1% on sulfur, and oil of this type can be obtained only from sources much farther away. The demand for such oil came at a time when petroleum tanker capacity already was in short supply. The closing of the Suez Canal, an unrepaired break in the Trans-Arabian Pipeline, and limitations imposed in 1970 by the Libyan government on production of oil in that country all contributed to much longer hauls by tankers in delivering oil to Western Europe, an area heavily dependent on oil. As a result, not enough tankers were available for shipping oil to the United States.

Further complicating the supply problem has been the uncertain future of the oil import program in the United States. This program, which has been in operation since the late 1950s, was designed to assure an adequate domestic productive capacity so that in times of national emergency the country would not be crippled by lack of foreign oil. Under its provisions, the amount of foreign crude oil that could be imported was limited to a percentage of the volume produced in regions of the United States where quotas were in effect. As a result of this policy, domestic oil sold for about $1.25 more per barrel than foreign oil.

Unlike oil, the U.S. coal industry always has maintained a relatively large excess production capacity, except during World War II. After the war, the industry realized that to compete in the energy market it would have to mechanize and increase productivity. Mechanization was expensive, and coal-mine operators sought guaranteed markets for high percentages of their production. This meant long-term contracts with utility plants, steel mills, and foreign importers. Because of these commitments, the historical excess capacity of the coal industry gradually declined. When shortages of other fuels developed, there was little slack left to supply the unexpected demand.

Coal exports, which had been expected to remain at about the same level as in previous years, rose by 25% during 1970. Because of the profitability of the export business, the industry made every effort to fill overseas orders. The coal that was exported, nearly all of it low in sulfur, was exactly what utilities were trying to buy to meet new air pollution standards at home. But because reserves of such coal near utility markets were not as plentiful as those of higher-sulfur coals, productive capacity to supply both overseas and utility markets was not immediately available.

The energy problems of 1969 and 1970 were intensified by several factors in addition to shortages of supply and productive capacity. Demand for energy, which historically had increased in the United States at a rate of 3.5% per year, rose during 1967-69 to an unexpectedly high 5.5%. Between this increase in demand, construction delays caused by environmental concerns, and the new sulfur standards for fuels, the United States is faced with the prospect of being unable to meet future needs unless appropriate remedies are found.

Northfield Mountain in north-central Massachusetts is the site of this giant hall, designed to be the powerhouse of the nation's largest pump-storage installation. More than 100 yards long and 10 stories high, it will contain four pump-generators that can raise 5,800,000,000 gallons of water to a reservoir on the mountain's top and generate one million kilowatts of power as the water runs down.

Meeting the demands of the 1970s

Decisions about energy require relatively long lead times. For example, building a nuclear-power plant takes five to six years, but planning and obtaining necessary clearances from local, state, and federal agencies can add four to six more years, even if there are no serious public objections to the construction. Tapping new domestic sources of natural gas will take time for exploration and development even if the search is successful. Construction of pipeline facilities to bring gas to consumers will add more time.

Reducing the demand for energy to achieve a balance with the supply would have far-reaching effects on economic growth rates and eventually on life styles. It may be possible to use less energy and have continued economic growth, but the historical makeup of the gross national product would have to be altered, and this too would require that life styles be modified.

Courtesy, POWER ENGINEERING Magazine

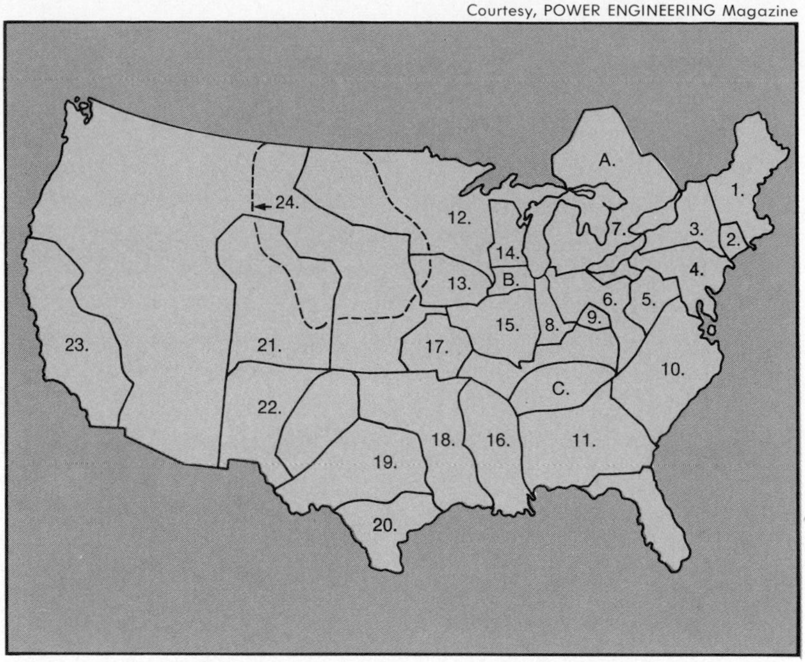

1. Electric Coordinating Council of New England
2. Connecticut Valley Power Exchange
3. New York State Power Pool
4. Pennsylvania-New Jersey-Maryland Interconnection
5. Allegheny Power System
6. American Electric Power Company
7. Michigan-Ontario Pool
8. Indiana Pool
9. Cincinnati-Columbus and Southern Ohio-Dayton Pool
10. Carolinas-Virginia Power Pool
11. Southern Company
12. Upper Mississippi Valley Power Pool
13. Iowa Power Pool
14. Wisconsin Public Service Corporation—Wisconsin Power and Light Company Pool
15. Illinois-Missouri Pool
16. Middle South Utilities
17. Missouri-Kansas Power Pool
18. South Central Electric Companies
19. North Texas Interconnected Systems
20. South Texas Interconnected Systems
21. Rocky Mountain Power Pool
22. New Mexico Power Pool
23. California Pool
24. Missouri Basin Systems Group

A. Hydro-Electric Power Commission
B. Commonwealth Edison
C. TVA

Major power pools in the United States consist of groups of utilities that interconnect their systems to achieve more economical and more reliable service. Such pools are also useful for long-range planning.

Methods to reduce the amount of energy consumed per unit of gross national product also are being studied as a means of meeting energy demand. For example, wider use of properly installed insulation could reduce significantly the demand for heating fuel. Less fuel would be consumed, too, if combustion equipment were better maintained. Selection of the most efficient fuel for each type of use also could reduce consumption with no sacrifice of economic growth. But such changes are difficult to accomplish quickly because the public has been conditioned to an abundance of low-cost energy and has not had to be concerned about its waste.

At least for the next several decades, the energy supply problem for the United States should not be a lack of physical resources. Although "proved" reserves of oil and gas are sufficient only for about 10 years, the country's top geologists believe that between 70 and 140 billion barrels of recoverable domestic oil are yet to be discovered (exclusive of Alaska). Even at the oil-consumption rate anticipated by 1980—19 million barrels per day—these reserves should last another 10 to 20 years. The ultimate supply of natural gas to be found in the United States exclusive of Alaska has been estimated at 1,000 trillion cubic feet. At the consumption rate anticipated for 1980—80 billion cubic feet of natural gas per day—this supply should last for 35 years.

Coal reserves are ample for many hundreds of years, but uranium reserves—producible at costs that would make them economically attractive using present nuclear reactor technology—would last only to the end of the century. As M. King Herbert, an energy geologist, has said: "Were electrical power to continue to be produced solely by the present type of light-water reactors, the entire episode of nuclear energy would be short-lived."

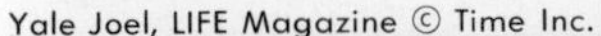
Yale Joel, LIFE Magazine © Time Inc.

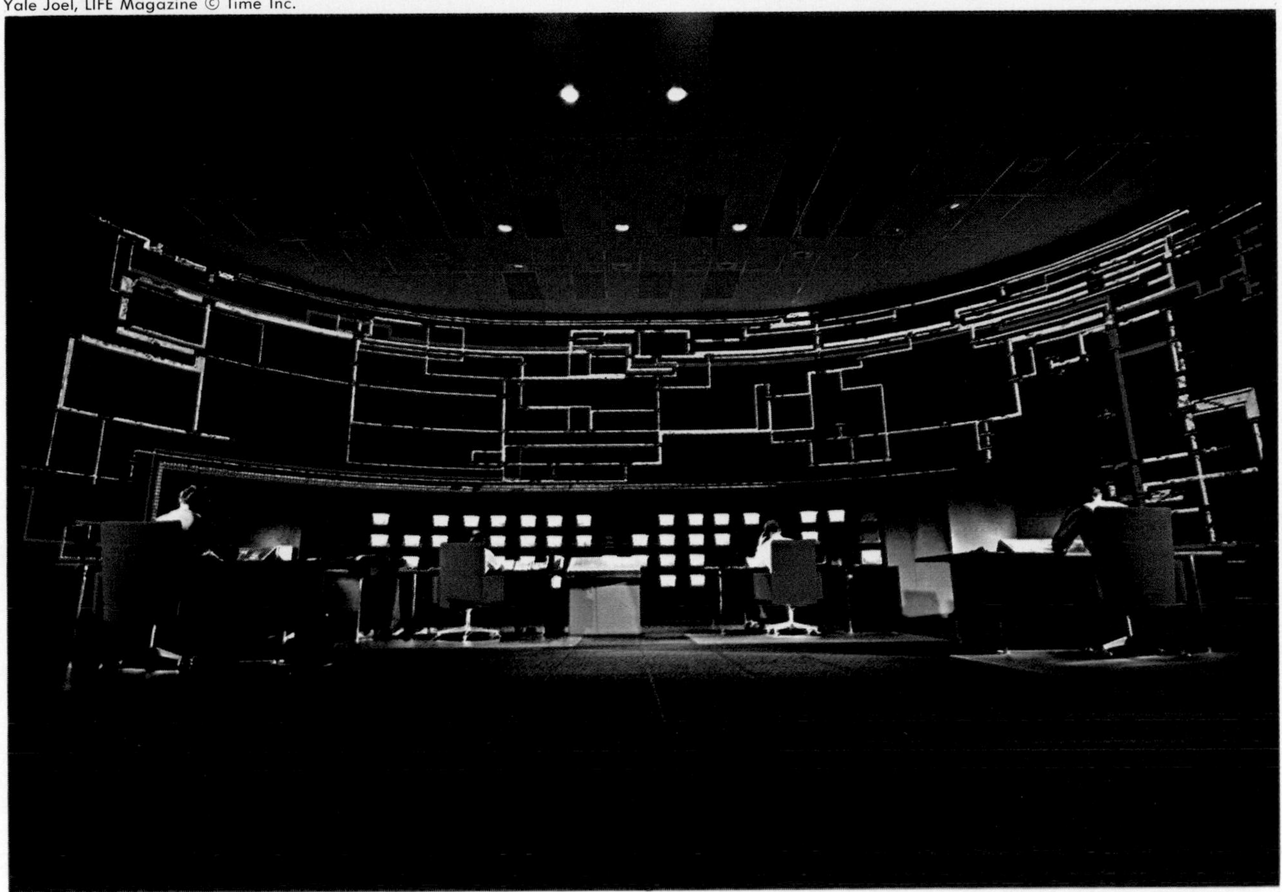

The control room for the New York Power Pool features an illuminated board that indicates where the energy is among the pool's eight systems. In emergencies the pool can import power over long distances from other pools.

The relative scarcity of oil and gas compared with known coal reserves suggests that at some time in the future, possibly before 1980 in some areas, the economically competitive production of gas and liquid fuels from coal might be possible, as might the production of liquid fuels from oil-bearing shale—all of which have been demonstrated on a large scale. Previously, such synthetic fuels could not compete economically with natural petroleum and natural gas, but as the cost of finding the latter increases, the relative economics could be reversed. The time when this might happen could be advanced appreciably if an intensive research and development program for producing synthetics were mounted.

Alleviating gas shortages

One method that can be used to alleviate potential fuel shortages in the United States is to develop additional gas reserves. This can be done by offering incentives to explore for more natural gas; increasing imports of gas from Canada; accelerating the leasing of federal lands both onshore and on the outer continental shelf; constructing pipelines to bring natural gas discovered in Alaska to the markets in the other states; importing liquefied natural gas (LNG) from Africa, Venezuela, or other overseas sources; and by constructing large plants capable of producing a synthetic natural gas from coal.

Courtesy, Linde Division, Union Carbide Corporation

This cross section of an underground superconducting cable shows the three niobium-plated copper pipes through which the electrical current is transmitted. The fourth, smaller tube holds liquid helium at —452°F. Because they eliminate practically all electrical resistance, such cables can handle power loads at least 25 times greater than the largest existing underground cables in the U.S.

What incentives, if any, will have to be offered to increase reserves is still to be determined, but there are some noncontroversial measures that could help. For example, speeding up administrative and legal procedures of the Federal Power Commission would resolve uncertainties for gas producers and transmission companies. This should increase the flow of gas to markets. Providing additional storage near the point of consumption to meet peak demands would help to assure customers of sustained service. Increasing research and development to find better exploration methods and to devise lower-cost drilling practices would encourage additional firms to enter the industry because the cost of entry would be reduced.

As for LNG, Britain has been importing this fuel for some years from North Africa; more recently, Japan has been importing it from Alaska. Now being considered are projects that would bring LNG from Algeria or Venezuela to the east coast of the United States. Moreover, additional research and development to improve LNG technology and the gasification of coal would reduce the cost of gas from these sources. If the cost of coal gasification could be reduced, it would provide a secure gas supply and also create a greatly expanded domestic coal industry with attendant economic benefits to the country.

Increasing the supply of oil, coal, and electricity

Because the oil supply situation in the United States is dominated by the uncertain future of the oil import program, advanced planning by the petroleum industry is difficult. A decision that "security of supply" was no longer a key issue might bring changes that would reduce domestic oil prices to the level of world prices. This would significantly affect the size and structure of the petroleum industry as well as the future rate and timing of investments in domestic oil exploration and development. But even if security of supply should continue to be a dominant factor in oil policy, there may be other ways to achieve security at lower costs than through the present oil import program.

Some of the same actions that have been proposed to increase gas supplies also are applicable to domestic oil supply. These include increased prices, more favorable tax treatment, continued limitations on oil imports, and less stringent air- and water-pollution standards. Short-term relief from the present tight supply—and higher prices—would occur if the Suez Canal were reopened, if the Trans-Arabian Pipeline were repaired, and if North African limitations on oil production are not imposed as they have been in the past. Over the longer term (two to three years), the construction of additional tankers would reduce transportation costs and provide more adequate oil supplies.

Another possibility is the development of Alaskan oil reserves to supply domestic demand. But because this would necessitate building a pipeline, it raises difficult environmental problems. Ecologists fear that the hot oil in the pipeline will melt the Alaskan permafrost, possibly causing soil erosion and landslides. Alternative above-ground pipeline transportation and other systems have been suggested as possible solutions.

Unlike the situation with oil and gas, coal reserves of all qualities in the United States are so large that the only major problems are those connected with productive capacity, labor supply, competitive pricing, and the length of time required to bring new mines into operation. For the short term, the ability to deliver coal to the user may be limited by the availability of freight cars, but with an assured long-term market this limitation will almost certainly disappear. Productivity was reduced temporarily when federal officials began enforcing the Coal Mine Health and Safety Act of 1969. As miners and management become familiar with the new rules and regulations, however, increased safety should result along with a return to yearly improvement in productivity.

Despite many problems facing the bituminous coal industry in 1970, production during that year was the highest since 1947. This increase was attained by the opening of many strip mines, which require smaller investments and can be brought into production quickly. However, as laws to prevent environmental damage from strip mining become more widely applied, the cost of coal from such operations will increase. More important, the growing concern by environmentalists that there is as yet no satisfactory way to reclaim hilly land that has been strip-mined may result in the extraction of more expensive coal from underground mines.

As for electricity, if its demand in the future can be estimated with reasonable accuracy, installation of adequate generating capacity should pose no insurmountable problems, even though long lead times are essential. It should also be possible, based on recent experience, to forecast load growth more accurately and thereby avoid a supply-demand squeeze.

However, siting problems for new plants are likely to become more difficult as the public, increasingly concerned over environmental matters, takes a greater interest in plant locations. Actions by local agencies, each of which has part of the responsibility for various aspects of plant siting, will have to be better coordinated to save time in the site-approval process. Suggestions have been made that a single agency be given responsibility for overall approval, and also that the federal government assume jurisdiction in any state that does not adopt the single-agency procedure.

Courtesy, Avco Everett Research Laboratory

Huge superconducting magnets, capable of producing a 40,000-gauss magnetic field, may help bring about the large-scale development of generating power by means of magnetohydrodynamics. Such a method is highly efficient and pollution-free.

Supplying energy to the year 2000

By 1980 major shifts may begin to take place in the energy supply-demand patterns that have characterized most of this century. Plans now being executed and those being formulated will have been transformed into plants and other physical facilities. But any modifications that may result from changed attitudes toward population and economic growth during the 1970s will not have had time to be reflected fully in new projects. Beginning about 1980, however, major changes in public policies and attitudes toward constraining population and

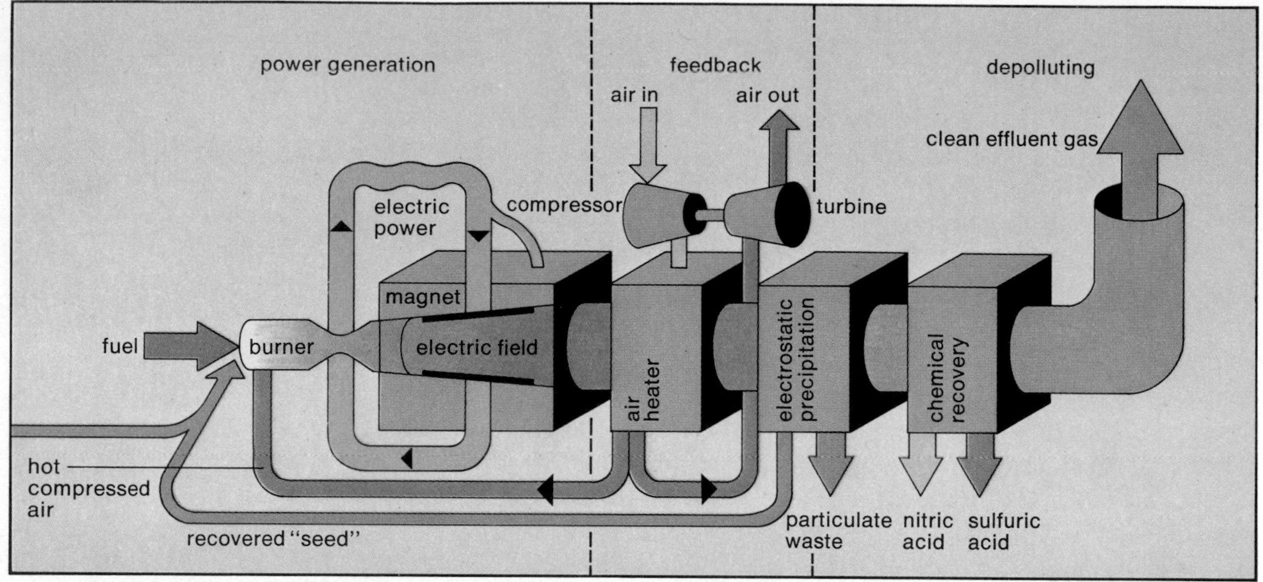

This flow diagram illustrates the relatively pollution-free production of electrical power by means of magnetohydrodynamics—generating electricity directly from a supersonic flow of hot, ionized gases. The fuel, along with a "seeding" of potassium crystals to raise the level of conductivity, is burned in a long chamber at 4,000°–5,000°F. This sends a high-velocity stream of conductive gases down a tube that is ringed by a superconducting magnet. There, current is drawn off from the gases by a series of electrodes. Part of the hot exhaust gas is fed back to the burner and part to an air turbine to produce even more power. An electrostatic precipitator recovers the potassium "seed" for reuse and also removes all particulate matter. A chemical unit recovers usable nitric and sulfuric acids, while the exhaust is emitted as clean gases, carbon dioxide and nitrogen.

traditional economic growth are expected to become increasingly important in energy planning.

Given any reasonable assumptions of population and economic growth, the energy needs of man probably could be supplied to the end of the century. By then, however, the world would be faced with depletion of many of its traditional energy resources. Assuring long-term supplies of energy, therefore, requires the development of new technology essentially free of resource restrictions.

As an interim measure, a policy should be vigorously pursued that would improve the efficiency of resource use to help extend the life of the nonrenewable resources. For generating electricity, the most promising new method is magnetohydrodynamics (MHD). This method usually involves heating a gas to 4,000° F or more, at which point the gas becomes ionized (electrically charged). When forced to move past a magnetic field, such a gas then becomes a conductor of electricity. When the technology for MHD is fully developed, scientists hope that it might increase efficiency in converting fuels to power from 40 to 50% and possibly even 60%. Not only would this conserve resources, but it also could help to minimize pollution.

MHD has been studied by nearly all industrialized nations, and in 1971 the Soviet Union reportedly was operating a large-scale demonstration unit using natural gas for fuel. But if MHD is to be employed extensively in the United States, coal will have to be used because it is the only fossil fuel with large enough reserves to last during the time that MHD would be most useful. Direct use of coal, however, poses many engineering problems that have yet to be solved, and a large investment in research and development would be required to overcome them. An alternative approach would be to gasify the coal and burn the gas as fuel, but with the present gasification technology, this would increase costs significantly.

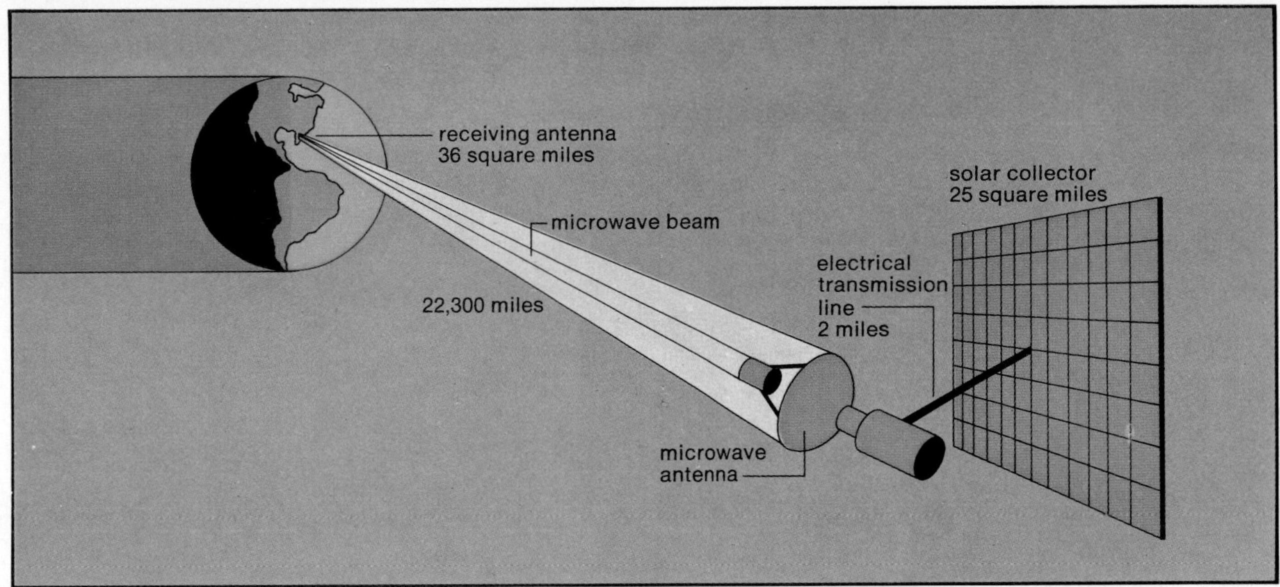

Energy from the sun may one day supply power to the earth. A solar panel on an orbiting satellite will collect the energy in sunlight. Solar cells then convert the energy into electricity, after which it passes through an electrical transmission line to the satellite's microwave antenna. The antenna will transmit microwave beams of electricity to a receiving station on the earth.

If MHD becomes commercially feasible, there should be beneficial effects on the environment. In advanced MHD concepts, thermal pollution would be either greatly reduced or eliminated and air pollution also would be reduced per unit of electricity generated. (An unresolved problem in MHD is how to handle the increased volume of nitrogen oxide that might be produced by the process.) Savings in fuel resources as a result of successful application of MHD also would have indirect environmental benefits. Because less coal would be needed for each unit of electricity generated, this means that less land would be disturbed by mining, less water would be polluted from acid mine drainage, and there would be fewer solid wastes in the form of ash or slag.

For the longer term there are several possibilities of a nearly inexhaustible energy supply. These include the breeder nuclear reactor, which produces more fissionable material than was originally supplied to it; fusion reactors; solar energy (enough reaches the earth each day to supply all the energy required for a greatly expanded world energy consumption); and controlled use of geothermal heat from the interior of the earth.

Although experimental work on breeder reactors began in the United States in the late 1940s, major engineering problems remain to be solved. Unless research is greatly expanded, present estimates indicate that a breeder may still be 10 to 20 years from commercial application. If the 20-year estimate is correct and if there is no major shift in energy growth rates, then there is only a small margin of safety if the breeder is to be relied upon to supply the energy requirements of the future. The U.S. Atomic Energy Commission (AEC) already has entered the "project definition" phase of its program to develop a liquid-metal, fast breeder reactor by awarding three contracts to outline steps that must be taken to arrive at a commercial breeder installation.

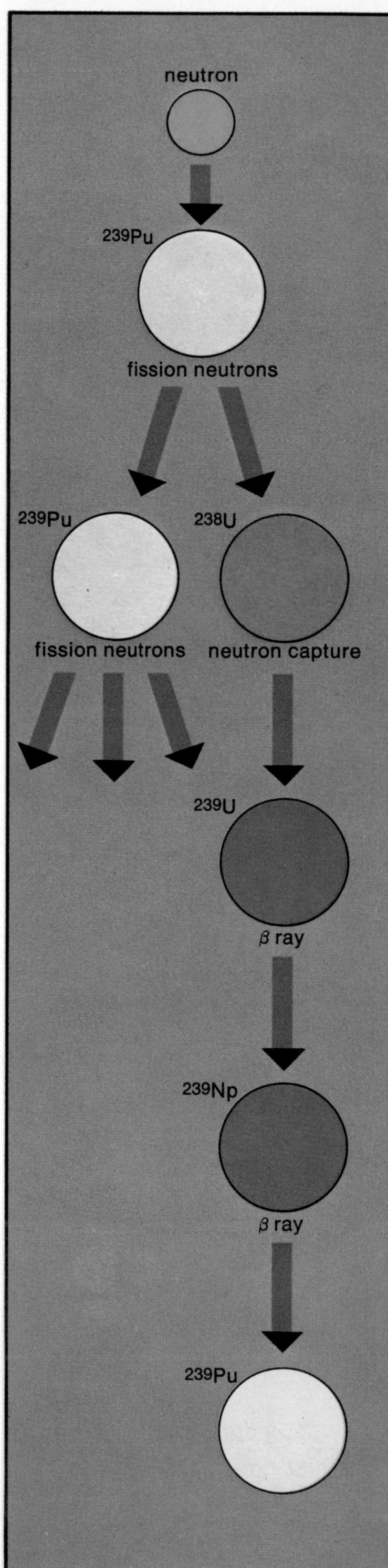

Even though there is worldwide interest in breeder reactors, the engineering problems are so complex that some attention has been directed toward development of "advanced converters" as a temporary alternative. These reactors can produce electricity with less fuel than conventional light-water nuclear reactors, but they are not capable of complete breeding. Concurrent with its research on advanced-converter reactors, the AEC began constructing a large, high-temperature gas reactor that employs the advanced-converter concept and started planning for several others.

Research on producing power by thermonuclear fusion also is being carried out in the United States and other countries. Although years have been spent on such research, the possibility of harnessing the energy released when two light atomic nuclei fuse into one heavy nucleus has not yet been demonstrated. Nevertheless, some optimistic fusion investigators hope to be able to demonstrate its laboratory feasibility sometime between 1975 and 1980. Should such a demonstration be made, the question of commercial feasibility and the economics of fusion would then have to be evaluated. Because of the uncertainties inherent in scaling up a laboratory model to commercial size, the mere demonstration of laboratory feasibility would in no way insure the ultimate use of fusion. It may be that an entirely different approach to that used to demonstrate feasibility in the laboratory will be needed if the engineering of a full-scale plant is ever to be accomplished.

Another potential source of power currently being considered is solar energy. Enough energy from the sun reaches us each day to provide many times the power consumed throughout the world. However, the complexity of installing energy-collecting devices, and the problems of transmitting, storing, and transforming the collected energy into conventional electricity so far has made its use uneconomic. Although small-scale, special-purpose devices have been developed for using solar energy, supplying large blocks of power with solar energy must await technological breakthroughs.

Energy from the heat contained in naturally occurring geothermal steam is another potential, but limited, power source. In those places where such steam is available, geothermal energy now is being used by some relatively small commercial installations, and more can be expected to be built in the future. This method of generating electricity is particularly advantageous because it not only uses the free heat in the magma of the earth but also because it is nearly pollution-free.

The number of locations where geothermal steam is available, however, are very limited, and probably represent only about 2% of the potential hydroelectric sources that could be tapped. But there may be many places where the hot magma is close enough to the earth's surface to provide the heat, if water could be pumped to it to be converted to steam. A feasibility study of this proposal is underway and, should it appear economically attractive, the potential of geothermal energy could be expanded greatly. (*See* Year in Review: EARTH SCIENCES, *Geophysics.*)

While there has been interest in the use of other energy forms, in most cases they are not as economically attractive as those already mentioned. Although many hydroelectric power sites remain to be developed, the total contribution of that source to the energy supply is expected to decline in importance. Tidal power also has been considered as a non-polluting energy source, but only a few small installations are in operation. Calculations of the maximum tidal power that could be harnessed indicate that it would be only about 1% of the power that could be produced hydroelectrically. Consequently, for the foreseeable future it will remain a source of energy only in a few favorable locations.

If the energy potential of the period from 1980 to 2000 is to be realized, it is obvious that new government-industry relationships will have to be developed. The magnitude of the research and development required to bring about the new technology that will be needed is larger than any single company can undertake, even if the risks were minimal. None of the new projects described can give such a minimum-risk guarantee, nor can they assure that, if successful, those who invested in the effort will be able to reap the benefits. Only through worldwide cooperation of government and industry will it be possible to supply the future energy needs of mankind.

Courtesy, Commonwealth Edison Co.

The Dresden nuclear power station near Morris, Ill., first went into production in 1960 with a unit capable of generating 200,000 kilowatts. In 1972 two more units, each designed to generate 809,000 kilowatts, are scheduled to become operative.

FOR ADDITIONAL READING:

Abelson, P. H., "Scarcity of Energy," *Science* (September 25, 1970, p. 1267).

Committee on Resources and Man, *Resources and Man* (Freeman, 1969).

Jensen, W. G., *Energy and the Economy of Nations* (Transatlantic, 1970).

Landsberg, H. H., and Schurr, S. H., *Energy in the United States* (Random House, 1968).

Mayer, L. A., "Why the U.S. Is in an Energy Crisis," *Fortune* (November 1970, pp. 74–77).

U.S. Library of Congress, Environmental Policy Division, *The Economy, Energy, and the Environment* (U.S. Government Printing Office, 1970).

U.S. Office of Science and Technology, *Energy and the Environment* (U.S. Government Printing Office, 1970).

A breeder reactor produces more fissionable material than it consumes. The process (opposite page) begins when a neutron interacts with the fissionable nucleus of plutonium-239. One of the high-velocity neutrons formed during this fission then continues the chain reaction (left branch), while uranium-238 can be used to capture another neutron. After this capture the uranium is transformed through two beta-decay processes to additional plutonium-239.

A Pictorial Essay

Grasslands of the World

by Lee M. Talbot

Man has created most of the world's grasslands, changed them for his own use, and left them severely depleted of their natural wildlife and vegetation. Much must be done if this vital resource is to be saved.

North American shortgrass prairie.

Dan Morrill

Lee M. Talbot

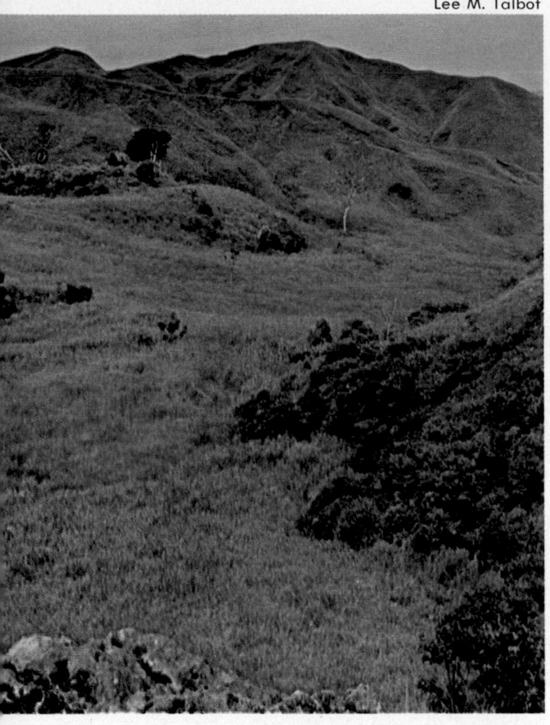

A mountain grassland on the island of Mindoro in the Philippines reveals the lush vegetation that is only one of the facets of this generally rugged, moist terrain.

Grasslands are immensely important to man. Their history and man's have been intimately associated for 2,000,000 years or more. Grasslands supported the grazing animals on which early man relied for food. They still supply most of man's meat and grain food crops, and they offer great potential for the increased meat and grain demands of the world's expanding human population. Yet, in spite of their significance to man and the rapidity with which they are being altered by his activities, grasslands have been very poorly known or understood.

Fortunately, this situation is changing. During the past 40 years or so, ecologists have come to recognize that the marked deterioration of wild grasslands, which almost automatically follows their pastoralization by domestic animals, is a dangerous waste of a valuable resource. Their attention has been directed toward the possibility of increasing protein production from many grasslands by using the indigenous wildlife, which is already well adapted to the environment, instead of attempting further management of nonindigenous animals.

As a result of worldwide recognition that basic ecological knowledge is required for effective long-term management of any living resource, the International Biological Program was established in 1967. Scientists from 70 nations are now involved in various projects, among which is an integrated research program by the United States to analyze ecosystems. Ultimately, all the principal biomes—the major biological communities or formations that include deserts, grasslands, tundras, and forests—will be studied, but the first effort, initiated in 1968, is concentrated on the grassland biome. The project is an exciting new approach to environmental biology because it involves large numbers of scientists from many disciplines and organizations. These workers are pioneering the type of effective methodology that is required for comprehensive research of grasslands throughout the world.

What is a grassland?

In numbers of genera and species, grasses are one of the largest families of flowering plants. The great food crops—wheat, rice, corn (maize), sugarcane, sorghum, millet, barley, oats—and the bamboos belong to the grass family. Geographically, grasses are probably the most widespread of all plants, covering about a quarter of the world's land mass. The treeless seas of grass on the prairies of North America were grasslands, as were the Russian steppes and the Argentine pampas. But so are high mountain meadows, the tropical savannas of Latin America, the grassy woodlands of Australia, and the acacia savannas of Africa.

All of these are biological communities where grasses are a dominant component. The species of plants in each area may be quite different, but the forms they assume and the ways they are grouped into communities and respond to common environmental conditions are remarkably alike. Among the environmental adaptations common to the grasses are fibrous roots, underground runners, growth from the base of leaves that are often at ground level, and quick response to

LEE M. TALBOT is affiliated with the Smithsonian Institution in Washington, D.C., as resident ecologist and field representative for international affairs in ecology and conservation. At the present time he is on leave to the Executive Office of the President, where he is senior scientist for the Council on Environmental Quality.

rainfall. These characteristics help grasses survive drought, fires, heavy grazing, and trampling, all of which are common conditions in the grasslands.

Grasslands range from communities of nearly pure stands of grasses to ones with relatively high densities of trees or other plants. Most such ecosystems have a grass layer of perennial bunch grasses. The ground surface between clumps of these grasses is often dotted with smaller perennial grasses, annuals, and small herbs.

This vegetation changes seasonally. Some species grow early in the wet season, while others respond later; showers during the dry season produce dramatic growth in some plants while leaving others dormant. Thus, the grass layer is extremely complex and efficient in utilizing ground surface, light, and the limited moisture. The upper, woody layer, if present, may be composed of a variety of woody plants that range from low bushes to tall trees and in density from isolated trees to thick stands.

Lee M. Talbot

Invertebrates play important roles in maintaining balanced environments. In the desert grass savanna of Kenya, termites build mounds like the one that stands about 10 feet high in the center of this arid landscape.

The dynamics of grasslands

Significant alteration of any of the components of a grassland usually changes the entire ecosystem. For example, climate sets the broad limits for growth of grass and other vegetation. But within any climatic zone, a series of grassland types may be possible, depending upon the interaction of other factors. Equally, many grassland types can exist in several climatic zones. Even within a given species of plant, size and vigor often increase with increases in the amount and reliability of moisture.

Generally speaking, the formation of grasslands is favored by a climate with marked wet and dry periods. The growth forms of many grasses are adapted to the moisture regime, and the dry period allows burning. Low and irregular precipitation, combined with high evaporation, produces vegetation that is particularly vulnerable to damage from heavy grazing and burning. Consequently, such areas are easily converted to grassland—and the resultant grasslands are equally vulnerable to degradation from the same causes.

The grassland soil itself is relatively dynamic. Its development and characteristics are affected not only by its parent material but also by topography, vegetation, climate, and the time during which these factors have operated. There is mounting evidence that, under most climatic conditions, if a forest cover is changed to grassland through cutting and burning and maintained as such, the forest soil will change toward grassland soil.

Effects of fire and grazing

Fires are characteristic of tropical savannas and of most other grasslands, where they suppress woody growth to maintain the open aspect of the land. Frequent or very hot fires tend to remove woody vegetation, thereby creating an open grassland; infrequent or cool fires often allow woody vegetation to increase. If burning is stopped, as it often is

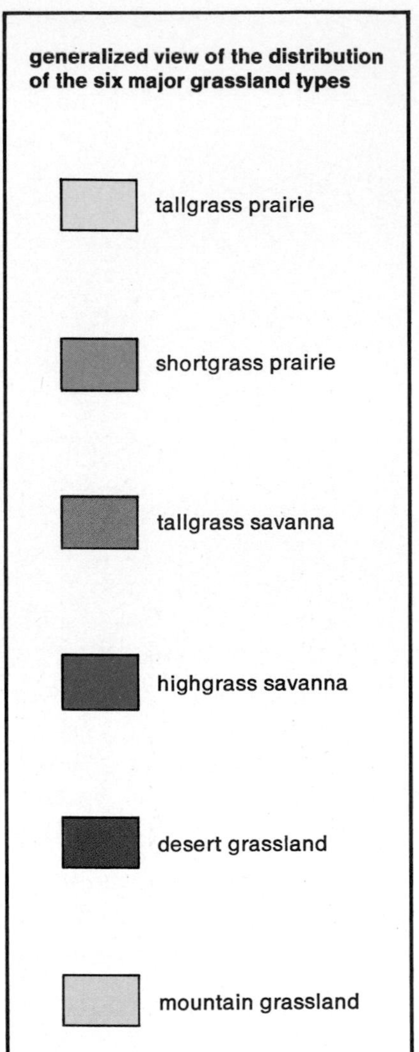

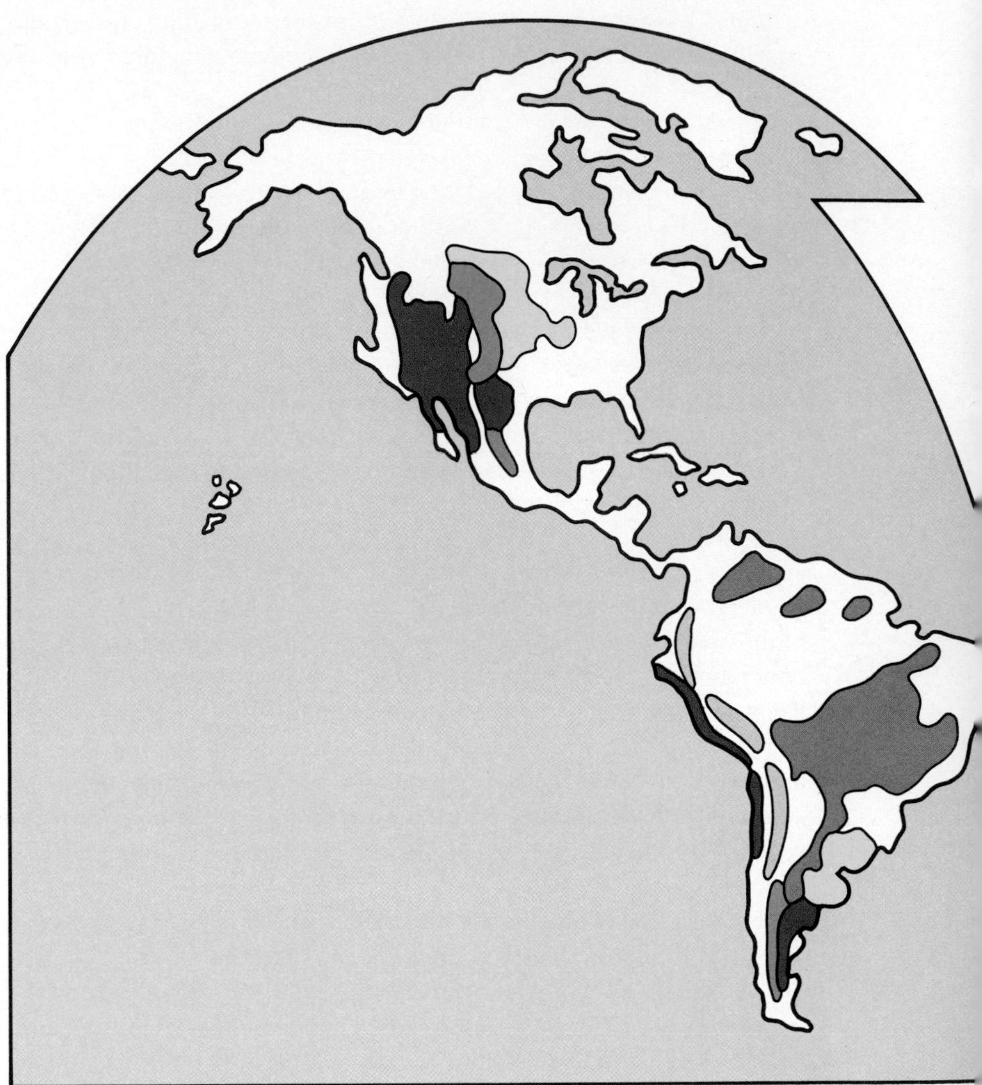

following settlement, a grassland may eventually revert to woodland; browsing alone usually will not maintain it.

For all practical purposes, man is responsible for grassland fires, which he starts to improve grazing, remove predators and pests, open new grazing land, or attract or drive game. It is now well accepted that human activities have created grasslands from the forests of much of the world or have at least significantly expanded and modified them. And it also is likely that archaeologists will learn that man's influence in changing grasslands is of far greater antiquity than is now believed.

Fires will not occur or will be too light to suppress woody vegetation in areas where grazing is so heavy that little or no fuel is left. Therefore, heavy grazing favors the increase of the woody components of savanna vegetation, and, when done by any single species, results in a disproportionate use of that species' preferred plants. Moderate grazing and browsing distributed over the available vegetation maintains a vigorous growth and a wide variety of plant species.

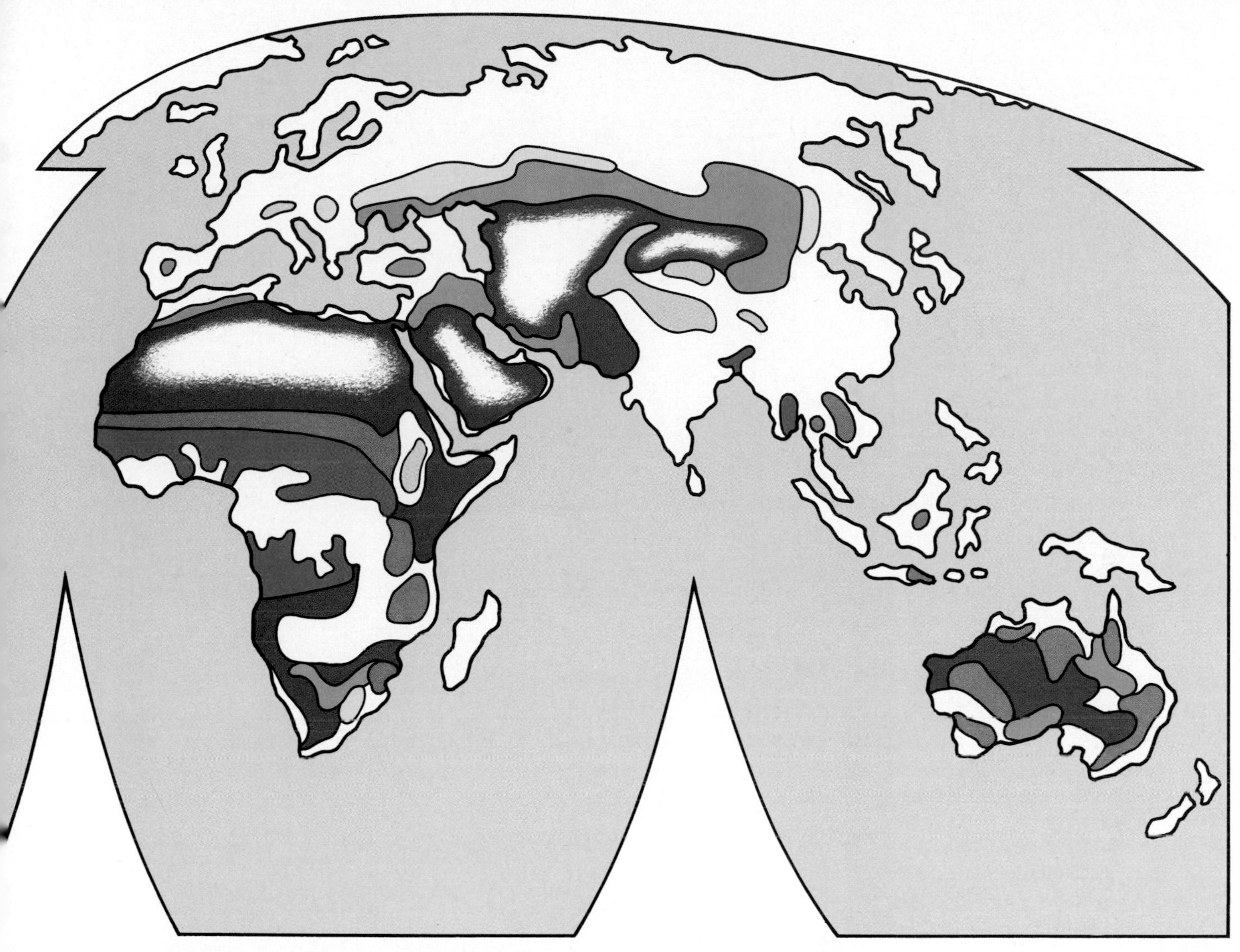

Grassland wildlife

Fire, and the alternating wet-dry seasons, put a premium on animals that can follow food and water and flee from fires and predators. Although there is a broad spectrum of grassland animals, ranging from invertebrates to elephants, the most conspicuous and well known are the ungulates, the hooved grazers and browsers. These animals have developed specialized tooth structures to deal with the silica content of grasses, specialized stomachs to digest them, relatively large size, and a hoof structure adapted for movement on open, dry plains. The other grassland fauna apparently have equally intricate roles to play in the ecosystem, but less is known of most of them than of the ungulates.

Wherever they occur, similar grassland types offer similar habitats for wildlife. Consequently, where there has been sufficient time, each type has evolved animals that perform similar functions in the ecosystem, even though the animals may have come from very different genetic stocks.

Dan Morrill

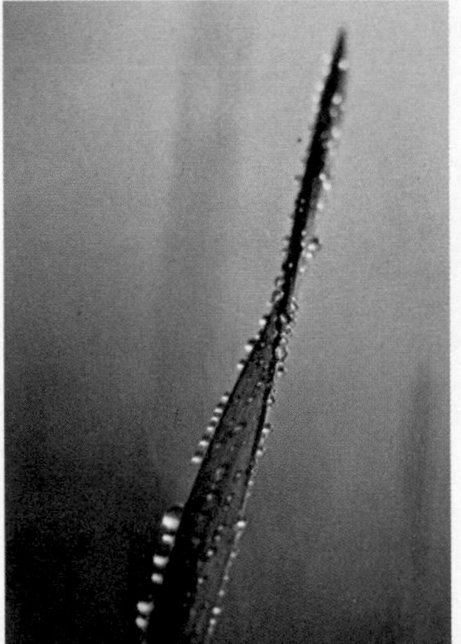

Lee M. Talbot

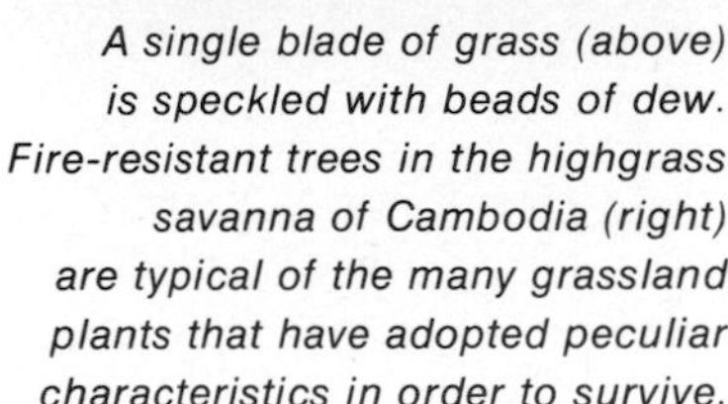

A single blade of grass (above) is speckled with beads of dew. Fire-resistant trees in the highgrass savanna of Cambodia (right) are typical of the many grassland plants that have adopted peculiar characteristics in order to survive.

Australia, for example, had no ungulates and evolved instead large grazing kangaroos, smaller wallabies, and a variety of other marsupials with functions similar to those of pigs, deer, cattle, sheep, goats, and horses, the farm and game ungulates of Europe. South America evolved a series of unique camels, including the guanaco, vicuña, and the domestic llama and alpaca, while Africa "specialized" in gazelles, along with a variety of other antelopes, giraffes, and horses. In Australia, the flightless emu fills the same ecological niche as the ostrich in Africa and the giant rhea in South America.

Major grassland types

Although grasslands can and do occur in great variety virtually everywhere in the world, six major types can be distinguished. The two most economically useful grassland types are found in the temperate zones. These are the tallgrass prairie and the shortgrass prairie or steppe, which, respectively, make up about 13% and 10% of the world's grass cover. The tropics have produced the two most extensive grassland types, tallgrass savanna (31% of all grasslands) and highgrass savanna (22%). About 18% of all grasslands belong to the desert grassland type; the remaining 6% are mountain grassland.

Now virtually taken over for livestock or cultivation, *tallgrass prairies* were all nearly pure grasslands dominated by a relatively few species of tall perennials growing on deep, rich loess (windblown soil). The eastern Great Plains of the United States, the pampas of the Río de la

Durward L. Allen

Man's tampering has frequently degraded a grassland's productivity rather than enhanced it. On this shortgrass prairie, the indigenous sagebrush, still visible on the right, was removed from the area on the left to permit planting of grasses for grazing. Instead, the broomweed plant took hold and now dominates the area.

Plata region of Uruguay and Argentina, the prairies of eastern Europe that include the Hungarian Basin and the Romanian plain, and the tallgrass Russian steppe were all tallgrass prairie.

Shortgrass prairie occupies the somewhat drier lands that are often transitional to deserts or dry woodlands. Its short bunch grasses are now used mostly for cultivation and rangeland, which is often degraded by overgrazing. The major shortgrass prairies are the North American prairie extending from Manitoba southward into Mexico, and the drier grasslands of Spain, southeastern Europe, and the Russian steppe.

Tallgrass savanna usually is dominated by three- to five-foot perennial grasses, often in dense stands with highly varied species. Where the tropical climate allows, these lands have been taken over for ranches or cultivation, and where protected, as in the parks of eastern and southern Africa, they support the world's greatest variety of plains wildlife. These grasslands were extensive in central, eastern, and southern Africa, north and east Australia, and scattered in a variety of forms across much of southern Brazil, Argentina, and Uruguay.

Highgrass savanna is found largely in high rainfall areas adjacent to the tropical forests from which it was derived. Now used for cultivation or grazing, this type was distributed in a broad band around the central African rain forests and in the great river valleys of southern Asia.

In arid places grasslands are characterized by a sparse assemblage of drought-resistant grasses, other such plants, and low trees or bushes. Major areas of *desert grassland*, or desert-grass savanna,

Lee M. Talbot

Steven C. Wilson, Entheos, Inc.

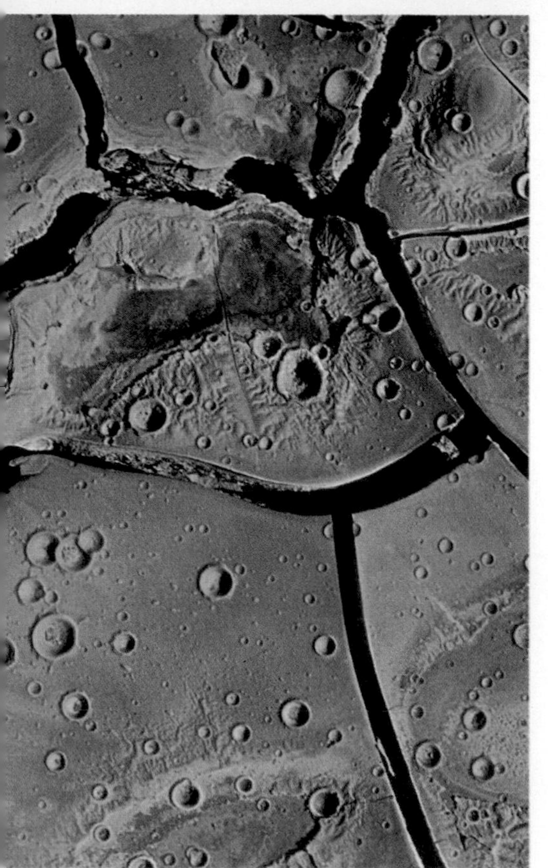

Steven C. Wilson, Entheos, Inc.

biomes include the sub-Saharan zones and arid eastern and southwestern portions of Africa, parts of the Middle East, central and western Australia, and southwestern North America.

The one major grassland type that occurs worldwide is *mountain grassland*. Although usually limited to relatively small areas above timberline, extensive areas occur in the plateaus of the northern Andes (the *jalcas* of Peru and *paramos* of Ecuador) and the high mountains of Asia. Its bunch or mat grasses are well adapted to the extreme temperatures and short growing season; seasonal livestock grazing is the area's only significant use.

The conservation challenge

Grasslands were not only one of the earliest habitats of man but, through cultivation, livestock grazing, hunting, fencing, and fire, they also have been one of the most completely altered. It is imperative, therefore, that we protect adequate examples of major grassland types throughout the world if the knowledge they may provide for human welfare and the biotic resources they contain for a burgeoning human population are not to be lost irrevocably. So far, however, the attempts to manage most grasslands have resulted in further degradation rather than in increased productivity.

An operating ecosystem cannot be reconstructed or studied after it is gone. If man is to derive maximum benefits from grassland resources, he must understand their ecology and their role in the world's biological systems. The studies now being conducted by the thousands

Lee M. Talbot

of scientists engaged in the International Biological Program are a start in this direction. The new type of multidisciplinary research that they are pioneering will be required in the future if we are to gain a true understanding of how the grasslands operate and of man's natural place within them.

FOR ADDITIONAL READING:

Carr, Archie, *The Land and Wildlife of Africa*, ch. 3, "Grasslands and Great Game" (Life Nature Library, 1964).

Thomas, William L. (ed.), *Man's Role in Changing the Face of the Earth* (University of Chicago Press, 1956).

U.S. Department of Agriculture, *Grass, The Yearbook of Agriculture* (U.S. Government Printing Office, 1948).

AUDIOVISUAL MATERIALS FROM ENCYCLOPÆDIA BRITANNICA EDUCATIONAL CORPORATION:

Films: *The Grasslands*; *Life in the Grasslands*; *What is Ecology*?

A complex interaction between living and nonliving elements is necessary to maintain a healthy grassland environment. The Serengeti Plain in eastern Africa (above) is being protected as one of the few remaining examples of a balanced, natural grassland. On the preceding page, an unusual African caterpillar (top, left) is one of the many insects that live out their life cycles in grasslands. The North American prairie dog (right) aerates the soil and enhances its richness as it burrows through the ground. Poor crop management or uncontrolled grazing of a grassland, however, can result in parched, useless earth (bottom, left).

Richard Keane

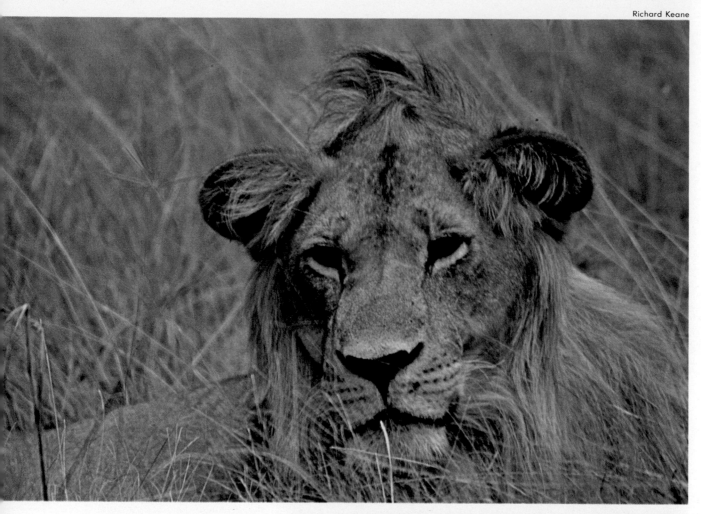

Grassland animals have developed unique devices for their survival. The light coloring of the lion (above) permits it to roam at will in open, sparsely covered country. The poor-sighted rhinoceros (right) is accompanied through the tall grass of an Indian river valley by egrets, which can warn it if danger is near.

Lee M. Talbot

Richard Keane

George B. Schaller

A wide variety of ungulates inhabit the African plains. A herd of gazelles (above) graze on open land. The wildebeest (left) stands paralyzed before its attackers, a pack of wild dogs.

Lee M. Talbot

Fire sweeps across the flat, wet terrain of a highgrass savanna (above), burning off woody growth that might otherwise cause the grassland to become a woodland. Guinea fowl of Africa (right), closely related to the domestic poultry of other continents, mill about a tallgrass savanna in Tanzania.

Ramkr from Pix Inc.

Massive herds of elephants, once a common sight in African grasslands, migrate to the green highlands. Unless protected from poachers in search of prized tusks, such a sight will soon be a thing of the past.

Lee M. Talbot

Many factors interact to produce great variety in the grasslands of the world. A lone tree (above) and a little moisture permit a brief change in an otherwise barren desert grassland in the mountains of Peru. Rice paddies on the island of Java (right) and oat fields in Siberia (opposite page, top) show man's efficient cultivation of the great grasses and contrast with an untamed African savanna at sunset (opposite page, bottom).

Y. Korolov, LIFE Magazine ©Time Inc.

Lee M. Talbot

The Science Year in Review: A Perspective by William D. McElroy

When we look back in future years, 1971 will probably be viewed as a year of transition for science—the beginning of a change from the old order to the new. Fresh currents are flowing, some of them conflicting, but the prevailing movement is toward a broader appraisal of science as a basic tool of service to society.

Increasingly, the national goals of the United States are expressed in terms of the quality of life. Degradation of the environment, disease, depletion of our resources, inadequate housing, and natural disasters are problems that must be solved. To do this, we have decided to reassess our customary ways of managing our national affairs. Science and technology have not been exempted from this general reassessment.

Science and social goals

Responding to the heightened urgency of public concern, the scientific community is moving to reaffirm its social responsibilities. Within that community, an increasing number of scientists are adopting an attitude that views science as a more positive instrument for social progress than ever before. While holding to the central need for fundamental disciplinary research, scientists are placing increased emphasis on interdisciplinary research focused on solutions of environmental, ecological, and social problems.

Another movement proposes that we subject technology to far more careful scrutiny to determine its social consequences. No longer are science-based technological innovations automatically considered improvements. Instead, their uses and possible misuses are beginning to be weighed carefully. Precautions are being taken, moreover, to eliminate unforeseen negative consequences. This is a relatively new development that requires, as in the case of scientific research related to societal problems, an underpinning of fundamental scientific research.

Part of the current unrest in the U.S. is directed against science and technology as threats to the quality of life. In large part, this view represents an overreaction to the situation, but it does point up the need to focus on three basic objectives. First, the pursuit of science as the essential element in national progress should be strengthened, not weakened; its potential exploited, not quarantined. Second, the citizen should have a better understanding of science so that he can assess and use scientific information and thus absorb it into the social fabric. Third, science should be flexible enough to play a stronger role in meeting the needs of society.

The efforts to achieve new and better goals for the quality of life, however, cannot be detached from the economic realities that affect the quantity of life. This, too, affects science. Not only is science dependent upon the economic health of society, but in developed countries the economic health of society in turn depends upon science and technology. The complexity of our economy is such that the real question is not whether we should be a science-based technological society—that is an accomplished fact—but rather what the future direction, methods, and pace of science and technology ought to be.

The need for more research

In moving toward a solution of the problems facing society today, we must realize that the situation is not comparable to the one confronting scientists during World War II or at the start of the space program. In those instances, much of the basic scientific knowledge was at hand; the greater part of the challenge lay in putting the knowledge to use, in developing the necessary technology. Today, there is woefully little basic knowledge that scientists can contribute to the solution of many current problems.

We know far too little about the interactions that occur within this biome we call the planet earth. We do not really understand the dynamics of our environment or the effects that technology might have upon it. We know little about the more subtle effects of pollution. We cannot predict with confidence the behavior of individuals or that of social groups and institutions. We are not in a position to assess adequately the relative costs and benefits to society of any technology or any course of action.

These are the dimensions of the challenge that lies ahead. They outline clearly the need for continued emphasis on basic research and for an increasing emphasis on the intelligent utilization of our scientific knowledge to help meet the problems of society.

Contents

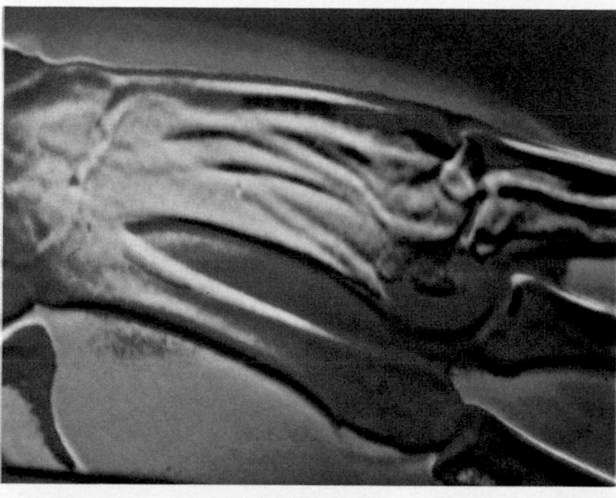

Agriculture

During the year, agriculture in the United States was strong and productive, firmly supported by a nationwide system of research and education to keep it responsive to the country's needs. Agriculture planned, produced, marketed, processed, and distributed one of the world's largest supplies of food and fiber for a domestic population of more than 204 million, and for millions more in other countries.

This past year also saw a reordering of priorities in U.S. agricultural research, with greater emphasis placed on some of the critical issues of the day. For example, the U.S. Department of Agriculture (USDA) concentrated on such vital problems as overcoming pollution of land, air, and water; eliminating poverty and lack of opportunity in depressed rural areas; developing in-depth knowledge of nutrition and teaching people good nutritional habits; providing high-protein foods for people of developing nations and helping them increase their food production; and insuring an adequate supply of food for future, fast-expanding populations.

Pollution problems studied. Environmental problems generated great interest and activity not only in the U.S. but in other advanced nations as well. Foremost among the actions in the U.S. was the creation of a new federal office, the Environmental Protection Agency, to spearhead the nation's war on pollution. This agency, headed by William D. Ruckelshaus, pulled together a variety of research, monitoring, standard-setting, and enforcement activities formerly scattered throughout the government. Its most important function was to establish and enforce environmental standards.

Agricultural research played a major role in dealing with a number of environmental problems. Scientists made progress in managing agricultural wastes and recycling some of them into useful products. In Albany, Calif., for example, researchers for the USDA's Agricultural Research Service (ARS) recently reported that high-pressure steam processing of wood chips made clean, washed wood fibers for manufacturing hardboard, plus a hemicellulosic liquor that could be concentrated for use as a molasses substitute in cattle feed.

ARS scientists in Arizona demonstrated the feasibility of "renovating" sewage water for human

Courtesy, R. R. Nelson, Pennsylvania State University

The effect of a new variety of fungus, Helminthosporium maydis, *that is highly virulent on corn with T-type cytoplasm is revealed in photographs of the 1970 southern corn leaf blight epidemic. Endangering more than 80% of the U.S. corn crop in 1970, the epidemic demonstrated dramatically that widespread reliance on one variety of corn can have devastating results.*

use. By utilizing plant-soil filters to clean secondary sewage effluent, they produced water that lost its identity as sewage effluent and looked and tasted like plain drinking water that was entirely suitable for human consumption.

Research advances helped to reduce air pollution from cotton gins, feed dryers, and other agricultural processors; select and breed pollution-resistant crops, ornamentals, and shade trees; and provide information on reducing or preventing soil from washing or blowing away at construction sites.

Research also developed ways to help reduce the use of persistent pesticides as well as to find other pest control methods. For example, federal and state agencies and industry were cooperating in one of the largest programs of its kind in agricultural history to determine if use of integrated control techniques can eliminate the boll weevil, the nation's most destructive cotton pest. The project was to provide the first large-scale coordinated testing of all the biological and nonpersistent chemical suppression techniques developed over the last decade. The massive tests were scheduled to begin in July 1971 and to continue for two years in Mississippi, Alabama, and Louisiana.

In other work, a research team at Michigan State University found that two insects—springtails and soil mites—break down DDT into harmless components, a decontamination mechanism that could be very important in the food chain. Scientists at Cornell University developed a method of killing more weeds with less weed killer, by using certain combinations of herbicides in order to obtain synergistic, or more powerful, response. This technique could greatly reduce the chemicals being added to the environment.

Rural life improved. Activities to improve the quality of rural living assumed greater priority in national planning for the U.S. this past year. The objective was to promote better housing, better health and health facilities, and more and better paying jobs in small towns and rural communities throughout the country. The states and USDA worked toward these goals by providing technical and economic assistance to benefit rural cooperatives, small businesses, and rural housing—and by conducting research of special benefit to the depressed rural areas.

One example involved the development of a low-cost building technique—surface-bonding of concrete block walls. This technique eliminated mortar between blocks, sealed the wall against leakage, strengthened wall joints, and provided an exterior finish, all in one operation. Research to provide low-cost housing emphasized self-help construction to utilize the spare time of low-income families, as well as the development of components that local people can produce economically in small plants.

Nutritional improvements sought. Studies on nutrition gained ground throughout the U.S. as surveys revealed the extent of malnutrition—among both those who had enough food but did not make the right choices, and those whose choices were limited because of low income. In the light of this information, USDA expanded its research and education programs in food and nutrition.

In education, USDA developed many new research-based publications to help homemakers choose and use food properly; disseminated nutrition information; expanded use of nutrition aides from the community to teach homemakers improved food habits; broadened the Food Stamp Program; and added nutrients in food distribution programs for low-income families and for infants, children, and pregnant and nursing women.

In research, USDA expanded its efforts to learn how individual nutrients were absorbed, transported, and utilized—and how they were affected by age, heredity, and environmental conditions. Greater emphasis was placed on measuring the nutritive values of various foods and finding out how these values were affected by household practices.

Many nations assisted. The more developed nations of the world, including the U.S., became increasingly involved in programs to help the poorer nations build up their agriculture to enable them to feed their growing populations. A special effort was also made to develop high-protein foods and to fortify existing foods to provide more nutritious diets.

Technical assistance came from a growing number of public and private sources—from state land-grant universities, USDA, private businesses, foundations, U.S. Agency for International Development (AID), and from the U.N. Food and Agriculture Organization and other international organizations.

Research assistance was provided in several ways—through Public Law 480 programs conducted in other countries with local currencies obtained through sale of agricultural commodities; through cooperative programs sponsored by USDA and AID; through USDA efforts to provide special high-protein and enriched foods; and, indirectly, through USDA's regular research program, the results of which can often be adapted successfully for use in the less-developed nations.

Recent studies by USDA and AID in Africa produced germ plasm for corn and sorghum that yielded four to five times more than indigenous

Courtesy, Bjorn Sigurbjornsson, FAO/IAEA

In this field at the Institute of Radiation Breeding in Japan, mutations are induced in plants to improve the quality of seeds for plant breeding. The tower in the center radiates gamma rays, which change the genetic structure of seeds, and the high wall surrounding it serves as a shield for the radiation.

varieties. Comparable studies in India produced higher yielding pulse crops (peas, beans, lentils). The desert locust plague was effectively stopped in East Africa, possibly for the first time in history, through the efforts of local control organizations and people from USDA, AID, and agencies from other countries.

In Great Britain, scientists made a discovery that could be, according to that country's National Research Development Corporation, one of the most significant of the century. Researchers there found a method for converting ordinary starch into a new high-protein food. The method produces a white powder containing 45% protein, nearly twice the percentage found in steak, and equal in quality to that of milk.

The international assistance effort was considered so significant an endeavor for mankind that the Nobel Peace Prize for 1970 was awarded to an agricultural scientist who devoted most of his life to such work. Norman E. Borlaug, working in Mexico for the Rockefeller Foundation, won the prize for his research in helping to develop and disseminate high-yielding dwarf wheats, universally described as the heart of the Green Revolution. (*See* Year in Review: HONORS).

Keeping agriculture productive. Agricultural research provided the technology farmers needed to continue producing efficiently. Advances were made in protecting crops against pests. Numerous resistant varieties were developed—for example, a tomato resistant to curly top virus, a potato resistant to leafroll and verticillium wilt, an experimental corn hybrid with genetic resistance to corn rootworm and earworm, and a wheat that is resistant to stem rust as well as other diseases.

Newer techniques developed through engineering research improved the handling and harvesting of agricultural products. Better methods for controlling pests and for conserving soil and water were basic to profitable production.

Advances in livestock research meant greater efficiency for producers and more wholesome meat products for consumers. One finding, at the University of Wisconsin, was a promising preventive for milk fever, which costs dairymen millions of dollars annually. Another, at the North Dakota agricultural experiment station, was the identification of 172 agents that cause abortions in cattle.

Remote sensing. As the world population continued to increase and the threat of food shortage continued to grow, such technology as remote sensing promised to open up new dimensions to alleviate world hunger and improve man's environment.

Remote sensing involves the use of special cameras and other sophisticated instruments in orbiting earth satellites. These instruments see and record invisible as well as visible light waves given off by objects on earth. The result is a picture that is far more revealing than one obtained by regular photography. Accurate interpretation of remote sensing pictures could provide rapid information on identification of crop species, crop health and production, range conditions, soils, major land uses, water resources, and pollution.

The U.S. National Aeronautics and Space Administration was funding most of the agricultural applications of remote sensing at Purdue University, West Lafayette, Ind., Oregon State University, and the universities of California and Michigan. ARS was coordinating this work and conducting work of its own in Weslaco, Tex.

The future. Yields from U.S. farms will continue to increase within the next few years as growers apply the latest research findings to their agricultural operations. Crops will be bred increasingly for resistance to diseases and insects. Greater stress will be placed on development of natural or biological methods of pest control, to be used singly or in combination with other methods.

Research will provide improved materials for irrigation and more automated systems. Weather extremes may be modified soon, based on some work already underway.

—George W. Irving, Jr.

Archaeology

Two notable trends in archaeology came into focus during the year. Archaeological sites that produce transportable art objects were being looted and destroyed so rapidly that archaeologists all over the world feared the greater part of our cultural heritage of antique objects still in the ground would soon be lost. In April 1970 a meeting of representatives from more than 60 nations was held in Paris, under the auspices of UNESCO, to work out some sort of international regulation, and a convention on the international trade in antiquities, produced at the meeting, was being considered by the consulting nations. In one sense, the antithesis of this trend was the rapidly growing interest in excavating, studying, and preserving the remains of one's own immediate history. In the United States, for example, this was reflected in the current very extensive excavation and preservation of historic sites, even including industrial sites of the Civil War period.

South and East Asia. During the year statements by Wilhelm Solheim, Kwang-chih Chang, and Chester Gorman appeared, summarizing recent developments in South and East Asian archaeology. They agreed in presenting a wholly new time perspective for the region and conclusive evidence for agriculture and pottery manufacture shortly after the end of the Ice Age. Apparently, the cold began to ameliorate about 12,000 to 10,000 B.C., when a culture known as Hoabinhian was already well established in Southeast Asia. With improving climate, this type of culture began to expand and soon achieved three revolutionary technological innovations—domesticated plants, pottery, and ground stone tools.

Early centers of culture. The best sequence of radiocarbon dates for Hoabinhian times is that from Spirit Cave in Thailand, where there are five distinct layers with datable organic material. The fourth layer from the top is on the order of 10,000 to 9000 B.C., but the lowest (fifth) layer, which had not yet been dated, was estimated to be on the order of 12,000 B.C. Solheim believes that ground stone tools from Australia, radiocarbon dated at 20,000 B.C., are of Hoabinhian origin. It is not known precisely when systematic cultivation of plants actually began in Hoabinhian times, but there is good evidence it had started by 10,000 B.C.

Jomon pottery in Japan has been radiocarbon dated as the oldest known (10,000 B.C.). However,

A bronze head from a 5th century B.C. shipwreck was one of many objects found by looters in the Strait of Messina. After the pillaging was reported to the Italian police, they recovered most of the objects.

David I. Owen with permission of Giuseppi Foti

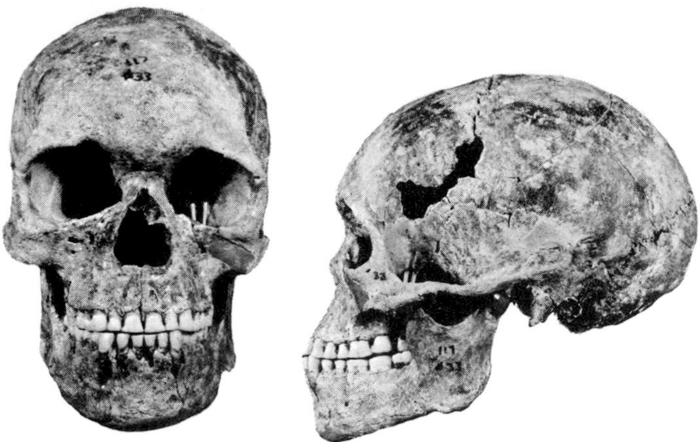

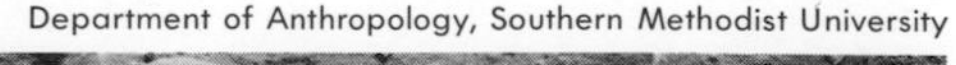

Department of Anthropology, Southern Methodist University

Excavation in the Upper Nile revealed a late Paleolithic skull (above) from Nubia, and (right) two skeletons from Jebel Sahaba, Sudan. Estimated to date between 12,000 and 10,000 B.C., these finds proved wrong archaeologists' former assumption that there were few Paleolithic remains along the Nile Valley. The recently completed work was done in connection with the salvage program in the Aswan Reservoir area.

Chang points out that it is cord-marked like Hoabinhian pottery, and that these two very early centers of pottery manufacture—as well as that found on the lower Wei Shui River and on the lower Yellow River in western Honan, north China—may all be linked in time as three early Far Eastern centers for the development of pottery and domestication of plants. There were no absolute dates for the presumed north China center, but all three centers seemed to have emerged in early post-Glacial times.

Unusual characteristics. There are other remarkable aspects of these early cultures. All appear to be of extraordinarily long duration. Hoabinhian culture persisted for at least 6,000 years at the Spirit Cave site and probably much longer in Southeast Asia as a whole. Jomon continued for at least 8,000 years, and the North China type must have been equally durable. There were changes, but at the current stage of investigation the nature of culture change in Asia appears to be very different from that in the Near East or America.

Gorman has pointed out that it has been quite impossible to classify collections of Hoabinhian artifacts on formal typological grounds, and that their most characteristic feature is the lack of well-defined types. Because of this he has analyzed the stone tools on the basis of edge damage, chipping technique, and other physical elements, often based upon microscopic examination. It is possible that the absence of well-defined artifact types will turn out to be characteristic of early material in East Asia as a whole.

Dating problems. It must be emphasized that the current conception of very early post-Pleistocene plant domestication and pottery manufacture in East Asia is based on the technique of radiocarbon dating. Without it there would be no way of recognizing the antiquity of these developments or of comparing, in time, similar developments in the Near East and Meso America. Unfortunately, none of the dates given above has been corrected for the known shift in the carbon-14 (^{14}C) inventory in the atmosphere. Thus, radiocarbon dates of about 6000 to 5000 B.C. are certain to be on the order of 1,000 year older, and those before 6000 B.C. may turn out to be much younger. (See *1971 Britannica Yearbook of Science and the Future,* Feature Article: DATING THE PAST.)

Oceania. Radiocarbon dates for early cultural horizons in the western Pacific Islands began to provide a time perspective for human occupation of Oceania. The 20,000 B.C. date for occupation of Australia, mentioned above, is supported by additional ^{14}C dates from Cannon Hill Station and Koonalda Cave, reported by Richard Shutler, and there are dates suggesting settlement of Australia as early as 35,000 years ago. There are ^{14}C dates from New Guinea, associated with a hunting-gathering economy, up to 24,000 B.C., but one must assume occupation of New Guinea at least as early as that of Australia, since New Guinea was part of the land bridge to Australia.

Settlement of north and central Oceania. A distinctive type of pottery known as Lapita, which is found in New Caledonia, Fiji, Tonga, and the New Hebrides, has been difficult to pin down in time. The ^{14}C dates are somewhat confusing, but the pottery would appear to have originated as early as 1000 B.C. and perhaps 1200 B.C. Technical problems with ^{14}C dating in the Pacific, particularly with shells in Micronesia and Polynesia, continued to plague attempts to reconstruct the history of settlement in the islands. Nevertheless, dates from Saipan and other sites indicated settlement of north and central Oceania as early as 1500 B.C.

Egypt. More evidence for very early domestication of plants came from the Upper Nile Valley as a result of the salvage program carried out in connection with the construction of the Aswan High Dam. Fred Wendorf, Rushdi Said, and Romuald Schild, summarizing some of their work on the late Paleolithic period (*Science,* Sept. 18, 1970), suggested that men were grinding grain along the Nile as early as 13,000 B.C.—4,000 years earlier than this activity was taking place in other regions of the Near East.

In June the Cairo newspaper *Al Ahram* reported that archaeologists had found in Egypt a mummy estimated to be almost 5,000 years old, "the most beautiful, most intact and most ancient" ever discovered. It was only the second time that a mummy had been found in its original burial ground, untouched by thieves.

The Near East. The whole question of the origin of domestication and settled village life received a "new look" with a discovery in the summer of 1970, announced by Robert Braidwood and Halet Cambel of the universities of Chicago and Istanbul. At the site of Cayonu in southeastern Turkey, evidence of the transition from the use of wild plants and animals to the domesticated varieties was found in the remains of a village with stone buildings. This could mean that men lived in villages before rather than after they had learned to domesticate plants and animals.

New evidence of early trading patterns. In southeastern Iran, C. C. Lamberg-Karlovsky of the Peabody Museum of Archaeology and Ethnology, Harvard, excavating at Tepe Yahya south of Kerman, discovered a proto-Elamite settlement dating from the 5th millennium B.C. Its significance lies in the evidence that the southeastern Iranians of this period were in contact with related people in the Persian Gulf (Bahrein), Baluchistan, and Mesopotamia and that the whole region was a trading area. Apparently, certain types of distinctive pottery and objects made from steatite, a kind of talc, are indicative of this trade.

Phoenician Iron Age discovery. In Lebanon, James Pritchard of the University Museum, Philadelphia, discovered the Iron Age strata of the Phoenician city of Sarepta at Sarafand in April 1970. This was expected to make possible a detailed study of a Phoenician city dating from the expanding period of Phoenician culture—a unique experience in Near Eastern archaeology. Identification of the site as ancient Sarepta (or Zarephath) is based on a combination of literary and archaeological evidence. References to it appear in the Papyrus Anastasi (13th century B.C.), I Kings 17 (9th century B.C.), and in cuneiform accounts of the campaigns of the Assyrian kings Sennacherib and Esarhaddon.

Painted tomb finds in Turkey. Machteld Mellink of Bryn Mawr (Pa.) College reported on the excavation of a second painted tomb in the district of Emali, west of Antalya (ancient Lycia), in Turkey.

A stone figure wearing an earring with a star of David in its center was recently excavated in Mexico, evidence in support of theory that Mediterranean people discovered America 1,000 years before Columbus.

Wide World

Courtesy, Israel Department of Antiquities and Museums

A Mycenaean charioteer vase, dating about the mid-14th century B.C., was discovered in a tomb in Tel Dan, Israel. The vessel is richly decorated with spiral motifs and linear patterns, but the rendering of the figure drawing is crude.

The first tomb, accidentally discovered in 1969, has painted scenes in the East Greek style of the late 6th century B.C. The second, probably dating from the first quarter of the 5th century B.C., has scenes of Graeco-Persian type. In Greece proper, most ancient wall painting has been lost.

America. The continuing debate on when man came to America was not advanced by a symposium held under the auspices of the San Bernardino (Calif.) County Museum and the University Museum, Philadelphia, held in San Bernardino in October 1970. The intention of the symposium, which included archaeologists and geologists, was to reach a decision regarding the results of excavations in an alluvial fan extending outward from the Calico Mountains in the Mojave Desert, carried out over several years by Ruth DeEtte Simpson under the general direction of Louis S. B. Leakey. The majority of the archaeologists present were unwilling to accept as human tools the flint objects so identified by Leakey, and a majority of the geologists believed that the deposit was much older than the 50,000–150,000 years estimated by the excavators. (See *1970 Britannica Yearbook of Science and the Future,* Feature Article: MAN IN AMERICA: THE CALICO MOUNTAINS EXCAVATIONS.)

Peru. In Peru, Richard MacNeish of Phillips Academy, Andover, Mass., and a team of North Americans and Peruvians discovered remains that profoundly expand the time and space horizons for ancient agriculture in America. During 1970 in the Ayacucho-Huanta region of highland Peru they excavated 12 stratified sites with a complete preceramic sequence extending from about 8500 to 1700 B.C., and, in another zone, two sequences dating from 9500 to 6000 B.C. and from 7000 to 2000 B.C.

One phase in this sequence, termed Jaywa, which has been found in five different ecological zones, appears to be the period of transition from hunting to plant collecting and may be a period of incipient agriculture and herding. Radiocarbon dates give an age for this period of about 6600 to 5500 B.C. A Piki phase, which follows Jaywa, produces definite evidence of agriculture and is dated at about 5500 to 4300 B.C.

The earliest remains in the area were found in Pikimachay Cave in different zones and levels with bones of extinct animals. Tentatively, the material is associated with two phases called Paccaicasa and Ayacucho in that order of time. There are bones of horse, sloth, possibly saber-toothed tiger, puma, skunk, extinct deer, paleo-llama, and possibly mastodon. Radiocarbon dates indicate that the period ended about 13,000 to 11,000 B.C. and may have begun as early as 23,000 to 21,000 B.C.

Mexico. In Mexico a great archaeological project, similar to the excavation and restoration of Teotihuacan, was under way in Mexico City, where the ruins of Montezuma's famed Aztec capital were being exposed by the construction of a subway. Ignacio Bernal had taken responsibility for a systematic study and preservation of very extensive and rich remains. Many sculptures and even some wood have been preserved in the old lake bottom upon which the present city rests.

Early man in East Africa. Both the origin of tool manufacture and the origin of *Homo* continued to retreat into the ever more distant past. In April 1970, Richard and Mary Leakey of the National Museum Centre for Pre-History and Palaeontology, Nairobi, Kenya, announcing the preliminary results of current excavations in the Koobi Fora area east of Lake Rudolf, described very good evidence for a tool-making hominid at 2.61 (±0.26) million years ago—800,000 years earlier than that found in Bed I at Olduvai Gorge in Tanzania.

The largest series of artifacts found at one site totaled 51 specimens. They had sustained very little weathering, indicating that they had been eroded out of the tuff in recent times. Five hominid specimens included two well-preserved and nearly complete crania. One, resembling *Australopithecus boisei* from Olduvai and found in a horizon below the volcanic tuff, has been dated at 2.61 (± 0.26) million years. The second, found about one mile distant but in the same geologic horizon, was notably different from the first and may be a very early representative of *Homo*. There is an expressed opinion that *Australopithecus* was not a tool maker and that an early form of *Homo* may have been a contemporary species which produced the tools in the same horizon.

The method used to date the Lake Rudolf site also represented a significant development. F. J. Fitch of Birkbeck College, London, and J. A. Miller of Cambridge recognized the possibility that the potassium-argon method for dating volcanic tuff and pumice overlying the deposit might give false dates because extraneous argon in the material may have come from preexisting rocks blown out of the volcanic vent. To avoid such an error, the gas being released from an irradiated sample of rock was analyzed over a series of heating steps toward complete fusion. Apparently these additional steps beyond the usual potassium-argon method lead to much more accurate and dependable dates for early periods.

A cult stand showing various figures of musicians was discovered at a site in Tel Ashdod, Israel, by a joint expedition of the Israel Department of Antiquities and the Carnegie Museum in Pittsburgh, Pa. It was estimated to date between the late 11th and early 10th centuries B.C.

Courtesy, Israel Department of Antiquities and Museums

Dating techniques. Persistent difficulties with the thermoluminescence (TL) method for dating fired clay meant that it was not as readily accepted and utilized as the ^{14}C method in its early stages. By 1971, however, continued improvements in the technique, worked out by Elizabeth Ralph and Mark Han of the University Museum, who first announced the method in 1965, and also by M. J. Aitken and S. Fleming at Oxford, had established it as a standard tool for dating.

Fine-grained pottery less than 3,000 years old could be dated by the TL method with more certainty than much older ceramics. Approximately 80% of the TL dates made on pottery of known age correspond with those known ages with a precision of ± 100 years. With much older, coarser, and more poorly fired ceramics, 30–40% of the TL dates differed more than this. Fixing a date through ^{14}C for these older specimens was also more difficult, however, so that part of the greater percent of error might not be real. One advantage of the TL method, which was already affecting the trade in antiquities, was its absolute dependability in detecting fraudulent terra-cotta figurines. There is no problem in identifying clays fired in recent years.

—Froelich Rainey

See also Feature Article: ASPHALT CEMETERY OF THE ICE AGE.

Architecture and building engineering

As in previous years, spectacular examples of the sheer quantitative power of the construction industry in Europe and America were matched by an equally generous outpouring of new techniques. A growing number of thoughtful people in the developed nations, however, could not help but regard these achievements as manifestations of an ominous paradox—the inexhaustible reservoir of technological innovation on one hand and the increasing inability of man to create a decent human habitation on the other.

New height records. Three structures brought to various stages of planning and construction during the year reached new records for height in their respective categories.

Polish television tower. The Polish government planned the erection of a television antenna tower that was to rise 2,100 ft and thus would become, at least for a time following its expected completion in 1974, the world's highest structure. Designed by the engineer Jan Polak and to be constructed at the town of Plock, near Warsaw, the tower will be a truss-like framework of steel tubes weighing no more than 550 tons. It will be guyed, or prevented from overturning under wind loads, by a system of steel cables attached to the framework at five different levels and radiating outward and downward to anchor points in the ground. Even more novel are the passenger and service elevators in the interior of the tower, which are to be operated by means of cogs fixed to their sides and engaging racks in the elevator guide rails—a vertical cog railway.

One Shell Plaza. At the opposite end of the scale for total mass was the Shell Oil Company's building, known as One Shell Plaza, in Houston, Tex., opened for occupancy at the same time that the Polish television tower was being designed. Standing 714 ft and 52 stories high, the Shell tower was the world's tallest concrete building, eclipsing the absolute height (645 ft) of the previous record holder, Lake Point Tower in Chicago. The Houston skyscraper, designed by the architectural and engineering firm of Skidmore, Owings & Merrill (Chicago office), is constructed in such a way that the loads are divided between an interior core and exterior bearing walls in the form of a dense framework of closely spaced columns and interconnecting beams.

Sears Tower. Far transcending these two structures in size and economic importance was the Sears Tower in Chicago, the immense skyscraper that Sears, Roebuck and Co. placed under construction in 1970 to serve as its international headquarters. Designed by the Chicago office of Skidmore, Owings & Merrill as a variation on the structural system embodied in the same firm's John Hancock Center (Chicago), the Sears building will establish a new record for absolute height at 1,450 ft, but will fall one story short of the 110 floors in each of the World Trade Center towers in New York City. Like the Hancock and World Trade structures, the Sears Tower is built as a rigid, steel-framed, tubular cantilever—that is, a structure anchored at one end (its foundations) and having the form of a hollow prismatic tube, the walls of which are rigid frameworks rigidly connected to each other. By building in this way, the designers can create the equivalent of a braced frame in which the outer walls provide all the resistance to the bending and shearing forces of wind. Thus the interior frame is restricted to carrying gravity loads, at a great saving in the total quantity of steel required.

The Sears Tower is unique in appearance by virtue of a novel variation on the setback principle, widely used in the skyscrapers of the 1920s. At the base the building consists of nine square prisms arranged in three rows, each 75 ft on a side, the three together thus forming a larger square of 225 ft. These prismatic tubes are column-free on the interior, being built up of peripheral columns only, and all of them rise together from the plaza level to the 50th floor, where the northwest and southeast prisms terminate. The remaining seven rise to the 66th floor, where the enclosures at the opposite corners end; a symmetrical cruciform arrangement of five prisms then extends to the 90th floor, where three are dropped, leaving the remaining two to continue to the top. This pattern of setbacks was designed partly on the basis of diminishing floor-space requirements, partly to reduce wind drift, and partly to maximize resistance to wind at the base of the building, where bending and shearing forces reach a maximum.

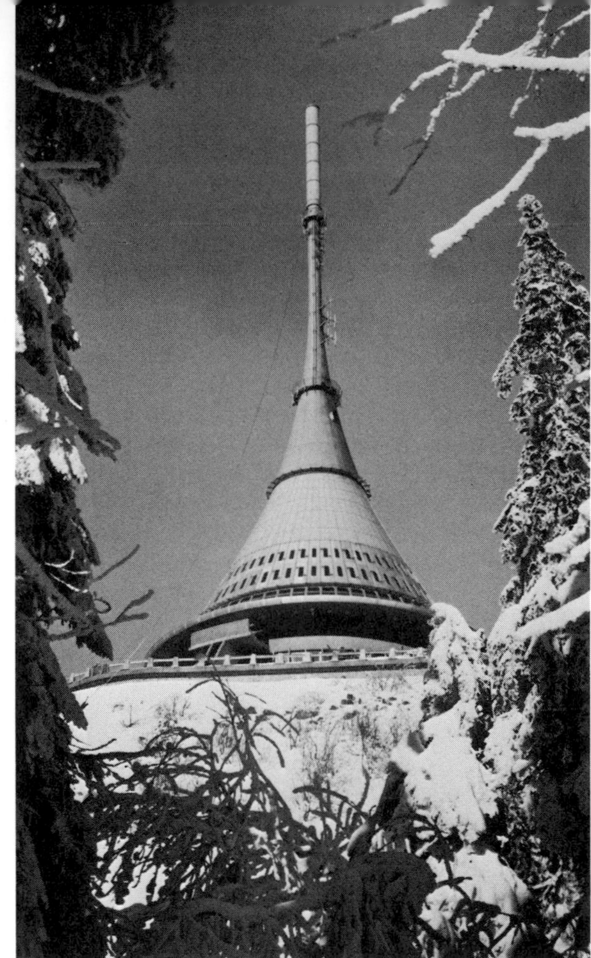

STAL-Hubacek

A 295-ft-high Czechoslovakian tower serves as a television transmitter and a tourist hotel. Plastics were used extensively for its diverse structural, electronic, and architectural needs.

The external walls of the nine prismatic elements are composed of massive corner columns and smaller intermediate columns spaced 15 ft from center to center, the columns and their associated floor trusses rigidly welded and bolted together to form what is called a Vierendeel truss, or a truss made up only of rectangular elements. The framing for the support of any one floor consists of trusses that extend between the walls in only one direction. This direction is held for two or three floors, then turned at right angles for the next several stories, a device that transforms the entire floor-framing system into a structure that serves not only to support floor slabs but to transmit wind loads along any cardinal direction from wall to wall. The total floor area of the skyscraper is 3.7 million sq ft, near the record for a single office building but still below that of Chicago's Merchandise Mart.

American Airlines hangars. A structural mode radically different from skyscraper framing appeared in two service hangars erected by American Airlines at the international airports of San Francisco and Los Angeles. The identical buildings, products of the architectural office of Conklin and Rossant and the engineers Lev Zetlin Associates, were the largest jet hangars so far constructed and the largest structures of any kind to be roofed by hyperbolic-paraboloidal shells of steel. This awkward mathematical term, ordinarily shortened to hypar, refers to a certain type of doubly curved, saddle-shaped surface that possesses great internal resistance to bending and consequently can support itself over long clear spans.

The individual hangar is divided between a central truss-framed core and eight hypar roof sections cantilevered from each side for the unprecedented distance of 230 ft, with a shell thickness of 0.09 in. The capacity of the hypar to sustain itself rigidly over this extensive free length arises from several factors, chiefly the double curvature and the twist of the geometric form, the welded stiffeners that extend along the length of each module, and the I-beams running along the ridges and the edges of the valleys. In addition to these elements, a series of longitudinal cables extend from the core trusses to various points along the hypars to take some of the tension, to prestress the shell, and to control the flutter that is a serious problem in structures composed of thin, flexible materials.

Sports stadia. Another hypar shell of record size in the customary material of reinforced concrete covers the sports stadium placed under construction in Milan, Italy, during the past year. The work of the architects Gilberto and Tommaso Valle of Rome and the engineer-contractor firm of Societa Italiana Condotte d'Acqua, the concrete stadium was planned as a shallow bowl turned up gracefully at the ends, with a seating capacity of 15,000. The 495-ft diameter of the nearly circular roof makes it the world's largest hypar and far outdistances its nearest competitor, the 221-ft-long hypar roof of the Edens Theater, opened in 1963 near Chicago. The Milan roof, however, is not a pure self-supporting shell; it consists of a lightweight concrete umbrella reinforced with wire mesh and resting on a thin sandwich of insulation fixed to a steel membrane, the whole complex in turn carried by a network of steel ribs. The composite construction of concrete on steel arises from the need to reduce the concrete density to a minimum in order to cover an area of about $4\frac{1}{2}$ ac without intermediate supports.

The building of stadia with roofing that does not obstruct sight lines has led to some of the most imaginative forms of reinforced concrete construction, a field in which French engineers have long been preeminent. Although they have been challenged in recent years by their Italian colleagues, they emphatically demonstrated the authority of their tradition in the Paris stadium known as Parc des Princes, scheduled to be opened for use in 1972.

Designed by the Paris architect Roger Taillibert as a soccer and Rugby stadium with a 55,000-seat capacity, the big structure presents a unique combination of prestressed concrete forms with a novel

Italian-born Architect Paolo Soleri proposed to house 170,000 people on less than half a square mile of land in his "Hexahedron," a model city that he designed.

Courtesy, Museum of Contemporary Art

Courtesy, Bethlehem Steel Corporation

The double-winged hangar at San Francisco airport is one of two identical structures recently erected by American Airlines. The largest structures to date to be roofed by hyperbolic-paraboloidal shells of steel, they will provide nearly six acres of clear floor space to service four jumbo jets simultaneously.

method of joining the precast elements that compose the various structural members. The columns extend upward and outward in 150-ft-high parabolic curves to carry the prestressed concrete roof beams that are cantilevered inward another 150 ft from the tops of the columns. Each column and beam pair consists of 12 precast segments that are joined end to end by an epoxy-resin glue—possibly the first case in which primary structural elements of the usual inorganic materials are joined with an organic adhesive. The separate column and beam segments are further tied together by prestressing cables that pass through openings in the concrete mass and extend throughout the length of the beam and column into the footing at the column base, in this way transmitting the stresses in the roof-supporting members to the foundation.

Plastics in construction. Organic substances in the form of plastics have been developed to the point where they may play a role in structural elements subject to relatively small loads. The most advanced applications of this kind appear in the combined hotel, restaurant, and television tower completed during the past year at Liberec, Czech., from the designs of the architect Karl Hubacek. From the ground up, the 295-ft structure consists of a glass-enclosed restaurant, a three-story hotel, an equivalent three-story height housing television-transmitting equipment, and a steel tubular mast, all supported on the interior by a double-walled cylindrical core of concrete.

The enclosure for the television equipment constitutes the novel feature: the curtain wall is made of laminated fiber glass panels laid like shingles, each one lapping over a small area of the one below it. These panels are fixed to a latticework of solid plastic rods tied at close intervals to a series of circumferential plastic rings, the whole system together forming a hyperboloidal truss of plastic materials. The lattice and the tie rods do not carry interior loads, which are sustained by the concrete core, but they must be strong enough to resist the bending and distorting forces of the wind. The fiber glass and plastic were chosen because they offer no electronic disturbance to or interference with television broadcasting.

A more extensive application of plastics appeared in an 11-story office building designed by International Environmental Dynamics and erected during the past year at Mountain View, Calif. The basic structural system is itself unusual, since the floors are hung from eight steel straps suspended from two reinforced-concrete core towers. The curtain walls are composed of prefabricated polyester panels reinforced with fiber glass, the windows of each individual panel being inserted into preformed openings lined with neoprene gaskets. The panels are held in place by being bolted to a light tubular framework. A similar kind of cladding was adopted for an office building of comparable character in Poole, Eng., completed shortly before the opening of the Mountain View structure.

The most remarkable exhibition of plastic construction would come with the completion of the Antioch College branch campus at Columbia, Md. (the main campus is located at Yellow Springs, O.). The various buildings reached the planning stage in November 1970, with Rurik Ekstrom of Columbia in charge of the design and the Research and Design Institute of Providence, R.I., as consultants. The various buildings of the small campus were planned for 200 students and were to be constructed as fiber glass domes, polystyrene plastic domes, and conventional highway trailers. All of them would be covered by a double-walled plastic vault 35 ft high at the crown and 210 ft square, supported by air maintained at elevated pressure by three blowers, and held in place by a grid of steel cables. This type of inflated structure was originally developed by the Goodyear Tire and Rubber Company of Akron, O., and was first used for a greenhouse built in Wooster, O. The entire Antioch campus was expected to cost only $250,000.

Prefabrication. The continuing need for low-cost housing the world over focused increased attention on the industrialization of building through the extensive use of prefabricated components. This technique has generally been confined to single-family houses, but in the past year the construction of two multistory buildings provided impressive demonstrations of the possibilities of large-scale applications.

A proposed system for residential construction is depicted in a project by SCAG (Structure for Change and Growth). The form can be changed constantly because parts can be added and replaced.

Courtesy, Helmut Schulitz, University of California, Los Angeles

Czechoslovak systems. An unusually extensive and thoroughgoing system of building with precast concrete elements was developed by Konstruktiva, a construction company in Prague, Czech. By using components designed and carefully cast as multiples of a basic module and precasting all primary structural components—columns, girder slabs, floor slabs, capital slabs, wall panels, stair towers, and elevator shafts—the company achieved a record by erecting a 17-story office building in 92 working days.

Cleveland public housing. A similar system of construction with precast concrete elements was adopted by the Cleveland (O.) Metropolitan Housing Authority in the building of a public housing group consisting of a 16-story apartment block and 38 town houses. The Cleveland system was developed by the architects of the project, Barbitta-James and Associates, and the two contractors, the Ohio Turnkey Company of Akron and the Tom-Rob Company of Cleveland.

Swedish innovation. Still another variation on building with manufactured components was developed by the Swedish manufacturer Karl-Erik Andersson. In a modular prefabricated house built up from a row of 10 × 27-ft manufactured units, he introduced a double-panel construction for walls, floors, and roof consisting of two steel plates separated by wood studs, the hollow space between the metal panels being filled with warmed or chilled air to heat or cool the house. Thin metal panels are excellent conductors of noise as well as heat, but the wood separators and the air sandwich proved to be an effective insulation in both cases. The chief advantage, however, is the provision of the highly efficient and comparatively inexpensive system of radiant heating.

Sewage treatment plant. Most sewage-treatment plants and incinerators are straightforward works of steel or concrete framing. One very large installation, however, represents a model example of the integration of a sewage-treatment plant with landscaped space and community recreational facilities. The North River Water Pollution Control Plant in New York City, recently placed under construction from plans prepared by three large New York engineering firms (Tippetts-Abbett-McCarthy-Stratton; Gibbs & Hill, Inc.; Feld, Kaminetzky and Cohen), will not only be one of the largest in capacity (220 million gal per day) but will also serve as the structural support for a recreational complex laid out on the scale of many city parks.

The plant will stand over the Hudson River on a 33-ac concrete platform, which in turn will rest on about 2,500 steel-lined concrete caissons sunk deeply into the bearing rock. The reinforced concrete structure itself will be an immense windowless building with a roof calculated for the unusually high loading of 400 lb per sq ft. This generous construction will allow the roof slab to carry a football stadium, handball and tennis courts, a 7,000-seat amphitheater, parking space, and a group of community buildings, all of them being planned by the New York State Park Commission in cooperation with various city and community organizations.

High-rise fireproofing. In the United States, a series of fires in high-rise buildings that were scrupulously designed and built according to the fire codes of the cities involved compelled a serious reappraisal of protective techniques commonly employed in multistory structures. The fatalities that occurred were a consequence of smoke rather than heat, and they indicated that there had been oversights in the traditional design of high buildings.

One problem had been known for years. A building may be built of incombustible materials, but furnishings, carpeting, personal possessions, and office supplies are usually readily combustible, and if enough such objects burn, a well-designed fireproof building may act like a furnace in the dissemination of heat and combustion gases. If the heat rises to a high enough temperature, high-strength steel will be rendered plastic, bringing about a swift collapse of roof or wall. The most ominous feature of recent skyscraper fires, however, was that the smoke and gases from a minor blaze could spread through shafts, stair wells, and hidden passages to trap people in distant rooms or elevators.

The New York Board of Fire Underwriters conducted an investigation in the winter of 1970–71 and issued a report prepared under the direction of W. Robert Powers, the board's superintendent. Its chief conclusions were plainly disturbing to the executive officers of the construction, steel, and elevator industries, for it was one weakness or another in their products that led to disaster. Fires on upper floors or in interior passages were not readily accessible to firemen, and buildings contained no means either to prevent the spread of the fires or to extinguish them. Elevators rose to centers of heat and became inoperable, resulting in the entrapment of occupants. The sprayed-on fireproof cladding that is universal for the covering of steel structural members proved far from adequate in the protection it provided.

The Board of Underwriters recommended the installation of automatic sprinklers throughout the entire height of a building, the adoption of encasement fireproofing such as hollow tile or applied concrete in place of the thin sprayed material, and the use of elevator call buttons that would not call a car to levels of concentrated heat and smoke. These were expensive devices, however, and the rapid escalation of construction costs during the past year made it unlikely that builders would take any decisive steps without a drastic revision of local building codes.

—Carl W. Condit

Astronautics and space exploration

Highlights of the year in astronautics and space exploration included the manned landing on the moon by the U.S. Apollo 14 astronauts in February 1971 and an unmanned lunar landing by the Soviet Luna 17 in November 1970. Three Soviet cosmonauts died during the landing of their Soyuz 11 spacecraft after setting an endurance record of 24 days in space. During their mission they had docked with the Soviet unmanned orbiting space station, Salyut, and had worked inside it.

Earth satellites

Earth-oriented satellites may be divided into three general categories: communications, earth-survey, and navigation. The first primitive versions of each type were launched in 1960. By 1971 each had developed greatly in performance, capability, sophistication, and promise for the future.

Communications satellites. The Communications Satellite Corporation (Comsat), a U.S. firm organized in 1964, strengthened its global capacity by launching in January 1971 the first of the Intelsat 4 satellites. Placed in commercial operation in late March, the satellite could handle 3,000–9,000 telephone circuits simultaneously. It was launched into geostationary, or synchronous, orbit (one that allows it to be constantly over the same portion of the earth's surface) over the Atlantic Ocean, joining five Intelsat 3 satellites, two of which were operating over the Atlantic, two over the Pacific, and one over the Indian Ocean. The Intelsat 4 has a cylindrical shape, is 17 ft long, weighs 1,075 lb in orbit, and has a life-time expectancy of seven years, compared with five years for the Intelsat 3s.

Two more Intelsat 4s were scheduled to be

Courtesy, NASA

An intelsat 4 is checked out prior to launching. The communications satellite, developed by the Communications Satellite Corp. (Comsat), is 17 ft in length and weighs 1,587 lb in earth orbit.

launched later in 1971, thereby making it possible for 77 countries to share in the operation of this global commercial communications system. With the addition of approximately 20 new ground station antennae that were expected to be completed in 1971, there would then be 70 such installations in operation.

On the military side, there were developments in both long-distance and tactical operations. Long-distance systems of the U.S. Department of Defense consisted of 23 operational satellites and 29 surface terminals. Two geostationary satellites of higher power were scheduled for launch in mid-1971. A tactical communications satellite (Tacomsat 1), in geostationary orbit over the Atlantic, was evaluated in 1970 to determine the feasibility of using such satellites for tactical operations. During the experiment 65 terminals were installed in mobile aircraft, jeeps, trucks, ships, and submarines. Because of the high power of the Tacomsat 1 terminal, antennae as small as one foot in diameter could be used.

Television transmission by satellites. In February and March 1971 Comsat and other U.S. firms made initial license applications for nationwide television transmissions by geostationary satellites. The Comsat proposal would employ three satellites larger than the Intelsat 4s, each with a capacity of 24 color television channels and utilizing an initial ground network of 132 stations.

Among the advantages that might be gained by increased development of communications satellites is the reduced cost of long-distance communications, both for the private consumer and for data transmission by industry. It is in the realm of television broadcasting, however, that such satellites perhaps offer their greatest promise of impact and general benefit. Current satellites require large, powerful receiving stations that transmit television by microwave or landlines to a local broadcasting station whose own transmitters then beam the television image to home receivers. Eventually, however, scientists may develop satellites capable of broadcasting directly to home receivers. Such a satellite must have high broadcast power and a large antenna. To meet these needs, the U.S. National Aeronautics and Space Administration (NASA) is currently developing the Applications Technology Satellite F (ATS-F). It will be equipped with a 30-ft-diameter antenna, the largest ever deployed in space.

Infrared photograph of the Texas-New Mexico border region, taken from an earth satellite, indicates denser crops in Texas. Such photography has proved useful in land exploration.

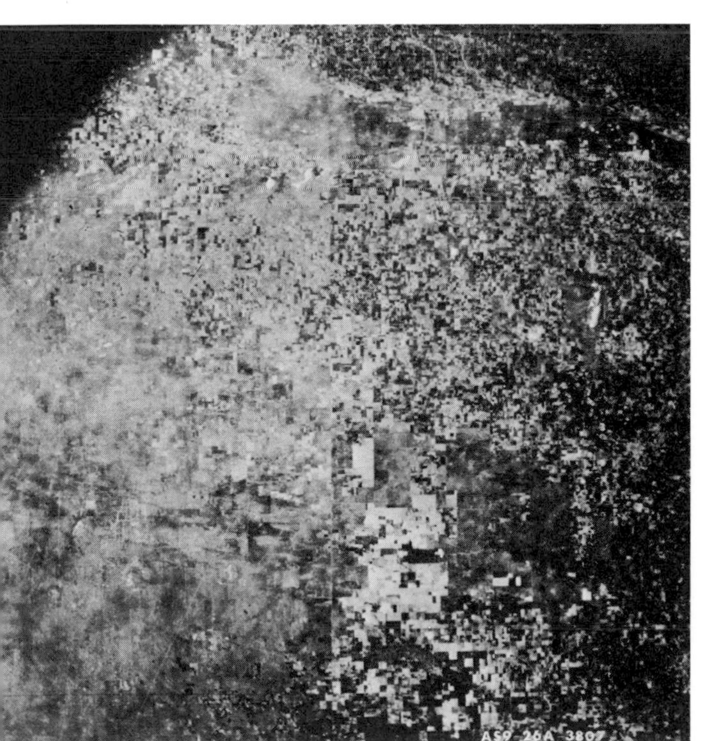

Courtesy, U.S. Geological Survey, Department of the Interior

Courtesy, NASA

The spherical satellite PAGEOS, to be used in the U.S. National Geodetic Satellite Program, undergoes an inflation test. As PAGEOS reflects sunlight falling upon it, it appears as a point of light in the sky.

In 1973 the ATS-F satellite is scheduled to be used in an attempt to provide mass education in India. Receiving television signals from an Indian transmitter, the satellite will then rebroadcast from its vantage point in space to approximately 5,000 villages. Each village, according to the plan, will have at least one low-cost, large-screen television receiver. The programs initially will deal with improvements in agricultural practices and methods of population control. The possible expansion of this experiment within India could lead to direct linkage of all of that nation's approximately 560,000 villages. Since interlinking television landlines do not exist and the country has insufficient funds to build them, the broadcast satellite holds immense potential as the most effective way of educating India's 524 million people.

European and Soviet communications satellites. France and West Germany embarked on a cooperative venture to develop and launch their own communications satellites. The program, known as Symphonie, calls for the launch of a satellite by the spring of 1972. The aim is to have the satellite in synchronous orbit in time to relay the broadcast of the 1972 Olympic Games from Munich, W.Ger., to Africa and the Americas. The space booster planned for the satellite is the Europa II launch vehicle, being developed under the direction of the European Launcher Development Organization. Total expenditure for the development by France and West Germany was estimated at $68.2 million.

The Soviet Union's approach to communications satellite systems remained quite different from that of the U.S. Beginning in 1965, the U.S.S.R. launched a series of Molniya 1 satellites in an orbit inclined about 65° to the Equator. The orbital path is not geostationary, but eccentric, moving closest to the earth in the Southern Hemisphere and reaching its highest altitude over the Northern Hemisphere. This path permits a Soviet satellite to travel for a relatively long time, about eight hours, over the portion of the earth visible to U.S.S.R. ground terminals. Because the Soviet craft are higher powered than corresponding U.S. satellites, relatively low-cost ground terminals can be used. Three such satellites, by proper spacing, can achieve 24-hour coverage within the Soviet Union. The Soviets also announced plans to launch a Molniya 2, operating at higher frequencies, and a Stationar 1 in a 24-hour synchronous orbit.

Earth-survey satellites. In this category are included meteorological, geodetic, earth resources, and reconnaissance satellites. Such satellites are designed to survey the earth with various sensors, both photographic and electronic, in order to obtain data that could not be gained by other means.

Weather satellites. In the U.S., developmental work on meteorological satellites is performed by NASA, while those spacecraft that are already operational are the responsibility of the National Environmental Satellite Service (NESS). NESS is a part of the National Oceanographic and Atmospheric Administration (NOAA) of the U.S. Department of Commerce. Formed in 1970, NOAA absorbed the activities of the Environmental Science Services Administration, which was abolished.

Meteorological, or weather, satellites are operational on a global basis. From a polar orbit, photographs of the earth's surface are taken and transmitted to earth receiver stations on command. Since the earth rotates constantly, total global coverage can be achieved in this way. Because weather stations cannot be built in many areas of the earth's surface, worldwide observation by satellites of weather and its movement is an important and significant contribution of knowledge. For example, tracking hurricane formation and movement has permitted advance warning of these storms, saving numberless lives and minimizing property damage.

As of April 1971, more than 500 Automatic Picture Transmission (APT) stations were in operation in approximately 50 nations. Each of these stations can receive facsimile printed transmissions of an area of about 1,700 sq mi. In the U.S., cloud pictures received from meteorological satellites by NOAA's Wallops Island, Va., station are processed for retransmission by geostationary Applications Technology Satellites (ATS) to local APT ground stations in Honolulu and San Francisco. Transmission of infrared nighttime imagery, which indicates ground, water, and cloud-top temperatures, was begun in November 1970.

Because the polar-orbiting meteorological satellites survey the same ground area only once every 12 hours, the geostationary ATS photographs are important for their ability to watch the development of severe storms continuously. Every 20 minutes, the ATS take and transmit photographs of the development and motion of severe storms. These are then monitored by the National Severe Storm Forecasting Center at Kansas City, Mo., and the National Hurricane Center at Miami, Fla. Some promising results have been obtained in attempts to correlate the history of cloud development with severe local storms and tornadoes in the central United States.

In late 1970, NOAA-1, the first operational "second generation" meteorological satellite, was launched. Twice the size and weight of the earlier TIROS satellites, it also included for the first time scanning radiometers to obtain direct readout and stored images of the earth's night side. Thus, global coverage at 12-hour, rather than 24-hour intervals, was provided.

Geodetic and earth-resources satellites. Geodetic satellites utilize the technique of simultaneous observation of a satellite from two or more points on the earth's surface. By means of this technique, the land masses of the world can be better measured and mapped. To this end, NASA in 1971 began conducting a year-long international cooperative program in association with the Soviet Union and several other nations.

Earth-resources satellites, as such, do not yet exist. However, a wide variety of techniques is becoming available that will eventually enable man to "see" and better appreciate the nature of the earth from the vantage point of space. Included among these techniques are systems built for the

"For I dipt into the future far as human eye could see, Saw the Vision of the world, and all the wonder that would be; Saw the heavens fill with commerce, argosies of magic sails, Pilots of the purple twilight, dropping down with costly bales; Heard the heavens fill with shouting, and there rain'd a ghastly dew From the nations' airy navies grappling in the central blue."
Alfred Tennyson

present meteorological and applied technology satellites; such new techniques as multispectral photography show promise of being able to detect narrow, selective bands of radiation in both optically visible and invisible regions. By mid-1971 the design of the first two earth-resources satellites, ERTS-A and -B, was complete, and construction was under way. Launching was scheduled for March 1972 and March 1973. Interested natural-resources scientists of all countries were invited to participate in analyzing and interpreting the data from these missions.

Expected to supplement such unmanned spacecraft in studying the earth's surface are manned orbiting laboratories. The first of three three-man, month-long flights in such a laboratory was scheduled for early 1973. During these flights astronauts will study, photograph, and in various other ways observe portions of the earth (see *Manned Space Exploration*, below).

Reconnaissance satellites. The U.S. and Soviet Union continued their reconnaissance satellite programs, but because such programs are highly classified for security purposes, no details about them have been released. It is known, however, that the spacecraft involved in these programs utilize extremely high-resolution photographic and narrow-band electromagnetic radiation surveillance techniques. In the last decade both countries launched more than 175 military observation, low-orbit, satellites; all payloads were recovered after a few days in orbit.

One additional type of U.S. military reconnaissance satellite deserves mention, the Vela. Its purpose is to detect nuclear explosions in the atmosphere and in space. Two Vela satellites were placed in orbit in 1970, the sixth pair in a series that began in 1963.

Navigation satellites. In the area of navigation satellites, the U.S. Navy Transit system continued to be fully operational and in use by the U.S. fleet throughout the world. Although commercial ships were permitted to use this system, it required a shipborne computer to calculate position from the Doppler shift of the satellite's signal and was thus probably too expensive for widespread public use. Future civilian systems were more likely to depend upon a signal transmitted from a ship via satellite to a shore-based computer, where position data could be determined and broadcast back.

The Soviet Union claimed to have an operational navigation satellite system, but released no details. NASA and the European Space Research Organization were holding joint studies concerning satellites that would be used for communications and for surveillance of aircraft flying in the North Atlantic air routes. Where today aircraft must be separated laterally by distances of about 120 mi, a satellite monitoring system could permit continuing "fixes" of aircraft position within a few miles. Studies and tests by the U.S. Federal Aviation Authority indicated that such a system could be in operation in the late 1970s.

—F. C. Durant III

Manned space exploration

During the past year, four manned space flights were undertaken, one United States lunar landing exploration and three Soviet Union earth orbital missions. Although the U.S. space budget was reduced by Congress, resulting in the cancellation of two planned Apollo lunar landing flights, plans were announced to start development of a new manned spacecraft called the shuttle. It would be used primarily as a transportation system to fly men and supplies to orbiting space stations, or to perform other short-term earth orbital missions.

Apollos 13 and 14. An extensive investigation was conducted to determine the failure that had caused the early termination of the Apollo 13 flight in April 1970. The investigation revealed that an electrical short circuit had occurred in one of the service module oxygen tanks. This caused wire insulation to burn and subsequently resulted in the explosion of the tank. As a result of this accident, significant design changes were made in the oxygen tanks in the remaining Apollo spacecraft. In

A comparison between Mercury-Redstone, the first U.S. manned spacecraft, and Saturn V-Apollo 15 reveals the advances made in space technology during the 10 years since 1961.

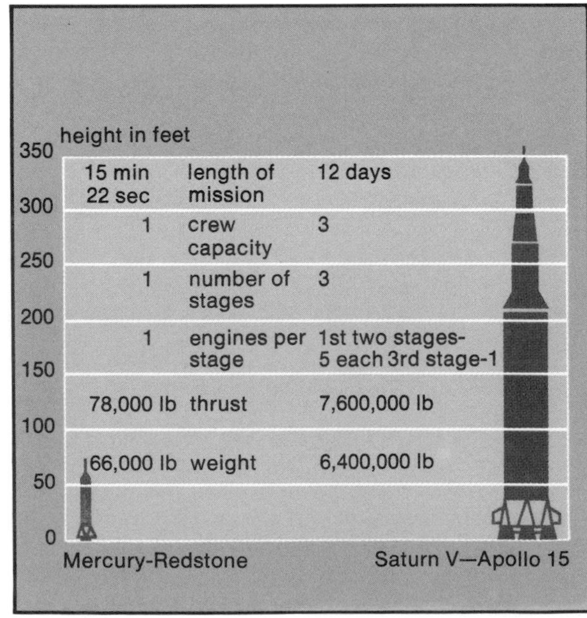

height in feet
350
300
250
200
150
100
50
0
15 min 22 sec
length of mission
12 days
1
crew capacity
3
1
number of stages
3
1
engines per stage
1st two stages-5 each 3rd stage-1
78,000 lb
thrust
7,600,000 lb
66,000 lb
weight
6,400,000 lb
Mercury-Redstone
Saturn V—Apollo 15

Adapted from "The New York Times"

addition, a third oxygen tank was installed in the Apollo 14 service module to provide a completely redundant supply in the event of a similar failure in future missions. The Apollo 13 incident caused a delay of approximately six months in the flight of Apollo 14.

Finally launched on Jan. 31, 1971, from Cape Kennedy, Fla., Apollo 14 was the third successful U.S. manned lunar landing flight. The mission was commanded by Alan B. Shepard, who in 1961 had become the first U.S. astronaut in space. Stuart A. Roosa piloted the command module and Edgar D. Mitchell the lunar module. The objectives of the mission were to explore a lunar highland area in an attempt to recover substances that may have formed part of the ancient lunar crust and subsurface material before it was scattered over the moon's surface by the impact of meteorites.

A Saturn V rocket placed the spacecraft and the third-stage Saturn IVB into earth orbit. After approximately 1½ orbits, the crew fired the Saturn IVB rocket to attain the velocity that would allow their spacecraft to escape from earth orbit and start the journey to the moon. The crew encountered the first problem of the mission while attempting to dock the command module with the lunar module. Roosa had difficulty in initially engaging the docking probe and latch mechanism to achieve a hard dock between the two vehicles. This was a significant problem because the docking mechanism would be required to function following the lunar landing phase of the mission and the return of the crew to the command module. After the dock was completed, the mechanism was inspected by the crewmen, and they determined that the mission could be continued.

Three days after the launch, the spacecraft was slowed to enter into a lunar orbital path. On Feb. 5, 1971, the lunar module was detached from the command ship and successfully landed in the predesignated area. While Roosa continued in lunar orbit in the command module, Shepard and Mitchell made two walks across the surface during their 33½-hour stay on the moon. On the first walk they set up the Apollo lunar surface experiments package (ALSEP), which contained five scientific experiments. These experiments were designed to permit scientists on the earth to study the moon's seismic activity, to detect particles that may be present in the lunar atmosphere, and to study the lunar subsurface structure. In addition, Shepard and Mitchell placed on the surface a passive experiment known as the laser ranging retro-reflector. With this experiment, scientists on the earth can study precise changes in the geometry between the moon and the earth.

On the two lunar walks the crew utilized a two-

Courtesy, General Electric Research and Development Center

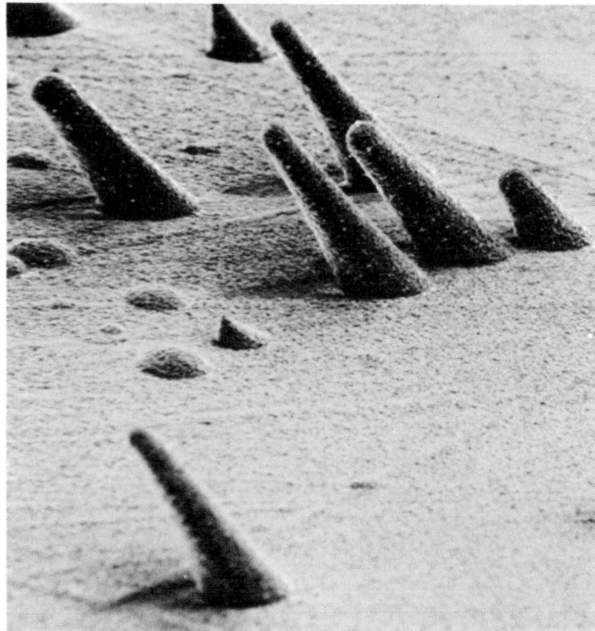

Silicone rubber replicas of etched tracks in a test helmet bombarded with cosmic rays are shown magnified 800 times. Astronauts on long missions will require additional shielding to protect them from these rays.

wheel transport vehicle called the MET. This unit was used to carry exploration equipment and scientific instruments as the crew moved about the lunar surface. The objectives of the second walk were to explore a highland area around the Cone Crater and to collect geological samples.

Shepard and Mitchell spent almost 9½ hours exploring the Fra Mauro landing site. They then used the ascent stage of the lunar module to rejoin Roosa. The crew made the return trip to the earth without a problem and landed on February 9 in the Pacific Ocean. The crew, lunar rock samples, and spacecraft were transported to the Manned Spacecraft Center in Houston, Tex., for a quarantine period until it had been determined that there was no biological hazard from the lunar material.

Apollo 15. On July 26 Apollo 15, the most ambitious lunar mission to date, lifted off from Cape Kennedy. Mission commander was David R. Scott, while Alfred M. Worden piloted the command module and James B. Irwin the lunar module. The launch and subsequent separation and docking of the command and lunar modules were performed smoothly, and, except for a faulty electrical switch that was later bypassed by manual operations, the journey to the moon was uneventful. On July 30 the lunar module, with Scott and Irwin aboard, separated from the command ship and landed on the moon at the foot of a mountain range and near the

Courtesy, NASA

On the moon's surface Alan Shepard, Jr., commander of the Apollo 14 mission, assembles hand tools for lunar exploration as he stands next to the modularized equipment transporter.

rim of Hadley Rille, a steep canyon. The landing was near the predesignated spot.

During their record 67-hour stay on the moon, Scott and Irwin took three excursions across the lunar surface. They were the first astronauts to do this in a vehicle, the 460-lb, four-wheeled lunar rover (see *J-series missions,* below). On their excursions, which covered a total of 17½ mi and lasted for 18 hours 37 minutes, the astronauts collected rock samples and photographed the canyon and mountain range. They discovered on a mountain slope crystalline rock that scientists believed might be a sample of the original lunar crust. Scott and Irwin also deployed near their landing craft ALSEP experiments similar to those of Apollo 14.

On August 2 Scott and Irwin, with 175 lb of lunar rocks, blasted off from the moon and rejoined Worden in lunar orbit. Before returning to the earth they deployed a 78½-lb "subsatellite" that was designed to remain in lunar orbit for a year collecting scientific data. During the return flight to the earth, which began on August 4, Worden left the command module for a walk in space similar to those taken by astronauts in the Gemini program. He did this to retrieve film from cameras that had been photographing the moon from the orbiting command module. On August 7 Apollo 15 splashed down safely in the Pacific Ocean.

J-series missions. The final three Apollo flights, 15, 16, and 17, were designated as J-series missions. Extensive changes were made in both the command module and lunar module to provide increased payload capability and extended lunar surface exploration time (from an original 36 hours to as long as 72 hours). These enhanced capabilities would permit extended exploration of the landing site, which was expected to result in an increase in the number and diversity of geological samples gathered and in the number of unusual lunar features investigated.

The command and service module was modified so that it could remain longer in lunar orbit and also could carry scientific instruments. The major changes were the addition of oxygen and hydrogen tanks to increase life-support-system and power-generation capabilities and the modification of one bay of the service module to carry scientific instruments. This bay had a door that could be jettisoned and also a data-acquisition system.

The lunar module was modified for a 72-hour stay on the moon's surface, by increasing the consumable supplies: water, oxygen, electrical power, and propellants. To increase astronaut comfort, the interior of the module was equipped with foldable seats and provisions for sleeping. In the module's descent stage, a stowage area was provided for the lunar roving vehicle (LRV). The space-suit life-support system was also modified to increase the operating life from 4½ to 7½ hours. This was accomplished by increasing the consumables—electrical power, water, oxygen, and lithium hydroxide (used for carbon dioxide removal).

A new extravehicular life-support system was developed to permit the command module pilot to walk in space and retrieve photographic film magazines from the service module science bay. Oxygen would be supplied to the pilot from spacecraft supplies through an oxygen/communication umbilical. In operation, a pressure-control unit would modulate the flow of oxygen to maintain an adequate supply to cool the crewman and maintain suit-operating pressures.

The four-wheeled jeep-sized LRV carried in the lunar module descent stage for deployment on the moon is an electrically powered vehicle capable of carrying two crewmen, communications equipment, and lunar-surface scientific and operational equipment. It weighs approximately 500 lb and is $10\frac{1}{2}$ ft long with a $7\frac{1}{2}$-ft wheelbase. The wheels of the vehicle, which has a top speed of 10 mph, were designed for compatibility with the lunar soil and rough terrain. The vehicle has a total payload weight of 970 lb, which includes the two astronauts, their life-support systems, and a scientific payload of 170 lb. The LRV was designed to have an operational life of 54 hours and to operate out to a radius of about 1.8 mi from the lunar module; this represents a traverse of approximately 45 mi. The LRV was to be equipped with a dead-reckoning navigational system that could determine the distance and direction between the vehicle and lunar module at any point in the traverse.

A device called the lunar-communications relay unit was being developed to provide communications during operation of the LRV. Stowed in the lunar module descent stage at earth launch and then placed on the LRV for surface operations, the relay unit was to be equipped with a self-contained battery power supply. Color television pictures would be transmitted from the LRV after the vehicle had been stopped. The television camera was to be mounted on a motor-driven gimbal system that could be directed from the earth to points of interest or to track the crew during the exploration periods. At the conclusion of the lunar mission, the ground-command television system might be used to provide pictures of the lunar module ascent.

Apollo J-mission science program. The J-mission science program was expanded to maximize the scientific-data return of the final Apollo missions. In the science bay of their service modules Apollos 15 and 16 would each carry nine of these experiments, while Apollo 17 was to have three additional ones. The data from these experiments would be processed through a scientific data system to the spacecraft telemetry system for transmission to ground stations. The orbital science experiments were planned to take place simultaneously with the 66-hour lunar-surface operations.

Among the instruments to be used in the experiment were gamma-ray, X-ray, and alpha-particle spectrometers. They had a common goal: to determine the types and distribution of selected elements on the lunar surface. The feasibility of conducting chemical surveys of the moon had been demonstrated by the Soviet Union in 1966, when a gamma-ray spectrometer was flown on Luna 9. By combining these three experiments on a single spacecraft, it was expected that the presence on the lunar surface of selected chemicals, such as uranium, thorium, and potassium, would be defined more clearly. An understanding of the chemical nature of the surface was considered a required step in unraveling the chemical processes that were active in the origin of the moon.

Closely tied to these spectrometers is the infrared scanning radiometer, which measures differences in lunar-surface thermal properties. From the results of this experiment, scientists believed that they could construct a thermal map of the moon that would provide clues as to the location of such features as volcanoes, areas of recent igneous activity, or the concentration of radioactive substances.

The moon does not have an atmosphere; however, the presence of trace quantities of gases may provide information about the origin of the moon. A mass spectrometer and a far-ultraviolet spectrometer were to be used to determine the distribution of the noble gases (helium, neon, argon, krypton, and xenon), carbon monoxide, carbon dioxide, sulfur monoxide, and water in the lunar atmosphere and to identify their sources, such as active volcanoes or areas emitting gas. The far-ultraviolet spectrometer was also to be used to study the distribution and movement of gases released during manned surface operations.

The remaining experiments would be directed toward a closer look at the layering and properties of broad areas of the moon's surface and subsurface up to more than a mile in depth. A biostatic radar experiment would provide data on the electrical properties of the lunar surface material. This information would complement that of a lunar-sounder experiment, which would be used to study subsurface features through the measurement of reflected radio-frequency signals.

An analysis of flight data from the Lunar Orbiter spacecraft resulted in a map of the near side of the moon that indicated the presence of gravitational anomalies and areas of the moon where there are large concentrations of dense matter lying below the surface. In the Apollo J-series missions, one transmitter-receiver facility operating at a wavelength of about four inches would be in the command module and another on a subsatellite to be launched from the service module into an orbit

60 naut mi above the lunar surface. By Doppler tracking of the command module and the subsatellite, minor changes in the orbital paths would be detected. From these changes scientists believed that the subsurface distribution of dense matter and associated gravitational anomalies could be determined and mapped. The subsatellite also would carry two additional experiments for the study of particles and gravitational fields around the moon.

Two new cameras and a laser altimeter would be carried in the service module as facility equipment. The high-resolution photographs that the cameras are designed to produce would not only be of direct aid to scientists in their study of the forms, distribution, and structures of surface features but would also enable the making of detailed maps of the lunar surface for future planning of its exploration. The laser altimeter would provide vertical control data with a resolution of about 6 ft at 60 naut mi for use in the study of surface features and in the preparation of the maps. A 24-in. panoramic camera was to have a lunar-surface resolution of approximately 9 ft. It was estimated that, on just one mission, approximately 10% of the lunar surface would be photographed.

Surface scientific experiments. The Apollo 15 and 16 missions were scheduled to carry ALSEP experiments that had been flown on previous missions. The Apollo 17 ALSEP, however, would contain five new surface experiments and would obtain and transmit information for up to two years. The mass-spectrometer experiment designed to provide information on the lunar atmosphere would study its dynamics, its changes in atmosphere from lunar day to lunar night, and its source.

A lunar ejecta and micrometeoroid experiment would provide data on the mass, speed, direction, and flux of cosmic rays that impact the lunar surface. It would also measure the flux and density of meteoroids that strike the lunar surface and measure the matter ejected from the surface as a result of its continual bombardment. The scientific significance of this experiment will be an understanding of the changes in the lunar surface caused by particle impact. With this knowledge, lunar scientists hope to be able to derive an understanding of the lunar-surface features and their evolution.

In a seismic profiling experiment four geophones (instruments designed to detect vibrations passing through rock) would be emplaced on the lunar surface, and the crew, using the LRV, would deploy eight explosive charges weighing as much as 8 lb apiece within a radius of about 1.8 mi. Each explosive would be manually armed to explode automatically at a prescribed time. This experiment was expected to provide data on the lunar subsurface to a depth of approximately 3,000 ft.

The objective of a heat-flow experiment would be to measure the net outward heat flux from the lunar interior. The equipment, consisting of two sensor probes inserted into holes drilled into the moon, would be connected to the ALSEP central station to transmit data and to receive commands. The data gathered by the sensors were expected to provide a daily thermal history of the moon and a profile of temperature as a function of depth below the lunar surface. Additionally, the sensors would be equipped with heaters for generating a known amount of heat to study thermal conductivity.

A tidal gravimeter was to be developed for emplacement on the lunar surface during one of the later Apollo missions. This complex instrument would be used to measure lunar tides and the free modes of vibration of the moon. A second purpose of the experiment would be the use of the moon as an antenna for propagating gravitational waves.

Three new traverse experiments were to be carried on Apollo 17. These experiments, each self-contained, would be deployed by the astronauts during their geological traverse. A traverse gravimeter would be mounted on the LRV to map the gravity characteristics of the area explored. At periodic LRV stops, the gravimeter would be actuated and would automatically measure and record the local gravity values to an accuracy of three milligals. The information gained from this experiment was expected to provide a more accurate determination of surface-gravity anomalies and of variations in gravity as a function of elevation of the lunar surface. This information would be of use in understanding the characteristics and origin of subsurface structural features.

In the surface electrical-properties experiment, a stationary radio-frequency transmitter and mobile receiver would be used to study subsurface structure. The transmitter would generate signals into the lunar surface at eight different frequencies. As the LRV moved about the surface, a receiver mounted on it would automatically record the time of the signals. Scientists believed that this would generate data that would help them determine layering in the lunar subsurface; this information, in turn, would contribute to the knowledge of the history of the moon. In addition, the presence of water in the lunar interior could be determined with this experiment. The presence of water on the moon is of great scientific significance because of its life-supporting qualities and because it would aid in an understanding of lunar-surface erosion.

Skylab. Development of the Skylab spacecraft and orbital workshop continued during the year.

Aviation Week and Space Technology

The Soviet manned spacecraft Soyuz 10 moves by rail (left) on its transporter-erector to the launch pad. Ready for launch (below), the white-shrouded spacecraft sits atop its booster rocket. The black rectangular sections hinged to the shroud pull it away from the spacecraft near the end of powered flight after leaving the earth's atmosphere. Launched in April, Soyuz 10 was the first Soviet spacecraft to dock with Salyut, the unmanned orbiting laboratory, and was similar to Soyuz 11, in which three Soviet cosmonauts died.

The third-stage Saturn IVB structure was being modified to serve as an orbital workshop. It was being equipped with two floors and the necessary support systems to serve as a science laboratory and as living quarters for a three-man crew. A Saturn V launch vehicle was to place it into near-earth orbit. Crewmen would then be transported to the workshop by an Apollo command module.

Three manned Skylab missions were scheduled, one of 28 days and two of 56 days. One of the prime objectives of the missions was the physiological study of long-term manned flight. Medical experiments planned for the Skylab program were designed to provide information on how the human body is affected by relatively long-term exposure to a weightless environment. Additional experiments included earth resources studies and the use of a solar telescope to study the sun. The first mission was scheduled for the spring of 1973.

Shuttle spacecraft. During the year the U.S. National Aeronautics and Space Administration established feasibility study contracts with several industrial teams concerning a shuttle spacecraft. These studies produced alternate designs for a reusable launch vehicle and orbiting spacecraft. This approach to spacecraft design was expected to reduce greatly the costs of manned spacecraft operations. Crews would man both the launch vehicle and the orbiting spacecraft at launch. After separation, the booster vehicle would be flown by the crew and landed like a conventional aircraft at a designated area; it would then be reserviced for another launch.

The orbiting shuttle vehicle would be flown by a

two-man flight crew and could carry up to 12 passengers. In addition, the spacecraft was being designed to carry large scientific payloads into earth orbit. The shuttle spacecraft would thus provide a low-cost transportation system to carry out earth-orbital space operations.

Soviet manned space flight. During 1970–71 the U.S.S.R. completed three manned space flights. Soyuz 9 was launched into earth orbit on June 1, 1970. Its two-man crew was commanded by Andrian Nikolayev, who had previously flown on Vostok 3; the second cosmonaut, Vitaly Sevastyanov, served as the flight engineer. The objectives for this mission were long-term medical and biological studies of the effects of weightlessness on the human body, scientific observations and photographs of the earth, and the study of the crew's ability to fly the spacecraft manually into changing orbital paths and to maintain spacecraft attitude.

The crew remained in orbit for 17 days, 16 hours, and 59 minutes to establish a record for the longest manned space flight. The flight broke the mark established by U.S. astronauts Frank Borman and James A. Lovell, Jr. Soviet authorities reported that the crew did encounter physiological changes as a result of the long flight. The crewmen stated that changes in their cardiovascular system had caused them to feel light-headed for the first 10 days of the mission and also required 4 to 5 days for them to readapt after landing.

On April 19, 1971, the U.S.S.R. launched an unmanned laboratory called Salyut into earth orbit. Four days later, Soyuz 10 was launched into orbit with a three-man crew. The commander was Vladimir Shatalov, a veteran of two previous flights. The flight engineer was Alexei Yeliseyev, who had flown on Soyuz 8, and the third crewman was Nikolai Rukavishnikov. The Soyuz 10 spacecraft docked with the Salyut laboratory on April 24, 1971, and the two remained coupled for 5½ hours. After two days in orbit, Soyuz 10 landed on April 25.

Soyuz 11, commanded by Georgi Dobrovolsky with Vladislav Volkov and Viktor Patsayev as crew (*see* OBITUARIES), was launched on June 6. The next day it docked with Salyut, and the cosmonauts entered the laboratory for an extended stay. Inside Salyut the cosmonauts exercised to test the effects of weightlessness and grew a "space garden" of various plants. After spending more than three weeks in the laboratory—a record time in space—they reentered the spacecraft and returned to the earth. Although the return flight seemed to be normal, the three cosmonauts were found dead in their seats. After an official inquiry the U.S.S.R. announced that a faulty seal in the spacecraft allowed a sudden drop in pressure, which caused the three to develop embolism and die.

International cooperation in space research. The U.S. and the U.S.S.R. conducted several meetings during recent years to discuss ways in which the two countries could formulate a cooperative space program. Discussions were begun on trying to develop a common docking mechanism that would allow spacecraft from the two countries to perform rescue missions for one another.

—Richard S. Johnston

See also Feature Article: COLONIZING THE MOON.

Space probes

Activity in space probes increased statistically and significantly during the year. In December 1970, scientists on earth for the first time were receiving data being telemetered from two bodies in the solar system, the moon, and Venus. It was a year of triumph for Soviet space scientists and one of frustration for their U.S. counterparts. The U.S.S.R. successfully launched three probes to the moon, two to Mars, and one to Venus. One attempt to launch a Martian probe ended in failure. Such activity seemed to indicate that unmanned probes to the nearer planets would continue to play a leading role in that nation's exploration of space in the future.

Boris Petrov, chairman of the Council on International Collaboration of Space of the Academy of Sciences of the U.S.S.R., revealed the new five-year plan for space (1971–75) in February 1971. It cited three areas of concentration: applications satellites (communications, meteorological, and navigational); earth resources satellites (manned and unmanned); and "a continuation of basic scientific investigations of the moon and the planets of the solar system."

The U.S., by contrast, lost its Orbiting Astronomical Observatory B during an attempted launch on Nov. 30, 1970, when its aerodynamic shroud failed to eject. A further setback occurred in the following May when Mariner 8 was lost on an aborted launching. Later in May, however, Mariner 9 was successfully launched on a journey to Mars.

Missions to Venus. On Aug. 17, 1970, at Tyuratam, the Soviets launched Venera 7 toward the planet Venus. In designing the spacecraft, engineers drew upon the experience of earlier probes, Venera 4 in 1967 and Veneras 5 and 6 in 1969, which reached the planet but lost their instrumented landing stages when they were crushed by the high pressure of the Venusian atmosphere. The instrumented lander of Venera 7 was built to withstand pressures up to 180 atm (1 atm [atmosphere] = 14.7 lb/sq in. at sea level) and to measure temperatures as high as 530° C. The necessity for providing a more massive structure for the landing pod accounted for an increase in weight of 100 kg over the previous Venera probes.

Two mid-course corrections were made by the probe's descent-stage engine on October 2 and November 17. The landing pod detached from the main stage on December 15 and began to sink rapidly through the Venusian atmosphere. During the descent, the velocity decreased from an initial 11.5 km/sec to 200 m/sec, while deceleration loads up to 350 *G*s were encountered. (1 *G* equals the force exerted by gravity on a body at rest.) The temperature within the stagnation region preceding the body reached 11,000° C. At a point about 60 km above the surface, where the pressure reached a value of 0.7 atm, a parachute system automatically deployed. The pod required 36 minutes and 32 seconds to reach the planet's surface.

Telemetry aboard the pod transmitted data for 23 minutes after landing. During that period, the temperature reported remained stable at 475° C (±20°). The pressure, computed from a mathematical model after the pressure sensor failed, was 90 atm (±15 atm). The density of the gases near the surface was approximately 60 times greater than that of earth.

Like its predecessors, Venera 7 consisted of two stages: the descent stage and the instrumented landing pod. The descent stage carried the landing pod and provided trajectory control. It had a sensor to measure cosmic radiation en route and telemeter data to the earth. Other sensors monitored pressures, temperatures, and voltages of components in both stages. The descent stage also had an orientation control system consisting of sun and star sensors and a cold-gas attitude-control system. An on-board programmer initiated separation of the landing pod and controlled other events.

Tass from Sovfoto

Lunokhod 1, the Soviet unmanned lunar explorer, was landed on the moon by the Luna 17 spacecraft on Nov. 17, 1970. The eight-wheeled vehicle, controlled from the earth and powered by solar energy, moved across the moon's surface taking pictures and analyzing samples of the soil. A U.S. reaction is shown below.

John Fischetti—"Chicago Daily News"

"Ours *will have sixteen wheels, lotsa chrome and play 'Yankee Doodle' in stereophonic sound."*

Adapted from "The Christian Science Monitor"

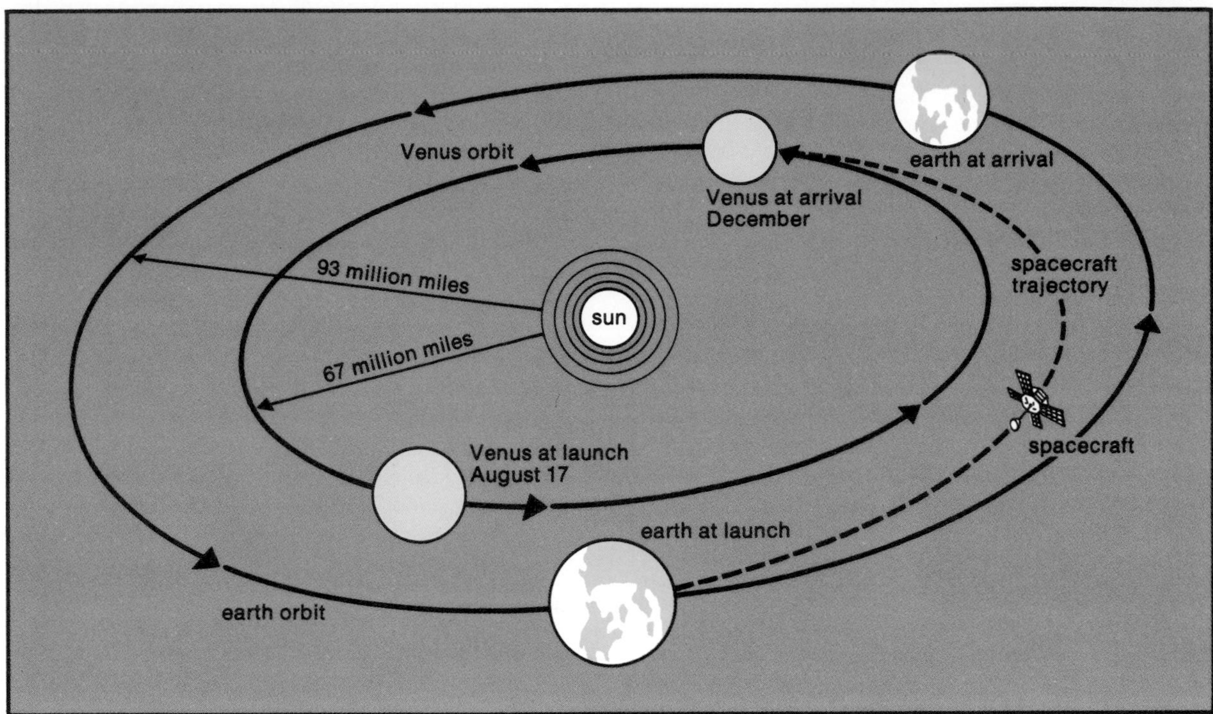

The Soviet unmanned space probe Venera (Venus) 7 traveled from the earth to its soft landing on Venus via the trajectory illustrated above. Because of its smaller orbit Venus had passed the earth at the time the spacecraft landed there, even though it had been behind at the time of the launch.

Electric power for radio telemetry and other operations was provided by chemical batteries that were recharged by solar cells in two panels kept facing the sun. This stage also had a high-gain, parabolic antenna and two small, low-gain antennae for the radio telemetry transmitters.

The landing pod was unlike its predecessors in that it had undergone considerable redesign while retaining the spherical shape with an offset center of mass that is characteristic of Soviet spacecraft designed for atmospheric entry. The container structure was strengthened considerably, and a new heat shield was developed for the external covering. In addition, instrumentation within the pod was provided with special mounting devices to damp the higher deceleration loads that were encountered. A more efficient temperature control system for the interior of the pod was also provided.

Probing the moon. The Soviets scored two technological triumphs with lunar probes that demonstrated a highly advanced state of space technology. In the field of unmanned, remotely operated vehicles they clearly equalled and surpassed the U.S.

The first Soviet probe to the moon during the year was Luna 16. Launched by a Proton booster from Tyuratam on Sept. 12, 1970, the two-stage probe achieved a near-circular parking orbit at 212.3 km. Only 70 minutes after launch the signal to fire the last stage engine was given, and the probe left orbit for a translunar insertion trajectory. Subsequently, one of two planned mid-course corrections was made by the probe's descent engine. On September 17, the craft went into a circular orbit 110 km above the moon. The orbit was reshaped twice by the descent-stage engine to produce a perigee of only 15 km. On September 20, the landing stage engine, and two smaller soft-landing engines broke the probe out of orbit and onto the Sea of Fertility. The coordinates of its landing site were 0°41′ S and 56°18′ E, about 100 km W of Webb Crater.

After a brief checkout of on-board systems, operations were initiated upon command of signals from the coordinating and computing center, near Moscow. A hollow-core drill was then deployed and driven into the lunar soil, being observed by a telephotometer. The drill reached a depth of 35 cm and extracted 101 g of lunar soil. The sample was then placed into the reentry body atop the probe and sealed. While the drilling was in operation, sensors on the craft telemetered to the earth radiation levels and temperatures of various components of the probe and the radiation level in its vicinity.

The upper stage lifted off the moon's surface on September 21, and no mid-course correction was necessary during the trajectory through near-lunar space. On September 24, the reentry package, in its spherical container, separated on command from the coordinating and computing center, and reentered the earth's atmosphere. Parachute deployment was triggered automatically by an on-board system. The reentry package landed about 80 km SW of Dzhezkazgan, and a homing beacon guided the recovery team to it in helicopters.

In May it was announced that 3 g of the lunar soil returned by Luna 16 would be given to the U.S. in exchange for samples returned by the astronauts of Apollos 11 and 12. Thus, scientists in each country would have the opportunity of analyzing material not otherwise available to them.

Zond 8 was launched on Oct. 20, 1970. Few details concerning the mission were released, and it was generally supposed that the probe was another instrumented Soyuz spacecraft such as Zond 7, launched by the Proton booster. The probe passed around the moon in a "free-return" trajectory on October 24. It took photographs of the lunar surface at the point of closest approach (1,200 km) and of the earth from varying distances, but, more significantly for later manned landings, Zond 8 was "to improve the reliability of on-board systems, units, and the design of the spacecraft." It reentered the earth's atmosphere on October 27 and landed in the Indian Ocean approximately 730 km SE of the Chagos Archipelago.

The Soviet Union produced an engineering tour de force in Lunokhod 1. On Nov. 11, 1970, the U.S.S.R. launched Luna 17 from Tyuratam with a Proton booster. After leaving its earth orbit, the probe underwent mid-course corrections of its lunar trajectory on November 12 and 14. On November 17, it landed in the Sea of Rains on the inner slope of a crater some 150 m in diameter.

The probe consisted of two stages: the descent stage, which was a modified model of that used with Luna 16, and the Lunokhod 1. The latter was a roving vehicle that weighed 756 kg and had a length of 2.218 m, and a width of 2.150 m. It had eight, independently suspended, wire-mesh wheels 5.1 cm in diameter, each driven by its own electric motor. It had two speeds forward and one in reverse. A hermetically sealed chamber contained the communications system, environmental control system, and power supplies (chemical batteries recharged by solar cells). Lunokhod 1 was remotely steered from the earth by a crew of three men in the Soviet Deep Space Communications Center, near Kerch, in the Crimea.

Lunokhod 1 was designed to be mobile during the lunar day and stationary at night. (The lunar day in this context is equivalent to about 14 earth days.) During the days the vehicle traveled about the lunar surface, crossing craters with slopes as great as 20°. It also crossed large rocks and extremely uneven terrain. At locations that looked interesting to scientists at the control center, the vehicle would be stopped and a variety of experiments would be made. During the nights, the Lunokhod 1 was parked, its solar cell panels folded up, and its instruments shut down. However, data on its internal environment continued to be telemetered to the earth. Temperatures inside the vehicle ranged between 15° C and 21° C, and pressures varied between 745 Torr and 780 Torr (1 Torr equals the pressure of 1 mm of mercury at 0° C and standard gravity).

Scientific instrumentation aboard the Lunokhod 1 included RIFMA (an acronym for Röntgen Isotopic Fluorescent Method of Analysis), for chemical analysis of soil; a penetrometer, for testing mechanical properties of soil; an X-ray telescope; a cosmic-ray radiometer; four television cameras,

Model of the Pioneer-Jupiter spacecraft awaits thermal testing. Two Pioneers were scheduled to be launched to Jupiter in 1972 and 1973, to pass within 100,000 mi of the planet. The nine-foot-diameter antenna will be pointed toward the earth during the mission.

Courtesy, NASA

for guidance and panoramic views; and a laser reflector supplied by the French Centre National d'Etudes Spatiales, in Paris. Determinations of the distance from the earth to the moon as well as the location of the Lunokhod 1 vehicle were made by laser beams projected to the reflector from the Crimean Astro-Physical Observatory near Yalta and the Pic du Midi Observatory in the Pyrenees Mountains.

A major scientific feat was accomplished by the radiometer on December 13. On that day, it recorded the results of a gigantic solar flare. At the same time, the same flare was detected and measured by Venera 7, approximately 300,000,000 km distant and on its way to Venus, and by Intercosmos 4, a scientific satellite in earth orbit. Thus, the energy and composition of the flare in space and time could be studied.

The Soviet probe continued to amaze scientists with its apparently unlimited lifetime in the lunar environment. The hardy vehicle survived eight lunar days and nights. In July 1971 it still was transmitting television pictures and telemetering data on its internal environment.

Return to Mars. U.S. hopes for a scientific coup in the unmanned exploration of Mars received a setback on May 8, 1971, when Mariner 8 was lost during an attempted launch from Cape Kennedy, Fla., on an Atlas Centaur vehicle. The Soviets also lost a Mars probe when it refused to leave earth orbit after having been launched from Tyuratam on May 10. It was designated Cosmos 419 and decayed from orbit on May 12. More successful were two additional Soviet probes launched by the Proton booster, Mars 2 on May 19 and Mars 3 on May 28. Mariner 9, the second of two U.S. probes that were to have orbited Mars, was launched successfully on May 30. All three probes were scheduled to arrive at Mars in late November or early December 1971.

Mars 2 was instrumented to detect and measure characteristics of the solar wind, cosmic radiation levels, and other radiation fluxes in the space between the earth and Mars. Three of the experiments were supplied by French scientists. Data from experiments were transmitted on a frequency of 928.4 MHz. The weight of the probe was 4,650 kg, a significant increase of 2,680 kg over that of Mars 1, launched in 1962. Speculation as to the meaning of this increase ranged from additional fuel to a small, remote-controlled, roving vehicle similar to the Lunokhod 1. A more likely explanation, however, was that the probe contained a soft-landing pod similar to that of Venera 7. Mars 3 was essentially the same as Mars 2, serving as a back-up craft in case of a catastrophe such as occurred with Mariner 8.

The 1,000-kg Mariner 9 was an elaborately instrumented U.S. probe, designed to take between 5,000 and 6,000 pictures of Mars over a three-month period; it was expected, however, that its lifetime would exceed that period by as much as nine months. On June 4 a mid-course correction was made to insure that the Centaur stage would not impact on Mars and possibly contaminate it.

The Mars orbit for Mariner 9 was originally to have had a perigee of 800 km and an apogee of 28,000 km. Had it followed this orbit, it would have been able to photograph variable features of the planet, such as the shifting polar caps and clouds, every five days. Thus, it could have studied the seasonal changes in certain areas that had been observed from the earth.

Following the abort of Mariner 8, however, the orbit for Mariner 9 was changed to permit it to accomplish most of the missions that would have been realized if both probes had been in orbit. The new orbit for Mariner 9 was inclined at 65° to the Martian equator (instead of a planned 50°) and had an apogee of 28,320 km and a perigee of 1,200 km. Thus, it should be able to photograph about 70% of the Martian surface. The new orbit also resulted in the probe's passing over variable features every 17 days instead of the originally planned 5.

Instrumentation in Mariner 9 was designed to examine the Martian surface by photography and by infrared measurement. The planet's atmosphere would be probed in the infrared and ultraviolet wavelengths as well as by the refraction of radio waves. The instruments were designed to become functional as the early winter appears in the northern hemisphere of Mars and early summer in the southern. Mariner 9 was scheduled to arrive at the planet at the peak of the "wave of darkening" in the southern hemisphere. This phenomenon thus would be observed at the peak of its intensity and could be compared with similar observations made in the northern half of the planet.

A wide-angle camera on board Mariner 9 could resolve Martian surface features as small as about 1,000 m in length, while a narrow-angle camera would be able to resolve features only 100 m long. The wide-angle camera was equipped with eight color filters that could be selected by command from the control center. In addition to photographing the Martian surface, the narrow-angle camera was designed to begin photographing sequentially the disk of the planet beginning at a distance of 1,609,000 km.

Another IMP. On March 13, 1971, the U.S. launched IMP (Interplanetary Monitoring Platform) 8 (Explorer 43). The 285.75-kg satellite was placed into a highly elliptical orbit with a perigee of 1,243 km and an apogee of 203,908 km. It contained 13 experiments, 12 scientific and 1 engineering (an on-board computer for processing some of the data from the other experiments). The scientific experiments were supplied by U.S. universities, the National Aeronautics and Space Administration, and the Atomic Energy Commission.

Six experiments were to measure energetic particles, especially solar and galactic cosmic rays. Two were designed to study solar plasma, while three others were to examine electric and magnetic fields in space. A radio astronomy experiment was designed to supplement data received from Explorer 38, launched in 1968, which measured very-low-frequency radio emissions from the sun, Jupiter, and other sources within the galaxy. One of the electric field experiments provided by the Goddard Space Flight Center, Greenbelt, Md., required four antennae, each 45 m long.

—Mitchell R. Sharpe

Astronomy

The universe in which we live is so large and contains such strange objects that progress in astronomy is marked by a series of disconnected new observations, insights, and explanations. Each explanation is followed by a new surprise that suddenly reverses previous priorities and interests. The year 1970–71 was marked by rapid growth of knowledge concerning objects which were not even known to exist five years ago (pulsars, infrared stars and nebulae, complex interstellar molecules). Few of the classical problems were solved; rather, they were temporarily bypassed as observational data established the existence of new types of objects for which not even the roughest model or explanation existed. Part of this new information came from successful experiments with satellites and rockets, above the earth's atmosphere, which opened new wavelength ranges for detailed observation. But a larger part continued to come from the increasingly sophisticated electronic technology that had so greatly increased the power of optical and radio telescopes.

The "cold universe," that of normal astronomical objects, contains stars that radiate most of their energy at optical wavelengths from 1,000 to 10,000 angstroms (one angstrom or Å is equal to 0.00000001 of a centimeter). In contrast, objects of the so-called "hot," or "high-energy," portion of the universe emit radio photons near 10 cm in wavelength, or infrared photons (0.1 mm), or X-ray photons (10 Å). They also produce high-energy particles, electrons or protons moving so near the speed of light that their mass increases enormously until they have energies of hundreds of millions of electron volts (an electron volt is the quantity of

A "black hole" in space is believed to be the remnant of a star that has collapsed into so small and dense an object that light cannot escape from it. Below left is an artist's rendition; at the right are satellite-determined locations (1967 and 1971) of a hole in the constellation Cygnus.

Helmut K. Wimmer, Hayden Planetarium

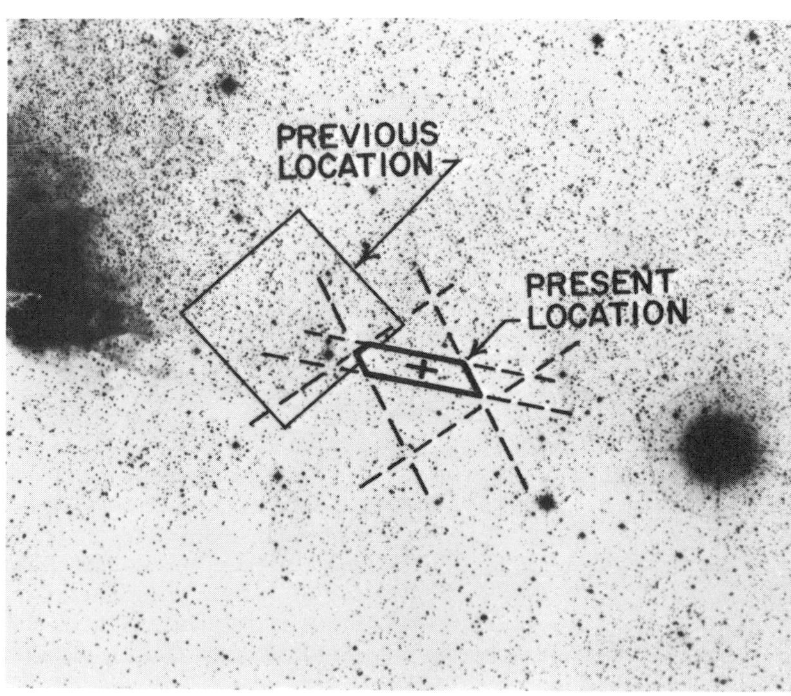

Courtesy, American Science and Engineering, Inc.

Courtesy, University of California—Berkeley

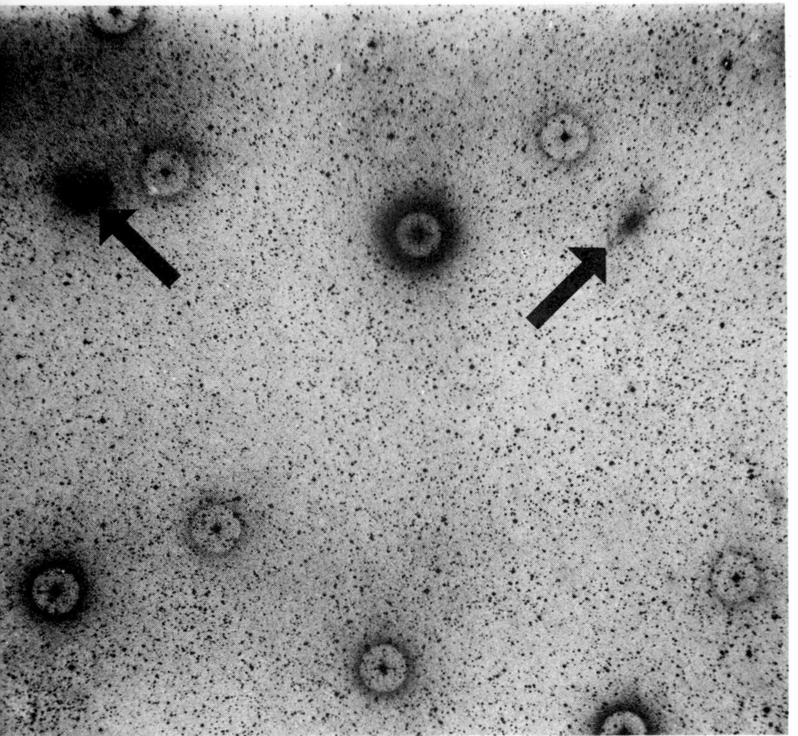

Maffei 1 and 2, two giant nearby galaxies, were not detected until 1967 because of obscuring interstellar dust. Discovered by Italian astronomer Paolo Maffei, they were identified as galaxies in 1971.

energy an electron gains in moving from one electric field to another whose potential is greater by one volt). The processes in the cold universe are mostly understood, although genuine new surprises occurred in recent months; the hot universe, however, grows ever stranger. In this account the planets, the sun, and normal stars are omitted almost completely because of the rapid pace of discovery of peculiar objects under extreme physical conditions.

Manned and unmanned space exploration continued to produce results of interest to astronomers. Rocks returned by the first successful Apollo mission to the lunar highlands revealed surprising further information about the great age of lunar surface samples. Scientists at the Jet Propulsion Laboratory in Pasadena, Calif., showed that ultraviolet radiation and the thin atmosphere on Mars could synthesize formaldehyde, acetaldehyde, and glycolic acid, possible building steps for life.

The rapid growth of major astronomical instruments in Europe was striking; the British were building an eight-km-long radio interferometer (an instrument used to measure wavelengths), the West Germans a 100-m dish for a radio telescope, and the Dutch a giant interferometer array. A 381-cm telescope for the European Southern Observatory was under construction in Chile with engineering help from the European Organization for Nuclear Research (CERN); an Anglo-Australian 381-cm telescope was being built in Australia, as was a new 254-cm instrument in Chile for the Carnegie Institution of Washington. Meanwhile, a study of the main problems of astronomy in the U.S.—its financial support and future—was underway with Jesse Greenstein as chairman. It was hoped that major forward steps in astronomical instrumentation would result from this planning effort.

Interstellar matter

About 1% of the matter in the Milky Way galaxy exists as a low-density gas composed largely of hydrogen and helium, either not yet aggregated into stars or else blown out of stars at the end of their evolutionary life. Observations with spectrographs established the existence in this gas of sodium, ionized calcium, titanium, and a few simple unstable molecules, such as CN and CH. Radio telescopes detected neutral and ionized hydrogen, helium, and carbon as well as hydroxyl (OH), water (H_2O), and ammonia (NH_3).

In the last few years, however, there was an unexpectedly rapid increase in the number of molecules found in interstellar gas and in their complexity. But the transitions from the lowest energy states of a molecule may not be at wavelengths convenient for observation. Moreover, molecules can absorb (or emit) energy by a variety of changes of configurations, such as rotation, vibration, and stretching or bending. To measure the wavelengths of the simplest of these energy transitions, chemists in the laboratory use a microwave spectrometer. Astronomical observation, on the other hand, not only requires accurate measurement of the wavelengths of the energy transitions of many molecules, but also a determination of the shift in wavelength caused by an isotope substitution. For example, CO (carbon monoxide) was found in gas clouds and proved very strong at radio frequencies. All possible isotopic compounds ($^{12}C^{16}O$, $^{13}C^{16}O$, $^{12}C^{18}O$, $^{13}C^{18}O$) therefore appear in the measurements and permit isotope ratio determinations for the first time for $^{16}O/^{18}O$. Although stars had indicated $^{12}C/^{13}C$ ratios which sometimes differed considerably from the terrestrial value, all interstellar clouds showed the terrestrial value. This was quite unexpected for the theory of the origin of the chemical elements in stars.

Abundance of molecules. Gas in space is normally in the atomic form. The gas density in space is very low and, at any temperature typical of gas clouds (usually either 100° or 10,000° K), the rates of chemical reactions are slow. Given enough time, however, stable compounds should be formed and any compound should be present in proportion to the abundances of its constituents. With the usual density of one hydrogen atom per cc, the concentration of CO could be as low as 0.0000001 the concentration of hydrogen, and could not be higher than 0.001. The expectation was that molecules should be rare, near the low limit. Optically detected CH and CN were both near 0.000001 the concentration of hydrogen.

But the study of certain peculiar high-density gas clouds, such as the Orion Nebula, M17, W3, W51, and the sources near the galactic center (Sagittarius B2 and Sagittarius A), showed that molecules are much more common than expected. The molecules are compounds of C, N, O, and H, and apparently all of the C, N, and O that possibly can be is bound in some molecule; some produce emission, others, absorption lines. However, the density of these molecules in space is very low, about one CO molecule per cc in the quite dense cloud W51, according to A. A. Penzias, K. B. Jefferts, and R. W. Wilson of the Bell Telephone Laboratories. B. E. Turner of the National Radio Astronomy Observatory found cyanoacetylene (HC_3N) at a density of about 10^{-2} per cc in Sagittarius B2; other molecules, such as formaldehyde (H_2CO) and methanol (CH_3OH), had similar or lower densities.

Nevertheless, in a dense ionized gas region (so-called H II region) like W51, the mass of CO observed is 125 times that of the sun. The other molecules are more rare, and there can be an unknown and large amount of molecular hydrogen (high ratio of H_2 to H) in dense clouds.

Significance of the molecules. The importance of the presence of complex molecules, however, is not measured by their rarity. The fact of their existence is of great importance, rather, to the theory of the formation of stars in dense gas and dust clouds. Furthermore, some of these molecules are essential to biological evolution, and might later exist on a planet near a young star. A search for SiO (silicon monoxide) gave negative results in the H II regions; astronomers expect the formation of silicates to be an early step in the growth of solids in a dense cloud. Interstellar dust grains should contain frozen gases (H_2O, CH_4, NH_3) and solids based on iron and silicon compounds. Orbiting observatory studies have suggested the existence of graphite in the solids.

The unresolved problem is whether the complex molecules found are the first step toward the condensation of solid grains with sizes up to a few thousand angstroms, or whether the solids provide the matrix on which complex molecules can form.

"Astronomy compels the soul to look upwards and leads us from this world to another."

Plato

For example, even ^2H is formed only at very high interstellar gas densities, but it would easily grow from H atoms on a solid grain. Star formation requires heat dissipation as a gas cloud condenses; dust grains and large molecules provide the mechanism by which heat is radiated away in the infrared.

The interplay of dust grains, large molecules, and dense gas is expected to be studied extensively, since infrared radiation has been detected in both small and large dense gas clouds, in dense H II regions like the Orion Nebula, in a planetary nebula (NGC 7027), and in infrared sources near recently formed stars like Hubble's variable nebula (NGC 2261). Old red giants, such as VY Canis Majoris, are imbedded in nearly opaque dust clouds, which radiate mostly infrared and which have expanding gas clouds, as found by their molecular lines at radio frequencies.

Preorganic complex molecules can be formed and trapped in collapsing gas clouds that will become stars and possibly planetary systems. Whether they will survive star and planet formation and cosmic-ray bombardment and still exist in the oceans of a new world is less certain. They cannot survive long in interstellar space against cosmic-ray and ultraviolet bombardment.

Stellar astronomy

Because stars are the major building blocks of galaxies, knowledge of their physical properties, composition, energy sources, and evolution is fundamental to the study of the galaxies. Their internal structure was supposed to be well understood, with one major unknown being the primeval ratio of atomic helium to atomic hydrogen (He/H). At the surface of young stars and H II regions, this ratio was found to be 1:10 from spectroscopic study. If a big-bang cosmology is taken seriously, this is expected to be the minimum ratio almost everywhere in the universe.

Solar theory challenged. Once formed, a star such as our sun can have its evolution calculated for the 4,600,000,000 years of its existence. Thus, elaborate computing-machine calculations by I. Iben of the Massachusetts Institute of Technology gave fairly reasonable models for the present sun. However, an experiment carried out for several years by R. Davis, Jr., to detect neutrinos emitted during the present energy-generation process in the sun led to serious difficulties with the models. Davis used 379,000 liters of cleaning fluid in a tank in a deep gold mine in South Dakota, where it was shielded from almost all energetic particles except neutrinos from the sun. A few neutrinos were expected to be absorbed by the chlorine in the tank. This reaction would then produce a radioactive isotope of argon, ^{37}Ar, which could be separated and counted.

The first predictions, based on well-known principles of nuclear physics and on the supposedly well-known composition and central temperature of the sun, foresaw a yield of seven atoms of ^{37}Ar per day. The present best value after several years of experimental work, however, is about 0.5 ($\pm$0.3) atoms per day. This large discrepancy suggests a radical modification of fundamental physics or drastic changes in the theory of the solar interior (an error in central temperature caused by wrong opacity theory, or a low helium content).

Mercury on hot stars. Stellar compositions are in general like that of the sun and earth. In regard to mercury, however, a startling observation was made during the year. On earth, mercury isotopes of weight 198 and 199 make up 27% of the total supply; isotopes 200 and 201 are 36%; 202 is 30%; and 204 is 7%. But at the surface of one hot star, G. W. Preston of the Hale Observatories found that isotopes 200 and 201 made up 16%; 202, 45%; and 204, 33%.

The enormous excess of 204 was quite unexpected and unexplained. In two other similar stars, M. M. Dworetsky, J. E. Ross, and L. H. Aller of the University of California at Los Angeles found isotope 204 even more dominant: 59% in one star and 97% in another. What type of nuclear process could so alter isotope ratios at the stellar surface? Intense bombardment by high-energy particles or neutrons is one possible answer.

Circular polarization in white dwarfs. A major discovery in stellar astronomy was the existence of circularly polarized light in four white dwarfs. Using delicate analyzing equipment, J. C. Kemp and J. B. Swedlund of the University of Oregon made the first observations; subsequently, J. R. P. Angel (Columbia) and J. D. Landstreet (University of Western Ontario) confirmed and extended Kemp's work.

White dwarfs are dying stars with a mean density one million times that of the sun (about 13.6 metric tons per cc); at their surface, the gravitational pull is 250,000 times that on the earth. Inside, they are "degenerate," that is, the electrons are relatively limited in their motion and position. Such stars lack nuclear energy sources and have crystallized in spite of central temperatures of 10,000,000–30,000,000° K. Normally, they lack hydrogen except at their surface; some show lines of only helium or carbon (products of nuclear burning).

It had always been assumed that white dwarfs were inert, and they also seemed to rotate relatively slowly. The discovery of circular polarization, however, altered this view, at least in regard to the inertness. The polarization is very weak, depends

NGC 1275, a Seyfert galaxy in the constellation Perseus, was found to be a strong emitter of X rays. Seyfert galaxies are dim in visible light but radiate great amounts of energy in the unseen wavelengths. Some astronomers consider them a link between ordinary galaxies and quasars.

Courtesy, Kitt Peak National Observatory

on wavelength, and in one star varies in 1.3 days (presumably the rotation period). The explanation for the existence of circular polarization, however, is based on these stars having an intense dipole magnetic field (one consisting of opposite magnetic poles), ranging from a few million gauss (the limit of detectability) to 30,000,000. (Gauss is the flux density of a magnetic field—that is, the amount of magnetic energy that passes through a specified area in a given time.)

Kemp elaborated this theory, which seemed the only plausible explanation, and showed that the first star observed, a faint white dwarf with unidentified weak lines, needed an intense field of 30,000,000 gauss. The sun and earth have fields near one gauss, and the magnetic stars have, at most, fields of 30,000 gauss. The suggestion was that if a star contracts, its magnetic field strength increases. Furthermore, since white dwarfs are essentially the exposed cores of red giants, it is possible that the intense magnetic field was produced in an earlier phase of evolution.

Most white dwarfs have much less intense fields than those with circular polarization, probably less than 100,000 gauss. All the white dwarfs with circular polarization so far observed have peculiar spectra; one has carbon bands, one very weak metallic lines, another no lines.

B. M. Lasker and J. E. Hesser of the Cerro Tololo Inter-American Observatory in Chile found that four other white dwarfs vary in brightness by about 1% in cycles of 212 to 1,638 sec. The variation is not absolutely regular in amplitude, and has several different periodicities. One of the light-variable stars has no spectral lines; another has helium, and two have hydrogen lines; none has circular polarization. The density of white dwarfs is so high that the natural oscillation frequency is about one second; hence, the longer periods observed suggest rotation or torsional oscillations.

X-ray sources

One of the most important successes of science in space was the *Uhuru* (the Swahili word for freedom) satellite, launched off the coast of Kenya. It contained a number of X-ray experiments and was the U.S. National Aeronautics and Space Administration's (NASA's) first Small Astronomy Satellite (SAS-A). By mid-1971 only part of the data from the satellite had been received and analyzed, but the first successes were remarkably interesting. Scanning our Milky Way, the satellite identified 25 X-ray sources in three hours of observation; many are concentrated near the galactic center. Also, known X-ray sources were much more accurately located. In addition, SAS-A confirmed earlier observations that the quasar 3C 273 emits X rays at a rate of 10^{39} w, slightly less than its optical radiation.

An X-ray source in our galaxy, Cygnus X-1, had by mid-1971 not been definitely identified with any star. The *Uhuru* satellite found its X-ray flux to vary in 12 min by a factor of 2. Superposed on this large change, the flux pulses by 25% in a fraction of a second. The source is apparently periodic on a short time scale, with a recent estimate of 0.292 sec. To change that rapidly, the emitting region must be smaller than about 960 km in diameter. It could be a rotating neutron star or a collapsed rotating black hole. (A black hole is an object—thus far only theoretical—that has such a strong gravitational field that not even light can escape from it.) No supernova remnant and no radio emission were found.

The satellite also detected rapid fluctuations and an X-ray flare in a source in the Southern Cross, and periodic variations in another X-ray source in Centaurus. The rate of discovery of such new phenomena was extraordinary, and it appeared certain that X-ray data would provide important clues to the nature of the high-energy universe. The highly promising early results from *Uhuru* were scheduled to be followed up by extensive optical and radio searches for these new objects. It was clear that X-ray emission involves either gas at tens of millions of degrees (thermal) or magnetic fields and high-energy particles (non-thermal).

There was much study during the past year of a known radio and X-ray source, Cassiopeia A. One of the first strong radio sources discovered, Cassiopeia A was also confirmed by *Uhuru* as a strong X-ray source. It is the remnant of a supernova that exploded in our galaxy about 300 years ago, the explosion still being visible as a group of small fast-moving gaseous filaments about 8,500 light-years distant in the Milky Way. Spectra of many of these filaments were obtained first by Rudolph Minkowski and then by S. van den Bergh. Moving knots of filaments show lines of oxygen, sulfur, and argon with typical sizes of a light-month. They move and change so rapidly, however, that a given object is barely recognizable after 10 years. Thus, supersonic motions build and dissipate small, dense regions in that short a time. These regions are condensations in an envelope expanding at 5,000 km/sec, presumably ejected when the supernova exploded with at least that high a velocity.

In addition to the filaments, van den Bergh found a number of nearly stationary nebulosities that move slowly but also vary in brightness and shape. They have spectral lines of hydrogen and nitrogen and an estimated total mass of 0.01 suns, much less than the interstellar gas that would have existed in the neighborhood of the supernova. Since their composition, in having strong hydrogen, is more normal than that of the moving filaments, it appears that the supernova composition was hydrogen-poor, in accordance with other work on supernovae of type II and V (massive exploding stars).

Radio astronomy and quasars

Two important developments in 1970–71 involved using interferometers to study the positions and motions of small radio sources, such as quasars.

Quasar position measurements. A group (G. Seielstad and others) at the California Institute of Technology's Owens Valley Radio Observatory used an interferometer to measure the position of the quasar 3C 279 as it passed near the sun in late 1969. A change of position is caused by the gravitational deflection of light produced by the general theory of relativity (the "Einstein deflection"). At the edge of the sun this bending was predicted to be 1.75 arcseconds, and the 1969 measurements gave 1.77 ± 0.20 (an arcsecond is equal to 1/3,600 of an angular degree).

Another, more precise, measure was made at the Jet Propulsion Laboratory by D. O. Muhleman and others. They used a 64-m and a 30-m antenna as an interferometer, to measure the separation of quasars 3C 279 and 3C 273. Radio waves from 3C 279 traverse the solar corona near the sun, producing some uncertainty due to coronal refraction; 3C 273 is much farther away from the solar limb. The deflections measured agreed with general relativity within 4%, with an uncertainty of about 10%.

In 1970–71, a much more elaborate attempt was made by I. Shapiro and a group at the Massachusetts Institute of Technology using a very-long-baseline-interferometer (VLBI). The entire width of the continental U.S. separated the two antennas of the VLBI. Shapiro's experiment, in October 1970, consisted of following 3C 279 and 3C 273 as the former passed by the sun; later, when the sun had moved, he remeasured their separation in the absence of the corona. He hoped to acquire enough information about the deflections produced by the corona to remove its effects, thereby obtaining a much more accurate value of the Einstein deflection.

At the October 1970 transit, unfortunately, solar activity was higher than in the 1969 experiment and seriously affected the 3C 279 positions. Thus, the accuracy of the results on general relativity was not immediately known. But Shapiro did find that the quasar 3C 279 was not a point source when viewed at a resolution of 0.001 arcseconds. It was, instead, a double source, and by repeating the study of 3C 279 four months later (when it was far from the sun), he found that the distance separating the two sources had expanded by 10%.

Suggestions of expansion had been found previously by A. Moffet and M. Cohen in other VLBI studies of quasars, many of which had small bright components about 0.001 to 0.003 arcseconds in diameter. The red shift (a shift in the spectral lines of certain celestial objects toward the red end of the spectrum because the objects are moving away from the earth) of 3C 279 is 54%, which corresponds to a distance of 9,000,000,000 light-years. Shapiro's angular separation gave the projected distance between the point sources as 20 light-years. If they are moving apart at a rate of 10% in four months, this corresponds to an apparent motion of six light-years in one year, a speed apparently six times that of light.

M. Cohen at the California Institute of Technology and others found that 3C 273, a quasar with a red shift of 16%, and NGC 1275, an explosive galaxy much closer to the earth than the quasar, also show apparent motions at very high speed, when VLBI observations are continued over some months. There is no need, however, to fear violation of Einstein's principle that nothing moves faster than light. For example, if a rotating searchlight illuminates different parts of an irregular distant cloud, the illuminated patches change and the disturbance appears to move faster than light. Similarly, if a quasar or a galactic nucleus explodes two clouds of high-energy particles outward at high speeds, the clouds will produce radio emission when they encounter magnetic fields, or stationary matter, and the moving front of the disturbance will appear, dependent on the geometry, to have very fast motion.

Clearly, more of the small-sized components in quasars and galaxies need to be studied by the VLBI technique, for various lengths of time, and with different resolving power. Some scientists do not believe that quasars are at the very large distances suggested by their red shifts; if they are nearby, the observed angular motions correspond to much smaller linear velocities and could be explained without relativity theory.

Distance to quasars. The controversy about the distance to the quasars continues. Their properties are so strange and their energies so large that it is reasonable to keep questioning distances derived only from the red shifts of their spectral lines. H. C. Arp of the Hale Observatories found several disturbing cases where a quasar of large red shift and a galaxy of small red shift were apparently close together, or even apparently linked by a spiral arm or other feature. Could quasars be blown out of galaxies and be subject to an unknown physical law producing a noncosmological red shift? If so, their distances and brightnesses would be less than they are now believed to be; the apparent faster-than-light expansion of 3C 279 would merely indicate that the distance assumed is too large.

According to G. Burbidge, such a spurious red shift may be produced in all very small and luminous nonthermal sources, even in the small nuclei of Seyfert or compact galaxies, perhaps by the interaction of light photons with whatever produces the nonthermal emission. A critical test of this idea would be the discovery of a typical quasar in a cluster of galaxies. J. E. Gunn of the Hale Observatories studied two faint objects in faint clusters of galaxies. He proved that one, called Tonantzintla 256, which is superposed on such a cluster, has the same red shift as the cluster itself (13%) and that it is slightly brighter than any other cluster member. The radio source, PKS 2251 + 11, which is quite starlike in appearance, was found by Gunn to be a conclusive case. At the 200-in. (508-cm) telescope at the Palomar Observatory, he measured its red shift to be 32% and found it super-

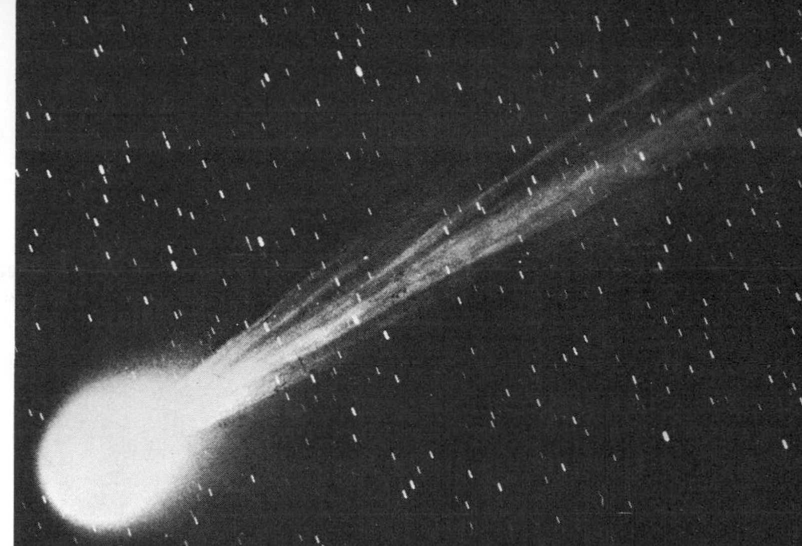

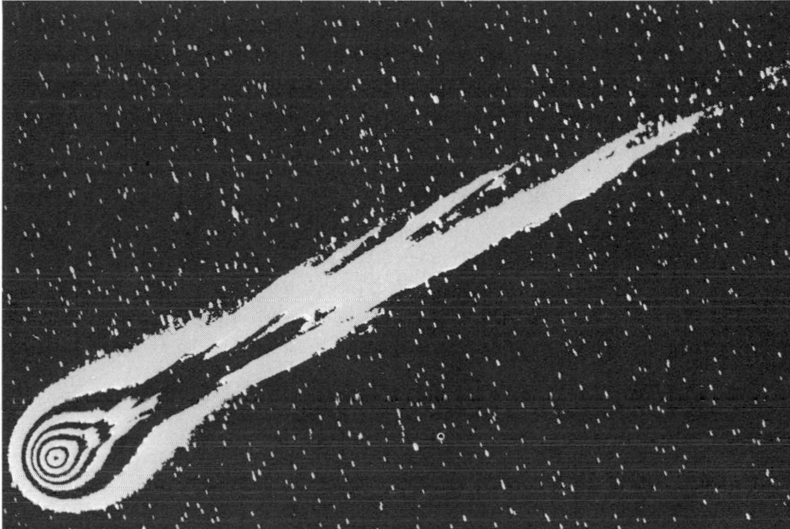

Keystone

A comet shows little light contrast when photographed with ordinary film (top), but the differing light concentrations within its head are revealed by the newly developed Agfa-contour method (bottom). This was expected to help astronomers better understand comets.

posed on a small, compact cluster of galaxies. Hydrogen emission and stellar absorption lines appeared in the galaxy that was the brightest member of that cluster. This galaxy was 28 arcseconds from the quasar and had the same red shift within 1%. The quasar was about 30 times brighter than the galaxy. This agreed with results for other typical quasars on the assumption that they are at distances given by their cosmological red shifts. Clearly, a few more such definite associations of quasars and galaxy clusters were needed to support definitively their cosmological distances.

—Jesse L. Greenstein

Atmospheric sciences

For many years, the relevance of the atmospheric sciences to matters of broad public concern seemed to be limited to research that promised to improve man's ability to forecast the weather. It was generally accepted that man could do little to affect the behavior of the atmosphere; the best he could hope for was to predict its vagaries accurately enough so that he could take precautions to protect himself from tornadoes, hurricanes, blizzards, and other violent assaults from the skies.

But as the 1970s began, it was becoming increasingly evident that human activities *could* affect the behavior of the atmosphere. Some of these effects can be achieved deliberately, as when clouds are seeded to dissipate fog, augment precipitation, or suppress hail. Others are inadvertent effects whose nature and extent can be understood only by extending our knowledge of the fundamental physical, chemical, and dynamic processes of the atmosphere, on scales ranging from the microscopic to the global. This realization caused an ever-greater involvement of atmospheric scientists in decision-making processes that confront broad and complex questions of public policy.

The atmospheric sciences and public policy. Early in 1971, the direct relevance of the atmospheric sciences to a major public-policy decision was evidenced by newspaper headlines about testimony at U.S. Congressional hearings on the civil supersonic transport plane (SST). Among the witnesses who appeared before House and Senate committees were atmospheric scientists who spoke both for and against continued government support of the SST program. James McDonald, an atmospheric physicist from the University of Arizona, testified that large-scale SST operations might increase the incidence of skin cancer by destroying atmospheric ozone, allowing more ultraviolet radiation to reach the earth's surface. However, William W. Kellogg, an associate director of the National Center for Atmospheric Research (NCAR), Boulder, Colo., favored development of two SST prototypes and attacked the validity of McDonald's skin-cancer hypothesis while defending his credentials as an atmospheric physicist. In the Senate hearings, television personality Arthur Godfrey and Sen. Gordon Allot of Colorado argued over whether or not jet aircraft contrails (streaks of condensed water vapor in the air) can trigger cloud cover over the Rocky Mountains in Colorado.

Although the final decision to end federal support of the SST program involved economic as well as environmental factors, there was no doubt that it was influenced by public pressure caused by fear that SSTs may damage the atmosphere. To many people, the SST episode clearly demonstrated that existing mechanisms for applying scientific knowledge of the atmosphere and other elements of the environment to public-policy decisions are not adequate, and that new roles for existing institutions and, perhaps, entirely new institutions must be developed to improve the linkage between environmental research and environmental policy. (For additional information on the SST, *see* Year in Review: TRANSPORTATION.)

Two presidential reorganization plans that went into effect in the fall of 1970 affected some major federal activities in the atmospheric sciences. The new U.S. National Oceanic and Atmospheric Administration (NOAA) absorbed the Weather Bureau —renamed the National Weather Service—the Environmental Research Laboratories, and other activities of the Environmental Science Services Administration, which became the core of the new organization. A second new federal agency, the Environmental Protection Agency (EPA), absorbed a number of federal activities concerned with the environment, including the National Air Pollution Control Administration, now known as the Air Pollution Control Office of EPA.

Perhaps the most important federal action affecting atmospheric matters was the passage of the Clean Air Amendments of 1970, signed into law on December 31. This legislation, extending and enlarging the Clean Air Act of 1963 and its subsequent amendments, was the toughest air pollution control law ever passed by Congress. It imposed rigid standards for the emission of hazardous materials, for air quality, for emissions from automobile engines, and for other elements of air pollution. Early in 1971, an EPA spokesman conjectured that a number of cities, including New York, Chicago, Los Angeles, Philadelphia, and Denver, Colo., might have to limit automobile traffic by 1975 to comply with the new air quality standards. In addition to spelling out standards and authority for

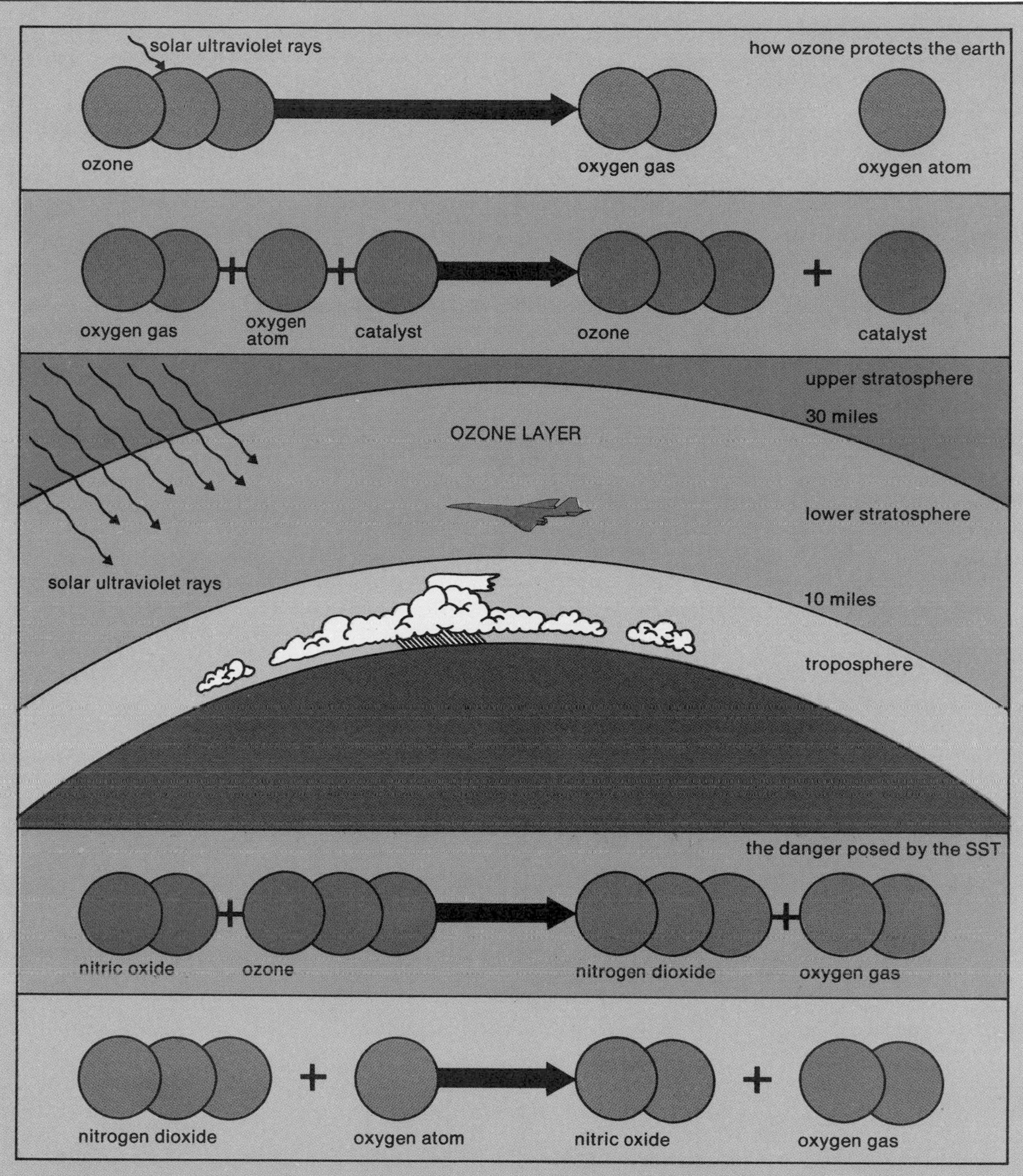

Supersonic transport planes (SSTs) may endanger life on the earth by breaking down the ozone layer in the lower stratosphere. Ozone, a faintly bluish gas created by adding a single atom of oxygen to the oxygen molecule, protects all forms of life by absorbing the sun's ultraviolet rays. Without this absorption barrier, according to some authorities, men and animals would be blinded and would develop skin cancer at an increased rate. (Top) The ultraviolet rays are absorbed by the ozone molecule, causing it to split. A catalyst, however, allows the ozone to be reformed. The SST might break down this process by adding from its exhaust nitric oxide that would split additional ozone molecules. These would not be reformed (bottom), and the ozone barrier gradually would be destroyed.

Adapted from "The New York Times"

air pollution control, this legislation also earmarked funds for scientific research on atmospheric pollution.

Observing and forecasting the weather. On Nov. 12, 1970, the coast of East Pakistan was struck by a cyclone that was described as one of the most deadly—if not the deadliest—storms ever to strike a coastal area. This cyclone, which is the same kind of tropical disturbance that is called a hurricane in the United States, may have killed approximately 300,000 people in the low-lying country along the Bay of Bengal.

Although there was some criticism of the warnings given by the East Pakistani meteorological service, an official of the U.S. National Weather Service who visited the stricken area said that the forecasters were not at fault. He said, "No action short of evacuation—which was impossible—could have made a substantial difference in the number of casualties."

Most of the victims of the cyclone were on low coastal islands that were flooded by the high tides pushed up in the Bay of Bengal by the powerful and sustained winds of the cyclone. In such a situation the benefits of improved weather forecasting are doubtful; if there is no way to escape from a storm, it is not particularly useful to know that it is on the way. One idea being explored to provide safe havens on the Pakistani islands is to build concrete-block school buildings atop 20-ft-high earth platforms set back from the coast. These school buildings could serve as storm shelters during cyclones, and early warnings of impending storms would then prove valuable.

In most places where people live, there are alternatives that can be employed if there is early and accurate weather forecasting. A Great Plains farmer can head for the storm cellar when he is informed that a tornado is coming; a rancher can herd his cattle to shelter and give them plenty of feed before a blizzard strikes; and a resident of a coastal area can move inland when he is warned that a hurricane is imminent. When these alternatives exist, the benefits of improved forecasting are obvious.

Although there is a popular tendency to remember all the times that the weather forecast turned out to be wrong, and to make fun of the weatherman, it is undeniably true that weather forecasting has improved over the past 20 years. Much of this improvement was achieved by using two valuable new tools: the computer and the meteorological satellite. During the past year perhaps the greatest strides were made in the application of new satel-

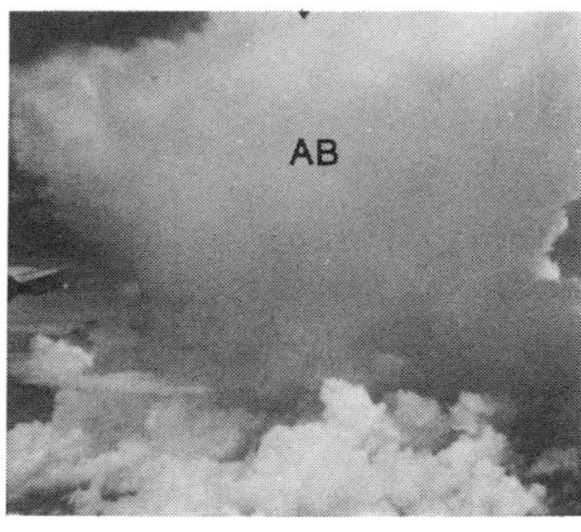

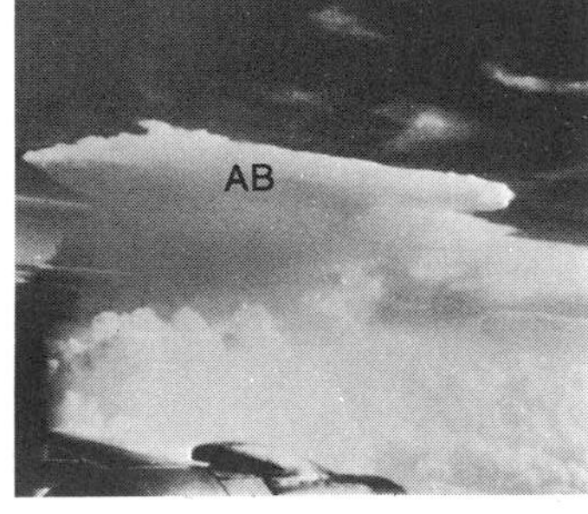

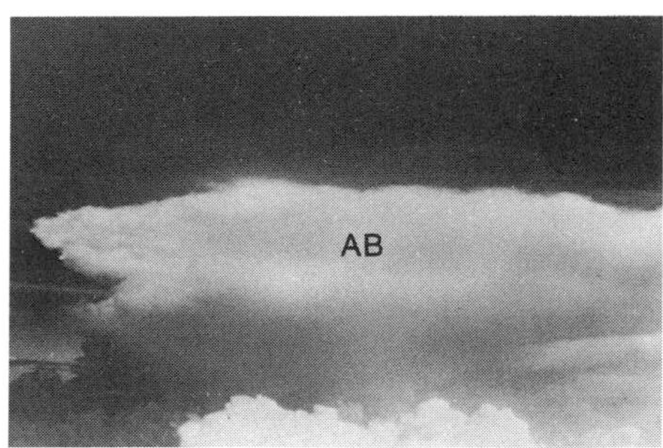

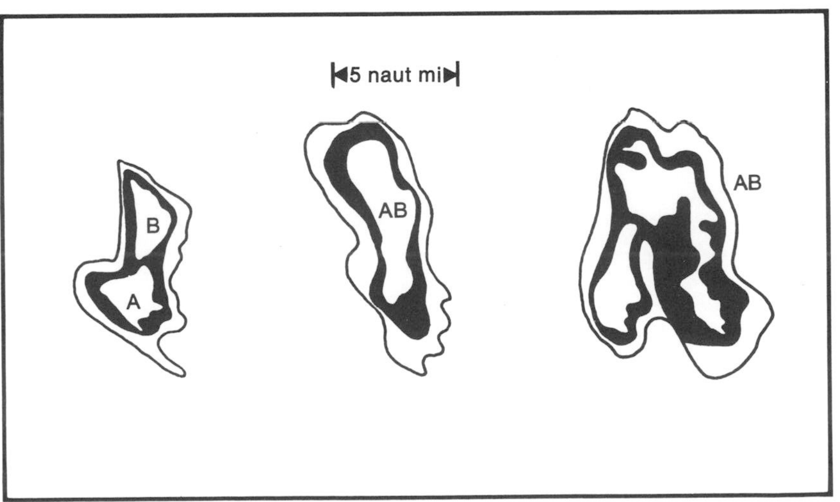

Seeding clouds to increase rainfall was attempted over southern Florida in 1970. Cloud A was seeded with silver iodide crystals, while Cloud B was not seeded. The two are shown (top, left to right) 6, 16, and 37 min after merging, which also increases rainfall. The radar pictures below show rainfall contours at (left to right) 5, 26, and 36 min after merging: inner white areas indicate over 0.68 in. of rain per hour; black, over 0.35 in.; and outer white, over 0.01 in. Nearby unseeded clouds that did not merge provided substantially less rainfall.

Courtesy, NOAA

lite technology to problems of operational weather forecasting. These satellites included ITOS-1, the first in a new series known as the Improved Tiros Operational System, and the Applications Technology Satellites (ATS). For further information about such satellites, *see* Year in Review: ASTRONAUTICS AND SPACE EXPLORATION, *Earth Satellites*.

During 1970, NOAA began transmitting ATS pictures routinely to two of its major weather centers—the National Hurricane Center in Miami, Fla., and the National Severe Storms Forecasting Center (NSSFC) in Kansas City, Mo. The satellite pictures enable the hurricane forecasters to spot storm systems on a global scale, while they are still in the formative stages half a world away. As a storm moves nearer to North America, it is probed by U.S. Air Force, Navy, and NOAA "hurricane hunter" aircraft. Satellite and aircraft data are analyzed, using an electronic computer, and the hurricane experts in Miami constantly update the portrait of each storm's vortex to tell whether or not it is increasing or decreasing in intensity. The Miami center provides guidance to four other National Weather Service offices with hurricane warning responsibility—San Juan, P.R.; New Orleans, La.; Washington, D.C.; and Boston. This improved system is designed to provide a hurricane warning time of 15 to 18 hours, including a full 12 hours of daylight, to permit people in threatened areas to take all possible precautions, including evacuation if the storm is extremely severe.

The NSSFC used the ATS photographs in an experimental program to evaluate their usefulness in predicting tornadoes. The severe storm forecasters analyzed the satellite pictures, along with upper-air data collected with weather balloons and radar, to watch for weather patterns that favor the formation of tornadoes. Big thunderstorm clouds were observed to expand explosively in the satellite pictures just before the onset of tornadoes. Although only the human eye can identify a tornado, the satellite pictures proved valuable in identifying areas where the probability of tornado formation is very high, a situation that calls for issuing a tornado watch to the public. A tornado warning is issued only when a human observer has actually sighted a deadly funnel cloud.

Deliberate weather modification. The technique that is at the heart of most attempts to modify the weather is cloud seeding with silver iodide or some other substance to influence the formation of precipitation from supercooled cloud droplets that are still liquid, even though they are colder than the nominal freezing temperature of water. Under the right circumstances, each tiny crystal of silver iodide that enters the cloud will serve as a nucleus around which ice crystals grow very rapidly, thereby taking up moisture from the surrounding cloud droplets.

Of the many weather-modification research projects taking place in the U.S. and other countries, three that were being conducted or planned during 1970 and the early part of 1971 represented good examples of the current state of knowledge in this field and how it is being applied to human needs.

Snow augmentation. In the winter of 1970–71, the Bureau of Reclamation of the U.S. Department of the Interior began a pilot project in weather modification in the San Juan Mountains of southwestern Colorado. This project was designed to test the effectiveness of systematic seeding of orographic (mountain-influenced) snowstorms in increasing the total amount of snow that falls from each storm. The ultimate goal was to increase the supply of water in the arid southwest by adding to the snowpack, the total accumulation of snow on the ground. This addition would, in turn, increase the spring runoff into the Colorado River, which supplies much of the water for human uses in the southwest, including the city of Los Angeles. Under pressure from people and organizations concerned with possible ecological implications of increasing the average snowfall in an area over a period of years, ecological studies were included in this project.

Rain augmentation. During the 1950s, commercial rainmakers conducted a thriving business in semiarid agricultural areas, seeding clouds for a fee. Sometimes their efforts were followed by rain, sometimes not. The scientific basis for their operations was somewhat uncertain, to say the least. But in recent months atmospheric scientists at the NOAA Experimental Meteorology Laboratory in Coral Gables, Fla., seemed to have established beyond reasonable doubt that, under certain conditions, cloud seeding can stimulate rainfall. In experiments in southern Florida in 1970, they tested a technique known as dynamic seeding. Large quantities of silver iodide were introduced into the tops of cumulus clouds by dropping pyrotechnic flares from aircraft. This seeding causes supercooled cloud droplets to freeze into ice crystals, releasing latent heat which causes the air in the cloud to rise. This causes a sudden explosive growth in the cloud, culminating in up to seven times as much rain as would have fallen from the unseeded cloud. The NOAA scientists also believe that seeding can cause two or more clouds to merge, producing a large cloud that will drop a great deal more rain than the total that would have fallen from the individual clouds.

In April 1971, at the request of the governor of

Courtesy, U.S. Department of Transportation

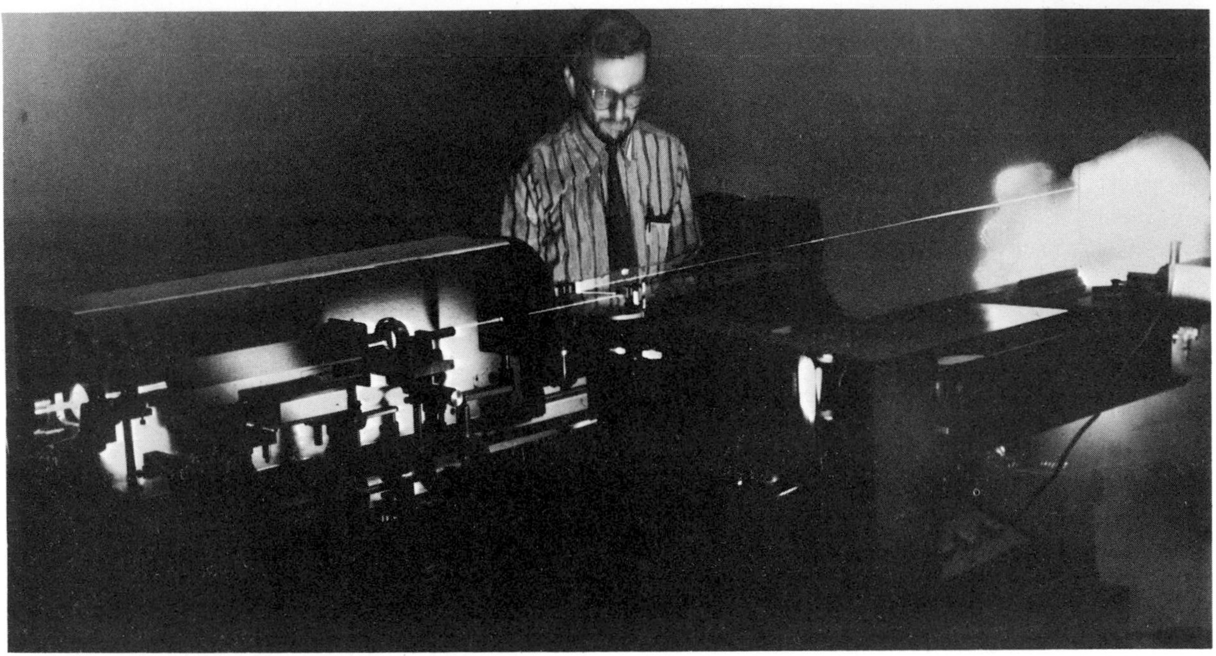

A laser beam is directed at a cloud of gas in a laboratory experiment designed to help develop a system for the U.S. Coast Guard to detect off-shore fog banks and provide better warning to ships. A laser beam transmitted from shore would trigger alarms in unmanned lighthouses upon reaching fog.

Florida, NOAA began a two-month experimental cloud-seeding program with a dual goal: to learn more about dynamic seeding and to try to help alleviate a severe drought that was afflicting southern Florida. The scientists emphasized that there were several unknowns in the experiment, and that their efforts were not likely to break the drought. Nevertheless, the possible benefits seemed likely to outweigh the uncertainties.

Hail modification. Hail is one of the most destructive atmospheric phenomena known to man. In the U.S. alone, the annual hail damage to crops and property runs from $200 million to $300 million, and in most years the cost of hail damage is higher than that from tornadoes.

In 1969, NCAR was asked by its federal sponsor, the National Science Foundation (NSF), to design a large-scale field experiment aimed at gaining a fundamental understanding of hailstorms and testing the feasibility of seeding them to reduce their destructive power. This experiment, known as the National Hail Research Experiment, will test the hypothesis that hailstorms can be seeded to make them exhaust their moisture in small hailstones or rain before large, destructive hailstones can grow. It will use radar, instrumented aircraft, ground crews, and other techniques to study the dynamics of hailstorms. Computer modeling techniques will also be employed to simulate the processes of hailstorm growth and hailstone formation. The field research area is in northeastern Colorado, in a region sometimes called "Hail Alley" because it has the highest hailstorm frequency of any location in the U.S. In the summer of 1971, a number of new systems were tested in the field, and successive summers were scheduled to be devoted to full-scale field operations aimed at studying hailstorms and testing seeding techniques. The experiment included researchers from universities and other groups, working under NCAR management with support from the NSF.

Atmospheric chemistry and climatic change. What is happening to the chemical composition of the earth's atmosphere over long periods of time, under the influence of both natural and man-made substances that enter the atmosphere from a variety of sources? Atmospheric scientists regard this as a critically important question that must be answered as soon as possible. It is possible that man-made atmospheric pollutants could alter the radiative processes of the atmosphere sufficiently to cause serious changes in worldwide temperature or weather patterns.

However, we can only speculate about such changes now, because adequate scientific information does not exist. No chemical sampling and analysis of the atmosphere has ever been performed on a comprehensive long-term basis. Until recently, few analytical techniques had been developed for measuring substances that are in the atmosphere in extremely small quantities, even though changes in the concentrations of such trace substances over long periods of time may cause critical problems for man.

In January 1971, the American Chemical Society published a list of 26 recommendations for high-priority actions that should be taken to clean the environment. The first recommendation concerning the atmosphere was that systematic measurements should be undertaken for a number of relatively long-lived substances in the atmosphere.

Both NOAA and NCAR began programs to make such measurements. The NOAA program called for establishing six stations to monitor carbon dioxide, the chemistry of precipitation, atmospheric attenuation of solar radiation, concentration of atmospheric dust at ground level, and vertical distribution of airborne particles. NCAR began developing an "atmospheric chemistry reconnaissance station" designed to require attention about once every two weeks. It would measure particle concentrations, gases that play a role in particle formation, carbon dioxide, carbon monoxide, and other substances in the atmosphere. NCAR chemists planned to set up five or six of these stations at such diverse locations as the Colorado Rockies, Alaska, Hawaii, Antarctica, and the Amazon jungle. They hoped that these stations eventually would be taken over by an international group, such as the World Meteorological Organization (WMO), and expanded into a full-time global monitoring network. Only by operating such a network and subjecting its data to rigorous study will man be able to assess the nature and extent of his influence on climate and other aspects of the atmosphere.

Global Atmospheric Research Program. During 1972 a great deal of effort by atmospheric scientists was expected to be devoted to developing detailed plans for the Global Atmospheric Research Program (GARP), an international undertaking aimed at gaining a far more detailed understanding than we now have of the global-scale physical behavior of the atmosphere. One end product of such an understanding should be the ability to make large-scale weather forecasts as much as two weeks in advance.

The broad outlines of U.S. participation in GARP were proposed in a plan prepared by the U.S. GARP Committee of the National Academy of Sciences. International GARP planning was centered in Geneva, under the joint auspices of WMO and the International Council of Scientific Unions. In the U.S. active planning was taking place in NOAA, where federal GARP activities were centered, and in the University Corporation for Atmospheric Research (UCAR), the consortium of 31 universities that operates NCAR. UCAR and NCAR worked to bring university scientists into active planning and organization for GARP.

The first large-scale GARP field program will be a study of tropical atmospheric processes in the Atlantic Ocean between Africa and South America. This project, scheduled for 1974, and subsequent larger experiments will utilize satellites, research aircraft, ships, balloons, and all of the most advanced tools of atmospheric research to examine the great fluid system that is the atmosphere of our planet.

—Henry Lansford

See also Feature Article: KILLER STORMS; A Gateway to the Future: CHALLENGING THE RESTLESS ATMOSPHERE.

Behavioral sciences

Why does man behave as he does? How did he become what he is? Can he change himself and the society he has created? During the year, behavioral scientists sought answers in investigations as diverse as the discovery of Paleolithic remains and the study of rage reactions in monkeys.

Anthropology

Since its inception 150 or more years ago, anthropology has taken as its province the study of mankind-as-a-whole, from its dim beginnings to its ever more complex present. During the past year, a well-organized interdisciplinary panel told an overflow session at the annual meeting of the American Anthropological Association that the profession needs to add systematic study of the future if it expects to provide understanding of the changing human career. Organized by anthropologist-philosopher Magoroh Maruyama, the panel provided guidelines for a new, realistic, and scientific perspective, presented a first book (*Human Futuristics,* Social Science Research Institute, Hawaii, 1971), and made preparations for a second session based on the book.

New studies. In its study of the origin and differentiation of the genus *Homo,* now represented by our single species, anthropology still learns from the living apes and monkeys. Because man is anxious to know himself, such popular writings as those of zoologists Konrad Lorenz and Desmond Morris and the playwright Robert Ardrey have been highly successful. Meanwhile, however, anthropologists all over the world quietly continue their exciting fieldwork with primate species in their natural habitat. Three new summaries of this work were *Social Groups of Monkeys, Apes and Men* by

Michael R. A. Chance and Clifford J. Jolly (1970); Colin Groves' *Gorillas* (1970); and *Functional and Evolutionary Biology of Primates,* edited by Robert Tuttle (scheduled for publication early in 1972).

Most of what was currently important to anthropologists continued to appear in conferences and special syntheses of knowledge. An example from a specialized field was Albert Dahlberg's publication of the second International Symposium of Dental Morphology, held in London (1971). In contrast to this was R. Narrol's and R. Cohen's 1970 *A Handbook of Method in Cultural Anthropology,* in which 40 anthropologists cooperated to describe the means by which anthropologists use disparate data to reach generalizations. Perhaps the most ambitious synthesis was the Smithsonian Institution's projected 20-volume *Handbook of American Indians,* by hundreds of scholars; begun with a series of conferences in 1970–71, it was to be published by July 4, 1976. Achieving fruition during the year was a major cooperative effort by UNESCO to summarize the state of cultural anthropology and the shape of its future. For this task, Maurice Freedman of Oxford University enlisted a wide spectrum of colleagues; publication was scheduled for 1971 or 1972.

Nevertheless, the truly significant events of the year were the unheralded studies by individual anthropologists who had learned something genuinely new in the field. Two contrasting studies illustrated both the quality and the variety: Joan Vincent's *African Elite* (1971), a brilliant "exploratory case study in societal change in Sub-saharan Africa"; and Fredrik Barth's still unpublished study of a hitherto untouched highland New Guinea tribe. Barth, of the University of Bergen, Nor., had previously worked mainly among pastoral peoples in the Middle East and was known for creative studies of social networks. Turning his attention to the cognitive side of human behavior, he literally "dropped in on" the small tribal group in upland New Guinea and succeeded in passing through the men's initiation rituals and deciphering a relatively complete cultural code.

Barth assumed quite a different role at a meeting of the Permanent Council of the International Union of Anthropological and Ethnological Sciences, where he was appointed to head a special working group to protect peoples and cultures from being destroyed. Meanwhile, in Chicago, early in May, the Smithsonian Institution's Center for the Study of Man took a long, positive step to mobilize anthropological studies of social problems that plague the species, among them overpopulation, environmental degradation, and the failure of education in a rapidly changing world.

In the U.S.S.R. anthropology was also "opening up" to modern problems. The author's visit with anthropologists in Moscow over May Day, 1971, indicated that (perhaps in common with colleagues in other sciences) the Soviet ethnologists were confident that they had unique and important contributions warranting a systematic program of translations into English. They therefore founded two new journals, in English: *Social Sciences* and *Social Sciences Today.* Moreover, they were preparing to publish an annual digest in English of the 50-odd books and monographs that they publish each year in the U.S.S.R. This practice also began in other Eastern European countries.

Sociology had become a recognized subject for study in the Soviet Union only within the past 10 years. Ethnologists, finding it necessary to reexamine their roles, concluded that they must be interdisciplinary and comparative, concentrating especially on the "ethnic" elements within the U.S.S.R. These elements, derived from the unique cultures developed by earlier self-contained social groups, still give character to individual parts of modern Soviet society, but they are rapidly being absorbed into the larger economy and body politic.

In the United States, the April 1971 inauguration of anthropologist Robert Hinshaw as president of Wilmington (O.) College signaled a serious possibility that a whole college would take an anthropological perspective. Its "Focus on Man" program corresponded to the interdisciplinary focus on environment that had attracted attention to the curriculum of the University of Wisconsin at Green Bay. As the year passed, it appeared that anthropology was indeed adding, to its traditional "little communities," the laboratory of the complex and changing modern world.

The Cueva Morín excavation. A remarkable excavation at Cueva Morín, a cave on the north coast of Spain, caused extensive reevaluation of previous conceptions about the Iberian Paleolithic sequence and added considerable detail to our knowledge of the lifeways of prehistoric man. The work was conducted by an interdisciplinary team led by prehistorians from the Prehistoric and Archaeological Museum, Santander, Spain, and the University of Chicago. The results of the first two seasons of fieldwork were published during the year; the final campaign, in 1969, was to be the subject of a later volume.

The sequence of intact prehistoric occupation strata at Cueva Morín, all dating from the Old Stone Age, begins with nine horizons belonging to the Mousterian tradition. One of these yielded two extremely important bodies of data. The first was an assemblage of deliberately worked bone artifacts far more abundant and varied than ever before reported from strata of this period. The

Excavations at Cueva Morín, a cave on the north coast of Spain, produced discoveries that added much detail to our knowledge of prehistoric man.

Courtesy, Dr. L. G. Freeman, University of Chicago

other was a straight wall of loose limestone rubble, dividing the cave interior into two "rooms"; all the recovered worked bone fragments were found in a mass on one side of this wall.

In the past, it had generally been thought that the industry marking the transition to truly Upper Paleolithic industrial complexes in France, the Chatelperronian (or Périgordian I), was absent from the Iberian Peninsula. At Cueva Morín, however, an intact occupation of clearly Chatelperronian character directly overlies the Mousterian strata. While the Morín Chatelperronian is very similar to its French counterpart, the nature of other occupations at the site makes it less certain that the affinities of the Chatelperronian are to be sought exclusively with the Périgordian. Morín yielded a series of very primitive Aurignacian occupations (Aurignacian "O" in the French sequence) with marked similarities to the Morín Chatelperronian. The discovery seems to call for a reevaluation of the evolutionary position of the Chatelperronian complex.

One of the early Aurignacian levels yielded several unique finds, including a dugout dwelling foundation, a posthole alignment, and three graves, two of which still contained bodies (or parts of bodies) preserved as soil shadows with three-dimensional integrity—pseudomorphs, or casts-in-the-round of decayed human bodies. Apparently, as the flesh disappeared, its place was taken by fine sediment, different in texture and sometimes in color from the surrounding earth. The importance of these finds demanded that they be studied in detail and preserved if at all possible. Since facilities for the purpose were readily available in the laboratories of the Smithsonian Institution in Washington, D.C., that organization offered to undertake the necessary steps for permanent conservation of the find if means could be found to transport the best-preserved burial (Morín I) from Santander to Washington. Thanks to the U.S. deputy undersecretary of the Air Force, a military cargo plane was dispatched to Santander, and the delicate task of transporting the burial, a 4,140-lb mass of easily crumbled sediment, was accomplished without mishap.

At the Smithsonian, it was possible to complete excavation of the burial under controlled conditions. Invaluable information was obtained about the burial practices of Morín's Aurignacian "O" inhabitants. The best-preserved individual seems to have been decapitated before or during interment and, coupling the puzzling absence of lower extremities in this grave with the charred condition of the lower limbs of a second body, he may have had his legs mutilated, perhaps by burning, at the time of burial. The body of Morín I lay on its left side, legs partly flexed, arms bent to bring the hands up in front of the chest and neck. The severed head was laid in front of the hands. Offerings of meat were placed atop the corpse. The grave was filled, and earth was heaped up above it to form a low mound, on top of which a fire was lit. The mound fill and an offering pit near the feet of the burial yielded abundant flecks of red ochre. Marks of the digging implements used were recovered at the bottom of two of the trenches. Enough of the outline of one arm was preserved to permit a Smithsonian anthropologist to estimate the stature of Morín I as over six feet tall.

From the evidence provided by this Aurignacian level, the excavators judged that the cave dwellers of the period were a very small group of people—perhaps not more than half a dozen—and that they utilized the cave for perhaps 10 to 25 years. Like their Paleolithic predecessors and successors, they made their living by hunting and gathering.

The Morín stratigraphic sequence continues with three more evolved Aurignacian levels, two Gravettian (Upper Périgordian) horizons, and a level each of Solutrean, Upper Magdalenian, and Azilian affinities; loose ceramic of the Early Bronze Age was found on the present-day cave floor. This sequence is as extensive and complete as any yet reported for Spain, and is much fuller than any yet known for the earliest Upper Paleolithic horizons in the Iberian Peninsula.

—Sol Tax

See also Feature Article: MARGARET MEAD: ANTHROPOLOGIST AND SOCIAL REFORMER; Year in Review: ARCHAEOLOGY.

Psychology

Each year the vigor with which behavior is studied seems to increase geometrically. The variety of problem areas being studied is represented by the following accounts taken primarily from basic research laboratories in psychology.

Monkey sign, monkey write. For centuries, language has been considered the exclusive ability of man. Indeed, several linguistic theorists, including N. Chomsky of the Massachusetts Institute of Technology and E. Lenneberg of Cornell University, recently reaffirmed that the acquisition of even the barest rudiments of language is quite beyond the capacities of nonhuman animals.

Clever demonstrations abound, to be sure. There is Tony the mynah bird who said, "Cogito ergo sum." We may recall the zoo raccoon that waited for the sign reading "Lunch is served" to be displayed before eating his midday meal. But these demonstrations are not sufficient to constitute a language system. In acceptable conversation a speaker uses words in regular relations to one another and to aspects of the environment. In reports upon the world about him, and in requests for information and items from the world, a speaker uses several words in many combinations in certain sequences. In addition, to prove that an individual knows the "language" rather than the particular combinations of words, it is usually necessary to show that he utters sentences that were not specifically taught to him. All previous suggestions of even meager animal communication have failed to meet these tests.

However, two independent and successful language programs were reported recently with chimpanzees. Taken together, they established rather convincingly that remarkable conversations may be carried on with at least one nonhuman species. Washoe, a wild-caught female chimpanzee, was taught American Sign Language (the manual communication used by the deaf in North America) by Allen and Beatrice Gardner of the University of Nevada. In the second program, David Premack from the University of California, Santa Barbara, constructed a special language written in plastic forms to converse with his chimpanzee Sarah.

Washoe. Washoe was raised in the relaxed setting of home (several homes in fact). It was necessary for all her human companions to master American Sign Language since, as far as possible, all conversation and interaction between Washoe and her trainers and between trainers was in this mode. Many different training techniques were employed. Imitation was encouraged, in which the trainers made a sign that Washoe would imitate, and then a second sign that Washoe would again imitate, for the reward of being tickled. Occasionally Washoe would be prompted by signing, "Speak up." In some instances outright instrumental conditioning was employed, as in the case of "more." When being tickled, Washoe tended to cover the area tickled by bringing both arms together, a crude approximation of "more." Then tickling was stopped and the trainers waited until Washoe again approximated "more."

While several signs were initially used only for specific activities, Washoe spontaneously generalized their use in novel situations. For example, following the second situation in which "more" was trained, a laundry-basket game, Washoe signed "more" in other activities, including feeding. By the time Washoe was four years old she had a vocabulary of about 80 words, using the strict criterion that a new word was counted after

A cat becomes enraged when an electrical current is applied to its brain by means of electrodes implanted there. Psychologists hope that such tests will help them gain a better understanding of the mechanisms of the central nervous system.

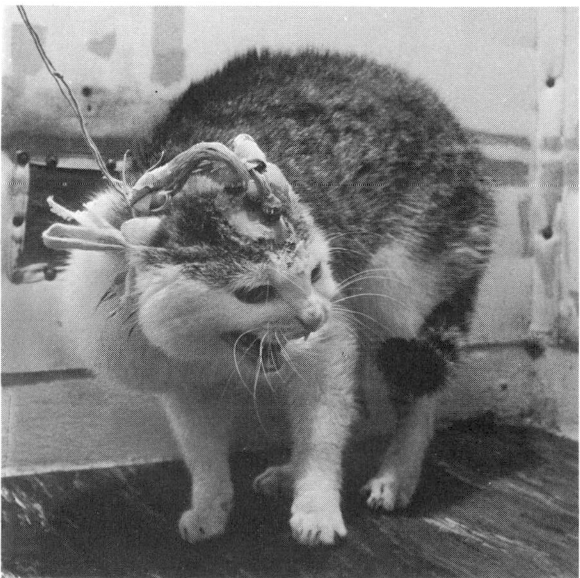

Authenticated News International

it had appeared appropriately and spontaneously each day for 15 consecutive days. As soon as Washoe had 8 to 10 signs, she began spontaneously to string them together. The Gardners observed 330 different strings of two or more signs. The sentence strings included "come gimme drink please" and "listen dog" (at the sound of barking by an unseen dog).

Sarah. While the Gardners tried to develop language in a naturalistic setting and to determine the extent and character of language attained in that setting, Premack was after a functional analysis of language and a set of procedures sufficient to train those functions to a chimpanzee. Sarah's language, consisting of colored, metal-backed, plastic form words, was written by placing the forms vertically (Sarah's preferred manner) on a magnetized slate board. By the time Sarah was seven years old, having been in training for two years, she used about 120 words. Vocabulary size was not the important objective; rather, the objective was to exhibit all the language functions with as small a lexicon as needed.

Initially, Sarah was given a piece of apple after she wrote "apple" on the slate. Then, she discriminated by writing "apple" for apple and "banana" for a banana. Later, with differing trainers, Sarah was to write "Mary apple" or "Randy apple," followed by "Mary give apple Sarah." Questions were answered by substituting a word for the "?" (interrogative marker). For example, the trainer would write "red? apple" ("red is what to apple") and Sarah would answer by replacing "?" with "color of." Note that "color of" is a class concept introduced after color names and object names are learned. Throughout, new words and new semantic and syntactic components were introduced in sentences with a single word missing or requiring replacement.

From the examples of Washoe and Sarah, it is clear that the rhetoric of grammar theories may rely no longer upon dichotomization among species. While the demonstration of language is clear in these cases, however, no one has claimed that equality exists between ape and man. More work is needed to specify the differences.

Chemical clues to schizophrenia? A chemical basis of schizophrenia has been sought for many years. Such an idea is attractive for several reasons. Some data suggest that schizophrenia is distributed in the population as genetic theory would predict and, of course, genetic expression must have biochemical nature. Furthermore, the identification of a responsible chemical would permit attack on the disorder by pharmacology. There have been many research efforts along this line, but most have been disappointing. However, Wyeth Laboratories psychologist Larry Stein and his colleague David Wise suggested another promising candidate, 6-hydroxydopamine.

6-Hydroxydopamine is a naturally although unusually formed variant of dopamine, which is systematically associated with nerve terminals in the brain. If dopamine is not converted totally into the usual substance, norepinephrine, and instead forms 6-hydroxydopamine, this toxic substance can slowly destroy the nerve endings.

The locus of action in the brain is quite important since these areas are involved with reinforcement, or reward, mechanisms, which in turn provide for goal-directed activity. Stein and Wise were able to show disruption of such activity in experimental animals after injections of 6-hydroxydopamine. They were also successful in creating catatonia, a highly specific form of schizophrenia, in rats by administering large doses of 6-hydroxydopamine with another chemical, pargyline. Chlorpromazine, a tranquilizer, is the drug of choice in treating schizophrenics, alleviating some of the more bizarre behaviors. Stein and Wise found that a program of daily doses of chlorpromazine apparently protects affected nerve terminals. They pointed out that a specific odorous substance occasionally detected in the sweat of schizophrenics can be derived from 6-hydroxydopamine.

Exercising a theory. One of the most stable theoretical distinctions in psychology has been that between respondent and operant types of behavior, most forcefully argued 35 years ago by B. F. Skinner. An example of respondent behavior is the classic case of Pavlov's dogs, who were presented with food while a bell was rung and eventually salivated in response to the bell alone. Operant behavior involves the performance of a designated act by the subject, as when a rat presses a bar to obtain food. The two types were differentiated according to anatomical basis, ability to be elicited, and the conditioning procedures applicable to them.

A monumental assault on this distinction began in 1968 with the work of Neal Miller and his colleagues at Rockefeller University, in New York City, who showed that such responses associated with the autonomic nervous system as cardiac action, blood pressure, intestinal motility, and renal action, formerly thought to be conditional only by respondent or Pavlovian procedures, could be conditioned by operant or instrumental procedures. Besides having important applications, this work destroyed the anatomical basis for the distinction.

Additional pressure on the theory occurred during the year in several similar experiments from independent laboratories on an altogether different phenomenon. A wide variety of vertebrate species demonstrated intense fighting against other animals and against inanimate objects upon presentation of painful stimuli, such as electric shocks, or when the reward used to reinforce behavior failed to appear. The effect was like the rage observed in humans upon being struck, or being fired from a job, or having a vending machine fail. The intensity of the attack was dependent on the intensity of the stimulus, and appeared to be independent of previous social experiences, motivational level, and other factors. It occurred unless the stimulus could be explicitly avoided. Thus, the shock-attack response appeared as a reflex reaction in which a painful stimulus produces a fairly stereotyped response that is relatively independent of normal learning experiences.

However, as was shown in several recent experiments, the shock-attack reaction may be suppressed by the same shock that produces attack. C. L. Roberts and K. Blase of Colorado College, R. Ulrich and his colleagues at Western Michigan University, and Nathan Azrin at Anna (Ill.) State Hospital all suppressed shock-attack with shock punishment. Azrin's study is representative. Monkeys received painful electric shock and attacked a rubber hose. With each unavoidable shock, a mean of 9 to 22 bites was observed by each subject. During the punishment phase, each attack bite was shocked. Even when the eliciting shock was 10 mA at 200 v, a punishment shock of 5 mA at 100 v was sufficient to suppress the attacking significantly. Thus, the operant (suppression by punishment) effect overrode the eliciting effect of the pain shocks. Accordingly, the experimenters questioned calling the shock-attack behavior "reflexive" and questioned whether elicited behavior can be considered as independent of its consequences.

—Daniel F. Johnson

See also Feature Article: INSTANT INTIMACY.

Botany

The past year was an exciting one for the plant sciences, as new methodologies produced interesting hypotheses and pointed the way toward the solution of many problems. Scientists sifted evidence from newly discovered fossils, from biochemistry, and from ultrastructure that pointed the way to a fresh understanding of the evolution of life on earth and its subsequent development.

Ancient records of life on earth. One of the most spectacular events of recent years was the discovery of fossils more than 3,000,000,000 years old, extending our knowledge of life on earth back more than 2,000,000,000 years. Formerly, the vast Precambrian era, which occupies nearly 4,000,000,000 years of the earth's estimated age of 4,500,000,000 years, had been thought to be devoid of life, or at least of any recognizable traces of life. By searching for the oldest known rocks and by studying very thin sections of these rocks with the electron microscope, scientists were able to demonstrate the presence of organisms at various levels in the Precambrian, and even in the very oldest rocks known.

The oldest fossils were located in rocks from the border region between South Africa and Swaziland, especially in cherts (an extremely fine-grained variety of quartz) of the Fig Tree series, which are more than 3,200,000,000 years in age. They consist of bacteria-like rods about 0.5–0.7 micron long and about 0.2–0.3 micron thick, with a wall about 0.015 micron thick, and spheroidal organisms about 17–20 microns in diameter (1 micron = 0.001 mm). In the Gunflint Iron formation along the shore of Lake Superior in western Ontario, about 2,000,000,000 years old, a more diverse assemblage of about a dozen kinds of very small fossil organisms was found. The most common of these are filamentous, about 0.6–1.6 microns thick, up to several hundred microns in length, and very similar to living blue-green algae.

One of the Ontario fossils consists of a bulb with a narrow stalk, surmounted by an umbrella-like structure. It was named *Kakabeckia*. Subsequently, similar, living organisms were discovered in soil

Electron microscopy reveals Eobacterium isolatum, *one of two primitive forms of life preserved in a Precambrian rock formation in South Africa. This fossil bacterium, found in cherts of the Fig Tree formation, is more than 3,200,000,000 years old.*

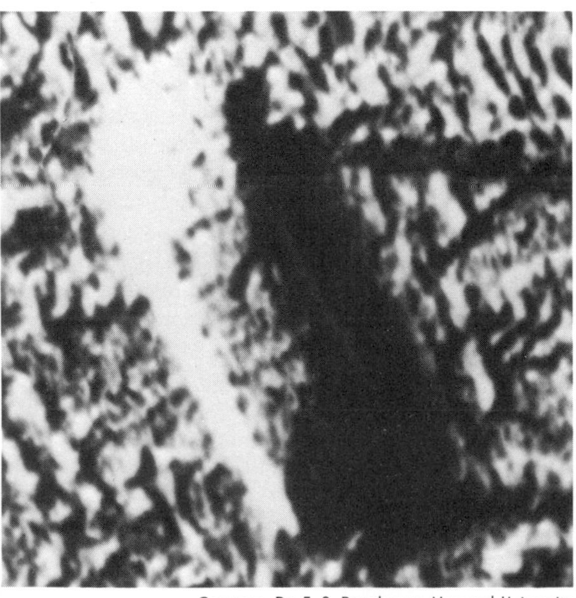

Courtesy, Dr. E. S. Barghoorn, Harvard University

samples being screened under atmospheric conditions similar to those on the earth during the early Precambrian: 25% methane, 25% ammonia, 10% oxygen, and 40% nitrogen. Whether these organisms are actually the same as the fossil *Kakabeckia* was not certain, but it is conceivable that this organism has survived locally in ammonia-rich soils for over 2,000,000,000 years.

Photosynthesis (the process by which green plants manufacture their own food) is probably at least 3,000,000,000 years old, as can be deduced from various kinds of chemical and structural evidence. Modern plants, in the process of photosynthesis, selectively accumulate carbon of atomic weight 12 in preference to its heavier isotope of atomic weight 13. In the atmosphere, there are normally about 99 parts of carbon-12 to one part of carbon-13, but plant parts are even poorer in carbon-13. The ratio of these two isotopes of carbon in the Fig Tree and Gunflint cherts is about the same as it is in modern plants, suggesting the existence of photosynthesis during the Precambrian era. Moreover, the hydrocarbons pristane and phytane, which are breakdown products of chlorophyll (the green photosynthetic coloring matter in plants), are found in the Gunflint cherts. Accumulations of calcium carbonate similar to those formed by modern blue-green algae are found in Rhodesian rocks some 2,700,000,000 years old, and in the Gunflint cherts. It is thought that the long-continued photosynthetic activity of these early organisms gradually changed the earth's atmosphere, decreasing the amount of carbon dioxide and increasing the amount of oxygen.

An even more diverse assemblage of organisms is found in Australian rocks about 1,000,000,000 years old—the Bitter Springs formation. These include not only blue-green algae, but also green algae and fungi. By this time, the outlines of modern groups of organisms begin to appear.

Fundamental differences in living organisms. It has become clear that the most drastic discontinuity among living organisms is not the historically recognized one between plants and animals, but rather the separation of those organisms with an organized nucleus (eukaryotes) from those without an organized nucleus (prokaryotes). The fundamental differences between these basic kinds of organisms was first suggested in 1941 by Cornelis van Neil and Roger Stanier of Stanford University, but it was not properly appreciated until recently.

Prokaryotes, which have been segregated as the kingdom Monera (or Mychota), include the bacteria and all bacteria-like organisms, such as spirochetes, rickettsiae, and blue-green algae.

Courtesy, Dr. E. S. Barghoorn, Harvard University

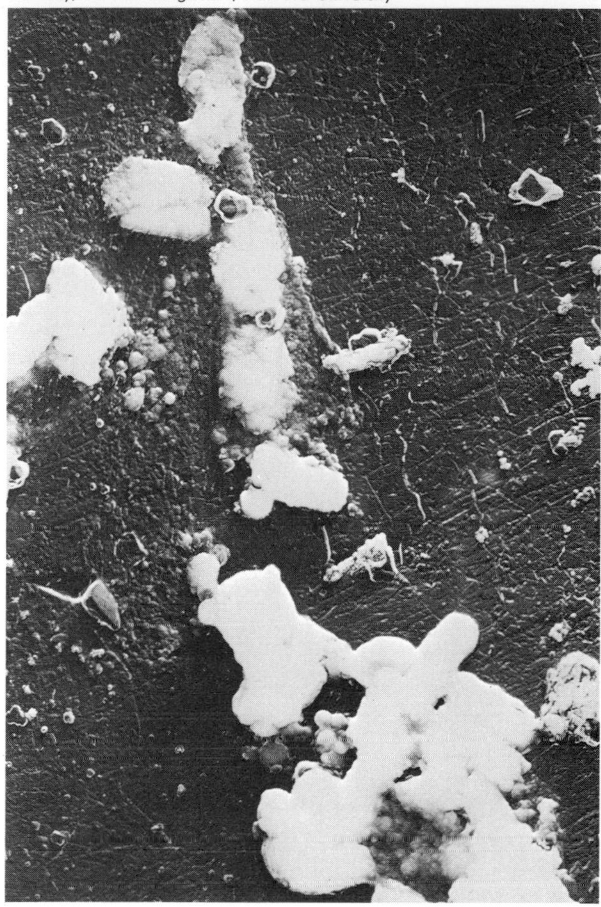

Fossil bacteria, 2,000,000,000 years old, are enlarged 30,000 diameters in an electron micrograph. These bacteria were located in beds of black chert exposed along the shore of Lake Superior in the Gunflint Iron formation in western Ontario.

Their genetic material, as far as is known, consists of a single, circular molecule of deoxyribonucleic acid (DNA). This molecule replicates itself, and the two daughter molecules attach to different points on the cell membrane, being carried in this way into the daughter cells. In the cells of prokaryotes, the ribosomes—the particles that are the sites of protein synthesis—are relatively small, with a sedimentation coefficient of 60–66 Svedberg units (a unit of time used to measure the sedimentation velocity of a protein or other colloidal solution in an ultracentrifuge for use in an equation for determining the molecular weight of a protein). Moreover, there are no nuclear envelopes, mitochondria, plastids, or flagella comparable to those found in eukaryotes (*see* below). Prokaryotes have small cells, comparable to only a portion of the typical eukaryotic cell, and these prokaryotic cells are solitary or united in chains or masses with no

Courtesy, Dr. J. William Schopf, U.C.L.A.

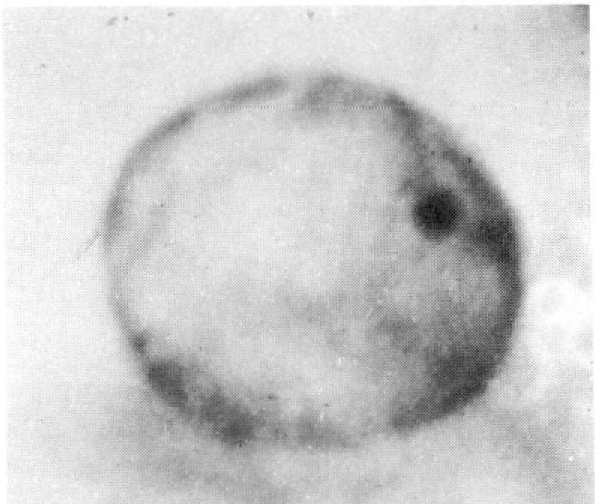

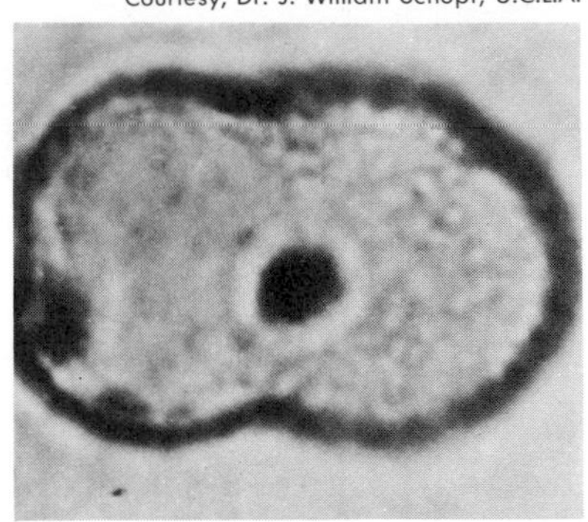

Stages in the cell division of a eukaryotic organism, a green alga of the genus Glenobotrydion, *are preserved in these fossils found in billion-year-old cherts of the Bitter Springs formation of Australia.*

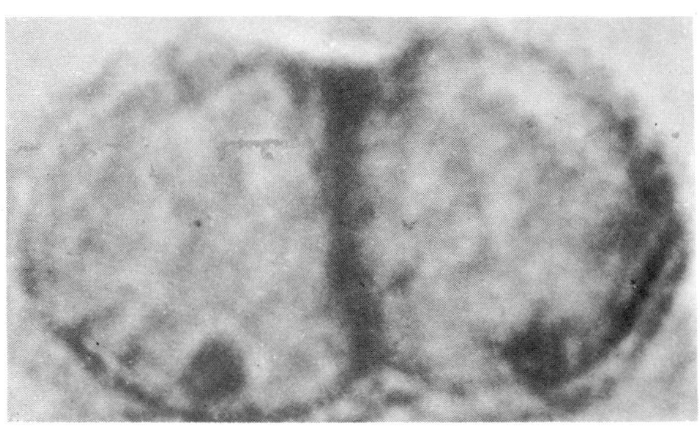

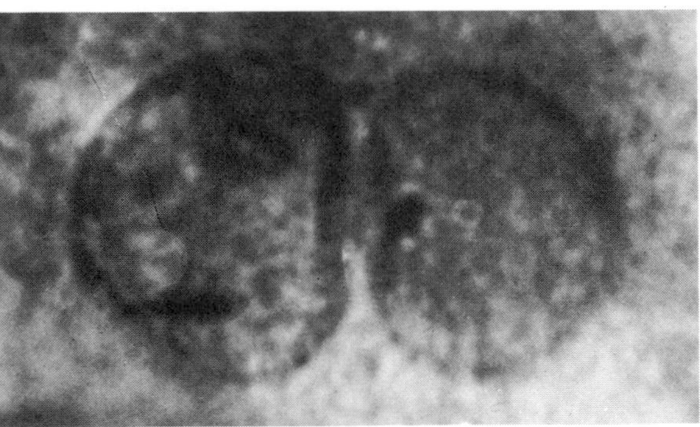

protoplasmic connections between the individual cells. Some are capable of moving by means of simple whiplike appendages (flagella) or by gliding.

Eukaryotes, which comprise all other organisms except viruses, have complex chromosomes in which much larger amounts of DNA are associated with histone proteins. These chromosomes are located in a nucleus, which is surrounded by a double membrane, the nuclear envelope; the chromosomes are partitioned to the daughter cells through the division process known as mitosis, which involves the formation of spindle fibers. All eukaryotes have within their cells mitochondria —complex organelles bounded by a double membrane in which aerobic respiration (of free oxygen) takes place—and some have plastids, also organelles bounded by a double membrane, in some of which (the chloroplasts) the process of photosynthesis takes place. The ribosomes within the cells of eukaryotes are larger than those in prokaryotes, with a sedimentation rate of 70–76 Svedberg units. Some of the cells of eukaryotes move by means of flagella or cilia (hairlike processes) in which there is a characteristic arrangement of nine tubules around the outside of the flagellum or cilium and two others in the center. The organizing structures of these flagella or cilia are similar to the centrioles, minute bodies that organize the spindle fibers in mitosis.

Modern classification of eukaryotes. There are many diverse groups of unicellular eukaryotes, which have been grouped as the kingdom Protista. These include various lines of what have been regarded as Protozoa (acellular or unicellular animals) as well as many groups of algae and fungi. One of these unicellular groups of algae, the dinoflagellates, was shown in 1969 to have chromosomes that differ from those of all other eukaryotes in that they do not contract or expand during mitosis and do not include any histone proteins. From some of these unicellular groups, evolutionary lines consisting of multicellular organisms have been derived.

Some of these are relatively simple organisms that can be thought of as colonial rather than truly multicellular. These include the sponges, in which certain cells in the colony are specialized in relation to certain functions, and also colonial green algae such as the spherical *Volvox*, in which many individuals similar to the unicellular *Chlamydomonas* are held together in a gelatinous matrix, each beating its pair of anterior flagella and thus moving the entire colony through the water.

Truly multicellular lines of evolution include the animals; the fungi; and the red, brown, and green algae. From the green algae have been derived the multicellular land plants. Each of these five groups of multicellular organisms can be traced back independently to unicellular ancestors, and there are still unicellular fungi (such as the yeasts and some water molds) and green algae, as well as a few unicellular red algae. In some schemes of classification, the animals, fungi, and plants (including land plants as well as the three groups of multicellular algae) are regarded as three distinct, fundamentally multicellular kingdoms of organisms. With the Protista and Monera, there are thus a total of five kingdoms, a system first proposed by R. H. Whittaker in 1969. In terms of nutrition, the animals ingest their food; the fungi absorb it, digesting it externally by means of enzymes; and the plants manufacture it.

Origin of the eukaryotic cell. The relationship between prokaryotic and eukaryotic cells has come to occupy a central position in attempts to understand the diversity of life on earth. Despite extensive studies with the light and electron microscopes, no forms intermediate between prokaryotic and eukaryotic cells have been discovered. It now appears likely that eukaryotic cells were derived from prokaryotic cells by a process of symbiosis (living together). For approximately the first 2,000,000,000 years for which fossils are found in the geologic record, all of the organisms are prokaryotes. The small size of these prokaryotes is the reason that they were not discovered until the advent of the electron microscope.

Mitochondria and plastids appear to be symbiotic prokaryotes which play different functions in the cells of eukaryotes. Both have a double membrane, the outer layer of which resembles the endoplasmic reticulum or membrane network in the cytoplasm of the cells in which they occur, and the inner layer of which resembles the cell membrane of prokaryotes. Both have characteristic forms of DNA and RNA, and the DNA is known to be present in at least some mitochondria in circular molecules similar to those found in the cells of bacteria. Both have their own characteristic ribosomes, with a sedimentation rate of 60–66 Svedberg units, like those of prokaryotes but smaller than those of the eukaryotic cells in which they occur.

Protein synthesis in mitochondria and chloroplasts is inhibited by the antibiotic chloramphenicol, as it is in prokaryotes but not elsewhere in the cells of eukaryotes. On the other hand, the antibiotic cyclohexamide inhibits protein synthesis in the cells of eukaryotes but not in their mitochondria or chloroplasts. In both mitochondria and chloroplasts, protein synthesis is initiated with the substituted amino acid n-formylmethionine, which is characteristic for bacteria. This amino acid is unknown elsewhere in the cells of eukaryotes and plays no role in protein synthesis there.

Photosynthetic algae are known to be symbiotic in a wide range of other organisms, including more than 150 genera of invertebrate animals. In these contemporary symbiotic relationships, all stages of assimilation of the symbiotic organisms into the cellular patterns of the host can be observed, including loss of the cell wall and various other morphological changes. Such relationships suggest the ease with which symbiotic relationships are established now, as they clearly have been in the past, the organelles maintaining a considerable degree of autonomy in their new role.

The appearance of eukaryotic cells a billion years or more ago seems to have been connected with the building of oxygen levels in the atmosphere to a point where prokaryotes with the properties of mitochondria could exist. Aerobic respiration is a far more efficient energy-releasing process than anaerobic respiration, to which earlier organisms had been confined, and it is likely that those relatively large prokaryotes that had the ability to ingest and to maintain "promitochondria" would have had a great advantage over their contemporaries. This would have been even more clearly the case had they been able to move, by virtue of their evolutionary acquisition of flagella or cilia, and to move through the water toward other organisms on which they could feed. Subsequently, and probably independently in several different lines, these early eukaryotes ingested and maintained other prokaryotes with the characteristics of chloroplasts, thus giving rise to several kinds of organisms that were capable of manufacturing their own food.

Among the characteristics in which these eukaryotic cells differed fundamentally from prokaryotic cells was a much greater ability to move and to respond to external stimuli. Their greater complexity probably also gave them the ability to differentiate from one another to a much higher degree than could their prokaryotic ancestors, thus paving the way for the appearance of multicellular organisms. A mass of prokaryotic cells adhering to one another must, by virtue of the simplicity of its members, remain an essentially homogeneous mass of cells, but in a similar aggregation of eukaryotic cells there can be specialization to perform the various diverse functions that must be combined to form a meaningful multicellular organism. In addition, the complex chromosomes of eukaryotes became involved in the process of true sexual reproduction, involving nuclear fusion, or syngamy, and meiosis (the nuclear division of germ cells giving rise to gametes or spores with half the full complement of chromosomes). In sexual reproduction there is a controlled release of genetic variability which makes evolutionary advance possible even in relatively long-lived, multicellular organisms.

Multicellular organisms appear in the fossil record about 650,000,000 years ago. They are, of course, much more easily detected as fossils, and were for years regarded as the first evidence of life on earth. All of these multicellular organisms were aquatic, and it is not until about 450,000,000 years ago that the first trace of terrestrial organisms is found. The level of oxygen in the atmosphere was still increasing, and it is likely that only then was it high enough to allow the occupation of the land. The reason for this is that oxygen (O_2) is in equilibrium with ozone (O_3), and ozone is the principal substance that shields the surface of the earth from ultraviolet radiation from outer space. Such radiation is highly destructive to living things and their genetic apparatus, and it is probable that their existence on the surface of the land and in the surface layers of the water could come about only when there was enough ozone in the atmosphere to protect them from these wavelengths of light.

—Peter H. Raven

See also Year in Review: AGRICULTURE; MICROBIOLOGY; ZOOLOGY.

Chemistry

Among the outstanding events in chemistry during the past year was the synthesis of the human growth hormone, a breakthrough that promises help for those thousands of persons afflicted with dwarfism. Other developments ranged from new techniques for disposing of used plastics, work on dissolving blood clots, and the use of lasers to gain information about the dynamics of electronically excited molecules.

Applied chemistry

Large molecules were the dominant theme in applied chemistry during the past year, as initial success in limiting the lifespan of plastics was reported and as the bright promise of enzyme engineering came into sharper focus. Innovations in the construction of polymers, and a promising role for small molecules in dissolving often-fatal blood clots were other products of basic work in chemistry.

Impermanent plastics. In a somewhat ironical turn of events, chemists began trying to find ways to shorten the lifetime of plastics. A few years ago, spurred by consumer complaints, they had found ways to stabilize plastics and keep them from yellowing and crumbling with time. Their success created "plastic pollution"—the highly visible plastic drinking cups, wrapping papers, and other packaging materials that litter the landscape and that will easily survive us and our children. The mounting concern with this problem was evidenced by the number of states—Hawaii, Massachusetts, New York, and others—that began restricting the use of plastic packaging materials or demanding that they be recycled or be sold as returnable items.

Disposal of plastics is not simple. They can be burned; but that adds to the pollution of the air and, in the case of some plastics such as polyvinyl chloride (PVC), produces corrosive fumes that can damage incinerators. Sanitary landfills are becoming increasingly impractical as cities run out of land to fill. A third alternative is to reverse the trend toward permanence and make plastics more vulnerable to sunlight, water, air, and microbes so that with time they are broken down to simpler materials that can be recycled by nature.

During the past year several schemes to provide plastics with a finite lifetime were reported. The trick, of course, is to have the plastic retain its desirable properties, but—like a good servant—disappear when no longer needed. Gerald Scott and his co-workers at the University of Aston in Birmingham, Eng., found materials, which, when added to several types of plastics, will reduce them to a powder within a given time. These materials capture the energy of a portion of the ultraviolet spectrum of sunlight, and then use it to break indirectly the carbon-to-carbon bonds that hold plastic molecules together. (A plastic is an example of a polymer, or chain-like molecule, created by linking smaller molecules like beads on a necklace. Proteins, carbohydrates, and nucleic acids are other examples of polymers.)

The actual chemistry involved in this breaking down process appears straightforward. The reaction of sunlight with the ultraviolet-sensitizing agents produces hydroperoxides, which, in turn, break apart; in doing so, they remove hydrogen atoms from the polymer chain. This hydrogen "abstraction" initiates the breakdown of the plastic, much as a building begins to crumble as crucial bricks are removed. In a typical experiment, a polyolefin packaging film of the type used to wrap meats and vegetables crumbled to dust after an 80-hour exposure to a lamp twice as intense as sunlight. Because ordinary windows screen out the portions of the ultraviolet spectrum to which these sensitizer molecules react, the plastic will remain stable if kept indoors. Telltale colors can be added to the formula to inform the user of a plastic that its disintegration is imminent. Scott reported that the properties of the plastic are unaffected by the addition of sensitizing materials, although manufacturing processes would have to be revised, inevitably adding to the consumer cost of the final product.

Several other efforts to find impermanent plastics were reported. For example, a method, somewhat similar to that developed at the University of Aston,

Courtesy, Monsanto Company

A burnable plastic bottle is being tested to demonstrate its special qualities. Made of a new plastic called Lopac, the container is designed to maintain its shape under high temperatures, although it will burn in conventional incinerators without polluting the air.

which hastens the breakdown of polyolefins such as polyethylene was being formulated by the Connecticut-based firm of DeBell & Richardson, Inc. Alternative chemical approaches to the problem were also being considered. For instance, many plastics must be protected from atmospheric oxygen by antioxidants built into their structures. If these antioxidants could be broken down in some controlled fashion, then the plastics they guard would also break down. Even the use of plastic-eating microbes was being investigated (in spite of jokes about disintegrating planes and disappearing clothes), with considerable attention given to the reluctance of microbes to react with plastics although they will devour paper, which has the same type of carbon-to-carbon molecular backbone.

Of course, incorporation of new material into common plastics, such as proposed by the chemists at the University of Aston, meant that the plastics would have to be screened for safety before they entered the mass market as packaging materials. And, in spite of the encouraging chemical advances, some wondered if the problem was being looked at right side on; that is, whether it was not the control of plastic litter rather than its disposal that should be emphasized.

Improved plastics. But while these chemists were seeking ways to delimit plastics, others were improving their versatility. A group at the University of Illinois reported that they had created the first polymer composed of a chain, or backbone, of nitrogen rather than the usual carbon atoms. This nitrogen polymer is a colorless and stable glass-like material whose possible applications should be considerable.

On a more immediately practical level, work at the Illinois Institute of Technology Research Institute indicated that plastic can be forged like metal. Forging, in which the material is first heated and then shaped under high pressure, is especially suitable for the precise manufacture of small plastic parts; products as thick as an inch can be forged in 20 to 30 seconds.

At the University of Michigan, Charles Overberger and his colleagues probed the still-unknown mechanisms of enzyme action by studying the chemistry of man-made polymers that share some of the same chemical groupings as these biological catalysts. Their aim was to use these simpler polymers as models to evaluate the effects of different chemical arrangements on the rates of reaction. From that basis they hoped to move toward an understanding of how an enzyme, essentially a protein, is able to catalyze the chemical reactions within living cells to rates thousands or even millions of times faster than can be done in a test tube, and to do so at normal body temperatures and pressures. Solution of this puzzle would not only be important to biologists but would also enable the creation of highly effective and versatile catalysts for the chemical industry, whose economic fortunes often depend critically on the efficiency of the catalysts used in their processes.

Enzyme engineering. But while enzymes remained an enigma, serious work was taking place in Israel, Britain, and, somewhat belatedly, in the United States, to find ways of applying them to a host of industrial processes. The possible applications are highly diverse: keeping chilled beer from clouding; speeding the production of mozzarella cheese; improving methods of manufacturing antibiotics, vitamins, steroids, and other medicines; and even simplifying artificial kidneys by providing enzyme-impregnated membranes to purify blood.

A critical problem in enzyme technology and engineering was finding suitable frameworks, or matrices, on which enzymes could be fixed, or immobilized, yet be free to act. Such "bound enzymes" would imitate the situation in living cells in which enzymes are fixed firmly in place by attachment to membranes. Recent work in several

Bill Wangell

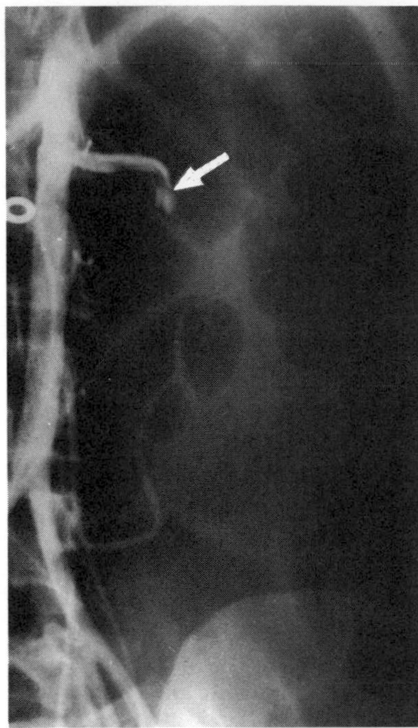

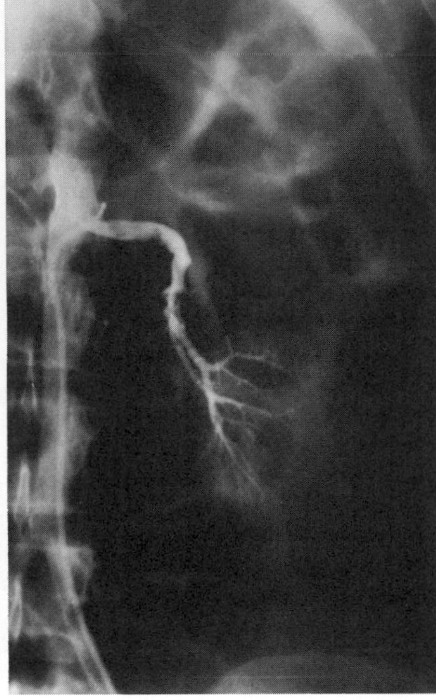

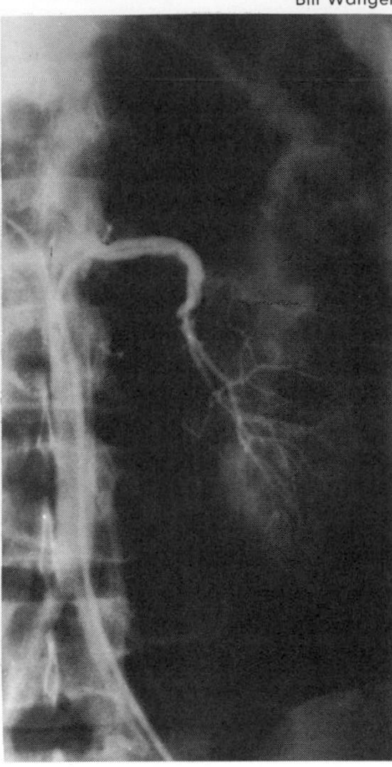

Urokinase, a chemical extracted from human urine, was tested on patients with pulmonary embolisms to determine its effectiveness in dissolving blood clots. The progress of the drug is observed directly in X rays. After the drug is administered to a patient with a blood clot (left), the clot begins to dissolve (center), restoring the patient's normal circulation (right).

laboratories—notably that of Ephraim Katchalski at the Weizmann Institute of Science in Rehovot, Israel—demonstrated that enzymes can be bound artificially without inhibiting their catalytic powers. Several binding agents can be used: organic chemicals; cellulosic, or paperlike, materials; and glass powder. The last appeared to be the leading candidate because of its inertness, rigidity, and resistance to breakdown by high pressure, microbes, and other chemicals. Some 40 to 50 enzymes were successfully bound to glass, usually through a chemical intermediary.

While much basic information remained unknown, enzyme engineering was by 1971 being applied in several ways. Enzymes were being used not only for their great catalytic powers but also for their choosiness in affecting specific molecules while ignoring others. This latter property results in an efficient reaction with fewer by-products. Also, compared to the microbes now used, enzymes are operative at relatively low temperatures, thus requiring less costly cooling processes.

A Japanese pharmaceutical company in Osaka began using immobilized enzymes on a limited scale to manufacture amino acids. Chemists at the U.S. Department of Agriculture used the enzyme glucoamylase, bound to a cellulosic membrane, to break down starch. Scientists at the University of Wisconsin prepared bound forms of the enzyme L-asparaginase, which breaks down asparagine, an amino acid apparently essential to the survival of certain leukemic cells. A possible future application is the use of "enzyme molecular screens" in which different materials—from foods to waste products—are simply passed through screens impregnated with various enzymes that degrade or convert them to other useful or less noxious materials. These screens may be used to break down carbohydrates such as starch into the variety of organic chemicals now obtained by the chemical industry from gas and oil. Pioneer work in this direction was done by W. R. Vieth of the Rutgers College, New Brunswick, N.J. Vieth bound the enzyme urease (which breaks down urea to ammonia and carbon dioxide) to a membrane and shaped the combination into a compact module that combines large surface area with small volume, enabling efficient exposure of various materials to the enzyme. A likely application of this type of system is the continuous treatment of by-product wastes of food processing.

Dissolving fatal blood clots. Chemists hoped that a merger of basic structural organic chemistry and decades of work on the chemistry of blood clotting might produce new agents for dissolving blood clots, considered to be major factors in fatal heart disease. Two enzymes, urokinase and streptokinase, were tested clinically as anticlotting agents. Both apparently work by promoting the appearance of plasmin, a natural clot-dissolving agent in the body. However, these materials might not work on very compact clots for the simple reason that their large molecular sizes prevent them from penetrating a small, compact mass. Partly for that reason, a number of smaller simple organic molecules—chloroform, urea, urethan derivatives, and, most recently, aromatic carboxylic acids—were being evaluated.

Chemists worked to relate the different structural arrangements and sizes of these types of molecules with their ability to dissolve clots. While these organic compounds proved to be effective for a limited time, they were rapidly inactivated in the body because they react with a ubiquitous body protein, albumin. If scientists could find a way to block this inactivation, they could produce materials whose structures can attack difficult, possibly lethal, blood clots with a minimum of side-effects on the patient.

—Norman Metzger

Chemical dynamics

Although many problems concerning the structural (or static) properties of molecules have been well understood for decades, the prediction of molecular reaction rates (chemical dynamics) still remains at a relatively primitive stage. Many features concerning the individual steps of a reaction sequence may be known, but quantitative answers to the question, "How fast does a given step proceed?" can rarely be made before experimentation.

The reader should be aware that the overwhelming majority of known chemical reactions do not occur in one single (elementary) step from reactants to products, but more commonly involve a fairly complex sequence of interrelated elementary steps. At present, an enormous reservoir of empirical data concerning reaction rates is available, and breakthroughs in theories on reaction rates may occur in the near future.

Electronically excited molecules. This article deals with a particular area of chemical dynamics that has made huge advances over the past years, the dynamics of electronically excited molecules. These molecules are produced when matter absorbs light.

The advantages of carrying out reactions by electronic excitation, relative to thermal excitation, are that one can control the molecules (or portion of a molecule) that are excited by controlling the exciting wavelength of light; one can also control the number of molecules excited per unit of time by controlling the light intensity. The lifetime of electronically excited states varies from about 10 sec to 10^{-12} sec, thereby providing an enormous dynamical range for study.

Laser applications. During the past year new laser technology made it possible to study, routinely, electronically excited molecules whose lifetimes are shorter than 10^{-6} sec. The most spectacular advances were those that allowed study of chemical events taking place in electronically excited states in time periods of less than 10^{-9} sec and as brief as 10^{-12} sec. The reader should be informed that, since nuclear movements (vibrations) occur in the range of 10^{-12} sec, and since normally observable chemistry involves rearrangement of nuclear configurations, scientists are approaching a theoretical limit for measurement of the fastest possible chemical events.

Two types of lasers are useful for the generation and study of excited molecules: continuously emitting lasers, which produce a continuous flux of highly monochromatic (single-wavelength) light; and pulsed lasers, which produce a burst of light in an extremely short time period. Although continuously emitting lasers probably represent the "purest," that is, most monochromatic, light in the universe, they suffer from the disadvantage that generally only one important emitting wavelength (which depends on the material of which the laser is composed) is available. However, in recent months continuously emitting "tunable" lasers, whose wavelength can be varied over a wide range, became available.

The development of the "mode-locked" pulsed laser perhaps offered the single most important technological advance in chemical dynamics of excited states. This type of laser generally consists of a standard pulsed laser (of, say, pulse duration of about 10^{-9} sec) and a crystal that "coordinates" the various phases of light from the pulse and then generates a series of very narrow pulses whose individual lifetimes are only about 10^{-12} sec. An important principle of chemical dynamics is that measurements on a system excited by a pulse cannot be analyzed until the pulse is completed. This is because the concentration of excited molecules is changing as the pulse is being absorbed and any measurement would, therefore, involve properties of the pulse in addition to properties of the molecules of interest. Thus, very short pulses from mode-locked lasers allow the chemical dynamicist to probe events that occur within about 10^{-12} sec, a region entirely unavailable to direct study before the advent of this type of laser.

The uses of pulsed lasers to study electronically excited molecules are rapidly expanding. The decay rates of upper vibrational levels of such

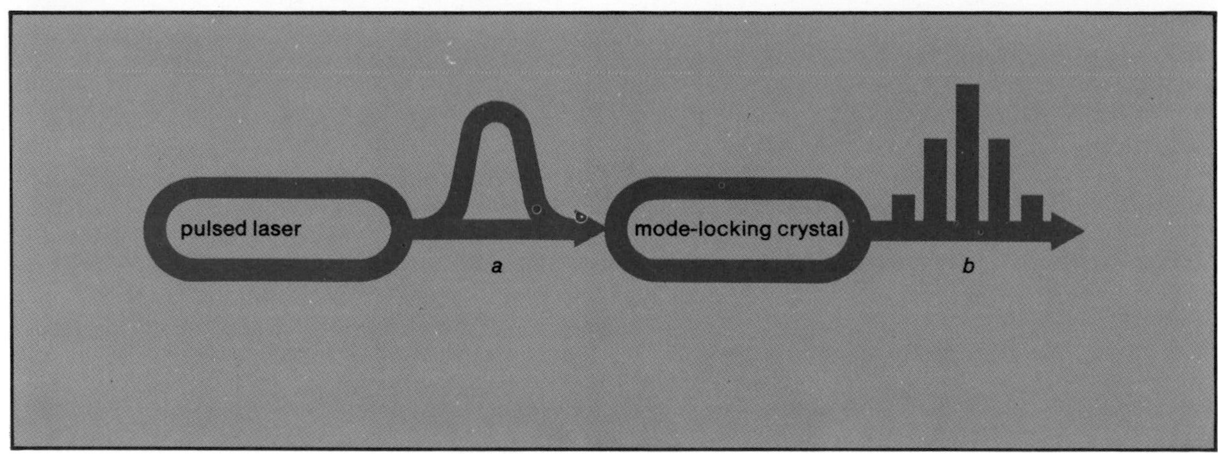

A standard pulsed laser produces a pulse, a, *of about 10^{-9} sec duration. The pulse then passes through a crystal that "coordinates" the various phases of its light and generates a train of very narrow pulses,* b, *whose duration is about 10^{-12} sec.*

organic molecules as azulene and benzophenone were measured by the mode-locked laser technique and found to be on the order of 10^{-12} per second. Emissions from upper electronically excited states, which are rarely observed, were discovered by the use of pulsed laser techniques.

Three states of a typical organic molecule are of special interest to chemical dynamicists: the ground state, that of the lowest energy (S_0); the lowest electronically excited state (S_1) whose total electron spin is the same as that of S_0; and the lowest electronically excited state (T_1) whose total electron spin is different from that of S_0. Light excites a molecule from S_0 to S_1, at which point it then has several choices for further decay: passage to T_1, reversal to S_0, and chemical reaction. If the molecule does decay to T_1, it has two choices for further decay: passage to S_0, and chemical reaction. Laser techniques allow direct study of all of these processes.

In addition to the above-mentioned intramolecular processes, intermolecular events involving electronically excited states are of major importance to the chemical dynamicist and also play a crucial role in photochemical processes. If energy can be conserved, excitation can efficiently "jump" from one excited molecule to another. The rate of this jumping can be measured by standard techniques for many systems, and mode-locked laser techniques have allowed the direct measurement of even the "ultrafast" energy transfer processes.

As an example of the latter methods, in dilute fluid solvents energy transfer between excited solute molecules and ground-state solute molecules is negligible because the excited molecule decays before it can encounter one in the ground state. As the solute concentration is increased, the probability that an excited solute molecule will encounter a ground-state solute molecule increases. By using a mode-locked laser, the lifetimes of the excited molecule can be studied as a function of concentration, and from this data the rate of intermolecular energy transfer between excited and ground-state solutes can be deduced.

Fluid solution studies. In addition to the advances made in chemical dynamics by the application of new laser technology, a large increase in the knowledge of rate processes was contributed by the study of the emission of electronically excited molecules in fluid solutions. The emission of light from S_1 is called fluorescence, while emission from T_1 is called phosphorescence. The rate of fluorescence is generally many orders of magnitude faster than that of phosphorescence. As a result, many molecules do not phosphoresce in fluid solution (because other processes compete favorably with phosphorescence for deactivation from T_1), but fluorescence is commonly observed. During the past year, however, scientists found that certain solvents, such as highly fluorinated organic molecules, allow phosphorescence to become measurable. Consequently, one may now study both the dynamics of S_1 and T_1 by studying fluorescence and phosphorescence, respectively.

Fluorescence lifetimes, which had been somewhat difficult to measure, became routinely available by means of a new technique called "single-photon counting." This new advance employs the principle of exciting a sample with a pulse of light and then measuring the time it takes for an excited molecule to emit a photon of light. It has proved a highly versatile and accurate method.

During the past year, the use of the single-photon technique for measurement of fluorescence and the use of fluorocarbon and other inert solvents for measurement of phosphorescence made it possible to measure a large number of important rate processes. For example, it was found that when an electronically excited molecule collides with another molecule in solution, the probability of energy transfer occurring is essentially unity, providing the energy transfer process is energetically going from a higher level to a lower one.

Photochemical reactions. Detailed analysis of systems by the techniques described above revealed that in many cases excited state aggregates result from the collision of an electronically excited molecule and a molecule in the ground state. Such aggregates (called exciplexes) are believed to play an important role in many photochemical systems. In some cases, these aggregates may serve as a "storage bank" for the electronic excitation; in other words, the aggregate may have a longer lifetime than the initially excited molecule.

The combined techniques of using pulsed excitation and light emission to study chemical dynamics possesses the unique quality of enabling scientists to excite particular molecules of interest and then use the emitted light as a "label" to follow the decay of the excited state; thus, the decrease in light intensity serves as a microscopic "clock" by means of which a scientist can gauge the dynamics of the excited state. The exact form of the emission spectrum and its variation with such parameters as solvent and temperature can also reveal specific information concerning the nature of S_1 and T_1.

As an example of the power of these combined approaches, a study of the chemical dynamics of fluorescence-quenching of norcamphor and its derivatives by the above techniques revealed that "electron-rich" molecules (molecules possessing an excess of negative charge at some site) specifically interact with the "edges" of the oxygen atom of electronically excited norcamphor. Furthermore, electron-poor molecules were found to interact specifically with excited norcamphor in the space above or below the C=0 function. From this behavior it can be deduced that excited norcamphor possesses electron deficiency near the "edges" of the oxygen atom (hence the preferred quenching by electron-rich molecules at the edges), and electron excess in the region of space above and below the C=0 function (hence the preferred quenching by electron-poor molecules).

At the year's end the theory of photochemical reactions remained in a relatively elementary stage. A number of promising developments had, however, taken place. It became possible to make a primitive, but revealing, calculation of the radiationless processes that occur from electronically

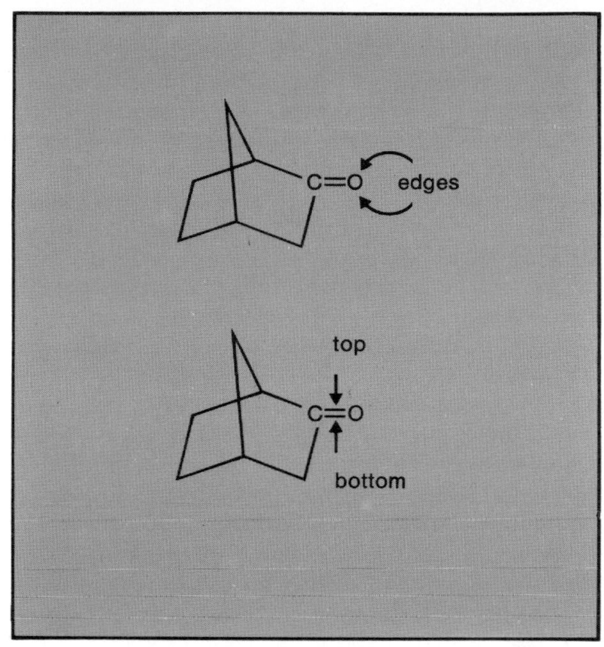

Norcamphor molecule has a carbon atom double bonded to an oxygen atom at one corner. When norcamphor is electronically excited, chemists found that "electron-rich" molecules interact with the edges of the oxygen atom (top) and that "electron-poor" molecules interact with norcamphor in the space above or below the C=O function (bottom).

excited states. Because photoreactions are themselves radiationless in nature, a general theory of radiationless processes would be of major importance to chemical dynamicists and should serve as an important aid in the understanding of the factors that control the rates of chemical processes from electronically excited states.

Future outlook. During the coming year chemists expect to see further development of laser technology that will be of use for chemical dynamics. As the number of "tunable" lasers increases and as their properties become more compatible with the requirements for the study of excited states, these devices can be expected to replace standard light sources. A major advance in technology awaits the use of mode-locked laser techniques as the excitation source for the single-photon counting device. Since the time scale that was made available by the ultrashort pulses of the mode-locked laser has barely been investigated, some exciting and novel chemical phenomena are expected to be revealed as intensive research in this area progresses. Finally, the application of the new laser technology, which is mainly being developed by physicists, to chemical problems offers unforeseeable but almost certain advances that will exploit the highly unique nature of these light sources.

—Nicholas J. Turro

Chemical structure

The establishment of the structure of a molecule is essentially the determination of its size and shape. The problem may vary from the determination of the bond angle between the hydrogen atoms in a water molecule (104.5°) to the determination of the structure of the myoglobin molecule for which J. C. Kendrew received the Nobel Prize for Chemistry in 1962 to the discovery of the double helix of DNA (deoxyribonucleic acid) for which F. H. C. Crick and J. D. Watson received the Nobel Prize for Physiology or Medicine in 1962. Of prime importance is the order of attachment of the atoms to each other.

Conformational analysis. Of nearly equal importance is the conformation of the molecule; that is, which of the many different arrangements of a given molecule that can arise by rotation around the bonds linking the atoms is most preferred. A molecule can be represented by a set of "tinker toys," in which the solid pieces or plastic spheres represent the atoms that constitute the molecule and the wooden sticks joining those pieces represent the chemical bonds linking the atoms. It is clear that if one puts together the pieces of a "tinker toy" set in a given order, there are still many different arrangements possible through simple turning of the wooden sticks. It is not necessary to take the model apart to obtain these different arrangements. In the same way, for almost any molecule, there exists a large number of conformations that can be interchanged by rotations around the chemical bonds. Actually, by such interchanging one is determining the three-dimensional shape of the molecule. This determination is often called "conformational analysis," and for their work in this field Derek H. R. Barton and Odd Hassel received the Nobel Prize for Chemistry in 1969.

The physical, chemical, and biological properties of a molecule depend critically on the conformation in which it exists in the resting state or at the moment of reaction. For example, many naturally occurring proteins have their large molecular chains arranged in complex folded and twisted shapes. The exact nature of the folding and twisting (the conformation) invariably plays a vital role in the function of the protein. Thus, the enzyme, lysozyme, which occurs in egg white, is so folded that a cleft is created within the molecule into which the substrates (a polysaccharide or a sugar molecule) will fit so they can be acted on chemically by the enzyme. As the enzyme is denatured (by heat, radiation, or chemical treatment), it loses its native conformation. The coils and pleats of the functional enzyme are converted into

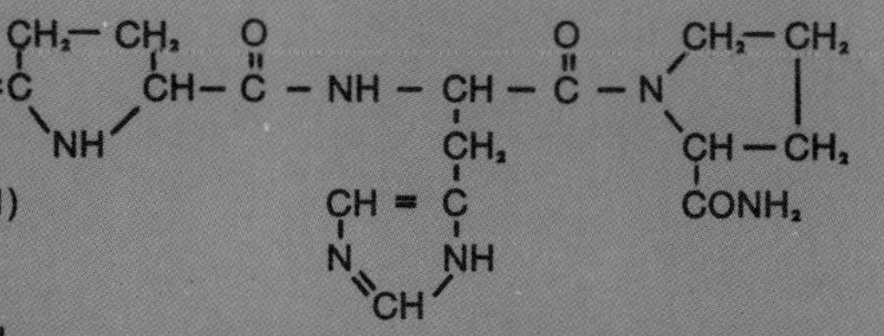

(11)

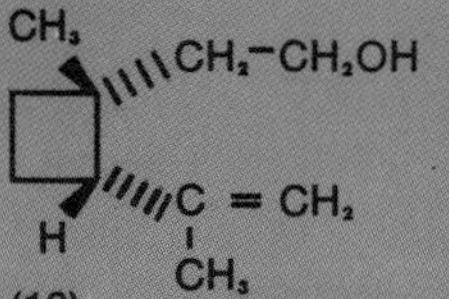

(12)

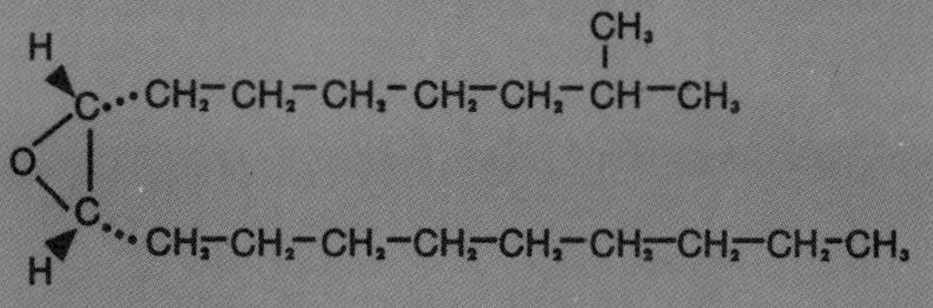

(13)

a straight zig-zag arrangement of the atoms in the denatured enzyme. This conformational change leads to complete loss of function.

Structure of enzymes. Proteins, which are composed of "building blocks" derived from amino acids, are used for many purposes within the body. One group consists of enzymes, the key catalysts that control all dynamic processes of living systems. In each living mammalian cell there are approximately 10,000 enzymes, and many of these are similar from cell to cell. The enzyme composition in a liver cell, however, is different from that of a muscle cell. Protein molecules are very large, having molecular weights ranging from 14,000 to 1,000,000.

Recently, the structures of a number of enzymes were reported. A group headed by Hans Neurath of the University of Washington reported the amino acid sequence of bovine carboxypeptidase-A. There are three forms of this enzyme: the α form has 307 amino acid residues, the β form 305, and the γ form 300. The three-dimensional model of carboxypeptidase-A had been previously established by William N. Lipscomb and co-workers at Harvard University on the basis of their X-ray crystallographic studies. Having learned both the amino acid sequence and the exact three-dimensional shape, these chemists were in a favorable position to speculate concerning the mechanism of action of this enzyme, which occurs in the pancreatic juice.

Also using X-ray crystallography, a group of scientists led by Margaret Adams at Purdue University, West Lafayette, Ind., resolved the structure of an intracellular enzyme, lactate dehydrogenase, which has a molecular weight of 140,000. Although the primary structure (the sequence of amino acids) remained in doubt, the secondary structure (the organization of the polypeptide chains into structural patterns, such as helices and sheets), the tertiary structure (the folding of the polypeptide chains), and the quaternary structure (the arrangements of different polypeptide chains with respect to one another) were solved.

Lysozyme, whose structure and bonding had been worked out by crystallography at the Royal Institution in London, recently was converted by two scientists at New York University to a lysosome; as such, it stores the latent enzymes in the cell, releasing them for attacks on invading viruses and other substances. The lysosome was prepared from three lipid (fatty) components (lecithin, stearylamine, and cholesterol) in which the lysozyme was encapsulated in the phospholipids.

Structure of membrane proteins and of hormones. Also, in the last year, myelin, a protective sheathlike protein that surrounds the brain and nerve tissue, had its structure determined. Edwin H. Eylar of the Salk Institute for Biological Studies, in San Diego, Calif., determined the amino acid sequence in the unfolded polypeptide chains containing 172 amino acids. This development should advance the experimental study of multiple sclerosis.

Hormones represent another group of proteins whose structures were recently established. Although the amino acid sequence of insulin had been known for some time to consist of two polypeptide chains, one containing 21 amino acids and the other 30 amino acids, the exact three-dimensional arrangements of the chains had not been determined. In the past year, a group headed by Nobel laureate Dorothy M. Hodgkin at Oxford University, established that insulin, which controls the level of glucose in the blood, forms a hexameric molecule existing in a triangular ring of three tilted football-like dimers around two zinc atoms. (A dimer consists of two identical molecules bonded in a single larger molecule.) The structural knowledge of this hormone should make possible a better understanding of diabetes.

In 1971 the structures of four mammalian glycoprotein hormones, which are produced in the anterior lobe of the pituitary gland, were identified. Each of the hormones contains an identical region composed of 96 amino acids which is combined with two carbohydrate groups. However, the second region of each hormone is different.

Not all protein hormones are large molecules. In late 1970 a group at the University of Texas Institute for Biomedical Research isolated and determined the structure of TRH, a thyrotrophin-releasing hormone of the hypothalamus (11). They found it to be a tripeptide derived from glutamic acid, histidine, and proline. Also in recent months Georges Ungar of the Baylor College of Medicine, Houston, Tex., isolated the first component material of what may be a system of molecular coding by which information is processed in the brain. The structure of this material was established as a sequence consisting of 15 amino acids. In Ungar's experiments, donor rats were trained to acquire a fear of the dark by receiving an electric shock when they took refuge in the dark. From 4,000 such donor rats Ungar isolated a chemical which, when injected into a mouse in 0.1-μg quantities, changed the mouse's normal preference for the dark into a fear of it. Since the number of possible peptides consisting of 15 amino acids is 20^{15} (there being 20 natural amino acids), the potential number of such peptides is well beyond that necessary for encoding all the information acquired by the brain in a lifetime.

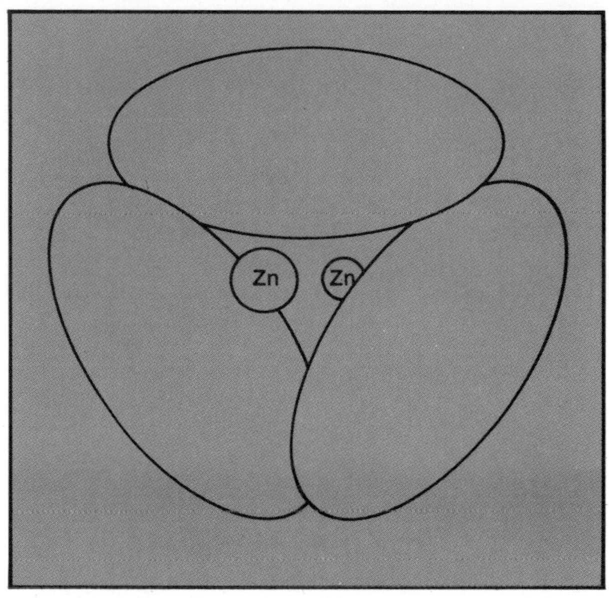

A model of the insulin hormone reveals that its structure is a triangular ring of three football-shaped dimers surrounding two zinc atoms.

Insect attractants. As people became more concerned with pollution of the environment, they were made aware that a long list of pesticides, such as DDT, would have to be replaced. As an alternative to the chemical control of pests, the use of sex pheromones, or attractants, gained much recent attention. Two groups, one at the Zoecon Corp. in Palo Alto, Calif., and the other at the U.S. Department of Agriculture's Boll Weevil Research Laboratory in State College, Miss., established the structure of the sex attractant for the boll

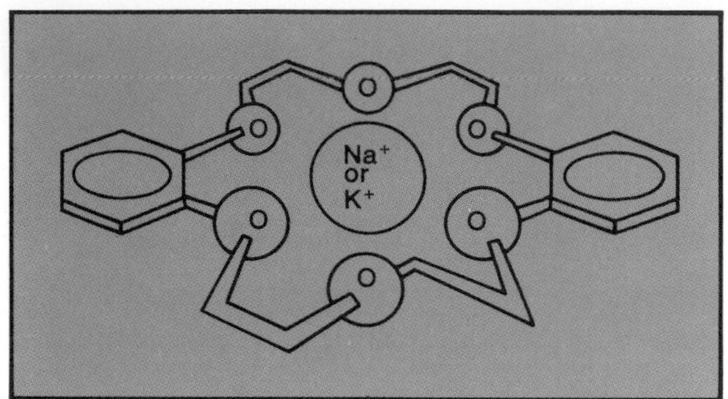

Dibenzo-18-crown-16, the first macrocyclic polyether made, is a compound with a large ring structure that contains ether-oxygen linkages. Compounds of this type coordinate with such alkali ions as sodium and potassium.

weevil, which does $200 million of damage to the cotton crop each year. The structure (12) was shown to be *cis*-2-isopropenyl-1-methylcyclobutaneethanol.

Also in late 1970, a group headed by Morton Beroza at the U.S. Department of Agriculture's Agricultural Research Center, in Beltsville, Md., identified the sex attractant emitted by the female gypsy moth to be *cis*-8-epoxy-2-methyloctadecane (13), which was active in laboratory assay in concentrations as low as 10^{-12}. Similarly, Wendel Roelofs at Cornell University showed that *cis*-8-dodecenyl acetate would attract the male of the oriental fruit moth, the oblique-banded leaf roller, the red-banded leaf roller, and the European corn borer. Obviously, traps baited with these sex attractants can eliminate a large number of insects without polluting the environment.

Another approach to the control of insects is to use the juvenile hormone which prevents insects from maturing by keeping them in successive larval stages until they die. The structure of this hormone was established to be methyl *cis*-10,11-epoxy-7-ethyl-3,11-dimethyl-*trans, trans*-2,6, tridecadienoate. Presumably, it or a simple analogue can be synthesized economically in the laboratory. In addition, 3,5-methylenedioxyphenyl ethers of 6,7-epoxygeraniol will prevent yellow meal worms and milkweed bugs from maturing. These artificial juvenile hormones are not insecticides since doses as high as 10 mg are not fatal. (See *1971 Britannica Yearbook of Science and the Future,* Feature Article: BATTLING THE BUGS, pages 50–63.)

Polyethers and anomalous water. An interesting class of compounds consisting of macrocyclic (containing a large ring structure, usually of 15 atoms or more) polyethers was discovered by C. J. Peterson. (A polyether is a thermoplastic material that contains ether-oxygen linkages, —C—O—C—, in the polymer chain.) The first of these compounds made was 2,3,11,12-dibenzo-1,4,7,10,13, 16-hexaoxacyclooctadeca-2,11-diene to which the name dibenzo-18-crown-6 was assigned. Compounds of this type coordinate with a large number of ions, including sodium and potassium. As the size of the crown and the substituents on it can be varied widely, a whole variety of interesting compounds can be made. These materials should aid in the analysis of organic substances, make ion-sensitive electrodes, and produce a number of new catalysts for a variety of reactions including polymerization.

Anomalous water (or polywater), which attracted much attention in recent months, was shown to be not an unusual form of water but ordinary water containing ionic impurities that cause it to have unusual chemical and physical properties.

—William J. Bailey

Chemical synthesis

The cherished goal of synthetic chemists is to duplicate in the laboratory the most complex molecules of nature. The joint efforts of R. B. Woodward and his co-workers at Harvard University and A. Eschenmoser and his group at the Federal Institute of Technology in Zürich, Switz., over the last several years to synthesize vitamin B_{12} (1) demonstrated this strong urge.

The less complex natural products, especially those with molecular weights below 1,000, are comparatively easy targets for the organic chemists of today. Because the continuing discovery of novel reactions and new reagents (substances that take part in one or more chemical reactions) is simplifying the task of the synthetic chemist, scores of naturally occurring molecules are being duplicated in the laboratory every year.

The quality of the synthesis of a natural product is no longer judged by the simplicity of the scheme or by the high yields obtained. It is not enough to achieve the gross juxtaposition of the right atoms; attempts must be made to assemble the molecule in a stereospecific manner (that is, to conform to a desirable molecular architecture). In what is described as an elegant synthesis, each reaction is planned with an understanding of the possible consequences, and a pathway is selected that provides the right intermediate as the major product at each step. In this respect the synthetic chemist is trying to approximate the extreme selectivity shown by enzymes in nature. An illustration of this attitude is provided by some of the recent syntheses described below.

Synthesis of cecropia juvenile hormones. Juvenile hormones play a regulatory role in the molting process of insects. They are of much interest because of the possibility of their application to the selective control of insect populations. The structure and stereochemistry (spatial arrangement of atoms and groups in a molecule) of two juvenile hormones (2 and 3) from the moth *Hyalophora cecropia* are known. The first synthesis, by a group at the University of Wisconsin, produced various isomers of the known juvenile hormones differing in geometry around the double bonds. (Isomers are chemical compounds that contain the same numbers of atoms of the same elements but differ in structural arrangement and properties.) Several subsequent syntheses also led to mixtures of geometrical isomers. (Geometrical isomers are those that differ in their spatial structure about the same plane through the molecule. The most common are found among compounds in which two atoms or groups may be on the same side—*cis* isomers—or opposite sides—*trans* isomers—of a double bond.)

Since then, alternative routes were developed at various laboratories that are increasingly selective regarding the stereochemistry. E. J. Corey of Harvard University developed a synthesis in which in a single step the basic juvenile hormone chain (7) was assembled from three components, (4), (5), and (6), specifically in the correct stereochemical form. The intermediate (7) could be transformed to either (2) or (3) without a change of geometry. The yields were good, and no complex separation of isomers was involved at any step. The juvenile hormones so obtained, however, were in the racemic (optically inactive) form, whereas they are optically active in their natural state.

Synthesis of quinine and indole alkaloids. M. Uskokovic and his group at the research laboratories of Hoffmann-La Roche Inc., Nutley, N.J., developed a practical synthesis of a key intermediate of predetermined *cis* or *trans* geometry. By using the right optically active stereoisomer of this intermediate, they synthesized a large family of related cinchona and indole alkaloids, such as (8), (9), and (10) in the exact form in which they are found in nature. (Optical stereoisomers have no plane of symmetry in the molecule and so are mirror-images that cannot be turned into a position of coincidence. Thus, compounds containing an atom to which four different atoms or radicals are bonded are optical isomers, gaining this name because one isomer rotates the plane of vibration of a beam of polarized light to the right and the other to the left.) Those isomers not occurring in nature also could be produced at will and their pharmacological activity tested. A number of the indole alkaloids available in quantity by means of this approach had been obtained previously from plant sources but in amounts inadequate for thorough biological evaluation.

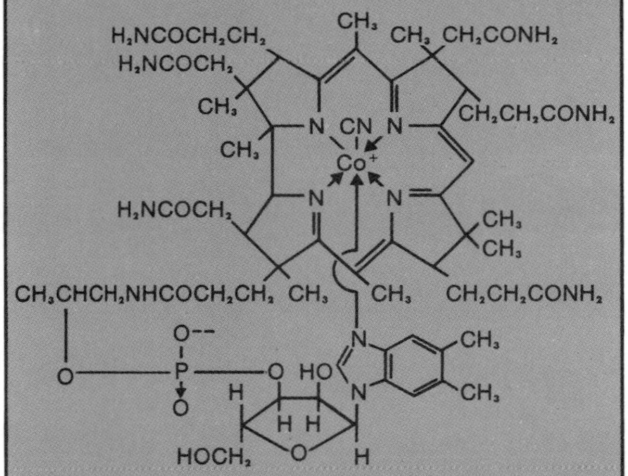

(1) cobalamin (vitamin B_{12})

Synthesis of peptides and proteins. Many biologically important compounds, such as enzymes and hormones, are proteins of various sizes. The task of isolating and purifying such compounds is usually arduous and time-consuming. For example, the thyrotrophin-releasing hormone obtained from the hypothalami of 165,000 hogs amounted to only 4.4 mg. However, during the past year G. Flouret of Abbott Laboratories described a convenient synthesis of this hormone in any quantity desired.

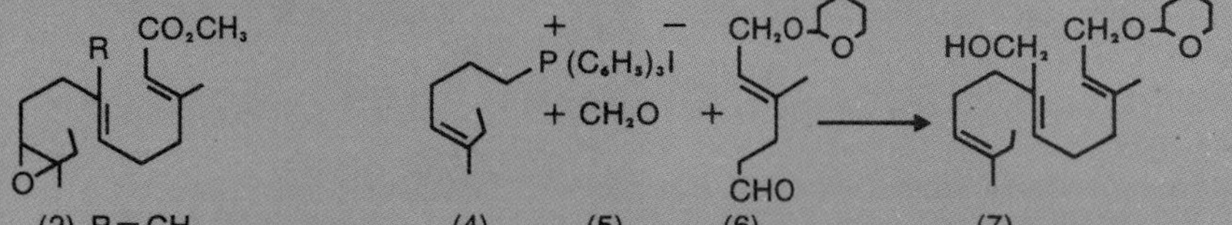

(2) $R=CH_3$
(3) $R=CH_2CH_3$
(4) (5) (6) (7)

(8) quinine

(9) ajmalicine

(10) tetrahydroalstonine

Thus, it will be possible for biologists to study this hormone extensively. Eventually, a synthetic peptide may be found to act as an antagonist of the natural hormone and permit the selective control of certain biological functions, including fertility.

Future trends. Although synthetic activity in the field of steroids seemed to be declining, antibiotics, especially penicillins and cephalosporins, were receiving increasing attention. Because a number of academic and industrial groups were involved in the synthesis of prostaglandins and peptides with known biological activity, important developments were expected in the near future.

Recent publications by E. C. Taylor of Princeton University and A. McKillop of the University of East Anglia in Great Britain established the usefulness of organothallium compounds for efficient organic synthesis. Organoboron and organoaluminum compounds continued to receive attention for the introduction of functional groups into alkenes, while increasing use was being made of organocopper compounds for synthesis. It was safe to predict that various organometallic compounds would be employed even more widely for simplified and stereospecific synthesis of complex structures.

—Ajay K. Bose

See also Feature Article: MATERIALS FROM THE TEST TUBE.

Breakthrough in chemistry: synthesis of human growth hormone

In one of the most important advances of the year, two California scientists succeeded in synthesizing the human pituitary growth hormone (HGH), showing it to be a single polypeptide chain that consists of 188 amino acid residues. The assembly of a protein of this size was an undertaking that would have been considered impractical even 10 years ago. But an ingenious technique developed by R. B. Merrifield of Rockefeller University changed the situation. In 1969 Merrifield described the successful application of this method to the synthesis of the enzyme ribonuclease, containing 124 amino acid residues. (See *1970 Britannica Yearbook of Science and the Future,* The Science Year in Review: MOLECULAR BIOLOGY, *Breakthrough in molecular biology: the synthesis of an enzyme,* pages 294–295.)

Choh Hao Li and Donald H. Yamashiro of the University of California's Hormone Research Laboratory in San Francisco applied Merrifield's technique in achieving the synthesis of HGH. The basic principle of the Merrifield method is to attach the carboxy (COOH) end of an amino acid to a polystyrene resin in the form of granules and then treat the resulting solid with a succession of appropriate reagents. The amino end (NH_2) of the first amino acid is then coupled with a second amino acid, and the cycle is repeated with other amino acids. Purification of the product at each step is achieved simply by washing the polymer granules clean of residual chemicals. After the desired peptide chain has been assembled, it is detached from the resin

Choh Hao Li (right) and Donald H. Yamashiro inspect some of the laboratory equipment involved in synthesizing the human growth hormone.

Bill Young, "San Francisco Chronicle"

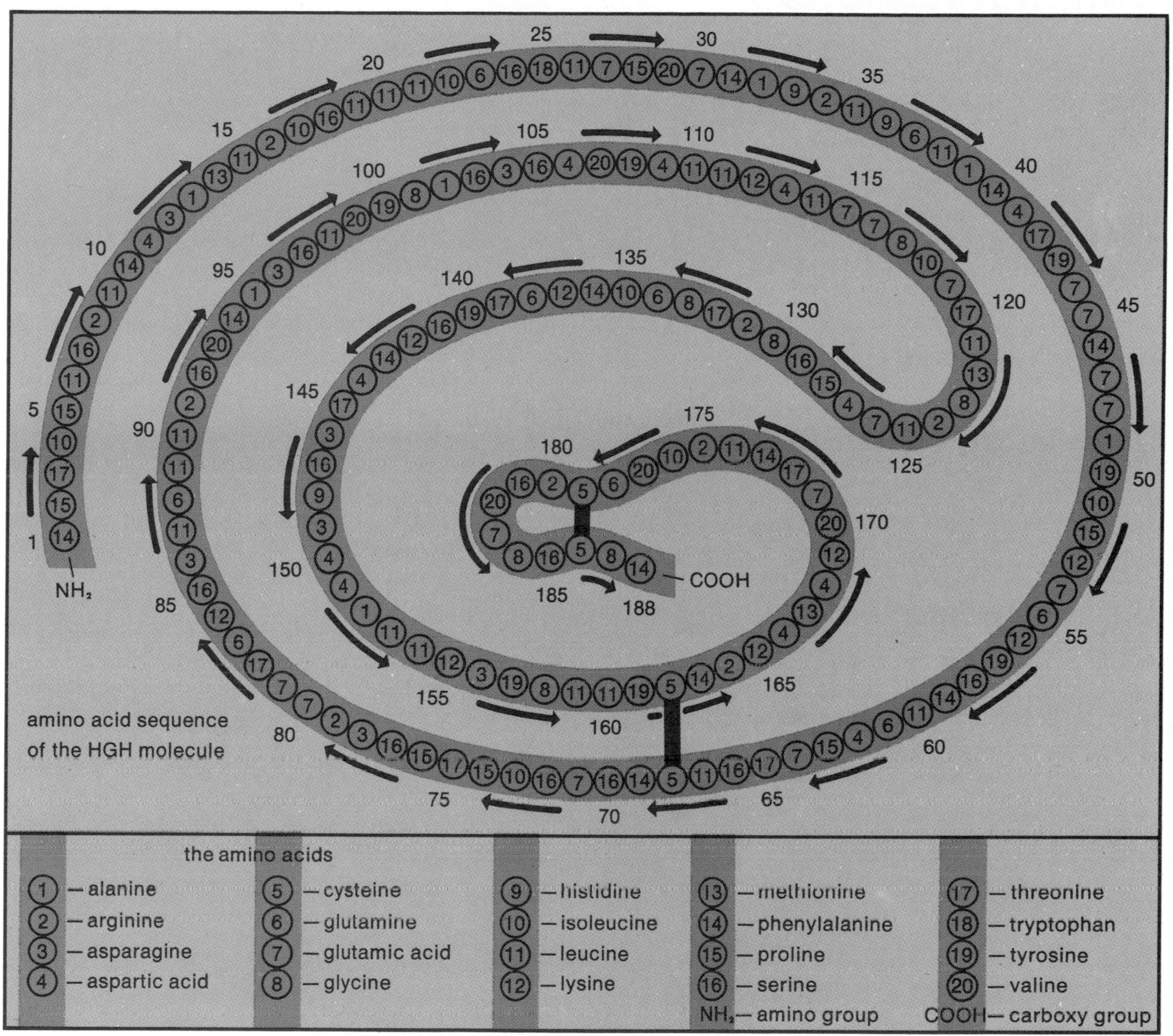

Structure of the human growth hormone, the synthesis of which was announced in 1971, is a complex chainlike molecule consisting of 188 amino acids. The dark bands indicate molecular bonds.

granules by a mild chemical reaction. These operations can be carried out with an automated machine with tremendous saving of labor and time. Li and Yamashiro borrowed such a machine and applied the peptide building cycle 187 times to phenylalanine. In this way they prepared a protein that exhibited to a significant degree the biological activities associated with the natural hormone.

The synthesis of HGH will make a relatively large supply of it available. Previously, the hormone could be obtained only by extraction from a person's pituitary—a pea-sized gland at the base of the brain—immediately after death, a method that yielded very small quantities. The increased supply was expected to help greatly in the treatment of dwarfism, which affected about 7,000 children born each year in the U.S. Dwarfs have an abnormally low supply of HGH.

Other possible applications of HGH include the stimulation of milk secretion by the mammary glands and the promotion of activity in the sex hormones—androgen in the male and estrogen in the female. Research on animals showed that HGH causes bone fractures to heal faster and that it also lowers the level of cholesterol in the blood. These latter effects, however, had not yet been demonstrated on humans. There was also some speculation that HGH might play a role in treating cancer.

Communications

The past year was an especially historic one for telecommunications. The dimension of sight was added to sound when video-telephone service was introduced commercially.

The demand for voice, data, and the new video-telephone services was expected to create greater demands on communications systems. Forecasters predicted that by 1980 communications capability would have to be double or triple that of 1971. The communications industries must also provide new services that did not exist in 1971 but that would be demanded in the near future.

During the year the significant developments in the science of telecommunications leaned toward fulfilling the new communications needs. The number of telephone calls increased significantly in the latest report on 1969 usage, and a number of new services, which would probably be commonplace by the end of the 1970s, passed important milestones. In addition to video-telephone service, there was increased interest in mobile radio and in domestic communications satellites. Other items of interest included the converting of printed text into synthetic speech via computers, more development work on solid-state lamps for future telephones, and the development of a semiconductor laser for a possible future communications system

A synthesized speech pattern is displayed on the television screen in the background. This synthetic speech is part of a new information-retrieval system being devised at Bell Laboratories to provide up-to-date information for a variety of purposes.

Courtesy, Bell Laboratories

that could be powered by ordinary dry-cell batteries. A new method for the commercial manufacture of materials for magnetic bubble devices was also announced. These devices may be used in future telephone switching systems and in computers.

Organizational innovations were also made during the year. In the United States, an Office of Telecommunications Policy was created; in Great Britain, 1970 was the first full year of a new arrangement for providing communication services. In the past, British telecommunication service had been run by the government; now it was to be handled by a private agency and operated for profit.

Telephones. As of Jan. 1, 1970, there were over 255 million telephones in the world, an increase of 7.3% in one year and the largest annual gain since 1956. Much of this worldwide increase was attributed to a few areas—Hong Kong, Greece, South Korea, and Brazil—where individual growth rates were more than double the world average. In conversations per person, an important measure of penetration into the social and business life of a country, the leaders were: U.S., 745; Canada, 710; Sweden, 650; and Iceland, 646.

As more telephones were placed in service, demands on the international telephone networks became greater. Overseas calling in the U.S. doubled in just three years—from 9.9 million in 1966 to 19.8 million in 1969.

Transmission systems. Communications satellites continued their important role in international communications. At the end of 1970, approximately 40% of all U.S. overseas traffic was handled this way. Three satellites were launched in 1970 by the Communications Satellite Corp., one for the Pacific area and two for the Atlantic. Unfortunately, the Pacific satellite failed to achieve its desired orbit.

A third satellite launched for use in the Atlantic in 1971 was a new design known as Intelsat 4. The most sophisticated communications satellite yet to be launched, it was capable of carrying television as well as telephone signals. A control station could reconfigure its antennas to provide spot or global coverage. The maximum capacity of about 6,000 telephone conversations would be obtained when all the satellite's power is concentrated into the spot-beam configuration.

A number of specific proposals were presented during the past year to the U.S. Federal Communications Commission (FCC) for satellites having domestic applications. Proposed systems ranged from one that could carry more than 10,000 message channels to one that would carry primarily television for distribution throughout the country.

Canada started construction of a domestic satellite system, and the Soviet Union continued to expand the Molniya satellite system it had used for a number of years. The Soviet system differed from that of the U.S. and Canada in that it used satellites in inclined orbits that had an apogee of 24,500 mi and a perigee of 300 mi. The U.S. satellites, on the other hand, were in circular orbits over the Equator at an altitude of 22,300 mi. Such a satellite would not be useful for the U.S.S.R. because so much of the country is in the northern latitudes.

A submarine cable capable of carrying more than 800 simultaneous conversations went into service between the U.S. and Spain. The new cable, plus the satellites, should provide adequate facilities to handle the needs of the early 1970s. By 1976, however, if present growth continues, systems of greater capacity will be needed. In 1970, therefore, development began on a submarine cable system capable of carrying 3,500 channels. This system would use a polyethylene-filled coaxial cable, 1.5 in. in diameter, with amplifying repeaters spaced every 5 naut mi. Work on a similar cable was also under way in Japan and Great Britain.

The Japanese were considering a submarine cable system to provide service between various parts of their country. They were investigating a system of several cables, with a total capacity of 25,000 channels, to interconnect Tokyo and Osaka. Installation of a submarine cable along the Japanese coastline might prove easier than cable or radio relay on land because the terrain is quite mountainous.

Field trials were begun in New Jersey on the new L5 coaxial cable system. It was designed so that eventually it would be able to carry 90,000 phone calls simultaneously when operated with a 22-tube coaxial cable.

The tempo of activity on the development of another high-capacity transmission system, called millimeter waveguide, picked up dramatically in the U.S., Japan, France, and the Soviet Union. In this system, a buried pipe about the size of a human wrist would someday carry 225,000 simultaneous telephone conversations. This is enough capacity to transmit, letter-by-letter, a full 24-volume set of an encyclopedia in one-tenth of a second.

Video-telephone service. As mentioned above, video-telephone service was introduced during the year in downtown Pittsburgh. By the end of the year it was also available in downtown Chicago and the southwest portion of Washington, D.C. Present plans called for several dozen more cities to be added by 1976.

The basic service provides a television camera and screen, approximately 5 in. by 5 in., on which

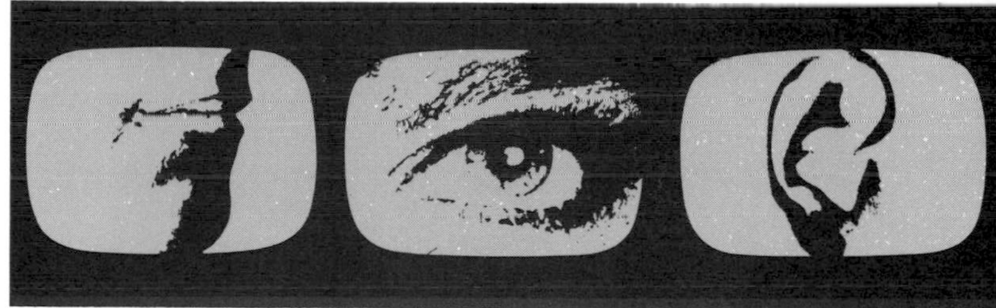

"Electric circuitry has overthrown the regime of 'time' and 'space' and pours upon us instantly and continuously the concerns of all other men. It has reconstituted dialogue on a global scale. Its message is Total Change, ending psychic, social, economic, and political parochialism. The old civic, state, and national groupings have become unworkable. Nothing can be further from the spirit of the new technology than 'a place for everything and everything in its place.' You can't *go* home again."

Marshall McLuhan

Courtesy, Bell Laboratories

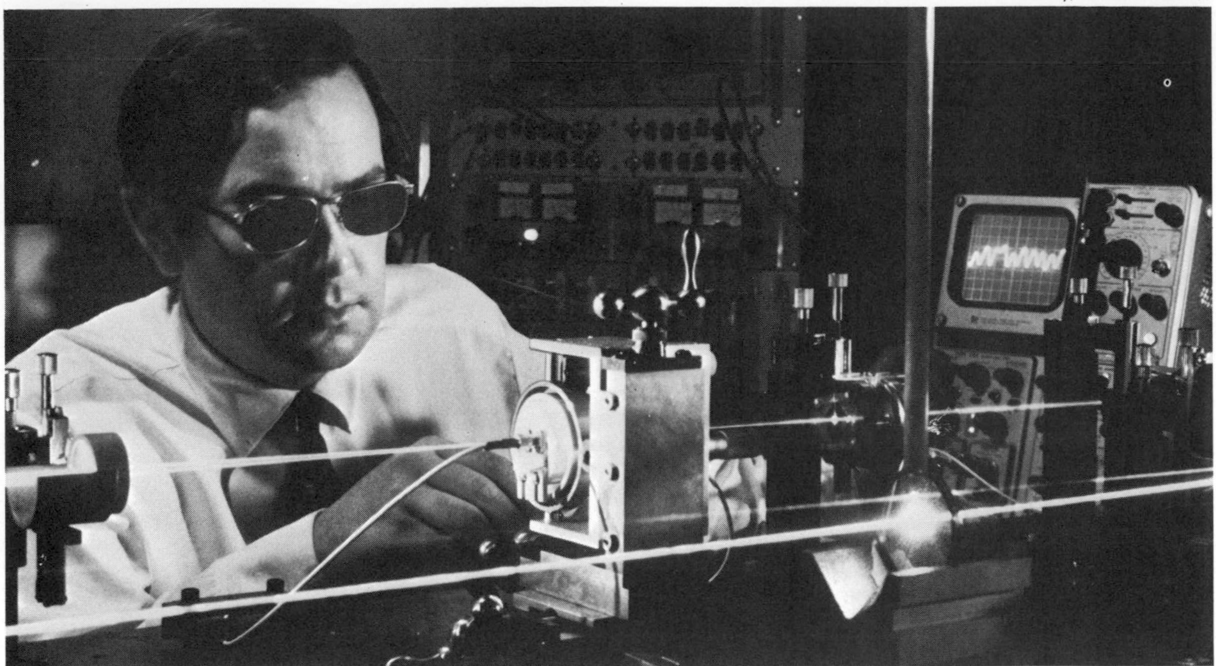

A billion bits of information are impressed onto a laser beam each second by a device called an optical modulator. This laser system, now in experimental stages at Bell Laboratories, eventually may be used to transmit television and telephone communications, replacing the traditional microwave or cable systems.

the user can see the party to whom he is speaking. A built-in speaker and microphone permit hands-free conversation. Not only will video telephone provide face-to-face communication, it also can be used for gaining access to a computer. The set's selector buttons allow the user to communicate with the computer and to view and manipulate data displayed on the screen.

Mobile telephones. The year was one of significance for mobile telephony. By authorizing the use of more frequencies, the FCC increased the capacity for this service from 33 channels to more than 500. The efficient exploitation of this increase will, however, require a major engineering and design job, and this service will, therefore, probably not be commercially available until the late 1970s.

An especially promising approach to developing mobile telephone technology was the "cellular" plan. In this, the range of the transmitters, both fixed and mobile, is limited to a small zone or "cell" so that the same frequencies can be used again in a nearby cell. The connection would go as far as possible over land lines and only short distances via radio to the mobile station.

Speech synthesis. In another interesting area of communications—man-machine interfacing—a computer was programmed to convert printed English text to speech. Speech researchers at Bell Laboratories had tried many methods for doing this, but they all had required human translation of each message into special machine terms. With the new computer, however, scientists were able to produce nearly natural sounding synthetic speech directly and automatically from ordinary English text.

In the experiments at Bell Laboratories, passages were typed and sent to a computer from a teletypewriter. The computer analyzed the sentence, assigned stress and timing to each word, and found a phonetic description of each word from a dictionary stored in the computer's memory. Mathematical descriptions of vocal-tract motions were then computed. They generated electrical speech signals that could be heard over a loudspeaker or telephone. This new information-retrieval system might someday assist a doctor who desires recitation of a page from a medical text, or help a stock manager seeking information about current inventory, or be an aid to the blind for programmed instruction.

In another development, a user desiring information from a computer library might call the computer from home and have the machine speak the information over the phone. The combination of synthesizing and the increasingly efficient storing of synthetic speech in a computer was expected to provide a new range of telecommunication services in the future.

Semiconductor light bulbs. Solid-state lamps made of synthetically grown gallium-phosphide crystal attracted increased attention in 1971. These lamps were expected to be more reliable, efficient, and economical than the incandescent lamps now used in telephones and other display boards, and might someday replace them.

Gallium-phosphide is a transparent, solid material that resembles amber. A small, suitably prepared crystal—0.015 in. on each side of the surface by 0.010 in. thick—will give off red or green light with almost no heat when a small electrical current is passed through it. The crystal's estimated life is more than 100,000 hr compared to 10,000 hr for incandescent lamps.

Gallium-phosphide lamps were expected to have a large potential application in business-type, multi-button telephone sets having illuminated push buttons that are used for various lines. These lamps were used to create number-display panels for telephones and "moving-light" display panels that can be used in combination with telephone handsets.

Lasers and magnetic bubbles. An important step taken during 1970 might affect the future of optical communications. Researchers developed a new semiconductor laser that is smaller than a grain of sand, can be powered by ordinary dry-cell batteries, and can run continuously at room temperature.

In the past, the heat generated in semiconductor lasers was too great to permit operation at room temperature for more than a fraction of a second. The new laser consists essentially of four layers of semiconductor material. In the design, two of the layers—each about 60 millionths of an inch thick—confine the laser light to a thin, central layer of the structure called the active region, which is about 20 millionths of an inch thick. The laser activity, which is effectively confined to such a small region, can be produced with much less current and therefore generates less heat during operation.

Work on the new magnetic bubble technology reached a major plateau this past year when scientists revealed a new technique that might be used to manufacture bubble devices commercially. The technique makes it possible to fabricate devices with more than one million bubbles per square inch. First announced in 1969, magnetic bubble technology may provide compact and inexpensive data-storage and processing devices for tomorrow's computers and telephone-switching systems.

Impact of telecommunications on society. As the world becomes smaller by virtue of rapid transportation, the need to communicate rapidly and conveniently becomes increasingly important. The key role of a reliable and dependable telephone system is especially highlighted during emergencies. During the 1965 power blackout in a northeastern area of the U.S., the uninterrupted operation of the telephone was credited with maintaining calm in a highly tense situation.

The coming decade was expected to see increasing reliance on the telephone. The recognition of its importance caused Pres. Richard M. Nixon to establish a special executive agency, the Office of Telecommunications Policy. Its job was to define the alternatives between certain broad areas of policy and to recommend the direction that should be taken.

Among the subjects being studied were the use of submarine cables and satellites for international communications, and the role of competition in providing special communication services. The close cooperation between U.S. government agencies, private telephone companies, and the various telecommunications agencies throughout the world should produce a worldwide system that will continue to provide service of increasing quality and quantity.

—Irwin Welber

Computers

During the last year a number of significant announcements of new computing equipment were made. By and large, however, the new developments were of an evolutionary, not radical, character, leading to faster, more reliable, more flexible, and more economical systems and devices. The United States continued to lead the world in numbers of computer installations although the 1970–71 economic recession, unlike earlier downturns in the U.S. economy since 1950, had a major dampening effect on new computing activity.

World computer growth. It was estimated that the number of computer installations in the United States had increased from 600 in 1950 to more than 70,000 in 1970–71. The computer population of the rest of the world was thought to be roughly half that of the U.S., but it was growing at a faster pace (a rate of about 25% during 1970) and some estimates indicated that it would approach the U.S. level in five or six years. Although computing activity in the major European countries was strong, Japan was actually second to the U.S. with approximately 6,000 installed computer systems. The number of computers in Japan was expected to multiply threefold by the end of 1974.

The Soviet Union's computer-use growth rate was estimated at only about 15% per year com-

Courtesy, Hughes Aircraft Company

Laser-cut suits roll from a new computerized device designed and built by Hughes Aircraft Co. Using a laser beam to cut patterns out of fabric for clothing, the computerized cutter promises to be a major advance in apparel manufacturing.

pared to a world average of about 20%. The Soviets conceded that they had a data processing gap with the West. In order to narrow this gap they planned to concentrate on the development of a few standardized models, thus reducing production costs. They hoped that by 1975 their annual computer production rate would be approximately 3,000 per year compared to their current 1,200 per year. The Soviets believed that their lack of progress in computer production had contributed to a slowing of their industrial and economic growth. Although the Soviet Union was extremely interested in buying computers, especially of the most advanced type, the U.S. Department of Commerce had not authorized such sales by U.S. manufacturers.

U.S. business environment. The economic recession in the United States resulted in a year of retrenchment for the computer industry. A number of small manufacturers went out of business, and, for the first time since the emergence of the modern computer, positions were hard to come by for programmers and other computer professionals. Computer sales activity in the U.S. in general slowed to a pace that was about 10% less than that of the previous 12 months, and there were also serious declines in software activity.

One major event among U.S. manufacturers was the formation, during the fall of 1970, of Honeywell Information Systems Inc., which resulted from the purchase by Honeywell Inc. of the general-purpose computer interests of General Electric Co. This merger was thought to have strengthened Honeywell's position in the highly competitive computer field.

One effect of the proliferation and rapid technological change of computers was the development of a market for used computers. The first public auction of secondhand computers and peripheral equipment took place in 1970.

Computer products and systems. A number of new data processing systems were announced in 1970. In many cases these were specifically designed for the remote-computing, integrated-system, large data-base needs that, it was thought, would characterize computing during the next decade. It has become customary to refer to each major technological improvement that has occurred in computing as characterizing a new "generation" of computers. In these terms, some of the newly announced machines were thought of as representing a "half-generation" improvement in technology.

Among the novel features of these new systems was the extensive use of monolithic integrated circuits. In some new models these circuits were even used for high-speed storage, replacing the customary magnetic core. Another new feature was the use of improved control storage devices to contain microprograms. Unlike most earlier such storage devices, these were more easily reloadable. Such a facility in a system allowed for the inclusion of microprograms, which are short programs that can be executed as though they are wired in. Microprograms are useful for the execution of comparatively simple functions that are called upon with great frequency, such as those in emulation (the imitation of one machine by another). The error-handling facilities were also improved in some new models. These improved facilities included means for the detection and correction of hardware and programming-system errors and, when an error was detected, the automatic retrying of certain processing unit operations.

Increasing attention was given in the past 12 months to the use of computers in a "sensor-based" process control environment, both in terms of new applications and in new machine design. In these applications, information obtained from various measuring devices during on-going processes was digitized and entered directly into the computer, which could then be used to provide the information needed, for example, in the automation of manufacturing operations or to control the course of laboratory experiments. Such computers could be tied directly to other large computers, making quite extended calculations feasible during the on-going process.

In the area of supercomputers, two new models were under development: Goodyear Aerospace Corp.'s Staran IV and Control Data Corp.'s Star. Staran IV employed parallel array modules that could provide up to 40 million operations per second by performing arithmetic and search operations on many parallel data streams simultaneously. Star was a string array processor that operated on "vectors," or strings of 32- or 64-bit words. These variable-length lists of operands were selected from main memory indirectly through a high-speed "scratch page memory." Operations were performed at great speed in assembly-line fashion.

At the other end of the scale, the minicomputers were booming in sales. There were approximately 50 models on the market. As with most computing products, costs were lower and improved technology provided faster memories. One significant use of minicomputers was to control peripheral equipment. By taking data management (one of the functional areas of a computer's operating system) out of the large central processor and putting it into an attached minicomputer, users could sometimes reduce the overhead costs associated with the routine processing of input-output operations. The number and uses of the small programmable desk calculators also grew at a substantial pace during the last 12 months.

The vision of this 14-month-old child is tested by a computerized instrument linked to electrodes on her head. The instrument records brain-wave patterns stimulated by the child's visual response to flashing lights, thus revealing how sharply the image was seen.

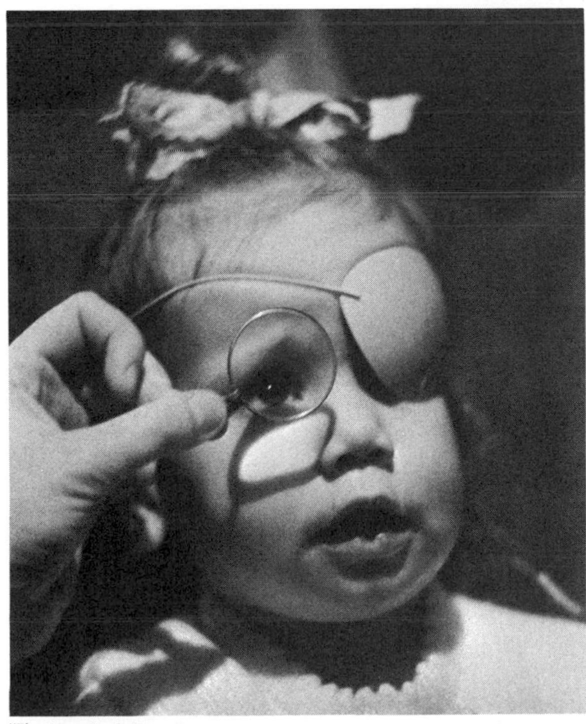

"The New York Times"

Microfilm was used more widely as a storage medium for both input to and output from computers. The number of installations making use of computer output on microfilm increased by about 70% during 1970. With facilities to handle computer input in the form of microfilm, a total picture-processing capability became feasible. For instance, fully automatic handling of engineering information from sketches to drawings was possible.

To handle the preparation of data for computer input more efficiently and economically, keypunch-to-tape and keypunch-to-disk devices were being used with increasing frequency. These devices permitted the direct entry of keyed information into off-line storage facilities and permitted the programmed editing and validation of data before it was entered into the computer, thus saving considerable central processor time.

Communications. Experiments, still quite tentative but promising, were under way in the establishment of networks of computers linked by communications lines. The computer network of the Advanced Research Projects Agency (ARPA) of the U.S. Department of Defense was an example of a large-scale, interconnected system. A goal of this network was to allow a person sitting at a console connected to any one of the computers in the network to use the hardware and software of any other computer of the network with the same facility with which he could use his own. The first transmissions between member installations in the ARPA network occurred in 1970 although the network was still in its initial stages as a research tool.

With somewhat related goals the Computer Sciences Corp. was developing its "Infonet" System, which would provide users the hardware facilities of more than a dozen different manufacturers. General Electric was working on an international information processing network that would make its services available through telephone lines to thousands of business firms.

The time-sharing service business, which provides users with terminal access to remote, shared computing facilities, was very active and highly competitive. Terminal products were becoming increasingly sophisticated, including in some instances programmable facilities and data files in the form of tape cassettes.

To allow greater competition in furnishing data transmission facilities, the U.S. Federal Communications Commission announced in May 1971 a policy of open competition in microwave and other specialized communications systems. This would allow any financially and technically competent company to offer such communication service, which was used primarily for data transmission, in competition with the existing common carriers.

Research. The search for devices that would lead to faster and cheaper computer systems continued unabated, yielding many promising preliminary results. Among the several competing technologies being considered to replace the use of magnetic cores for the building of high-speed computer memories were integrated circuits, charge control devices, and magnetic bubbles. Magnetic bubble domains were being experimented with, and significant advances in creating such physical configurations were made during the past 12 months. A 10,000-bit magnetic bubble register requiring very little power was developed at Bell Telephone Laboratories with eventual costs of substantially less than one cent per bit predicted.

Research continued in the development of a type of transistor that would be capable of much faster switching and higher frequencies. One promising configuration was the "Schottky-barrier" field effect transistor of MESFET (metal semiconductor field effect transistor), which was capable of operation at a frequency of more than 17×10^9 Hz. Another device under study was the Josephson tunneling cryotron, which was a superconducting switching device with extremely fast switching of less than a nanosecond (one billionth of a second).

Theoretical research in the field of computer science included efforts to develop a useful theory that would provide a better understanding of the complexity of computations, programs, and machines, and also to develop theoretical constructs that could be applied to verifying the correctness of programs. Activity also continued on the analysis of algorithms and on heuristic methods of artificial intelligence. All results along these lines were fragmentary and of little import in the practice of computing. Some computer scientists believed that the current period of gathering evidence about computational phenomena was a necessary, preliminary stage to the establishment of a broad scientific theory that would be applicable in making optimal use of computers.

Computer applications. The number of successful new applications of computers continued to be legion. In the U.S. financial markets a number of steps toward the greater utilization of computers took place. More than 3,000 subscribers with terminals located at brokers' offices began to make use of the National Association of Securities Dealers' Automated Quotations System. A member of the U.S. Securities and Exchange Commission proposed restructuring the securities industry into two computerized markets that would include an amalgamation of existing stock exchanges as well as the over-the-counter market. Nine major banks in New York City began sharing a computerized communications network to automate and expedite the handling of clearing house interbank payments. This represented the first employment of "electronic money" within the commercial banking system. The handling on a real-time retrieval basis of the individual accounting necessary with the daily exchange of four million personal checks in the New York Clearing House, however, was still a distant prospect.

Considerable attention was also given to problems related to ecology. Computers were being

Courtesy, Bell Laboratories

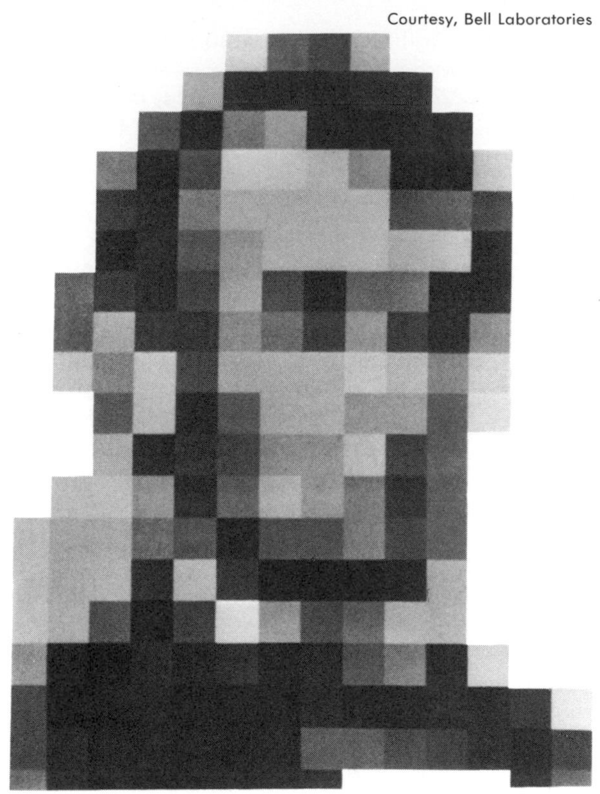

The familiar face of Abraham Lincoln has been blurred by a computer as part of an experiment to learn the minimum amount of visual information a picture may contain and still be recognizable. This study may prove to be useful for computer storage of pictures.

applied to the collection, storage, and retrieval of information on the toxic effects of substances being introduced into the environment; they were also used in environmental surveillance systems to monitor pollutants and their effects, and in environmental control systems to establish and implement governmental strategies for pollution reduction.

Such applications were of special interest to the computing fraternity, which was becoming more sensitive to its social responsibility. The 1970 annual meeting of the Association for Computing Machinery included a discussion of the social consequences of an increasingly computerized society. Consumer advocate Ralph Nader, in a key address to the Association, suggested an "Information Bill of Rights" that would guarantee individuals the right to examine government-maintained records of their personal matters. Concern with the citizen's right to privacy in view of ever-expanding data bases was continually expressed in the media and by public officials.

But there were some disappointments in the development of computer applications. The degree of progress anticipated in a number of areas simply was not achieved. To cite some examples: (1) Much thought in the past had gone into the planning of total management information systems that would permit an effective and, to a large extent, automated scientific approach in the management of large, complex organizations. Systems that were produced in many cases did not come anywhere near the expectations of their originators. (2) A decade ago there was considerable optimism about the possibilities of using computers for language translation. This area was far from achieving the promise seen for it. (3) Computer-assisted instruction was a field in which there continued to be substantial research, but it had yet to achieve in compelling fashion the successes originally anticipated. (4) The potential of the computer in medicine was still not widely accepted in the medical profession. Only about one out of eight hospitals had a computer, and of these about 90% were used exclusively for business applications. (5) For 10 years the U.S. Post Office Department had sponsored the development and testing of optical character readers to be used in reading and sorting letter mail, but only about 15% of total mail volume was being handled by optical character readers.

It was thought likely, however, that success would eventually be achieved in all of these areas. Large, complex computer systems often required years for successful development. It was not surprising, therefore, that in the brief history of computing a number of such systems have not yet achieved the maturity, economy, and infallibility required to be successful. Perhaps economic considerations offered the most serious difficulties. For example, it was doubtful that computer-assisted instruction would be very successful until very cheap terminals offering audio and visual facilities were available together with well-developed, carefully tested programmed course material. It seemed highly probable that these would be available in the future. To illustrate continued progress in another of these areas, management information systems were evolving into reality, albeit at a pace much slower than originally expected.

—Frank S. Beckman

See in ENCYCLOPÆDIA BRITANNICA (1971): COMPUTER.

Dentistry

Perhaps the most critical challenge confronting the U.S. dental profession at the beginning of 1971 was to ensure that dental care would not become a "stepchild" or be totally ignored if a national health insurance program were passed by the U.S. Congress. In assessing the implementation of such a program, the American Dental Association (ADA) considered such issues as the uneven distribution of dentists across the country in relation to population, the cost of dental care (which had increased less compared to other living expenses), the fulfillment of future dental manpower needs (which appeared to be jeopardized by the financial crisis in dental education), and the continuing growth of private prepaid dental care plans.

In dental research, progress was achieved in a number of areas. New dental uses for the laser were investigated, and more effective topical fluoride solutions were found that could augment water fluoridation—the major weapon in dentistry's armamentarium to combat tooth decay.

ADA policies on national health care. The multitude of national health legislative proposals strongly attested to the prospect that Congress would eventually make a decision on a national health program. For some time, the ADA has given signal attention to this matter by appointing its own special task force to review long-standing ADA policies on national health care and to formulate recommendations for a new policy statement. The task force was to present its final report and recommendations to the ADA House of Delegates at its annual session in October 1971.

Meanwhile, the ADA took a dim view of some of the proposals, particularly those which purported to be complete health programs and yet excluded

Courtesy, Dr. Leonard B. Shulman, Harvard University

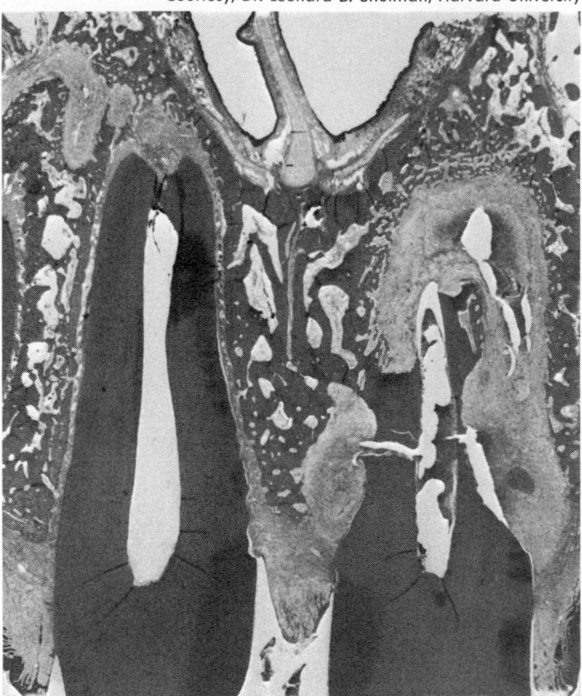

Distinct differences appear in tissue sections from tooth transplants in a rhesus monkey. The autograft (left), a transplant from the same monkey, remains almost intact. The allograft (right), a transplant from another monkey, displays extensive damage.

dental care. ADA spokesmen said that dental care was an integral part of total health care and that no national health proposal could be considered sound or responsive to public need if it did not include a realistic dental program, particularly one for children.

The ADA specifically criticized U.S. President Richard M. Nixon's Family Health Insurance Plan. According to the ADA, the plan not only ignored the need for dental care but, by eliminating Medicaid, apparently would cancel such dental care as was now available to several million needy children. The ADA also took a negative position on the administration's proposal for a National Health Insurance Standards Act since it failed to include dental coverage.

It also criticized all of the Congressional health proposals that involved the use of federal funds for personal health care for those individuals who could afford to pay for it. The ADA once again stressed the need for a national dental care program for children, emphasizing that nearly 50% of all children under 15 years of age had never visited a dentist. The ADA urged the administration and Congress to recognize that dental care was essential to the total health of the nation.

Tooth-decay research. One of the most promising areas of dental research in the past year was in the use of topical fluoride treatments as a decay-preventive measure. A Harvard University School of Dental Medicine research team reported that a liquid-base resin sealant applied to the teeth of teenagers following the topical application of fluoride prolonged the degree of concentration of the fluoride on tooth surfaces and increased the decay-prevention ability of the topically applied fluoride. In their experiment, the resin sealant was applied to all tooth surfaces on one side of the mouth in 50 children. The teeth on the other side were not sealed.

After the first six months, the research group found consistently higher concentrations of fluoride in the resin-sealed teeth in comparison with those that were not sealed. Comparison of the decay rates between the treated and untreated teeth showed that the untreated had four times the number of cavities.

Phosphoric acid treatment. A dental scientist at the Forsyth Dental Center School for Dental Hygienists in Boston, Mass., found that pretreatment of teeth with a phosphoric acid solution prior to topical fluoride application could increase decay resistance. Children receiving the pretreatment and an acidulated phosphate fluoride (APF) topical application developed 26% fewer decayed and filled teeth and 32% fewer decayed and filled surfaces over a one-year period when compared with children who received only the APF treatment. Previous research had shown that a major portion of the fluoride deposited in enamel from topical treatments washes away during the first 24 hours and that only modest amounts are retained permanently.

New fluoride agents. Several new topical fluoride agents appeared to offer greater protection against tooth decay than the commonly used sodium fluorides, the International Association for Dental Research meeting in March 1971 in Chicago was told. One of the new fluoride agents was titanium tetrafluoride, which acted in two ways to protect tooth enamel: chemically, by promoting a decrease in enamel solubility and sustaining its effect; physically, by providing a "glaze" on enamel surfaces. The fluoride content of enamel surfaces treated with the solution was consistently higher than that of enamel treated with fluoride compounds now in wide clinical use. Two other solutions, acidulated ammonium fluoride and an ammonium silicofluoride solution, were also described as being more effective in depositing fluoride in the teeth than the conventionally used sodium fluoride.

Fluoride-containing orthodontic cement. A fluoride-containing cement used to bond braces to teeth might substantially reduce the threat of decay to those teeth, according to University of Iowa researchers. They reported that, after a one-day coating with orthodontic cement containing 10% stannous fluoride, the teeth held about twice the normal amount of fluoride in their outer enamel layers. After a seven-day coating, the outer enamel layer contained about three times the normal amount of fluoride. This process could ease a decay problem for children whose braces were fitted loosely or whose adhesive cement was washed away.

Fluorides in teeth transplants. Fluorides were also found to be effective in prolonging the survival of teeth transplanted between individuals. Four Harvard University researchers discovered that concentrated phosphate fluoride was highly effective in inhibiting tooth root resorption, the process that leads to the rejection of teeth transplanted between individuals. Studies on Rhesus monkeys on whom the fluoride solution was used in teeth transplants revealed that there was significantly less root resorption and that the existence of the bottom third of the root was prolonged by 27%. The same fluoride solution was to be used eventually for human transplant experiments planned at Harvard and the Massachusetts General Hospital.

New theory on origin of dental decay. Research conducted by two dental scientists at Loma Linda University, Loma Linda, Calif., offered a new insight into the development of dental decay. From their studies, Ralph R. Steinman and John Leonora concluded that a slowdown in the speed of internal tooth fluid movement resulting from a high sugar diet might be an early indication of the eventual development of decay. The fluid movement within each tooth is a transport system flowing from the tooth pulp or nerve through tiny channels in the dentin to the covering enamel. The scientists theorized that resistance to dental decay might be systemic and that complex physiological processes are involved in maintaining the health of the tooth. This theory differed strongly from the commonly accepted concept that decay is a local phenomenon on the surface of the tooth.

Periodontal disease research. Emanuel Cheraskin, a dental scientist at the University of Alabama School of Dentistry, reported that periodontal disease could be almost totally prevented if treatment were started early in life. He recommended that the preventive procedures performed for adults should actually be started on children. Periodontal disease is found in virtually all adults in some form and is frequently found in children. The disease accounts for most of adult tooth loss.

Cheraskin also believed that prenatal influences might play a role in the development of periodontal disease. Studies on rats revealed that rodents kept on various prenatal and preweanling diets showed marked differences in post weanling development of calculus. Calculus is the hard substance that clings to the tooth and irritates the gums, eventually causing gum disease.

A U.S. Army dental scientist reiterated the concept that the human body's reaction to microbial agents was a factor in the inflammation of chronic periodontal disease. Lieut. Col. John E. Horton of Bethesda, Md., noted that saliva, as well as deposits from pockets at the base of the teeth of patients with moderate periodontal disease, stimulates certain white blood cells called lymphocytes to enlarge and divide. Terming this change "transformation," he pointed out that such enlarged lymphocytes were found in inflamed areas where they take part in various immune reactions. He believed that this process caused a sensitization to bacteria in the patient and that this sensitization was probably a major factor in contributing to the chronic inflammatory lymphocyte infiltration of the gum tissues.

Dental laser uses. As dental researchers continued their investigation of the laser beam, it appeared almost certain that the device would find

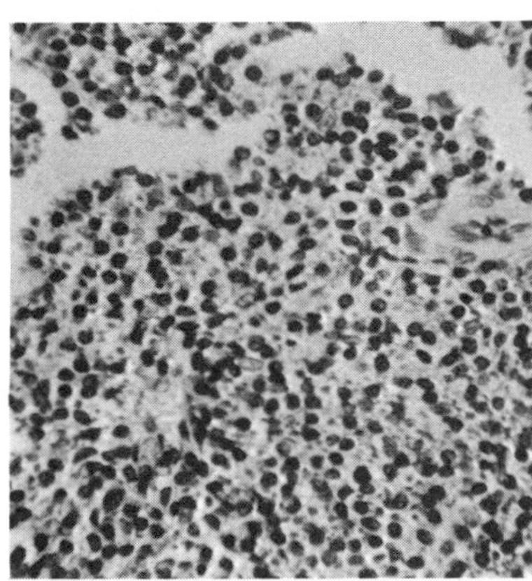

Courtesy, Dr. John E. Horton

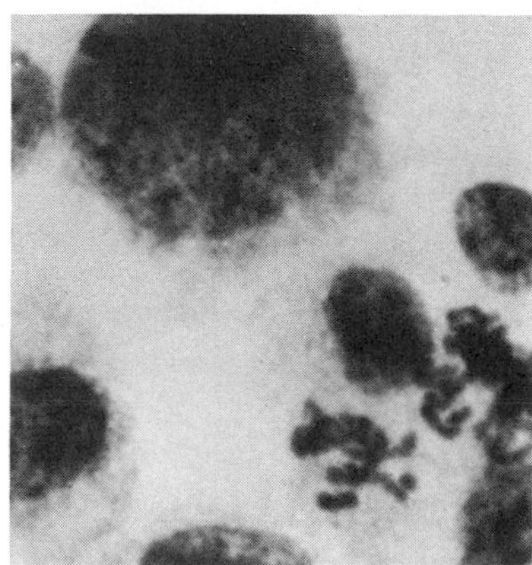

The belief that the human body's reaction to microbial agents was a factor in the inflammation of chronic gum disease was supported by recent research. Far left, inflamed human gum tissue is shown magnified, revealing the presence of lymphocytes (white blood cells). Left, an enlarged lymphocyte (center) is shown in an immunizing reaction, a process that was thought due to sensitization by bacterial products to be a key factor in creating inflammation.

a place in the dental practice of the future. A unique experiment utilizing a "mini-lab" inside a gold bridge worn in a human mouth raised speculations that laser treatment of teeth might prevent cavities. In this experiment, dental scientist Ralph Stern of the University of California at Los Angeles School of Dentistry took bits of laser-treated and untreated enamel from freshly extracted teeth and placed them inside tiny windows of a gold bridge in the mouth of Reidar F. Sognnaes, a co-investigator in the project. The bits of enamel were shielded from toothbrushing by a thin gold plate to trap food debris and facilitate early decay in the enamel samples. At the end of four weeks, the bits of enamel were removed and examined. The samples not treated by lasers exhibited the typical chalkiness of beginning decay while the laser teeth showed no signs of decay.

As the result of another laser experiment, Thomas E. Gordon, Jr., a dental researcher in Orlando, Fla., predicted that laser welding might eventually replace conventional soldering of dental bridges and appliances. It was found that the laser process could reduce the time required to fabricate a joint for the appliance from over one hour by the shortest current conventional methods to an average of four minutes.

Oral cancer. Dental researchers were investigating the possibility of developing an oral cancer risk index based on carcinogenic (cancer-triggering) agents to forecast which persons would be most susceptible to oral cancer. Such a risk index could be derived from epidemiological studies aimed at examining variable factors that may bear a relationship to the development of oral cancer. These findings were based on a 10-year study of nearly 200 oral cancer patients at the Philadelphia General Hospital in Philadelphia, Pa. The study centered on systemic diseases and such variables as age, sex, and race of the patient, consumption of alcohol and tobacco, and oral trauma.

A U.S. Air Force dentist, Lieut. Col. Arden G. Christen, believed that the general public was beginning to look to the dentist for help in quitting the smoking habit. He felt that "the recommendations of a trusted and knowledgeable nonsmoking dentist could be a strong motivating force" in influencing patients to stop smoking. Such a step was important, he emphasized, in view of the fact that "there is a well-established link between tobacco use and these diseases: cancer of the lower lip, oral cavity, esophagus, larynx, and lung; hairy tongue, leukoplakia, periodontal disease, nicotin stomatitis (smokers palate); chronic bronchitis; coronary disease, emphysema, peptic ulcer and peripheral vascular disease."

—Lou Joseph

Earth sciences

Destructive earthquakes in Peru and California, new information about lunar rocks and plate tectonics, and the unforeseen hydrological effects of the Aswan High Dam were among the highlights in the earth sciences during 1970–71.

Geology and geochemistry

The most important developments in geology and geochemistry during the past year concerned lunar geology, the history of continental drift and the ocean basins, and environmental geology.

Lunar geology. Luna 16, an unmanned Soviet spacecraft, soft-landed on September 20 on the moon's Sea of Fertility, obtained samples, and blasted off to return safely to the U.S.S.R. It was the first unmanned spacecraft to land on another body in the solar system and return to the earth.

In February 1971, the manned U.S. Apollo 14 spacecraft landed on the moon at Fra Mauro, which had considerably more rugged terrain than the sites of the two previous Apollo landings. Astronauts Alan Shepard and Edgar Mitchell brought back 96 lb of lunar soil and rock. While on the moon, they dug a trench 18 in. deep, revealing a layer of the lunar regolith (unconsolidated material overlaying the solid rock in place), and seismically determined that the regolith is 25 ft thick at that site.

Based on the data collected over the past several years, the following history of the moon is emerging. The moon formed about 4.6 billion years ago as a separate planet which originated from the primeval solar nebula during the development of the solar system. The early accretionary period of the moon was relatively short, so that by the end of 20 million years it was a hot, solid body with its approximate present size.

Carbon compounds, including amino acids of probable nonbiogenic origin, developed early in the moon's history and are trapped in rocks billions of years old in trace amounts. No evidence suggests that the moon was ever inhabited by life before the astronauts arrived. No liquid or solid water ever seems to have been present on the moon's surface, and its rocks contain no significant amounts of chemically trapped water. Similarly, the rocks are oxygen-deficient, which suggests that at no stage in its history did the moon ever have an appreciable atmosphere.

Early in lunar history, perhaps from the heat generated during the accretionary period, relatively light-colored, low-density igneous rock melted and then erupted to form the lunar highlands. As the lunar surface solidified, craters formed by mete-

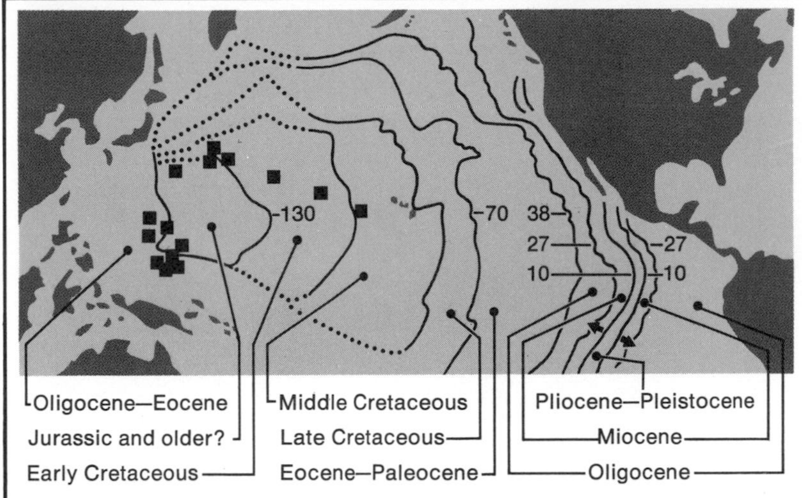

Map shows the spread of the West Pacific floor as it moved westward from the East Pacific Rise upwelling. Numbers indicate the ages (in millions of years) of the dividing lines that demarcate eras of geologic time. The black squares indicate the drilling sites of the "Glomar Challenger." This ship used a sharp bit (below) that was guided into an undersea exploratory core hole by a high-resolution scanning sonar probe. The bit was guided into a cone that marked the site of previous drilling. This was the first time that a drill could be directed back to a previous exploration.

orite impacts began to be recorded. The early rate of impact cratering was about 300 times as great as at present, but as the solar system was gravitationally sucked clean of debris, the rate and size of cratering declined.

Radioactivity heat, probably from traces of uranium and thorium, caused the moon's interior to heat up, forming magma (molten rock) at depths of 200 km or more; the lunar core melted so that for a time the moon had its own magnetic field. Lava erupted intermittently 3.7–3.3 billion years ago into the maria depressions and over much of the lunar surface. Then the moon cooled, developing a generally solid interior again as the radioactivity diminished. The lava-filled maria became scarred by meteorites. Small amounts of lava escaped from time to time along fractures, and some gaseous emanations may still occasionally leak out. At some relatively late date, perhaps as recently as 700 million years ago, the moon was gravitationally captured by the earth. The interior of the moon is now fundamentally inert and probably has been so for three billion years. The moon lacks a magnetic field, has no convection currents to cause drifting of the surface or the crumpling of mountain ranges, and has no atmosphere to weather the igneous rocks so as to yield sediments and to produce the metamorphic rock types that have originated on the earth.

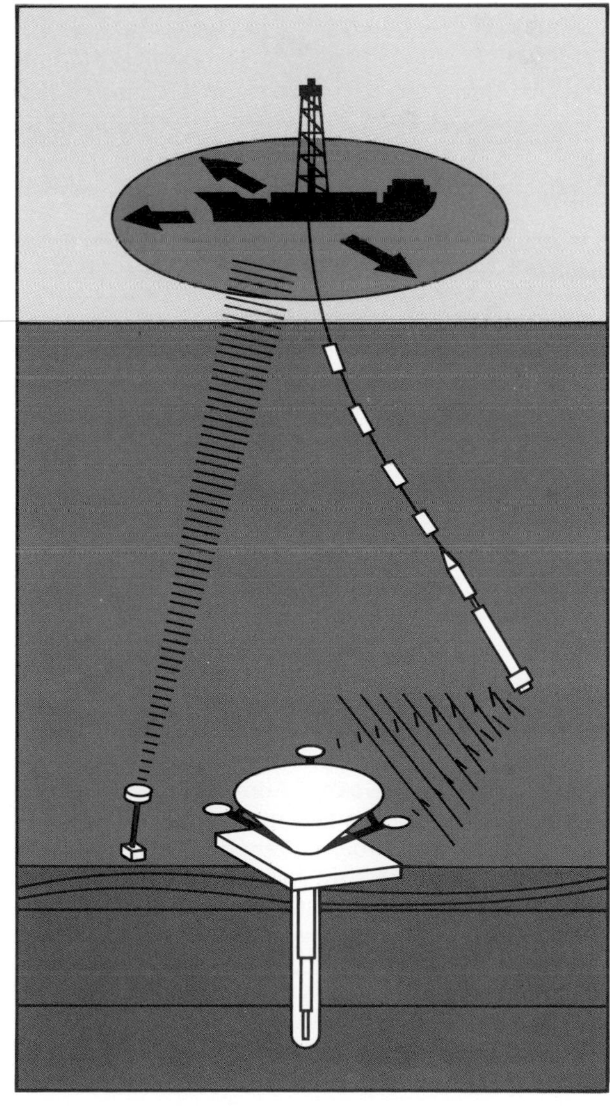

Geochemistry. Laboratory studies of tholeiite basalts heated to 1,300° C and subjected to varied confining pressures up to 40,000 atmospheres were used experimentally to simulate the mineral composition deep in the mantle of the earth. Under reasonable assumptions of confining pressure and temperature, basalt changes with increasing depth to garnet granulite and then to plagioclase eclogite. Melting occurs at conditions corresponding to a

UPI

A violent explosion of hot lava and magma occurs during an eruption of Mt. Etna, Europe's largest active volcano. Because during previous periods of activity Etna has proved capable of causing great destruction, farmers from the nearby village of Sant' Alfio prayed before the rumbling mountain for a miracle to stop the flowing lava that was devouring their vineyards and hazelnut groves.

depth of 75 km. These data seem fairly consistent with seismic velocities and densities that can be determined in the upper mantle. Another group of workers, however, believed that peridotite is a more probable rock type for the mantle. Peridotite dikes which occur in some regions probably represent intrusions that made their way upward from the mantle.

In other work a new radioactivity dating method using lutetium-176 and hafnium-176 isotopes and mass spectrometer techniques was perfected. The procedure can be applied to rocks one billion years old.

A group of researchers from the U.S. National Aeronautics and Space Administration (NASA) reported the first positive identification of extraterrestrial amino acids, obtained from a meteorite that fell near Murchison, Austr., on Sept. 28, 1969. The five amino acids found in the meteorite did not appear to be of biological origin. This discovery demonstrates that amino acids, the basic building blocks of proteins, had formed by 4.5 billion years ago in a system outside the earth.

Plate tectonics. The idea of continental drift has been intermittently in vogue since 1910, but only since 1968 has it been recognized that the ocean basins are more actively involved in the driving mechanism than are the more visible continents. By 1971 scientists realized that the earth's upper layer, consisting of the crust and upper mantle to a depth of about 100 km, is divided into six relatively rigid, migrating plates that have spread apart and come together to form many of the major geological features on the earth. Because continental drift is only part of the story, "plate tectonics" is a more appropriate term to describe this process.

The scientific world of the West was enthusiastically accepting the idea of plate tectonics. Soviet geologists, however, continued to prefer the theory that vertical movements of the crust are the major driving forces of geologic processes, rather than the horizontal migration of crustal blocks called for by plate tectonics.

In accordance with plate tectonics, the Pacific Ocean is spreading apart along the East Pacific Rise, much like the sea-floor spreading that is occurring along the Mid-Atlantic Ridge. The oldest parts of the West Pacific, which emerged and spread westward out of the East Pacific Rise, contain the most ancient fossils, as well as the greatest age for the basaltic rocks that occur directly beneath the sedimentary cover.

Marine geology. Studies of marine geology continued to be dominated by the finding from the deep-sea cores obtained by the "Glomar Challenger" cruises during the last year to the Gulf of Mexico, Atlantic Ocean, and Mediterranean Sea. For the first time, technology was developed that allowed the drill to be removed, a worn-out bit replaced, and the same hole then reentered to continue drilling to greater depths. A site approximately 480 km northeast of Cuba was drilled because it theoretically should contain the earliest sediments deposited after the Atlantic sea floor spread open by continental drift; Jurassic Period (130–180 million years ago) sediments there were found to overlie the suboceanic basalt, a result consistent with the postulated breaking apart of Africa and North America just after the Triassic Period (180–225 million years ago). The total thickness of oceanic sediment there was only 315 m, indicating extremely slow deposition during the last 160 million years. These are the oldest sediments known from any ocean basin; in contrast, rocks exceeding three billion years are known on different continents.

Attempts to increase the supply of the scarce metal manganese continued during the year. Manganese nodules have been found in nearly every ocean area, but large concentrations generally lie at great depths. In order to reach them Tenneco, Inc., developed a deep-sea dredge device that acts like a giant vacuum cleaner.

Paleontology. The discovery in 1969 of the first Antarctic reptile fossil, *Lystrosaurus,* was augmented in 1970 by the uncovering of a 10-in. *Thrinaxodon,* a 200-million-year-old Triassic mammal-like reptile. It was the first complete reptile skeleton from Antarctica. Because similar fossils have been found in South America, Africa, and India, they provide the strongest evidence so far obtained that these regions were joined together as a single land mass during the Triassic Period.

—John M. Dennison

A huge dome of lava boiling upward to a height of 50 ft appears during an eruption of the Kilauea volcano in Hawaii. Kilauea's latest series of eruptions began in May 1969. Since then it has produced approximately 250 million cu yd of lava and shows no signs of stopping in the near future.

UPI

Geophysics

The decade of the 1960s saw the development and wide acceptance of the concept of sea-floor spreading and continental drift. The lateral mobility of the earth's outer layers, the principal feature of the new concept, is a sharp departure from the formerly prevailing view of the earth, in which movements were considered to be mainly up or down. An understanding of the forces that cause the lateral movements, however, has not yet been achieved. While testing of the sea-floor-spreading hypothesis continues, a new international project is being launched to gain that understanding.

The destructive power of earthquakes was impressively demonstrated in Peru and California during 1970 and 1971, respectively. The Peruvian earthquake ranks as the most destructive shock in the history of the Western Hemisphere. An estimated 70,000 Peruvians were killed and 50,000 injured. The earthquake was assigned a magnitude of 7.75 on the Richter scale. The magnitude-6.6 earthquake that struck on the fringe of the Los Angeles metropolitan area caused between $500 million and $1 billion damage, and killed more than 60 people. A disaster unprecedented in the United States nearly occurred when a dam in the Los Angeles area partially collapsed. Had the dam failed completely, tens of thousands of people would have been killed.

Deadly earthquake in Peru. On May 31, 1970, a large earthquake shook the coastal region of central Peru. The valley of the upper Santa River, which parallels the coast between the 14,000-ft Cordillera Negra on the west and the 22,000-ft Cordillera Blanca on the east, was devastated; 90% of the buildings in the valley were seriously damaged or destroyed. Destruction was on a large scale primarily because the principal building material there is adobe.

The most spectacular and deadliest effect of the earthquake was a massive landslide that occurred when the side of a peak in the Cordillera Blanca was shaken loose by the earthquake waves. The mixture of rock, mud, ice, and water roared down upon the towns of Yungay and Ranrahirca at a

Wide World

A house in downtown Lima, Peru, lies in ruins after a severe earthquake struck a portion of the country on May 31, 1970.

speed estimated at about 250 mph. There is evidence that the debris rode on a cushion of compressed air, much like a hovercraft, to reach such a high velocity.

The occurrence of a large earthquake in Peru came, of course, as no surprise. Peru lies in the well-delineated circum-Pacific seismic belt, and has a long history of strong earthquakes. The zone of intense earthquake activity along South America's west coast results from the collision of the westward-drifting continent with the eastward-spreading floor of the East Pacific Ocean.

California earthquake. Approximately 100 earthquakes of magnitude 6 or larger occur each year in the earth. Most of them are situated in remote areas, or at great depths, and receive little notice except from seismologists. On Feb. 9, 1971, however, a magnitude-6.6 earthquake was centered near San Fernando, Calif., on the northern fringe of the Los Angeles metropolitan area. The destruction resulting from this moderate-sized shock gave a grim warning of what Californians will face when the next truly major earthquake of magnitude 8 or greater strikes. (*See* THE TREMBLING EARTH.)

The San Fernando earthquake occurred when the San Gabriel Mountains were thrust upward and to the southwest a distance of three to five feet, overriding the San Fernando Valley. Thrust-type fault movement had also been responsible for the magnitude-7.7 earthquake near Bakersfield, Calif., in 1952. In contrast, strike-slip faulting, where movement is horizontal, predominates in the San Andreas, the major fault system in California. The site of the San Fernando earthquake had not been regarded as a particularly hazardous one; in fact, the fault was not even plotted on the California geologic map.

The Los Angeles area was well-instrumented with strong-motion seismographs. These devices record the accelerations experienced by the ground and buildings during an earthquake, information that is critical in designing structures to withstand tremors. The extensive data obtained in Los Angeles will undoubtedly lead to revision of the building codes concerning earthquake-resistant design. Upgrading of design criteria will probably be called for because the accelerations measured were unexpectedly large. At Pacoima Dam, located near the earthquake's epicenter, the horizontal acceleration reached 110% that of gravity. Accelerations of 30% gravity were measured 20 mi away. Many buildings and dams are designed for accelerations of only 10 to 15% gravity.

Geodynamics Project. Much of the international cooperation that was necessary in developing the sea-floor-spreading hypothesis was coordinated under the auspices of the Upper Mantle Project (UMP). The progress during the ten-year UMP has been acclaimed a revolution in the earth sciences.

To maintain the momentum of research generated under the UMP, and to focus attention on some of the important questions raised by it, a new international cooperative program, the Geodynamics Project, was initiated. This project will focus on movements at the surface and within the upper portions of the earth's interior. Its principal goal is to encourage research on the dynamic history of the earth. During this project an understanding of the forces that move continents and build mountains may be achieved.

Lunar seismology. The deployment of seismometers was an important activity during the Apollo 11, 12, and 14 landings on the moon. These instruments have detected moonquakes, meteoroid impacts, and impacts from spent rockets.

Two major discoveries resulted from the lunar seismological studies. One is that seismic signals on the moon are unusually complex and of a long duration. An interpretation of this observation is that seismic waves are not as strongly attenuated on the moon as on the earth, and that waves are strongly scattered by heterogeneous material near the moon's surface.

The other important finding is that most moonquakes occur near times when the moon is closest to the earth, indicating that the strong tidal stress at these times may cause the tremors. A similar correlation between quakes in the earth and tidal stresses has been sought by numerous scientists, but by mid-1971 they had not yet achieved convincing success.

The astronauts on the Apollo 14 mission set out a seismic experiment to determine the structure of the moon at shallow depths beneath the surface. On command from the earth, a mortar on the moon projects a shell, whose explosive impact is then detected by the lunar seismometers. From the velocity of the shock waves scientists can deduce the elastic properties of materials buried beneath the moon's surface.

Power source of the future? Electrical power is mostly produced by the burning of coal or oil, hydroelectric facilities, and nuclear reactors. There are problems associated with all of these sources in fulfilling long-term power needs. Coal and oil resources are limited, and burning them pollutes the atmosphere. Practically all suitable sites for hydroelectric plants in the U.S. have already been developed. Nuclear reactors cause thermal pollution, produce radioactive wastes that are difficult to dispose of, and generate political controversy. There is a possibility that a significant amount of our future power needs could come from geothermal energy, the heat inside the earth, which brings with it none of these problems. In 1971 the exploration for and development of this resource was underway, guided primarily by geophysical techniques.

The combining of a gyroscope with a TV camera in borehole photography enables geologists to measure accurately the dip and direction of a strike. The compass direction is determined by the relation of the seam's strike (shown in squares) to the north (top); the direction of the dip of the seam is indicated by dots.

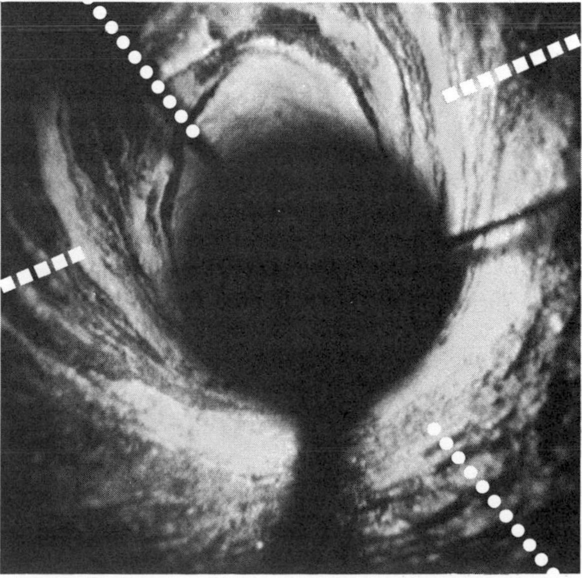

Courtesy, International Underwater Contractors, Inc., ENGINEERING NEWS-RECORD

Hot water from inside the earth has been used for centuries to heat houses in some countries. The use of geothermal energy to produce electricity was first accomplished in Italy, at Larderello, in 1904. Geothermal energy was first used for power in the U.S. in 1960, when production of electricity started in The Geysers area about 70 mi N of San Francisco. By 1971, an 82,000 kw facility was operating there. Over the next few years it was anticipated that production would be raised to about 300,000 kw, sufficient power for a city of a half million people.

A potential geothermal power resource exists where groundwater can migrate to the vicinity of a heat source, such as a hot intrusive rock mass, and thereby rise to a depth accessible to drilling. A fracture system is required to provide sufficient permeability. Several characteristics of this structure can be delineated by geophysical methods. For example, the presence of the hot water may be revealed at the surface by an anomalously high measure of electrical conductivity. Earthquakes may be associated with the fracture system. Structural characteristics can be detected by magnetic, gravimetric, and seismic methods. Remote sensing methods, such as radar imagery and infrared photography, may also prove useful. During the next few years these various techniques will be critically evaluated for their usefulness in geothermal power development.

—Robert M. Hamilton

See also Feature Article: ENERGY CRISIS, *Supplying Energy to the Year 2000.*

Hydrology

The origin of rivers, springs, rain, and the water filling the lakes and oceans has probably been a subject of speculation since before the earliest written records. The science of hydrology, however, did not come of age until 1674, when Pierre Perrault clearly demonstrated that the rainfall over one of the tributary basins of the Seine River, in France, was more than enough to account for the stream flow out of the basin.

Hydrology has its roots in the engineering analy-

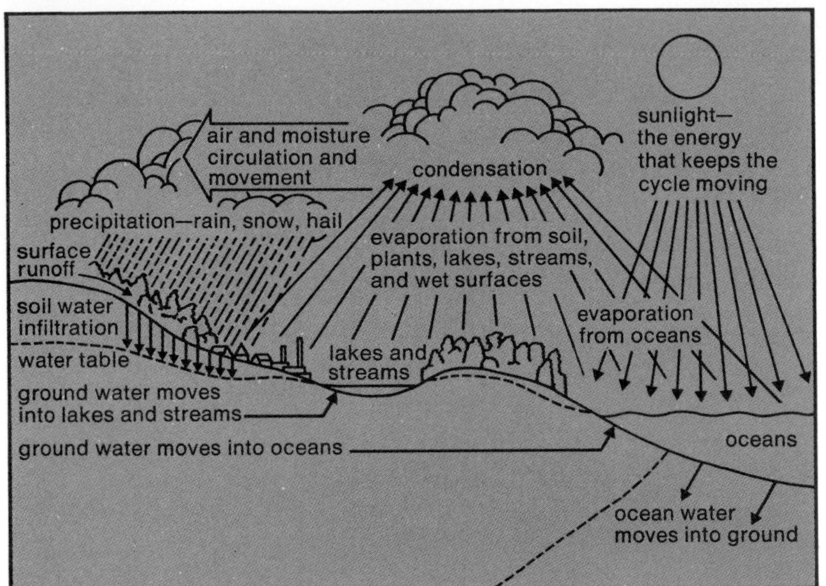

The hydrological cycle is continuous, though irregular, circulation of water from clouds to the earth and back to the clouds again. All phases operate simultaneously but at different rates in different places.

sis of stream flow, but in the past 100 years, with increased understanding of the significance of ground water, soil moisture, and the roles of rain, snow, and ice, its scope has grown. The pervasive circulation of water is commonly referred to as the "hydrological cycle." Today, the most widely accepted definition is that used by the International Hydrological Decade (IHD): "Hydrology is the science that deals with the waters of the earth—their occurrence, circulation, and distribution on the planet; their physical and chemical properties; and their interactions with the physical and biological environment, including their effects on human activity. Hydrology covers the entire history of the cycle of water on the earth."

In practice, however, hydrology is not quite so broad as the IHD definition indicates. Atmospheric moisture is within the purview of the atmospheric physicist and meteorologist, and the water of the oceans is primarily the concern of the oceanographer. Hydrology, however, is not entirely restricted to the water regimens of land areas. Because of the importance of precipitation, evaporation, and the movement of moisture through the atmosphere, the hydrologist also is concerned with these processes along with the oceanographer and the atmospheric scientist. Hydrology overlaps with geochemistry and geology in its interest in the absorption of water by rocks undergoing metamorphic change and in the release of water from rocks, from magmatic liquids and gases, and from crystals as they are exposed or ejected at the surface.

Ganges delta flooding. The outstanding physical hydrological event of 1970 was the catastrophic flooding of the Ganges River delta and adjacent areas around the perimeter of the Bay of Bengal in East Pakistan on November 12–13. A cyclonic storm coincided with spring tides to produce abnormally high storm surges. Along the main path of the storm, nearly 12 in. of rain fell on November 12. From the combined effects of high rainfall and the storm surges generated by driving hurricane-speed winds, the outer islands of the delta were completely submerged under 20 ft of water, and an area of about 3,000 sq mi on the mainland was flooded or devastated. The loss of life was estimated at up to 300,000, and the destruction of property, public facilities, livestock, and irrigated lands at hundreds of millions of dollars. This event was called the 20th century's worst natural disaster.

Effects of the Aswan High Dam. What may be a man-made hydrological disaster, although it was not so intended, is the Aswan High Dam on the Nile River in the U.A.R., dedicated on Jan. 15, 1971. Among the largest dams in the world, it will impound a reservoir which, when full, will be 300-400 mi long, cover about 2,000 sq mi, and contain about three times as much water as Lake Mead, the reservoir behind Hoover Dam on the Colorado River. The main purposes of the Aswan High Dam are to provide irrigation water for more crops to feed the growing population and 10,000,000,000 kw of electrical power for new industrial and municipal uses that will give the people productive work. The dam also was designed to provide flood control by eliminating the peak of the annual floods of the Nile.

The reservoir is not yet full, but already several unforeseen, or perhaps neglected, consequences are making themselves felt. Because the reservoir is not filling as rapidly as expected, power production is behind schedule. Bilharzia, a widespread, debilitating intestinal and urinary disease carried by snails, has spread because annual floods and annual droughts no longer restrict the upstream migration of the snail. Also, the U.A.R. is now forced to produce or buy inorganic fertilizers to make up for the loss of the annual enrichment of the soil by silt that in the past was left behind by the receding flood waters. This silt now settles in the bottom of the reservoir.

Another unforeseen consequence is the accumulation on the land of inorganic salts in sufficient amounts to take some agricultural areas out of production; these salts formerly were flushed out by the high flood waters. Finally, the sardine catch in nearby parts of the Mediterranean Sea is less than one-third of what it was when the flood waters dumped their annual loads of nutrients into the waters fringing the Nile Delta. The long-term balance between the increased productivity expected from water and power development and the losses resulting from environmental changes remains to be seen.

International Hydrological Decade. On the international scene, the most significant advances in hydrology during the past year were made in connection with the International Hydrological Decade. During 1970, the Decade's sixth year, the IHD program was reoriented to put increased emphasis on the practical needs of developing countries, as recommended by the Mid-Decade Conference held in Paris in December 1969. The scientific program of the past year, carried out almost entirely through national investigations, can be summarized to some extent by a partial list of international symposia and conferences held in 1970. These included: Great Lakes Research (U.S.); Hydrobiology (U.S.); World Water Balance (Great Britain); Ice (Iceland); Hydrochemistry and Hydrobiochemistry (Japan); Hydrometry (West Germany); and Representative and Experimental Basins (New Zealand). Most of these meetings were organized by the International Association for Scientific Hydrology as contributions to the IHD.

The International Field Year for the Great Lakes. On the North American continent, the U.S. and Canadian national committees for the IHD together initiated the International Field Year for the Great Lakes. Planning for the Field Year began in 1966, and the Field Year itself, a period of intensive data measurement, was scheduled to begin in April 1972 and to last well into 1973. The long-term objective would be to provide more reliable and more comprehensive information and understanding of the natural hydrological phenomena of the Great Lakes as a basis for the improvement of

The recently completed Aswan High Dam was built to overcome unpredictable cycles of flood and drought that plagued the economy of the U.A.R. This technological triumph, which provides flood control and abundant and cheap electricity, simultaneously creates, however, unexpected ecological problems.

UPI

their management. The program contains five main components: terrestrial water budget, atmospheric water budget, energy budget, circulation patterns, and nutrient cycle.

Planning and guidance for the Field Year were under the general supervision of the Joint U.S.-Canadian Steering Committee, while actual research operations were to be conducted by U.S. and Canadian agencies and institutions. Principal U.S. agencies contributing to the program were the National Oceanographic and Atmospheric Agency, the Environmental Protective Agency, and the U.S. Geological Survey; Canadian agencies included the Canada Centre for Inland Waters, the Canadian Meteorological Branch, and the Ontario Water Resources Commission. Many universities in both nations were involved in individual projects.

Role of basic research. Recent hydrological investigations continued to concentrate on problems of water quality and the application of computer-based analytical techniques to those and other problems of water resources. The emphasis was on practical problems rather than long-term research. The increased interest in statistical approaches to hydrological problems, as well as the emphasis on the practical, meant that a proportionally smaller part of the research effort was being expended on understanding the basic physical and chemical phenomena involved in the transit of water from one place to another and from one phase to another. Practical needs must be met, and increasingly sophisticated statistical and computer-based procedures are useful in many scientific as well as engineering applications, but further so-called "pure research" is also needed if we are to continue to improve our understanding of the physics and chemistry of water in order to resolve problems in the future.

—Leo A. Heindl

Education, Science

A basic trend in science education during 1970–71 was an increasing concern with environmental and ecological factors that affect the earth and its population. In the United States particularly, a sustained interest was evidenced during the past 12 months in environmental education. Educators at all academic levels were becoming active in getting their students involved in, and in amassing relevant data about, society's most pressing environmental problems: pollution and population.

This concern also was apparent in other areas. The U.S. National Science Foundation (NSF) and the U.S. Office of Education both gave financial support to the development of environmental science curricula. The U.S. Office of Education-supported ERIC Clearinghouse on Research in Science Education, in Columbus, O., expanded its coverage to include not only mathematics but also environmental education. The information gathered at the University of Maryland's Science Teaching Center for their *Eighth Report of the International Clearinghouse on Science and Mathematics Curricular Developments* (to be published in 1972) reflected considerable ecological and environmental activity. Concurrent with this development, curriculum projects in the "hard" sciences were developing new integrative formats with the social sciences.

This concern for environmental education also was apparent in countries other than the United States. The Organization of American States (OAS) sponsored an international meeting on environmental education in Washington, D.C., during the fall of 1970. Delegates from a large number of OAS member countries as well as professional organizations concerned with the relationship of science and society discussed the current developments and shared plans for future educational programs. Other groups in numerous countries had similar meetings on regional and local projects.

Science curriculum development. Science curriculum development work in the U.S. and in other countries maintained the rapid momentum that had characterized the past 12 years. Although environmental and ecological subjects were being added to a number of projects, basic subject matter of a more traditional nature was still being modified and made more appealing to student interests. A number of the more elaborate U.S., British, and Australian science and mathematics curriculum projects were being translated, adapted, and implemented in other countries, but new projects on a smaller scale were also being undertaken in some of the developing nations. In Africa a number of countries began projects on their own or developed joint projects with other nations that had similar educational needs. The Scottish Integrated Science Project and the West African Primary Science Project (produced by the Educational Development Corp. in the U.S.) were both being used on an experimental basis in various African countries. Another project receiving widespread attention was the British-originated, Nuffield Foundation-sponsored project in biology, chemistry, physics, and mathematics, which was being tested in a number of countries, including the U.S. and several in the Commonwealth.

UNESCO continued to be particularly active in developing science curricula. The UNESCO-sponsored Chemistry Teaching Project team, working out of Bangkok, Thailand, completed its writing effort and was involved in implementing its products in a number of Asian countries. The UNESCO Biology course writing team in Africa completed its major phases and was taking up more ecologically oriented topics.

UNESCO sponsored a number of conferences designed to initiate more curriculum projects in integrating the sciences. One of the most successful was a two-week-long Asian regional conference held in the Philippines in August 1970. One of the topics discussed was the improvisation of inexpensive equipment for science teaching, an important consideration for developing countries strapped for educational funds.

One U.S. curriculum project nearing completion was a three-year sequential science project for junior high schools known as the Intermediate Science Curriculum Study. The curriculum, which was being written at the Florida State University, integrated science and mathematics and utilized teaching procedures paced to the individual student.

Centers for the development of science education. The number of science teaching centers continued to grow throughout the world. Israel expanded its center and new, similar teacher-aid centers were begun in the Philippines, Malaysia, Thailand, and in South America, and Africa. The number of science teaching centers in the U.S., which was already substantial, also expanded. One impetus for the increase was a new NSF policy of awarding substantial financial grants for five-year periods to clusters of university science departments that developed comprehensive programs for improving science education in their geographic areas. Six such grants were awarded during the past 12 months.

One of the most important grants was to the state of Delaware to make major improvements in the teaching of science and mathematics. NSF contributed funds to this vast project along with the Delaware State Board of Education, Delaware's three universities, and E. I. du Pont de Nemours & Co. Similar grants were to be made in the future by NSF if such cooperative efforts and fund-sharing arrangements could be worked out.

Science education publications. UNESCO was completing revision of the *UNESCO Sourcebook for Science Teaching,* with a release date anticipated for late 1971 or early 1972. Substantial changes were made in updating and changing the earlier edition under a U.S. team coordinated by the director of the University of Maryland's Science Teaching Center. The sourcebook was UNESCO's largest selling publication and was published in 21 languages.

The latest in UNESCO's Division of Science

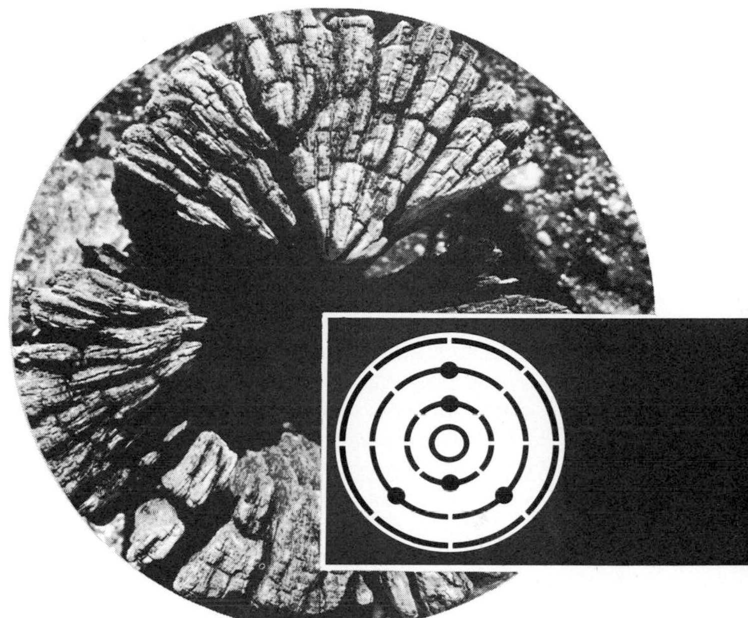

"The stumbling way in which even the ablest of the scientists in every generation have had to fight through thickets of erroneous observations, misleading generalizations, inadequate formulations, and unconscious prejudice is rarely appreciated by those who obtain their scientific knowledge from textbooks."
James Bryant Conant

Teaching "Trends" series of books, *Trends in Integrated Science,* was released in June 1971. It included papers presented at the Varna, Bulgaria, conference on that topic in 1968 as well as other articles summarizing curriculum projects of an integrative and interdisciplinary nature.

The U.S. National Science Teachers Association (NSTA) announced that it would produce a third professional journal (in addition to its *Science and Children* and *The Science Teacher*) entitled *The Journal of College Science Teaching*. The journal, aimed at NSTA's ever-increasing college-level membership, was to be edited by Leo Schubert, chairman of the chemistry department of American University, Washington, D.C. The first issue was to be released in October 1971.

Work continued in the U.S. during the past 12 months on the preliminary edition of the *Guidebook to Improvised Science Teaching Equipment Worldwide*. The guidebook was to include descriptions and actual construction plans for producing inexpensive equipment for student investigations in introductory biology, chemistry, and physics classrooms, with primary distribution aimed at developing countries. The preliminary edition was to be refined by feedback information from around the world and a final, revised publication eventually released.

In the U.S. science textbook business, most of the activity revolved around revisions of earlier editions of NSF-sponsored curriculum projects. For example, a third edition was published of a high school text developed by the Physical Science Study Committee (PSSC) called *PSSC Physics*. The text had been the first resulting from an NSF-sponsored science curriculum project, which had begun its work in 1956, a year before the first Soviet satellite launching accelerated U.S. interest in improving science teaching at all levels. In other science areas such as chemistry "public domain" rights went into effect and, consequently, three different commercially produced texts were available for the NSF-supported CHEM Study course.

Professional science teaching organizations. NSTA, the largest of the science teaching societies in the United States, held its convention in April 1971, in Washington, D.C. The convention theme—"Decisions in Science Teaching-Selection, Implementation, and Evaluation"—attracted the largest crowd in the NSTA's 27-year-old history with more than 7,000 NSTA members attending. It also attracted a substantial group of exhibitors of science teaching materials, in contrast to the smaller numbers of such exhibitors attending the conventions of the more typically science-oriented organizations.

—J. David Lockard

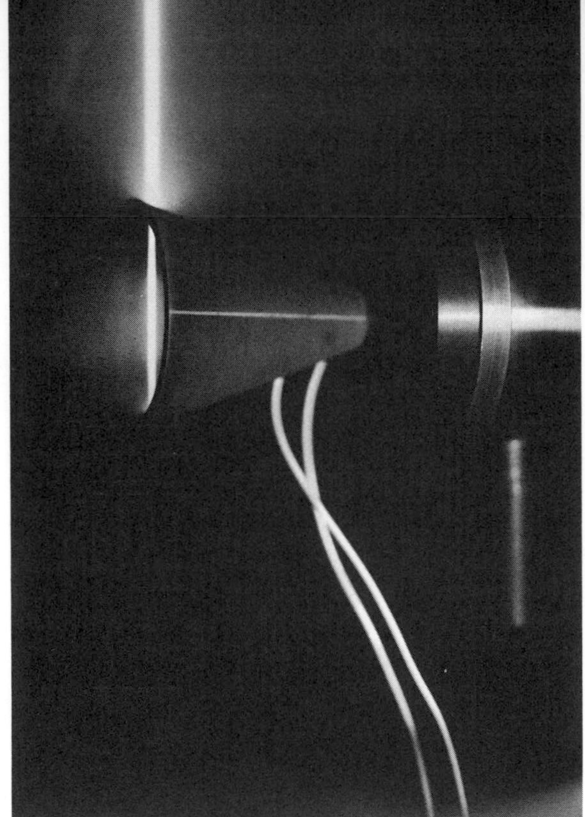

Courtesy, Cornell Aeronautical Laboratory

An electron beam is passed through an exhaust gas flow as part of a study to measure the temperature and density of rocket-exhaust flows. Information derived from the study will be used to predict the performance of rockets that control the attitude of satellites.

Electronics

The term "electronics" originally denoted what is now called "electron physics," that branch of physics concerned with behavior of charged subatomic particles. The present-day use of "electronics" as meaning a branch of modern technology dates back only to 1930, when a magazine by that name was begun in New York. During the next 20 years uses of electronic devices were limited largely to the telecommunications and entertainment industries. (Scientific instrumentation and control of certain industrial processes were the most important exceptions.) Even the tremendous expansion of electronics that followed the invention of the transistor in 1947 and the introduction of transistor-like devices into the manufacture of electronic computers were not far removed from communications. This endeavor gradually became known as the "information" industry, which now has come to encompass such widely divergent applications as technological aids to education, to traffic control, to the graphic arts, and to health care.

As the 1970s arrived engineers were becoming

increasingly aware that they could not continue to consider themselves the passive tools of society, doing its bidding without regard for the consequences, which would remain the province of "others." Now the technologists were finding that they were part of the "others," those who were being asked to make society's decisions and who were importuned, held responsible, and in extreme cases vilified for failing to be more outspoken about the possible adverse effects accompanying their achievements.

In the face of this new consciousness of the social effects of technology, the trickle of engineers concerned with such matters grew to a mighty stream. Engineering periodicals reflected changes in attitudes as they opened their pages to discussions of technological aspects of societal problems. Electronics engineers did not lag behind. They found no shortage of areas into which electronics had expanded beyond the traditional fields of information.

Electronics and the health sciences. Few fields typify the current drive of technology toward social relevance better than the search for applications of electronics to the healing arts. Yet this is not altogether a new development. Electrophysiology traces its beginnings to the 17th century; the employment of electronic devices in medicine goes back at least to the first use of X rays by Röntgen in 1895; and techniques such as electrocardiography (EKG) and electroencephalography (EEG) that have long since found general acceptance in the diagnosis of abnormalities in the heart and brain would be unthinkable without the faithful reproduction afforded by electronic amplification. But new prominence was given to medical electronics in 1970 by a combination of factors ranging from the new emphasis on socially relevant uses of technology to the sudden resurgence of the U.S. electronics industry's interest in a field that promised to offset decreasing military and space expenditures.

Medical electronics had limped along as best it could for decades, hampered by technical and economic problems of the sort that characterize many emerging areas of technology: lack of a mass demand, a rapidly advancing scientific base, and unusual marketing conditions. Despite such difficulties, the field had chalked up some remarkable successes.

Automated health services. The success of some prepaid group health plans, coupled with the opposition of many U.S. physicians to a comprehensive governmental health service, led some observers to predict that what ultimately would emerge would be a system of health care and services underwritten by the government but dependent on physicians in private group practice. Since the supply of physicians was not keeping up with the rising medical care needs of the population, it was safe to predict that such a system would also come to depend more heavily on technology than in the past. Auguries of things to come are voluntary schemes such as the Kaiser Foundation Health Plan centered in the San Francisco Bay region, whose managers not only continued to pioneer in the utilization of electronic technology but also operated the program along sound economic principles, passing on part of the savings resulting from group practice to participating physicians.

One feature of such programs is the multiphasic clinic, in which routine medical tests are administered by a succession of paramedical personnel to a group of patients who go through the several stages of a periodic health checkup according to a predetermined schedule. Of prime importance to the multiphasic clinic is the associated processing and storage of the data resulting from the examinations. The data are registered and transcribed by means of electronic computers. Techniques were being developed that render increasing portions of this process automatic—from simple counts of pulse and respiration to the chemical analysis of samples taken from the patient and even the examination of visual patterns, such as blood smears and X rays.

Automated clinical laboratories. Since many chemical laboratory tests are quite simple (such as nitrogen or glucose analyses) and also highly repetitive, considerable advances toward automating them were being made. The rewards are high: more rapid service and diagnosis, identification of abnormalities by automatic comparison with normal bases, and the presentation and recall of data in the context of past performance to establish trends. Partially automatic blood analysis was already available in many hospitals, but the automation of even simple blood counts (relative numbers of different red cells and white cells) continued to present considerable difficulties.

The automation of clinical laboratories was the subject of intense research and development in 1970. Among projects under way in the U.S. were the following: (1) high-resolution systems to identify and analyze approximately 130 components of urine; (2) a rapid clinical system in which specimens are analyzed automatically, the data recorded by a spectrophotometer and displayed on a cathode ray tube, and the results calculated and printed out by a computer; (3) the monitoring of physiological functions, both during acute illness and in periodic checkups; (4) computer identification of about 50 microorganisms associated

John Launois from Black Star

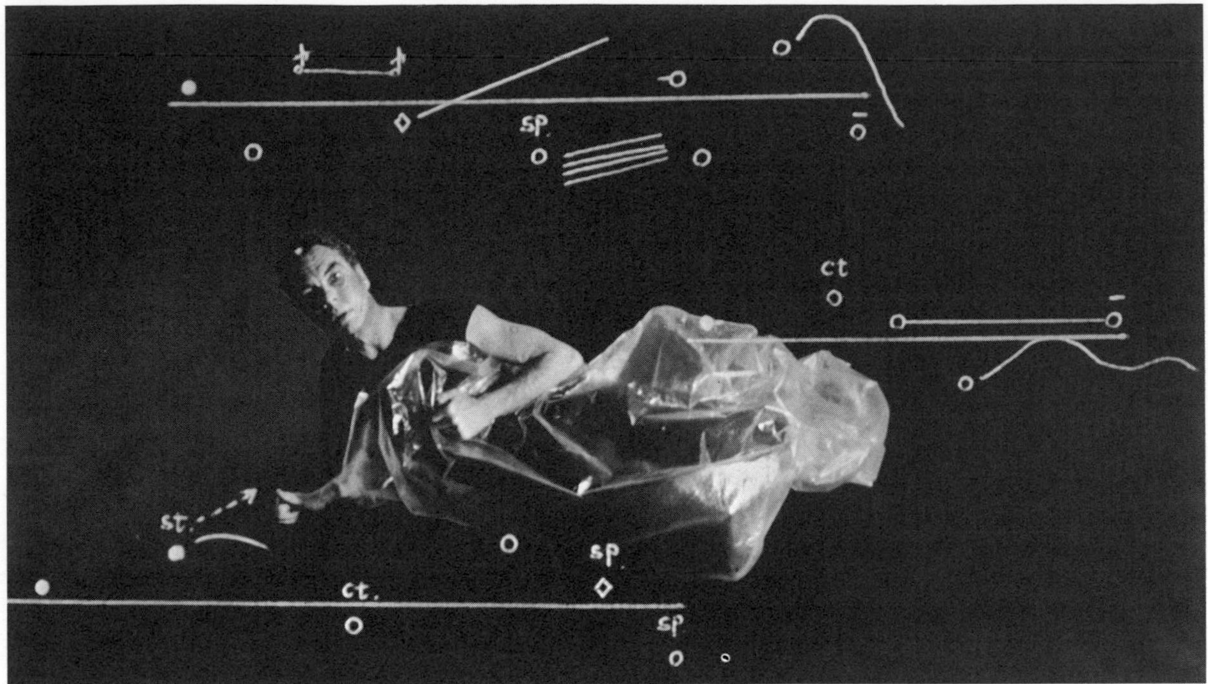

Computer-prepared notations for an electronic music score and its accompanying choreography are superimposed on a photograph of dancer Merce Cunningham. The use of computers in musical synthesis has greatly expanded the range and control of composers as well as eliminated the tedious process of building tones by means of an array of generators and other instruments in a tape studio.

with various diseases; (5) probing by laser microbeams so fine that they can penetrate a single cell and measure concentrations of various trace elements inside it; (6) multiple tests on a single drop of blood to accommodate patients who cannot tolerate much blood loss; and (7) automatic identification of possibly malignant cells in the Papanicolaou smear for cervical cancer and the electronic display of suspected cells for the pathologist's special attention.

The full realization of these projects remained in the future, but several instruments embodying at least some of the principles were available commercially. The earliest versions comprised components originally intended for other uses and relied on general-purpose computers; but at least one company (Berkeley Scientific Laboratories) led the way in evolving systems designed specifically for the biochemical or medical laboratory, with just as much instrumentation and computer capacity as was necessary to collect, process, and present the requisite information and no more. Moreover, because personnel unskilled in the electronic arts would operate the equipment, an effort was being made to make its manipulation no more complex than that facing, say, a travel clerk making a computerized airline reservation.

Another aspect of the application of electronics to clinical procedures is the part electronic devices play in diagnostic techniques, such as exploration of tissues by ultrasonics. Although ultrasound itself does not depend on electronic phenomena, the method of producing it—and, in the present instance, displaying it—requires electronics.

The most spectacular use of ultrasound in medicine was in visualization of internal tissues, such as the eye or the brain, by a technique reminiscent of sonar (the underwater equivalent of radar): pulses of sound energy are sent into the body and their reflections are mapped on a picture tube. The technique yields displays of higher contrast than are obtainable with X rays, which are not as useful for soft tissues; moreover, ultrasonic mapping avoids the hazards of X radiation to the patient.

During 1970, the use of these "ultrasonograms" in the diagnosis of brain tumors and eye disorders remained largely experimental and was nowhere in routine clinical use, but the technique was attracting increasing interest and was being successfully extended to such fields as cardiology, abdominal exploration (kidney stones show up particularly clearly), and obstetrics and gynecology—though a warning was sounded by an expert who pointed out that there was little information concerning possible adverse effects of sound energy (such as a collapsing effect called "cavitation"), especially on such sensitive organisms as embryos.

Therapeutics. Another use of ultrasonics was therapeutical, by application of deliberately large and concentrated doses of sound energy to destroy malignant concentrations, to cure the vertigo-inducing Ménière's disease of the inner ear and parkinsonism, and to deliver energy for heating of deepseated tissues and joints in the treatment of rheumatism and arthritis.

A better-established and more directly electronic method to accomplish the last goal was use of high-frequency electromagnetic energy. Portable radar-like transmitters had been commercially available for the purpose for some time and had brought relief to thousands of patients. In 1970, however, many physicians set them aside in the wake of the well-publicized realization that such devices could also do harm to certain tissues through undetected overexposure, which could lead to eye cataracts and other aberrations. Indeed, the devices were in many respects similar to the "microwave ovens" that utilized the same principle for rapid cooking of food, and physicians were increasingly reluctant to expose their patients to unknown hazards in the process of relieving painful spots by microwave heating.

In another biomedical application of electronics, an attempt to reduce the immensely complex waveforms produced by the human brain to a relatively simple presentation resulted in the successful completion of a stereo electroencephalophone—a device by which the EEG signals from the four quadrants of the skull are translated into two pairs of audible notes, with each pair sent to the observer's two ears through headphones. Moreover, the spatial relationships were preserved in the presentation, so that a listener got a subjective impression of listening to brain waves from the particular portion of the skull where they originated. The same technique also proved to be applicable to the sort of heart signals obtained in "vector" EKG, a technique that likewise depends on the preservation of spatial relationships for successful interpretation.

The therapeutic applications of the electroencephalophone remained highly speculative. Included among them was the fascinating possibility of feeding back his own brainwaves to a patient under treatment for such conditions as insomnia or heightened irritability. Considerable progress had been made previously in a related technique, in which the waves were analyzed by a computer almost instantaneously to determine which of the several EEG components were present and the patient was "trained" to evoke the component associated with a relaxed state by watching for a light that came on whenever he had been successful. Replacement of the simple on-off device (the light) by a signal that is actually related to the brainwaves being produced at the time and that enters the patient's consciousness painlessly through the medium of his own ears would seem to open up entire additional dimensions for future clinical exploitation of this technique.

Electronics and the graphic arts. An example of electronics applied to a completely new field is phototypesetting. In 1970, the superiority of computer-driven cathode ray tubes over older methods of preparing copy for offset printing was established firmly, and one printing firm after another sought to take advantage of the new invention.

In this application, type characters are "painted" on the face of a cathode ray tube by an electron beam in accordance with signals processed by a computer, which can supply a limitless number of styles and sizes. The result is photographically transferred to film, either by writing a line at a time and advancing the film, or by displaying and photographing an entire paragraph or page at once.

Not only is this method of phototypesetting vastly more versatile than the more conventional

A new process for fabricating complex electronic displays was announced in 1971. Tiny lamps called light-emitting diodes are combined in rectangular arrays. Left, a typical array is shown on a piece of newspaper. Right, the diodes can be activated in various combinations to make visible the 64 standard computer characters.

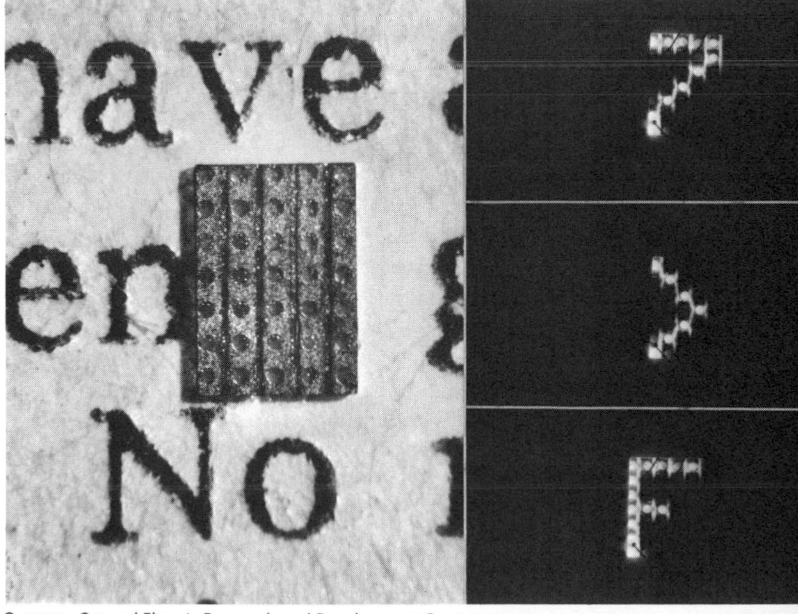

Courtesy, General Electric Research and Development Center

metal-casting machine (which provides at best eight fonts and a limited selection of sizes), but it is also at least 10 times as fast as the best metal linecasting machine. Digital storage of type fonts and computer drive of the cathode ray tube made it possible to produce a phototypesetter capable of producing 6,000 characters per second.

This example has been included as a demonstration of how the involvement of electronics can revolutionize entire industries. It also serves to illustrate the new concern of the professional engineer for the responsible use of his talents. But it must not be thought that this concern is entirely of recent date. As far back as the 1929 edition of the *Encyclopaedia Britannica,* Alfred Douglas Flinn wrote:

> The engineer is under obligation to consider the sociological, economic and spiritual effects of engineering operations and to aid his fellowmen to adjust wisely their modes of living, their industrial, commercial and governmental procedures, and their educational processes so as to enjoy the greatest possible benefit from the progress achieved through our accumulating knowledge of the universe and ourselves as applied by engineering. The engineer's principal work is to discover and conserve natural resources of materials and forces, including the human, and to create means for utilizing these resources with minimal cost and waste and with maximal useful results.

—Charles Süsskind

See also Year in Review: MEDICINE: *Biomedical Engineering.*

Environmental sciences

"Everything is connected," wrote the British novelist E. M. Forster early in this century. Ecologists have long been saying the same thing, though less succinctly, and everyone caught up in the environmental movement this past year has learned its truth—sometimes the hard way. For it was a year when "the ecological conscience," trumpeted abroad by the festivities of Earth Day, April 22, 1970, began to take root in some very rocky soil.

Ecology, the precise, often mathematical, study of the relationships of living things to their environment, was given a concrete meaning as *the ecology* (which meant anything from a local stream and its frogs to the entire planet), and as each new threat and each new hammerblow to the environment was identified, the economic and political ramifications of being in favor of *the ecology* became painfully clear.

The ecological movement changes. On April 18, 1971, almost exactly a year after the event, one of Earth Day's young organizers, Stephen Cotton, wrote that "the idea that students had discovered a new supercause was largely a myth," inspired by a few energetic workers in Washington, D.C., and inflated by the national media. Indeed, the organized student ecology movement, if it really ever existed, failed to make it through the summer. As early as the fall of 1970 it was clear that students were not the core of the environmental movement; only a few large campus groups remained active, such as those at the University of California (Berkeley) and the University of Michigan.

Yet Earth Day may well have accomplished its main task. While campus groups faded quickly, students' energies were often drafted by other citizens in community action groups. Earth Week 1971 (April 18–24) was more practical and less dramatic than the original Earth Day; for example, citizens' groups devoted themselves to bottle recycling projects. Environmental law societies gained a new vigor or sprang up anew in more than 50 law schools as student lawyers effectively took up the cudgels of environmental law.

More important, the message had been spread far and wide. It was hard for anyone anymore to be unaware of the ecological havoc man was creating. (*See* Feature Article: THE SENSES BESIEGED.)

New publications began, ranging from the hip, radical, and angry to the technical, industrial, and nervous. Many of them, such as *Earth Times,* an ecological cousin of the hip newspaper devoted to music, *Rolling Stone,* disappeared almost as fast. A British magazine called *The Ecologist* seemed the most thorough and reliable of the new eco-journals, but it was difficult to find in the United States.

Quantities of books were published on the subject—many merely "quicky" reiterations of the problem, but some, like Richard Saltonstall, Jr.'s *Your Environment and What You Can Do About It,* thoughtful and authoritative attempts to lead citizens to appropriate and practical action. The most typical publication in the year following Earth Day, however, was the ecological tip-sheet. Published in paperback-book form or often just mimeographed (on white paper because dyes foul the water) and distributed by the hundreds of thousands, these collections of dos and don'ts were directed at the person who seemed to be the ultimate polluter (or, at least, the most visible), the nation's chief consumer, the average housewife.

Life-style and the embattled housewife. The washday miracle, it turned out, was a catastrophe. Of all the phosphorus dumped into the Great Lakes, for example, 70% came from municipal sewage-treatment plants, and 70% of that amount originated with household detergents containing phosphates. By encouraging the growth of algae that use up oxygen in fresh water, phosphorus hastens the natural process known as eutrophication by which lakes and rivers "die." To avoid such consequences, environmental activists urged housewives to abandon phosphate detergents in favor of soaps that would not harm the nation's waterways.

In April 1971, Indiana followed the lead of two New York counties (Suffolk and Erie) and became the first state to ban high-phosphate detergents. By January 1973 no detergent with more than 3% phosphate content would be sold in Indiana; during the interim year, 1972, a phosphate content of only 12% was to be allowed.

It seemed likely that more states would follow suit, as a pattern began to emerge of increasingly strict standards imposed on a variety of pollutants, such as automobile emissions, the loads put on local sewerage systems, and the thermal emissions from power plants. Because it was also likely that the consumer would pay virtually all the costs resulting from such antipollution measures, the real question was: what will happen to the high resource-using American way of life?

Another urgent environmental problem involved the disposal of ever-increasing quantities of solid wastes. A cry went up for recycling these materials. Newspaper reclamation bins sprang up and were immediately oversubscribed. Centers for the collection and recycling of metal cans were also established. A technique was announced for using old bottles to build highways. But by mid-1971 there was little visible progress on any national or local scale in dealing with solid waste; it was still just garbage, to be put out the door and taken away in the ancient manner, while municipal governments desperately sought new places to put it. (*See* Feature Article: BURIED IN AFFLUENCE.)

Toxic substances. Consumers were warned that dangerous amounts of heavy metals and new chemical poisons were turning up in such staples as tuna fish and swordfish. First, it was mercury, then cadmium, arsenic, and lead. When one includes contamination by the insecticide DDT, which had been officially banned in many areas but was still being used on a large scale nonetheless, there seemed to be no safe food.

At the annual meeting of the American Chemical Society in March 1971, members cited item after item of dangerous chemicals in the environment. In April the U.S. Geological Survey reported that in most of the 720 urban and rural areas it had studied, the amounts of "new" chemical pollutants in the water did not exceed U.S. public health standards. In 12 urban areas, however, the report said that cadmium and arsenic in the water did exceed health standards. In May, swordfish was officially described as containing dangerous amounts of mercury and a New York woman who had eaten large quantities of it over a period of months was found to be suffering from mercury poisoning.

Michael Lloyd Carlebach from Nancy Palmer Photo Agency

Deprived of its natural water and plagued with the worst drought in years, the Everglades is devastated by fire. Years of diverting the natural water flow for land development has upset the ecological balance, threatening the survival of the former swampland.

As a result of such situations, the Senate began hearings on the proposed draft of a toxic substances bill that the administration had sent to Congress. This measure, which some scientists said was too weak, requested authority to determine the effect on human health and the environment of every new chemical before it is released. Environmentalists were pointing out, however, that we do not yet understand the synergistic effects of pollutants; that is, the extent to which one pollutant at a "safe" level will interact with another at a "safe" level to produce an amplified and lethal pollutant—as is the case with dust particles and sulfur dioxide in the air, which, when combined, cause severe lung damage.

Citizen action and the citizen's chief tool. Environmentalists claimed several major election victories in 1971. In Idaho, for example, an ad hoc group had formed to oppose the development of molybdenum mining claims in the remote, wild White Cloud Mountains, on land held by the U.S. Forest Service. The governor of the state opposed the conservationists, who were joined by such national organizations as the League of Conservation Voters, the political arm of the newly formed Friends of the Earth. The governor was not reelected, and while other issues were also involved, his defeat did demonstrate to environmentalists the viability of old-fashioned political activity. (Usually, however, in environmental issues the "villains" are not so clearly identified.)

Following the approach of Joseph Sax, a professor at the University of Michigan, a great deal of legal activity was aimed at establishing the right of citizens or private groups to bring suit even though their own economic interests were not at stake. Such a suit was filed by the Sierra Club against the federal government to prevent Disney Enterprises from building a resort on Forest Service land at Mineral King in the Sierra Nevada mountains. While the right of a private group to file such a suit was still awaiting a final decision from the Supreme Court, legislation was introduced in Congress in the summer of 1971 to make it legal for citizens to file class-action suits.

Another emerging pattern involved the many environmental lawsuits being directed against the federal government. The strategy for this action was to make the courts compel the administrative agencies of government to enforce existing laws or to follow appropriate administrative procedures.

Federal government initiatives. Many citizen suits took advantage of the 1970 National Environmental Policy Act, which created the Council on Environmental Quality (CEQ) and required federal agencies to prepare detailed environmental impact statements before taking such actions as licensing a nuclear power plant, building a highway, and so forth. The law did not state whether these "102 reports," as they are called, were to be made public. After bitter complaints by conservationists, however, the CEQ decided to publicize the reports 90 days before any final administrative action was taken.

Thus, in April 1970 conservationists were able to argue in court that federal agencies had not followed proper administrative procedures—that is, they had not adequately taken into account the environmental impact—when they granted permission to build an oil pipeline across the Alaskan tundra. The pipeline was delayed by a preliminary injunction.

In this instance, as in others, conservationists found an ally in the Environmental Protection Agency (EPA). Organized in October 1970 under the National Environmental Policy Act and headed by William D. Ruckelshaus, formerly of the Justice Department, EPA moved quickly to establish more stringent standards for air and water quality and the use of chemical pesticides. In the Alaska pipeline case, Ruckelshaus disputed a Department of the Interior report that said the pipeline was basically safe; in March 1971 he urged Interior to consider an alternate route through Canada.

Many of EPA's efforts were designed to force states and municipalities to set and enforce higher standards. For example, seven large power plants were planned or under construction on the shores of Lake Michigan; as a result, a severe thermal pollution problem was forecast. In March 1971, three of the states and the utilities involved agreed with EPA on temperature standards for the lake's waters and the means to maintain them. The fourth state, Illinois, announced that it would build no more nuclear power plants on the lakeshore.

EPA could and did intervene directly in some situations. In a ruling early in its existence, the agency set a deadline for a Union Carbide plant in Marietta, O., to stop polluting the air. The company answered that in order to comply it would have to close the plant and throw a number of men onto the unemployment rolls. Although this matter was later resolved (the company agreed to use low-sulfur fuel), 19 other plants did close or threatened to do so when faced with a similar order. This drew attention to a problem that will occur and reoccur: economically marginal businesses and plants may not be able to afford the installation of pollution-abatement equipment or to change plant processes. As increasingly strict pollution standards become more widely enforced, there are certain to be many cases involving economic hardships.

The federal government also demonstrated during the year its willingness to "waste" considerable investment of public capital. A case in point was the Cross-Florida Barge Canal, which conservationists strenuously opposed. This project by the Army Corps of Engineers involved building a canal across the breadth of Florida. It was already well under way when, in January 1971, Pres. Richard M. Nixon announced that, for environmental reasons, construction was to stop—in spite of the nearly $50 million that had already been spent on the project.

Economics, environment, and the SST. The choice between economics and environment was

During Earth Week an animal graveyard was built in Bronx Zoo to commemorate 225 animals that have become extinct since 1600. Wildlife experts estimate that at least 75% of the extinct creatures are victims of man's careless action against the environment.

Courtesy, New York Zoological Society

not always so clear. In March 1971, the U.S. House of Representatives and Senate voted to cut off further funds for the development of the supersonic transport (SST), a project that the administration supported just as strongly as it had the establishment of the government's staunchest defender of the environment, the EPA. Environmentalists, who had raised the anti-SST hue and cry years before on the grounds of noise pollution and other ecological considerations, claimed a historic victory. In May, the House reversed itself, but the Senate again voted the SST down—this time apparently for good.

It was clear that several other considerations besides environmental ones were involved in the defeat of the SST. They included:

The egalitarian—was public funding of the SST merely subsidizing the rich few who travel abroad?

The practical—what good was it to reduce air travel by a few hours if the time saved is immediately used up in getting from a remote airport to one's real ground destination?

The economic—was the SST the best way of bailing out a fiscally troubled aerospace industry, and was *that* more important than developing a rational, overall transportation policy for the entire nation?

Everything seemed very much connected, indeed tangled, in the demise of the SST, but several consequences for the future were perceptible in the immediate aftermath:

1. Fourteen thousand men, many of them highly skilled technicians, were thrown out of work. Legislation was immediately introduced to help retrain them to find jobs. More importantly, hearings were under way in Congress on the possibility of converting much of the energy and talent of the aerospace industry to such environmentally critical matters as housing and urban mass transportation.

2. The SST's demise gave a boost to those in and out of government who had been calling for "technology assessment," that is, the attempt to determine in advance the real effects of new technologies on society and the environment, rather than merely doing something because technologists have found out how to do it. (See *1971 Britannica Yearbook of Science and the Future,* Feature Article: PRIORITIES FOR THE FUTURE: GUIDELINES FOR TECHNOLOGY ASSESSMENT, pages 292–311.)

3. The present state of ecological knowledge is embarrassingly inadequate; as in other cases, both sides of the SST controversy had been armed with their own equally vehement ecological facts and predictions.

For additional information on the SST, *see* Year in Review: TRANSPORTATION.

The most complex ecosystem. When a question arises about that most complex of all ecosystems, the whole earth—as was the case in the arguments over the SST's possible effects on the world's climate—the state of the ecological art is far from satisfactory. Part of progress, however, is the assessment of ignorance. In July 1970, anticipating the United Nations Conference on the Human

Courtesy, General Electric Research and Development Center

An engineer tests a laboratory model of a proposed new cooling tower in an environmental chamber in which an atmospheric inversion is simulated. The cooling tower, designed to accomplish penetration of such inversions, may provide an effective means for thermal exhaust disposal.

Environment scheduled for 1972, the Massachusetts Institute of Technology sponsored a month-long meeting of scientists and professionals to assess the global impact of civilization's effluents. The results of the meeting, published later in the year under the title *Man's Impact on the Global Environment,* was one of the most significant books of the year, especially in its forthright discussion of our common ignorance.

Also known as the SCEP report (Study of Critical Environmental Problems), it warned that SSTs *might* have a severe impact on regional (not global) climate owing to increases of up to 10% in atmospheric water vapor in areas of high traffic, leading to increased cloud cover. SSTs would also, the report said, produce dangerous increases in particles (sulfur dioxide, hydrocarbons, and soot) in the atmosphere. In other areas of concern, the SCEP report:

- found that the amount of atmospheric oxygen was remaining constant, and was thus not a problem;
- saw increasing "heat islands" over cities from thermal power output with possible effects on regional climates;
- warned of severe potential problems if DDT and other chemicals as well as heavy metals such as mercury spread throughout the land, oceans, and living things.

All of the findings and recommendations of the SCEP report reiterated the need for more research, more data, and more monitoring of the global environment. Even predictions of such matters as how much heat man will add to the atmosphere by the year 2000 through the generation of power were hampered by lack of precise knowledge of how much he is adding right now.

Technology to the rescue. Much of the information SCEP scientists lacked will begin to be provided starting in 1972, when the U.S. National Aeronautics and Space Administration launches its first earth resources technological satellite (ERTS-A). Coupling earth-orbiting technology to the newly developed techniques of remote sensing, such as infrared photography, ERTS missions would be capable of rapid, wide-scale monitoring of the earth's natural resources.

The ERTS satellites would detect gross geologic structures, evidence of sick forests and food crops, soil types, land-use patterns, tell-tale marks of oil seeps, fish migrations, evidence of groundwater, rates of melting from mountain snows, signs of pollution, and even trends in urban development. When such information is rapidly disseminated and plugged into sophisticated mathematical models of large ecosystems manipulated by computers, it might one day provide the basis for intelligent planning and use of the planet.

Meanwhile, scientists were beginning to develop other technologies that might eventually relieve some of the impact of man on earth. For example, power companies were acceding to the demand for cooling towers. Generally speaking, for every three units of heat produced by a power plant, one becomes electricity and the other two are given off as waste heat—called thermal pollution or thermal effects, depending on who is talking. In conventional and nuclear power plants, most of this waste heat is carried off by cooling water, usually into rivers and streams, where the added heat has been proved to be destructive to aquatic life.

In March 1971 General Electric Co. announced a design for a new type of cooling tower. This six-story doughnut-shaped structure, the company claimed, would cool the water used to cool a power plant and eject the waste heat in such a way that it would penetrate temperature inversion layers overhead, thus ventilating the atmosphere.

The problem of growth. The above mentioned efforts are technological Band-Aids, some were saying, as they looked into the future and saw a dying planet. The central dilemma of power illustrated their point. To fuel its growing economy, the U.S. (and other industrial countries) needs to double its power-generating capacity every decade. Aside from various kinds of pollution that may or may not be technologically contained, and aside from the specter of an unprecedented proliferation of power plants dictating an arbitrary land-use policy for the nation, there is a very fundamental problem. According to some current estimates, fossil fuel resources will be exhausted in 30 years or so. Even with the development of breeder reactors, the cost of uranium ore by then will have skyrocketed. (*See* Feature Article: THE ENERGY CRISIS.)

So, during the year after Earth Day, ecologists and some economists were beginning to discuss seriously the mystifying ramifications of a "non-growth economy," seeking desperately for some way to visualize a world that one day would use less rather than more of the earth's natural resources.

—James K. Page, Jr.

See also Feature Articles: GRASSLANDS OF THE WORLD; WASHING OUR DIRTY WATER; Year in Review: FUEL AND POWER; MARINE SCIENCES; ZOOLOGY.

Several seals take refuge aboard a buoy in San Francisco Bay after two oil tankers collided, polluting the bay with massive oil slicks. The increasing frequency of oil leakage seriously threatens the marine environment and poses a major pollution problem.

UPI

Foods and nutrition

Have we made any significant progress this past year in the world's battle against hunger and malnutrition? Will there be fewer famines in the world in the next few years as a result of the Green Revolution, which has greatly increased the supplies of badly needed cereal grains in many less developed countries?

The answer to each of these questions is a qualified "yes." However, the battle to supply food to the rapidly growing world population is far from won, and we are still in a very precarious position.

The overall goal of providing enough nutritious food for all the world's population is extremely difficult to reach. Good nutrition for any individual, especially one in a less developed country, is affected by many important forces other than the actual food supply, with population density and family planning being the most important. Other factors include commercial food processing and retailing practices; medical and health standards; climate and the environment; transportation and storage facilities; and good nutrition education.

The 3,700,000,000 people on earth today consume about 4,000,000,000 pounds of food (dry basis) per day. This would fill 1,000 railroad trains of 50 cars each. Of course, these numbers would be about three times larger if the amount of water ordinarily present in foods was included.

In the United States in 1970, approximately $114 billion was spent on food, including food eaten away from home (about a fifth of the total). In the U.S. and Canada, the overall food bill amounted to about one-sixth of total personal income. People in most of the less developed countries spent a considerably greater proportion of their total income or total work effort on food for their immediate family; figures of from 40 to 60% were not uncommon.

In spite of this vast amount of money and effort devoted to obtaining food, hunger and malnutrition were still widespread in many less developed countries, with hundreds of millions of persons being affected.

There was abundant evidence in the U.S. to show that even there malnutrition existed in many forms, though extreme hunger was rare. In 1970 roughly 10–15% of the U.S. population, or 20 million–30 million persons, had some clearly evident form of malnutrition, such as inadequate intake of food and nutrients, anemia, goiter, underweight, or severe obesity, or suffered from nutrition-related disorders such as cardiovascular diseases, alcoholism, osteoporosis (softening of the bones), and poor or nonexistent teeth. An estimated 10% more were in borderline situations, though they might show no outward evidence of malnutrition.

Food consumption patterns in the U.S.—1970. The causes of malnutrition in the U.S. can be better understood by looking at the food consumption habits of the general public. Data provided by the U.S. Department of Agriculture (USDA) showed that in 1970, on an annual per capita basis, approximately half of the average American diet consisted of: 102 lb of white table sugar (sucrose); 53 lb of purified fats (mainly butter, lard, margarine, and vegetable oils); 100 lb of wheat flour ("white flour"); 14 lb of corn sugar (including that in corn syrup); and 7 lb of milled rice. This total of 276 lb of highly processed foods provided at least half of the energy requirements of the American public but only a small fraction of most of the essential nutrients, such as amino acids, vitamins, and minerals.

The other half of the average diet in 1970, calculated on a dry basis from figures released by the USDA, included: 92 lb of meat, poultry, and fish; 14 lb of eggs; 46 lb of milk; 18 lb of various fruits; 49 lb of various vegetables, including potatoes; 21 lb of whole wheat, cereals, and cornmeal; and 19 lb of peanuts, beans and miscellaneous foods (259 lb total).

Nutritionally speaking, a person can do quite well on such a total diet provided the foods are selected with some knowledge of their nutritional value. Unfortunately, half the American public was eating greater amounts of the highly processed foods than the "average" person. It was those persons who had most of the nutritional problems. From a nutrition point of view, furthermore, it appeared that the food habits of Americans were getting steadily worse instead of better. For example, in 1970 the average American consumed more soft drinks (32 gal) than milk (22.6 gal), a reversal of the ranking just five years earlier. About 10% of most soft drinks is table sugar. Thus, a 12-oz bottle supplies 1.2 oz of sugar (about two heaping tablespoons) and practically no other nutrients, whereas milk is a nearly complete food. The average American was eating (on a dry basis) more white sugar than eggs, fruits, and vegetables combined. In addition, the per capita intake of many important vitamins and minerals had declined steadily in recent years.

Developments in foods: the year of the consumer. There was a marked increase of public interest in foods and nutrition in the U.S. in 1970–71, partly as a result of the 1969 White House Conference on Food, Nutrition, and Health and partly because of widespread recognition of the existence of malnutrition, especially by the communications media. This interest was stimulated by the current concern about ecology and the environment, by

trends in food production
and population in the less developed regions

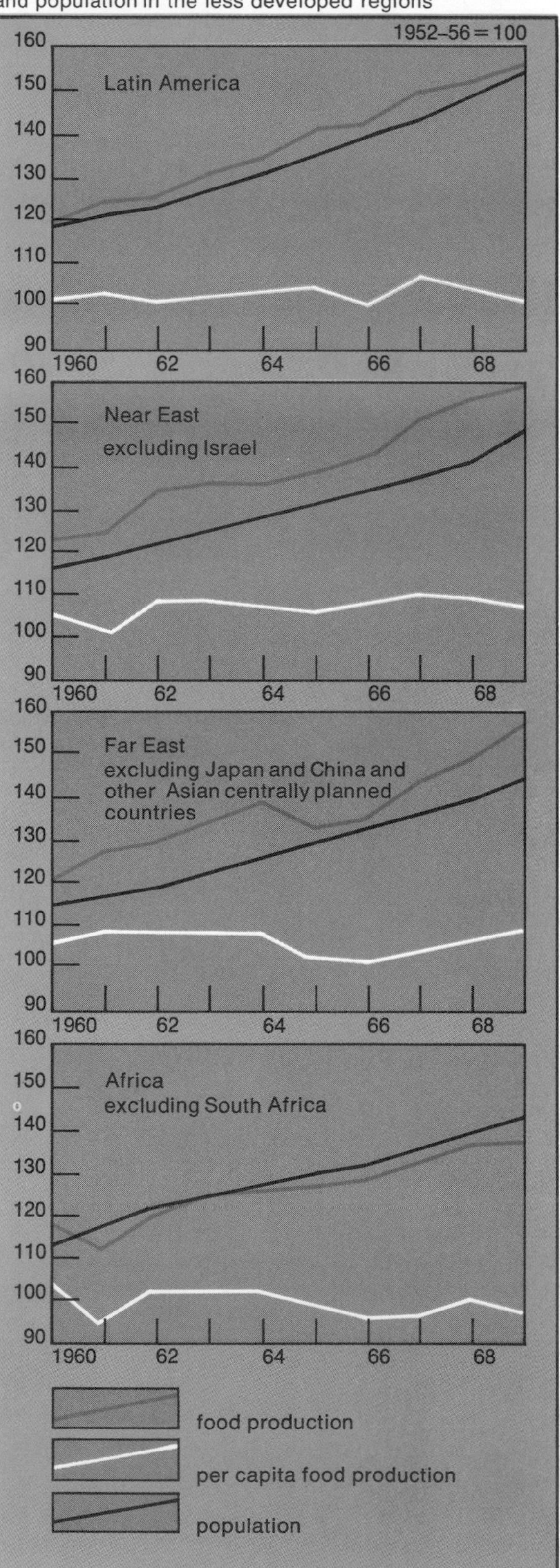

popular anxiety over the quality of the food supply, and by the recent claim that large amounts of vitamin C can prevent the common cold (*see* below).

In the U.S. a countrywide "Nutritional Awareness Campaign" was started in 1970, sponsored by the Food Council of America representing all segments of the $114 billion food industry, with the aim of better informing the consumer about nutrition matters. Much attention also was being given by various groups to the development of more informative labels on food packages and cans. A reevaluation of all currently used food additives was being made by the federal government, and there was particular anxiety over pesticide residues, such as DDT, and residues of possibly harmful trace minerals, especially mercury. In 1971 nearly all swordfish was removed from the U.S. food market as a precautionary measure because of its relatively high content of mercury—over 0.5 parts per million. The mercury content of other marine products was being monitored although, for the time being at least, such products appeared to be safe.

In 1970 Robert B. Choate, Jr., a nutrition crusader of Washington, D.C., made national headlines when he challenged the U.S. breakfast cereal industry in connection with what he termed the poor nutritive value of prepared breakfast cereals and misleading advertisements aimed at children. Though he did not take into account the nutrients naturally present in breakfast cereals or the milk generally consumed with such products, the incident served to point up the marked increase in public concern over the nutritional value of common food products—a concern with which most food scientists and nutritionists were in full agreement.

Many new types of food products were being developed by the food industry in response to changes in consumer demand. In 1970 "aseptic canning" was perfected, making it possible to market such canned products as flavored puddings. Advancements also were made in food-fortification procedures, food packaging, flavors and colors, freeze-drying, microwave cooking, dehydration, and in food-storage techniques. Of special interest was the increased commercial use of "textured proteins"—vegetable proteins (usually soybean) modified in such a way that they can be used in meat products or as an economical substitute for meat.

Nutrition advances. Probably the most far-reaching discoveries of the past year (largely unheralded by the press) were in the area of essential trace minerals—elements needed in the diet of animals in very small amounts, often one part per million or less. Using experimental rats and chickens, nutritionists at the USDA Agricultural Research Center, Beltsville, Md., and the Veterans Administration Hospital, Long Beach, Calif., discovered the dietary need for nickel, vanadium, and tin. One must go back to 1957 and 1958, when selenium and chromium were discovered to be essential in the diet, for a finding of equal significance in basic nutrition.

This discovery raised the number of inorganic trace elements known to be essential in the diet of an animal organism to 12 (the others were zinc, iron, copper, iodine, manganese, chromium, selenium, cobalt, and molybdenum). Though the need to include them in the human diet had not been demonstrated as yet, these trace elements could assume special importance in relation to world nutrition problems. Their presence in food depends to a great extent on their being absorbed by plants from the surrounding soil. As far as is known, plants do not require the minerals themselves, but act as carriers only.

Rapid advancements were made during the year on the chemistry and metabolism of vitamin D. Biochemists at the universities of Wisconsin and California (Riverside) and at Cambridge University in England isolated new and more active forms of vitamin D that are more effective against rickets and other bone disorders than vitamin D itself. Of great interest medically was the indication that such compounds are able to overcome certain vitamin D-resistant forms of rickets and certain rare metabolic diseases in man.

In April 1971 the climax of a 30-year search for pacifarin was announced by Howard Schneider, formerly of the American Medical Association in Chicago where much of the work was done. Pacifarin, found in certain batches of whole wheat and dried egg and produced by bacteria, protects mice infected with mouse typhoid. It was identified as enterobactin (a compound of known structure), recently discovered by biochemists at the University of California, Berkeley. The uses of this compound in human medicine were not yet known.

Several major reports on the effect of diet on circulatory diseases of man became available. Increasing recognition was being given to evidence that pointed to a variety of causes of heart disease and indicated that diet is not always an overriding factor. Five different types of lipid distribution (Types I-V) have been recognized in the blood of persons who have high levels of blood lipids (hyperlipoproteinemia). Dietary treatment varies with the type. In some instances a reduction in total calories, or in sugar consumption, is more useful than a reduction of dietary saturated fats or cholesterol. There was as yet no proof that dietary changes, beyond total calorie balance, have anything more than a minor role in preventing cardiovascular disease in most individuals with normal levels of blood lipids. As far as the eating habits of the general public were concerned, however, it still seemed reasonable to recommend the avoidance of rich diets with large amounts of saturated fats, cholesterol sources, and excessive carbohydrates (especially table sugar).

Does vitamin C, consumed in large amounts, prevent the common cold? This was the conclusion reached by Linus Pauling of Stanford University, holder of two Nobel prizes, whose book *Vitamin C and the Common Cold* (published November 1970) precipitated a worldwide controversy. According to Pauling, a very high level of vitamin C intake (as much as 5–15 grams per day, or about 100 times the usual requirement) might be able to prevent the common cold. Though he had strong support from the rapidly growing health food industry, most nutritional scientists were unwilling to accept his conclusions without further evidence.

A breakdown of the per capita food consumption patterns in the United States during 1970 indicates the vast need for an Expanded Nutrition Education Program (ENEP). Calculations were made from U.S. Department of Agriculture figures.

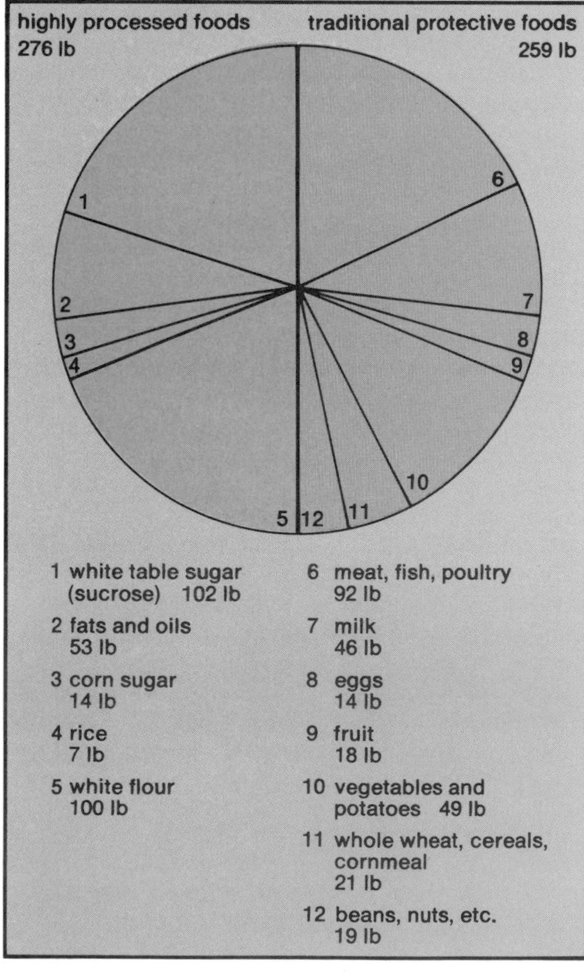

Advances were made during 1970 in understanding the possible role of diet in alleviating many aspects of diabetes, alcoholism, and dental problems. Improved techniques for measuring the nutritional status of individuals were discovered, including the use of hair, nails, blood, and urine as readily available indicators of nutritional deficiency.

Not all nutrition discoveries could be classified as "good," however. Studies made during the year on the extent of malnutrition throughout the world confirmed the existence of severe malnutrition not only in the economically deprived nations but also in the so-called developed countries. In April 1971 the U.S. Department of Health, Education, and Welfare announced that anemia and serious vitamin deficiencies had been found in 25% of 83,000 persons surveyed in a 10-state area. The poor were twice as likely to have low or deficient indicator levels of malnutrition as those above the poverty level, and blacks were four times as likely to be nutritionally deficient as whites with similar incomes.

Nutrition education. Significant advances were made in many countries during the year in the development of techniques useful in the nutrition education of the consuming public. It was recognized that when an unlimited variety of foods is available, an informed public will at least tend to choose better diets than an uninformed public.

Jean Mayer of Harvard University and other nutrition leaders questioned the adequacy of the well-known "four food groups" pattern (meats, vegetables and fruits, breads and cereals, dairy products) as an educational tool. It was agreed, however, that this pattern was still useful in teaching the meaning of the "variety of good food" that nutrition educators recommend.

Several millions of dollars were spent in the U.S. during the year on an Expanded Nutrition Education Program (ENEP), sponsored by the Cooperative Extension Service of the Department of Agriculture. With these funds persons from low-income communities were given training so that they, in turn, could provide guidance to other persons in their neighborhoods. Elsewhere in the world, "mother-craft centers," designed to further nutrition education, were established in Haiti, the Philippines, and parts of South America.

Obesity, which continued to plague many people in the developed countries, is a form of malnutrition caused by excessive consumption of calories. Some degree of success in treating obesity was

Courtesy, William J. Shoemaker, Massachusetts Institute of Technology

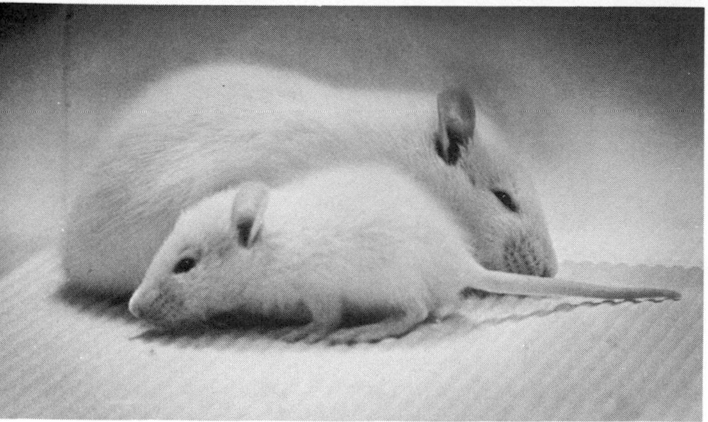

Experiments with rats provide evidence of the direct relationship between nutrition and early brain development. The smaller rat, nursed by a nutritionally deprived foster mother, showed a decrease in the necessary amounts of neurotransmitter substances in his brain. His adequately fed littermate was normal.

being achieved by group therapy programs. Especially effective in the U.S. were TOPS (Take Off Pounds Sensibly) and the Weight Watchers.

The future. With about 10,000 research publications on foods and nutrition being published each year, many important new discoveries could be expected. In the near future, significant advances in the enrichment and fortification of common foods with low-cost vitamins, minerals, and vegetable proteins in order to combat malnutrition were likely, as was the wider use of "fabricated" or manufactured foods, including complete meals containing low-cost vegetable protein fortified with many nutrients.

More studies could be expected on the metabolic role of nutrients and on the relationship of good nutrition to resistance to infection, and to normal behavior, mental health, and development of persons of all ages. Special attention would be given to the production of new food sources, especially "single-cell" protein sources, and to the question of food additives. For the immediate relief of malnutrition, improved food-distribution programs and school feeding programs would be essential.

—George M. Briggs

Fuel and power

The United States in the early 1970s continued to lead the world in the per capita consumption of energy. In 1970 the average American consumed energy equivalent to about 60 barrels of oil. Western Europeans and citizens of other advanced industrialized countries consumed about half that amount; persons in the less developed nations consumed less than 10 barrels each. But while U.S. energy requirements increased at average annual rates of 3 to 4% in recent years, the energy needs of other nations were advancing even more rapidly. Energy demand in the non-Communist world outside the U.S. grew at annual rates in excess of 6%.

In general, the per capita use of energy rises in accord with the growth of a national economy and the increased average income per person. In the four years from 1966 through 1970, however, the per capita consumption of energy in the U.S. increased even faster than the gross national product. This can be attributed to accelerated growth in home and automotive air conditioning, expanded use of electrical heating and other appliances, increased use of industrial energy for the processing of raw materials, greater use of passenger vehicles, and less improvement from the application of technology to better the efficiency of fuels usage.

However, despite the rapid increase in energy consumption, the U.S. and the other highly industrialized nations of the world continued to depend to some extent on less developed countries for their fuel supplies. Western European nations obtained about two-thirds of their total fuel requirements from the oil-exporting countries of North Africa and the Middle East; Japan was dependent on foreign sources for more than 90% of its fuels; and the U.S. imported about 25% of its petroleum, largely from Canada and South America.

The combination of accelerating demand for fuels in the major consuming countries and the growing dependence on oil-exporting countries for supplies resulted in several confrontations between these sides during 1970. As a result, there were transportation and supply stringencies, an increase in the share of oil concessionaires' profits paid to the governments of the producing countries, a rise in consumer countries' oil prices, and a growing international concern over a developing energy crisis. (*See* Feature Article: THE ENERGY CRISIS.)

During 1970, the major energy concern in the U.S. and other industrialized countries continued to revolve around problems associated with meeting the enormous demands for fuels and power, while at the same time preventing further degradation or detrimental effects on the environment. As much as any other activity, the supply of fuels and power to the economy has an impact upon the environment at each step in its production, conversion, distribution, and consumption. Too often in the past, fuels were provided on a competitive economic basis by producers who developed re-

Courtesy, National Coal Association

A continuous mining machine tears coal from a mine seam with whirling teeth made of steel. The loosened coal drops to the floor, where rotating steel arms load it into shuttle cars or conveyor belts for further handling. The development of this new machine has helped revolutionize the underground mining of coal.

sources without paying adequate attention to the harmful effects of their activities on air, water, scenic locations, and other common "free resources." Belatedly, there has developed an awareness of the necessity to prevent abusive use of these resources. Potential health hazards and other detrimental environmental effects have led to public concern about air, water, and thermal pollution; nuclear radiation; prevention or reclamation of land damage; and related environmental matters.

The deep concern expressed by many segments of society toward energy supply and environmental problems was reflected in several governmental actions and hearings on the current and future situation. U.S. Pres. Richard Nixon directed the Council on Domestic Affairs to investigate energy problems, and at least 10 energy studies were conducted or are in progress in various executive departments and agencies. Several Congressional committees held hearings to investigate the situation, and more than 50 bills relating to various aspects of the energy crisis in the U.S. were introduced in the 91st Congress. Included were measures affecting electric power, nuclear energy development, oil- and gas-well drilling and production in the outer continental shelf, coal-mine health and safety, land reclamation after strip mining of coal, oil imports, development of Alaskan oil production on the North Slope and the need for oil and gas pipelines, development of oil shale lands, leasing of geothermal steam lands, price controls on natural gas, and research on solar and tidal energy.

The National Environmental Policy Act, signed early in 1970, was one of the most significant of the many U.S. laws enacted pertaining to the environment and the fuel and power industries. The law established a national policy on all environmental matters, created a new Council on Environmental Quality, and placed new responsibilities and requirements on all federal agencies to take environmental factors into account in their decisions on the development of fuel resources on all public lands, the siting and operation of powerplants and other large facilities, and all other activities under their jurisdiction.

Petroleum and natural gas

During the past year the world's petroleum and natural gas industries faced new challenges in meeting soaring demands, in adjusting to changed political and economic conditions in oil-producing and exporting countries, in developing new and diverse oil supplies, and in conforming to environmental requirements. In the U.S. the industry was faced with more stringent environmental considerations, a decline in excess producing capacity and greater dependence on foreign imports, and a worsening natural gas supply situation.

Worldwide oil demand in 1970 averaged about 45 million barrels daily, nearly 10% more than that of 1969 and a marked increase over the average growth rate of 7.7% annually during the 1960s. The U.S. continued as the dominant oil-consuming nation, accounting for almost one-third of the total.

New demands by producing countries. Political and economic conditions in the international oil industry underwent profound changes. The balance of power in oil transactions switched from the buying to the selling countries. Producer countries forced oil concessionaires to give up their historical prerogative of setting export oil prices and negotiated an increase in their share of oil profits. For example, Libya won concessions from oil companies by restricting oil production, thereby causing a disruption in oil supply patterns. The cutback in Libyan production, at a time when Middle East pipelines were not permitted to operate and the Suez Canal continued to be closed, caused a worldwide tanker shortage. The necessity to haul oil from the Persian Gulf the long way around the

Cape of Good Hope to Europe increased the average delivered price of crude oil to Western Europe and Japan by about 40%. Price increases at U.S. ports were even greater, and many companies were not able to import their allowed quotas.

Following the Libyan success, a number of oil-producing countries in the Persian Gulf region (Iran, Iraq, Kuwait, Qatar, Saudi Arabia, and one of the Trucial States), negotiating through the Organization of Petroleum Exporting Countries, concluded an agreement with oil companies that gained for them a significantly higher revenue per barrel of crude oil produced. The five-year agreement, signed on Feb. 14, 1971, was expected to increase the producing government's share by 30 cents per barrel in 1971 and by about 50 cents in 1975. Much of this price increase was expected to be passed on to the consuming countries. Even more increases seemed likely because Algeria, Libya, Nigeria, Venezuela, and other producing countries continued to press demands for greater shares of oil profits and control of oil operations.

Oil exploration and development. The industry continued exploration for and the development of new oil sources. The most notable success was the discovery of large oil reserves in the North Sea, where nine large oilfields were found off Norway, Great Britain, and the Netherlands. Other important discoveries were made in the offshore area of Indonesia, in Iraq, and in Ecuador.

The U.S. petroleum industry struggled to develop new domestic oil and gas resources in the face of new, more stringent environmental protection requirements. The development of the reserves on Alaska's North Slope was a case in point. In order to develop and exploit this large new reserve, operators proposed construction of an 800-mi, 48-in.-diameter trans-Alaska pipeline as the most effective and economical way of transporting the oil from the Arctic fields to an all-weather port on Alaska's southern coast. Owing to environmental and ecological concerns, however, a series of events virtually halted continued development drilling and delayed the construction of the pipeline indefinitely. Since the proposed line traverses large permafrost areas and regions of high seismic

Below, cooling towers at the Keystone Steam Electric Station in western Pennsylvania serve the station's two 820,000-kw generating units. The towers, 325 ft high and 247 ft in diameter at the base, are natural-draft towers of hyperbolic form. Most of the height functions to produce a strong updraft of air inside the tower which meets warm water from the plant falling through a draft near the bottom of the tower. Breakup of the water by fill inside a tower is shown at the right. Falling water repeatedly hits pieces of fill and breaks up into small drops and films which present a large surface area to the moving air in the tower. Cooling takes place by evaporation of a small amount of the water. Clouds of nonpolluting vapor emerge from the towers when they are in operation.

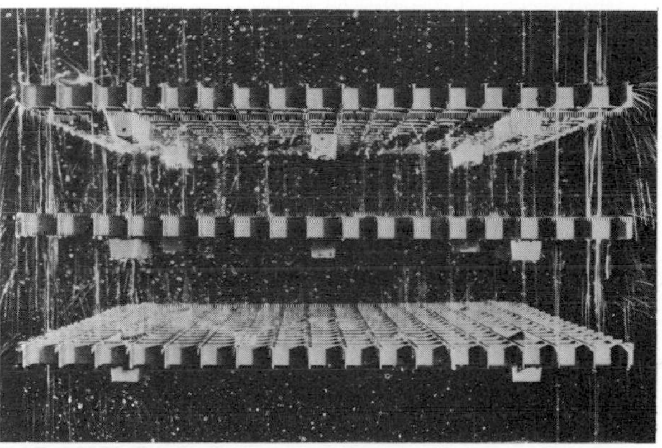

Courtesy, Ecodyne Corp., Fluor Cooling Products Co.

Courtesy, Pennsylvania Electric Company

activity, it poses construction problems and dangers to the environment, especially in the event of a rupture. The U.S. Department of the Interior undertook a detailed investigation of the planned pipeline and issued strict regulations for its construction.

Another problem involved land claims by Alaskan natives for the property along the pipeline route, which remained unresolved. Settlement of those claims was expected to require legislative action by Congress. Also, certain environmental groups obtained a court injunction against the issuance of a construction permit, based on the fact that the required right-of-way is wider than authorized under existing leasing acts.

Meanwhile, the U.S. was becoming increasingly dependent upon foreign sources for about 25% of its oil supply. Because in past years the country had a large excess oil-producing capacity, oil imports were limited to specific quotas in order to maintain this capacity for use in a national emergency. However, as demand increased, the domestic industry was not able to discover new oil reserves and develop new capacity at rates sufficient to maintain the margin of excess. Not only did the excess capacity disappear but the U.S. in 1970 could not produce enough domestic oil to meet its demand.

As a result of the disruption of normal oil supply patterns, crude oil imports into the U.S. from foreign sources declined in recent months. Accordingly, U.S. Pres. Richard Nixon ordered increases in the amounts of oil that could be imported from Canada. Also, he ordered the Department of the Interior to increase well production in the federally owned portion of the outer continental shelf.

Natural gas. The natural gas supply situation in the U.S. also continued to worsen. In 1970, for the third consecutive year, natural gas consumption exceeded new additions of reserves. A potential gas shortage was confirmed by this sharply declining availability of new gas supply to meet growing demand and by the inability of gas suppliers to guarantee delivery of gas needed to supply new industrial customers.

In an effort to improve the supply situation, the U.S. government took a number of measures, such as rate hearings by the Federal Power Commission (FPC) to consider better economic incentives for gas producers. On Dec. 15, 1970, the government held a federal lease sale of tracts in the Gulf of Mexico that had great potential for new gas discoveries. The record sale was worth $845 million and ended a two-year moratorium on oil and gas lease sales in the outer continental shelf that had been imposed because of the Santa Barbara, Calif., oil spill.

Natural gas is a clean, nonpolluting fuel that supplies about one-third of the total U.S. energy requirements. Shortages of natural gas will cause an increase in the use of less desirable fuels and may jeopardize environmental objectives. In the near future, there are no large sources of low-sulfur fuels available to substitute for natural gas in the quantities needed and at competitive costs. Increased imports of pipeline gas from Canada and liquefied natural gas from other foreign gas supply areas are the most probable sources of immediate supplemental supplies. Over the longer term, however, the technological development of low-cost, pipeline-quality synthetic gas from coal, oil shale, or heavy tar sands will be required to meet the demand.

Estimates of the quantities of undiscovered natural gas resources in the U.S. exceed the level of the demand for it for the balance of this century. However, the natural gas industry must improve its recent performance in finding and developing these resources. If the accelerating natural gas demand is to be filled, it will be necessary to develop new supplies at annual finding rates nearly twice as great as accomplished from 1965 to 1970. Increased economic incentives to gas producers, in the form of higher wellhead gas prices, are considered essential by many industry and government officials to stimulate the needed exploration and development effort. The FPC, which regulates the price of natural gas sold in interstate markets, was continuously studying the price situation.

Coal

During 1970, the U.S. coal industry raised its output to the highest level in nearly 25 years. Buoyed by exports of 74 million tons and a strong domestic demand, bituminous coal and lignite production reached 590 million tons, approaching the record-high levels of slightly more than 600 million tons annually that occurred in the mid-1940s. Because coal remained the most important fuel for generating electric power, electric utilities accounted for 62.4% of all domestic coal consumed in 1970. Despite continuing problems in meeting antipollution regulations, other environmental restraints, labor shortages and unrest, shortages of railroad cars, and growing competition from nuclear and other competitive fuels, U.S. coal demand rose 2.3% in 1970.

Supply and demand patterns. The strong export demand for high-quality U.S. coals of recent years continued in 1970. Paced by Japan, the demands by industrialized nations increased U.S. coal and coke exports during 1970 by about 26% to an estimated 74 million tons, valued at over $1 billion.

"London Daily Express" from Pictorial Parade

There was a worldwide shortage of coking coal during 1970 that was reflected in sharply increased U.S. exports. This shortage resulted from an unprecedented rise in world steel production coupled with a reduction in foreign metallurgical-grade coal-mining capacity and the depletion of large coal and coke stockpiles. The increased export demand caused a sharp increase in export coal prices. The average export value rose to $13.40 per ton in 1970, an increase of about 30% over 1969.

Continuous advances in technology increased coal productivity nearly threefold and reduced costs, permitting coal to remain competitive with other fuels during the past 20 years. In 1970, however, the industry was faced with increased costs that offset productivity gains. Among these were the expenses incurred in complying with new environmental, health, and safety requirements.

Effect of antipollution and safety laws. Antipollution laws and regulations require the use of low-sulfur fuels (sulfur content of 1% or less) in many areas. Although the U.S. has large resources of low-sulfur coal, most developed reserves are of the high-sulfur varieties. Furthermore, most low-sulfur resources are located in the western states, far distant from the major consuming markets. Because it is not yet feasible to remove enough sulfur from coal to meet strict antipollution requirements, there is an urgent need to develop a commercially viable method of removing sulfur dioxides from combustion gases or to develop new methods of combustion, or processes for converting high-sulfur coal to low-sulfur gas or liquids. These objectives were pursued by government and industry researchers in 1970. Preliminary results of research by the U.S. Bureau of Mines showed that coal can be converted to a fuel having a sulfur content as low as 0.2%. Also, the Office of Coal Research obtained encouraging results on research to produce a low-sulfur power plant fuel.

Mandatory safety standards under the Federal Coal Mine Health and Safety Act of 1969 became effective on March 30, 1970. Three months later the mandatory health and respirable dust provisions of the act also became effective. To implement the act, the Bureau of Mines made plans to expand greatly its inspection and enforcement staff. A federal Mine Health and Safety Academy was established to train inspectors, supervisors, and other enforcement personnel. Although compliance with the new regulations resulted in lost

A massive atomic reactor in the first completed portion of the Novovoronszh nuclear power station in the U.S.S.R. is prepared for operation. This section alone is estimated to have a generating capacity of 575,000 kilowatts.

Courtesy, Los Alamos Scientific Laboratory

Technicians install the first of three sectors of the quartz torus of Scyllac, an experimental machine designed to investigate the possibility of controlling nuclear fusion. A continuous ring of aluminum coils (foreground) will deliver the energy simultaneously from 3,000 energy storage capacitors to the torus.

production in some mining areas, its overall effect was not measurable by mid-1971. Some small mines closed voluntarily soon after the act became effective, and others closed, in whole or in part, for various periods of time. It was anticipated that any lost productivity as a result of the new health and safety requirements would be temporary and largely offset by the coal industry through the adoption of improved technologies and operating practices.

Electric power

Electric power consumption continued as the fastest growing sector of U.S. energy demand. In 1970, the national demand for power generated by electric utilities exceeded 1.5 trillion kw hr, about 7% more than that of 1969. During the late 1960s and into the 1970s electricity demand grew more than twice as fast as total energy demand. This rapid growth required a doubling of generating capacity in the past decade. This was too rapid for the electric utility industry, which was not able to expand output fast enough to meet peak demands at all times of the year in all areas of the country. Blackouts, voltage reductions, and other curtailments or interruptions of electric power service to homes, offices, and factories were the results.

Along with this increased need for power, there developed a growing concern with the environmental degradation and potential health hazards associated with power generation. This was reflected in shortages of fuels of low-sulfur content that could meet more stringent air-quality standards and in delays in the approval of construction and operating permits or licenses for new plants.

Sources of electric power. Although hydroelectric plants provide an important share of electric power, fossil fuels remained the major source of centrally generated power. In 1970, fossil-fueled plants produced 82.5% of net generation in the U.S.; hydroelectric plants accounted for 16.2%; and nuclear power plants produced only 1.3%.

Traditionally, utilities have used the most economical fuels available to generate electric power at central plants. During the 1960s, coal accounted for about 65% of the fuel used, natural gas for about 28%, and oil for all the remainder except for an insignificant percentage by nuclear fuels. The relative shares of fossil fuels used for power production changed in 1970, with coal's share slipping to 55% while natural gas remained at about 28%, and oil increased its share to about 16%.

The growth of nuclear power and limitations on the sulfur content of fuels caused a shift from low-cost, high-sulfur coal to residual oil that had a lower sulfur content but was more expensive. Most of this oil had to be imported into the U.S., its use made necessary by the short supply of natural gas and of domestic low-sulfur oils. Most low-sulfur coals are of the metallurgical grades used in steelmaking and are more costly than steam coals. They are in great demand for export markets, and available supplies are limited. Thus, all fuels used in power generation were in short supply, and there was a marked increase in their prices.

Nuclear-electric power. Although nuclear power plants had produced only a negligible portion of U.S. electric power by 1971, they were expected to provide a much greater share in the future. In 1970, there were 16 operable nuclear plants having a generating capacity of 5,300 Mw. However, an additional 55 plants were under construction and others were on order. All together these plants comprised nearly 94,000 Mw of generating capacity, an amount equivalent to one-third of the total existing U.S. capacity.

Delays were experienced in the completion of all types of electric power plants, but those for nuclear plants were the most serious and were mainly responsible for the inadequate generating capacity in certain areas of the country. These delays were caused by late delivery of reactors and other major plant components, longer construction times than planned, and public opposition to nuclear plants because of possible adverse radiation hazards and other environmental effects. These factors and the attendant increased costs caused some utilities to defer plans for new nuclear plants. In 1969, only 8 reactors totaling about 8,000 Mw of capacity were ordered. In 1970, however, there was an upsurge in new orders as 16 plants with a total capacity of nearly 17,000 Mw were announced.

Although nuclear plants do not produce air pollutants such as sulfur oxides and particulate matter, they do pose other environmental problems. These include potential radiation hazards due to plant emission during operation and to the permanent disposal of radioactive wastes from spent fuels, and thermal pollution caused by the discharge of waste heat into bodies of water. Heat discharges from the present type of nuclear plants exceed those from fossil-fuel plants by about 40%. Public concern over these environmental problems led to opposition to the siting and licensing of such nuclear plants. These vital concerns must be resolved if the growing demand for electrical energy and the requirement that it be provided without excessive environmental degradation or health hazards is to be brought into reasonable balance. The outcome will have an important impact on the future growth of nuclear power.

The U.S. Atomic Energy Commission and electric power companies gave top priority in their nuclear research and development efforts to the fast breeder reactor program. The goal is to develop a reactor that will "breed," or produce more fissionable material than it consumes so that it can refuel itself and another reactor of equal size every seven to ten years. U.S. and foreign researchers concentrated on developing a fast breeder reactor of the liquid-metal type.

Continued research on the extremely difficult technologies needed to obtain controlled fusion produced some encouraging results. Scientists at the Los Alamos (N.M.) Scientific Laboratory activated a large experimental test machine called the Scyllac and demonstrated its capacity to contain repeatedly an electrified form of gas, heated to 15,000° F, for as long as 30 microseconds. The demonstration, considered a milestone in scientific achievement, was one of many steps needed eventually to sustain a fusion reaction. Fusion would permit the direct conversion of nuclear energy into electricity in almost limitless quantities.

Outlook

The immediate outlook is that fuel and power concerns will claim a greater share of the public's attention in the U.S. and other industrially advanced nations. The worldwide demand for all forms of energy is expected to continue undiminished in the months ahead. These growing demands for all types of fuels will cause supply problems that will probably result in shortages of fuels in certain places at times of peak requirements. Interruptions in power service, such as blackouts, brownouts, and voltage reductions, will occur more frequently and cause greater inconvenience to the consumer. Environmental and ecological problems associated with the extraction, processing, transportation, and utilization of fuels and energy resources will attract greater notice. Increased production, transportation, and environmental protection costs will be passed on, and the consumer will pay higher prices for the fuels he consumes.

The near-term outlook for the specific fuels and energy sources is as follows. (1) Petroleum and natural gas will remain the principal fuel supply. Oil, especially that from foreign sources, must be relied upon to fill the gap between demand and supply created by shortages or restrictions on coal, natural gas, and nuclear capacity. (2) Coal demand will continue to grow, but its future will depend substantially on how successfully it can develop commercial technologies to overcome environmental and social objections to its use. (3) Natural gas demand will accelerate because of its clean-burning characteristics, but new supplies are insufficient to fill all market demands. (4) Delays in new plant construction will continue to affect adversely the rate of nuclear capacity development.

—James A. West

Honors

The following major scientific honors were awarded during the period from July 1, 1970, through June 30, 1971.

Aeronautics and astronautics

Daniel and Florence Guggenheim International Astronautics Award. The International Academy

Wide World

Luis Leloir

of Astronautics gives an annual award for individuals who have made an outstanding contribution to space research and exploration through work done in the preceding five years. The 1970 Daniel and Florence Guggenheim International Astronautics Award was presented to the Soviet cosmonauts Andrian Nikolayev and Vitali Sevastianov, for the experiments performed during the June 1970 flight of Soyuz 9, which constituted a milestone in manned space flight. The Soyuz 9, aloft 17 days and 17 hours, greatly expanded the knowledge of medico-biological effects of space conditions on human organisms.

Founders Medal. The National Academy of Engineering each year presents the Founders Medal to honor outstanding contributions by an engineer both to his profession and to society. The 1970 Founders Medal was awarded to Clarence L. Johnson, senior vice-president of Lockheed Aircraft Corp. in Burbank, Calif., for his contributions to aeronautical engineering, including the designing of advanced aircraft. His best-known work included the designing of the Hudson bomber, the Constellation and Super Constellation transports, the P-3B, the T-33 trainer, the F-90, and the JetStar.

Astronomy

Benjamin Apthorp Gould Prize. First awarded in 1971 by the National Academy of Sciences, the $5,000 Benjamin Apthorp Gould Prize went to Elizabeth Roemer, professor of astronomy at the University of Arizona, in recognition of her distinguished contributions in the field of cometary astronomy. In particular, Roemer was cited for her consistent successes in first sighting comets as they return toward the sun. Since 1952, she and her associates have been responsible for first sightings of 51 periodic comets as they journey from outer space toward and around the sun.

Helen B. Warner Prize. Presented annually since 1952 by the American Astronomical Society, the Helen B. Warner Prize is given in recognition of a significant contribution to astronomy during the preceding five years. The 1970 award was made to John N. Bahcall, theoretical physicist at the California Institute of Technology, for his contributions to cosmology. In the area of theoretical work Bahcall participated in an experiment designed to explore the temperature and density of the nuclear core of the sun by detecting and recording solar neutrinos. In a second area Bahcall examined light from distant quasars, which enabled him to determine the density and composition of interstellar gas and dust.

Henry Draper Gold Medal. The National Academy of Sciences awarded the 1971 Henry Draper Gold Medal to Subrahmanyan Chandrasekhar of the Laboratory for Astrophysics at The University of Chicago, for his major contributions to theoretical astrophysics, and particularly for his studies of the structure of stars, their evolution and atmospheres, and the dynamics of gravitational interactions with other stars in clusters or galaxies.

Biology

John J. Carty Medal. Established in 1930 and supported by the American Telephone & Telegraph Co., the award is administered by the National Academy of Sciences. The 1971 John J. Carty Medal, with an honorarium of $3,000, was presented to James D. Watson, Nobel laureate biochemist at the Biological Laboratories of Harvard University, in recognition of his outstanding accomplishments in molecular biology. Watson concentrated his efforts in determining the role of ribonucleic acid (RNA) in the synthesis of protein, and with his colleagues provided the most precise available characterization of the ribosome particles in which protein synthesis occurs and identified the small portion of RNA that carries genetic information.

Louisa Gross Horwitz Prize. Established in 1967 by the College of Physicians and Surgeons of New York, the Louisa Gross Horwitz Prize is administered by Columbia University and is awarded annually to recognize outstanding basic research in biology or biochemistry. Sharing the 1970 prize of $25,000 were Albert Claude of the Free University of Brussels, George E. Palade of The Rockefeller University in New York City, and Keith Porter of the University of Colorado. The three scientists were cited for their "important contributions to our knowledge of the functions and fine structure of cells." It was noted that the three recipients had been associated with the Rockefeller Institute from 1946 until 1949 for a period of critical interaction in their work.

U.S. Steel Foundation Award in Molecular Biology. Administered by the National Academy of Sciences, the U.S. Steel Foundation Award in Molecular Biology may be given annually for a recent notable discovery in molecular biology by a young scientist. The $5,000 award for 1971 was presented to Masayasu Nomura, co-director of the Institute for Enzyme Research at the University of Wisconsin, in recognition of his studies on the structure and function of ribosomes and their molecular components. Nomura demonstrated that parts of bacterial ribosomes may be disassociated or broken down into their components—ribonucleic acid (RNA) and proteins—and then reassociated to regenerate fully active particles. This discovery was vital to the study of ribosome structure and function.

Chemistry

Garvan Medal. The American Chemical Society each year presents the gold Garvan Medal and a $2,000 honorarium to a U.S. woman chemist selected for her distinguished service to chemistry. The 1971 award was given to Mary Fieser, a research fellow in chemistry at Harvard University, for her research, and as co-author with her husband, Louis F. Fieser of Harvard, of a series of outstanding textbooks and other publications relating to the field of organic chemistry. Mary Fieser's citation stated that generations of organic chemists have learned the rudiments of science from the Fieser text and the Fieser laboratory manuals.

Ipatieff Prize. Under the trusteeship of Northwestern University, the Vladimir N. and Barbara Ipatieff Trust Fund was established in 1943 to support the Ipatieff Prize, to be presented every three years to recognize experimental work in the field of catalysis or high pressure. In 1971 the $3,000 prize was awarded to Paul B. Venuto, research associate at Mobile Research and Development Corp. in Paulsboro, N.J. Venuto was cited for his extensive research on catalytic agents, during which he uncovered a broad and unusual area of information on how catalysts function. His study of zeolite (aluminosilicate) catalysts elucidated the chemical nature of the active areas on catalyst molecules.

Nobel Prize for Chemistry. The Royal Swedish Academy of Sciences selected Luis Frederico Leloir, professor at the Institute of Biochemical Research in Buenos Aires, as recipient of the 1970 Nobel Prize for Chemistry. Leloir, a French-born Argentinian, was the first Latin American to win a Nobel chemistry prize. The $78,400 award was given for Leloir's discovery and isolation of sugar nucleotides (a substance needed for transforming one type of sugar into another), and for the subsequent research on the functions of sugar nucleotides in the biosynthesis of carbohydrates. His work was cited as fundamental to the understanding of the workings of the human body.

Perkin Medal. Presented annually by the American Section of the Society of Chemical Industry, the Perkin Medal is considered to be the highest honor bestowed for outstanding work in applied chemistry in the United States. The 1971 award was given to James Franklin Hyde, senior scientist at the Dow Corning Corp., Midland, Mich., for his role in developing the group of synthetic substances known as silicones. Hyde's 40 years of pioneering work resulted in transforming a blend of silicon, oxygen, carbon, and hydrogen into the versatile family of silicones used in aircraft and space suits, in surgery, and in kitchen appliances.

Priestley Medal. Established in 1922, and presented each year by the American Chemical Society, this high honor is in recognition of distinguished services to chemistry. The 1971 recipient of the gold medal was Frederick D. Rossini, professor of chemistry and vice-president for research at the University of Notre Dame. Rossini, an authority in the fields of chemical thermodynamics and thermochemistry, conducted extensive research at the National Bureau of Standards in order to achieve standards of precision requiring immense care and accuracy in purifying compounds, and also to measure their physical and thermochemical properties.

Roger Adams Award in Organic Chemistry. Established in 1959 and sponsored by Organic Syntheses, Inc., and Organic Reactions, Inc., along with the Division of Organic Chemistry of the American Chemical Society, the Roger Adams Award is granted for outstanding contributions to research in organic chemistry. The 1971 recipient of the $10,000 award was Herbert C. Brown, R. B. Wetherill Research Professor of Chemistry at Purdue University, Lafayette, Ind., for his work in the discovery of the hydroboration reaction. Brown's discovery opened a new era in chemical synthesis characterized by the utmost simplicity. As co-discoverer of the borohydrides, he helped add a new dimension to synthetic organic chemistry, and to the field of organometallic chemistry.

UPI. Wide World

Sir Bernard Katz (left), Julius Axelrod (center), and Ulf von Euler shared the Nobel Prize for Physiology or Medicine.

Earth sciences

Howard N. Potts Medal. The Franklin Institute in Philadelphia, Pa., awards annually the Howard N. Potts Medal for distinguished work in science or the arts. The 1970 recipient of the gold medal was Jacques Yves Cousteau, French underwater scientist, inventor, research pioneer, and photographer, for his extensive pioneering in marine biology, oceanography, ocean engineering, and especially for his development of the "Aqua-Lung" —a key to man's undersea research capability. Cousteau, co-inventor with Emile Gagnan of the Aqua-Lung, began his work on the apparatus in June 1942 and eventually demonstrated that man could be freed from the old heavy rubberized suit with its helmet and surface hose connections, and could instead enter the water for long periods of time and have complete mobility in swimming, descending, climbing, and working.

Nobel Peace Prize. The Nobel Committee of the Norwegian Parliament awarded the 1970 Nobel Peace Prize to Norman Ernest Borlaug of the United States, for his great contribution toward creating a new world situation with regard to nutrition. Borlaug, an agricultural expert and a director of the Rockefeller Foundation, heads a multination team of scientists conducting extensive experiments with new grain types at the Rockefeller Foundation's International Maize and Wheat Improvement Center of Mexico. Borlaug's development of high-yield grains resulted in the Green Revolution, referring to the use of improved wheat and rice strains, and more efficient fertilizers and irrigation methods for the production of larger crops. The $78,000 award citation stated that Borlaug's improvement of wheat and rice plants had created a technological breakthrough making it possible to abolish hunger in the developing countries in the course of a few years.

Penrose Medal. The Geological Society of America awards the Penrose Medal each year for research in pure geology. The 1970 medal was presented to Ralph A. Bagnold, a British engineer of Edenbridge, Kent, for his work in sand dune formations and movement. Bagnold, long in his country's military service, pioneered in research concerning the physics of water-transported sediments on sea beds and rivers. His sediment transport studies have been applied by geologists and engineers to deal with problems of flood control, bank erosion and levee construction, beach and harbor erosion, and especially in working with airborne particulate matter pollution.

Van Cleef Memorial Prize. Established by Eugene Van Cleef, professor emeritus of geography at The Ohio State University, in memory of his wife, the Van Cleef Memorial Medal was presented for the first time in 1970. The award, for outstanding research in the applied as well as theoretical aspects of urban studies, was presented to John R. Borchert, professor of geography at the University of Minnesota. Borchert's interest in urban geography led to his writing more than a dozen articles covering various phases of the subject; and six years as Urban Research Director of the Upper Midwest Economics Study deepened his knowledge of the many facets of urban morphology.

Vetlesen Prize. Administered by Columbia University, the Vetlesen Prize was established in 1959 by the G. Ungar Vetlesen Foundation to honor leading scientists in geophysics. The 1971 award was presented to three geophysicists: S. Keith Runcorn of the University of Newcastle Upon Tyne in England, Allan Cox of Stanford (Calif.) University, and Richard R. Doell of the U.S. Geological Survey. Runcorn, who received half of the $25,000 prize money, was cited for his pioneer work in the use of remnant magnetism in rocks to determine the past positions of the magnetic poles, and to reconstruct the movement of continents. Cox and Doell, who divided the other half of the $25,000, were honored for demonstrating that the earth's magnetic field reverses its polarity at irregular intervals. The timetable of reversals explained the striped magnetic patterns on the ocean's floors and confirmed the theory of continental drift.

Medical sciences

Albert Lasker Medical Research Awards. The Albert and Mary Lasker Foundation each year bestows a number of awards for medical research. The 1970 Albert Lasker Clinical Medical Research Award, with its $10,000 honorarium, was awarded to Robert A. Good, pediatrician, anatomist, immunologist, and chairman of the department of pathology at the University of Minnesota, for his 25 years of clinical research, and specifically for his feat of transplanting bone marrow cells into immunologically deficient children.

The 1970 Albert Lasker Basic Medical Research Award, also with a $10,000 honorarium, was presented to Earl W. Sutherland, a physiologist at Vanderbilt University School of Medicine, Nashville, Tenn., for his work in elucidating the function of cyclic AMP (adenosine 3′, 5′-monophosphate), which appears to regulate hormone activity throughout the body.

James B. Herrick Award. The American Heart Association presents the annual James B. Herrick Award for distinguished achievement in the advancement and practice of clinical cardiology. The recipient of the 1970 award was Eugene A. Stead, Jr., professor of medicine at Duke University School of Medicine, Durham, N.C., for his research accomplishments, his leadership in medical teaching, and for his role in developing leaders in medicine. Stead's research pursuits included work on plasma volume, on methods for measuring blood flow, on the uses of sodium nitrate to study circulatory collapse, and on the absorption of sulphanilamide as an index of blood flow in the intestine. He developed a method for determining the protein content of tissue and utilized the catheterization technique for studying blood flow to the splanchnic area—the heart, kidneys, and brain.

Joseph Goldberger Award in Clinical Nutrition. Sponsored jointly by the American Medical Association and the Nutrition Foundation, Inc., the award commemorates the U.S. Public Health Service physician who uncovered the cause and treatment of pellagra in the early 1900s. It is given to stimulate medical investigators in advancing the frontiers of public and personal health, and to honor physicians who have made important contributions to the knowledge of nutrition. The 1971 prize was presented to Robert E. Hodges, professor of internal medicine at the University of Iowa College of Medicine, and to John E. Canham, colonel and commanding officer of the U.S. Army Medical Research and Nutrition Laboratory at Fitzsimmons General Hospital in Denver. The two physicians were cited for their collaborative work on projects to further define minimum dietary requirements in humans for vitamins A and C, and to gain deeper insight into how the body responds both to decreasing body supplies of the vitamins and to restoring them.

Nobel Prize for Physiology or Medicine. The Swedish Royal Caroline Medico-Chirurgical Institute awarded the 1970 Nobel Prize for Physiology or Medicine to three scientists, all working in the field of research on the movement of nerve impulses from nerve endings into glands, muscles, and other parts of the body. The $78,400 prize was divided among Julius Axelrod of the United States, Sir Bernard Katz of Great Britain, and Ulf Svante von Euler of Sweden. Axelrod, professor of pharmacology at the National Institute of Mental Health at Bethesda, Md., was honored for his discovery of a new metabolic route by which the body carries intravenously administered drugs, and the effects of drugs on the nervous system. During the course of his work Axelrod found the mechanisms that the body uses to regulate the production of noradrenaline, a neurotransmitter. Katz, professor of biophysics at University College in London, was cited for discoveries concerning the way a transmitter chemical substance called acetylcholine is released from the nerve terminals at the nerve-muscle junction under the influence of the nerve impulses. Von Euler, professor at the Royal Caroline Institute in Stockholm, was recognized for his 1949 discovery that noradrenaline serves as a neurotransmitter at the nerve terminals of the sympathetic nervous system.

Samuel A. Talbot Award. The Institute of Electrical and Electronics Engineers gives the annual Talbot Award to a young (under 30) investigator who has made a creative contribution to medicine or biology using engineering techniques. The 1970 Talbot Award was presented to William S. Rhode of the University of Wisconsin for his investigative work involving the measurement of the motion of the inner ear using the Mössbauer effect, a radioactive method for measuring extremely small quantities. Rhode confirmed and extended the work of previous investigators, and also revealed two completely new phenomena: the first, a mode of vibration peculiar to high intensity; and the second, an intensity-dependent nonlinearity, which opened up a new phase of auditory research.

Physics

Dannie Heineman Prize for Mathematical Physics. Endowed by the Heineman Foundation for Research, Educational, Charitable, and Scientific

Purposes, Inc., the Heineman Prize is presented by the American Physical Society and the American Institute of Physics. The 1971 award of $2,500 and a certificate was given to Roger Penrose of the department of mathematics, Birkbeck College of the University of London. Penrose was cited for his contributions to general relativity, including new mathematical techniques, new conservation laws, and his theorem on singularities in space-time, each being represented by an outstanding publication in the field of mathematical physics.

Enrico Fermi Award. The U.S. Atomic Energy Commission presents annually the $25,000 Enrico Fermi Award in recognition of outstanding scientific or technical achievement related to the development, use, or control of nuclear energy, and to stimulate creative work in the development and application of nuclear science. The 1970 award was given to Norris E. Bradbury, head of the U.S. Atomic Energy Commission's Los Alamos Laboratory, for his key role in helping revolutionize nuclear weaponry and for his work in developing peacetime uses of atomic energy.

Franklin Medal. The Franklin Institute of Philadelphia presents its highest award, the Franklin Medal, annually for recognition of those workers in physical science or technology whose efforts have done most to advance a knowledge of physical science or its applications. The 1970 Franklin Medal was awarded to Wolfgang K. H. Panofsky, director of the Stanford Linear Accelerator Center at Stanford (Calif.) University, for his work in the creation of the giant, two-mile-long accelerator that demanded a 20-fold jump in energy, a 100-fold jump in power, and a 500-fold jump in length to achieve a research tool for investigating the elementary particles that are the basic building blocks of the universe.

J. Robert Oppenheimer Memorial Prize. The University of Miami Center for Theoretical Studies awards annually the Oppenheimer gold medal for important contributions to mathematics, theoretical physics, chemistry, biology, or the philosophy of science. The 1971 recipient of the prize was Abdus Salam, director of the International Centre for Theoretical Physics at Trieste, Italy, for his contributions to quantum electrodynamics and to elementary-particle physics.

Nobel Prize for Physics. The Royal Swedish Academy of Sciences designated two European physicists as recipients of the 1970 Nobel Prize for Physics. Sharing the $78,400 award were Louis Eugène Félix Néel of France, and Hannes Olof Gösta Alfvén of Sweden. Néel, of the University of Grenoble, was awarded the prize for fundamental work and discoveries concerning antiferromagnetism and ferrimagnetism in the iron oxides known as ferrites, which are magnetic but do not conduct electricity. Alfvén, who was teaching half a year at the University of California at San Diego, and the other half of the year at the Royal Institute of Technology in Stockholm, was cited for his discoveries in the area of magnetohydrodynamics, which included the Alfvén waves and their application to the physics of plasma or ionized gas.

Wide World

Norman Borlaug

Science journalism

AAAS-Westinghouse Science Writing Awards. Administered by the American Association for the Advancement of Science, the AAAS-Westinghouse Science Writing Awards, with honorariums of $1,000, are presented each year in three categories. The categories and 1970 winners were: (1) General magazines, to Tom Alexander, assistant editor of *Fortune*, for his article, "The Wandering Continents," which appeared in *Nature/Science Annual*, 1970. (2) Newspapers with over 100,000 daily circulation, to Jerry E. Bishop, staff reporter with the *Wall Street Journal*, for his article, "Taming the H-Bomb: Physicists Edge Closer to Controlled Release of Energy From Fusion," which appeared in the *Wall Street Journal* Dec. 3, 1969. (3) Newspapers with daily circulation under 100,000, to Patrick R. Cullen, reporter with the Palm Beach (Fla.) *Post*, for two series and three articles on pollution and its dangers in Florida, published between January and September 1970.

AIP-U.S. Steel Foundation Science Writing Award. The American Institute of Physics, cosponsor with the U.S. Steel Foundation, annually presents the AIP-U.S. Steel Foundation Science

Writing Award in Physics and Astronomy to stimulate distinguished reporting and writing of advances in those two disciplines. The 1971 recipient of the $1,500 award was Kenneth F. Weaver, assistant editor of the *National Geographic* magazine, for his article entitled "Voyage to the Planets," which appeared in the August 1970 issue of the *National Geographic*.

Howard W. Blakeslee Awards. The American Heart Association presents annually the Howard W. Blakeslee Awards for outstanding reporting in the cardiovascular field. The five 1970 award winners were: (1) Bill Sidlinger, district editor of the *Hutchinson* (Kan.) *News*, for his three-part series of articles entitled "Newsman Faces Life Instead of Death," which related his own experience in undergoing open-heart surgery (*Hutchinson News*, February 1970). (2) Patrick Young, staff writer for the *National Observer*, Silver Spring, Md., for his article in that weekly, "New Technique Holds Out Hope to Heart Patients" (*National Observer*, Dec. 29, 1969). (3) Lou Adler, anchorman-reporter for WCBS News radio 88, New York City, for coverage of developments in the cardiovascular field. (4) "The Heartmakers," a one-hour program on the National Educational Television Network on Nov. 5, 1969. (5) "The Pros and Cons of Exercise," a half-hour documentary presented over station WAGA-TV, Atlanta, Ga., on Feb. 6 and 7, 1970.

James T. Grady Award. The American Chemical Society presents the annual $2,000 James T. Grady Award to recognize, encourage, and stimulate outstanding reporting directly to the public to increase materially the public's knowledge and understanding of chemistry, chemical engineering, and related fields. The 1971 recipient of the award was Victor Cohn, head of science reporting for *The Washington Post*, for, among others, his articles on the scientific aspects of the Apollo 11 and 12 moon flights of 1969.

UPI

Louis Néel

Hannes Alfvén

Wide World

Miscellaneous

National Medal of Science. The United States government's highest award for distinguished achievement in science, the National Medal of Science, is presented annually by the president of the U.S. to those persons who in his judgment are deserving of special recognition by reason of their outstanding contributions to knowledge in the physical, biological, mathematical, or engineering sciences. The nine 1970 recipients of the gold medal were: (1) Richard D. Brauer, professor of mathematics at Harvard University, for his work "on conjectures of Dickson, Cartan, Maschke, and Artin, for his introduction of the Brauer group, and for his development of the theory of modular representations." (2) Robert H. Dicke, professor of physics at Princeton University, for his "fashioning radio and light waves into tools of extraordinary accuracy, and for his decisive studies of cosmology and of the nature of gravitation." (3) Barbara McClintock, Carnegie Institution of Washington, for "establishing the relations between inherited characters in plants and the detailed shapes of their chromosomes, and for showing that some genes are controlled by other genes within chromosomes." (4) George E. Mueller, senior vice-president of General Dynamics Corp., New York City, for his "many individual contributions to the design of the Apollo system, including the planning and interpretation of the large array of advanced experiments necessary to insure the success of this venture into a new and little known environment." (5) Albert Bruce Sabin, president of the Weizmann Institute of Science in Rehovot, Israel, for "numerous fundamental contributions to the understanding of viruses and viral diseases, culminating in the development of the vaccine which has eliminated poliomyelitis as a major threat to human health." (6) Allan R. Sandage of Hale Observatories for "bringing the very limits of the universe within the reach of man's awareness and unraveling the evolution of stars and galaxies—their origins and ages, distances, and destinies." (7) John C. Slater, professor of physics and chemistry at the University of Florida, for "wide-ranging contributions to the basic theory of atoms, molecules, and matter in the solid form." (8) John A. Wheeler, professor of physics at Princeton University, for his "basic contributions to our understanding of the nuclei of atoms, exemplified by his theory of nuclear fission, and his own work and stimulus to others on basic questions of gravitational and electromagnetic phenomena." (9) Saul Winstein (d. Nov. 23, 1969), who had been professor of chemistry at the University of California, Los Angeles, in recognition of his "many innovative and perceptive contributions to the study of mechanism in organic chemical reactions."

Marine sciences

The National Oceanic and Atmospheric Administration (NOAA) was established during 1970 as a part of the U.S. Department of Commerce. With the exception of aspects of the ocean sciences related to national defense, and the interest of the National Science Foundation in university studies and meteorology and oceanography, NOAA encompassed all of marine science and meteorology under its administration.

Pollution. The omnipresent problem of marine science remained pollution, and not much real progress in reducing it appeared to have been made during the past year. The magnitude of the problem was documented even more starkly as the Baltic Sea, Japanese waters, and portions of the coast of California were shown to have high amounts of pollution. Plans for reducing pollution were taking shape, however, and concrete results were expected to be evident soon.

Methyl mercury and DDT. Two kinds of chemicals seemed to follow a similar pattern in becoming concentrated enough to damage higher forms of life. They were the decomposition products of DDT, such as DDE, and various organic compounds of mercury.

By 1971 DDT derivatives had spread worldwide. Soluble in oils and fatty tissues, these compounds were found to be increasingly more concentrated in the bodies of living creatures as one moves up the food chain through phytoplankton, zooplankton, small fish, large fish, and finally birds and mammals. DDT derivatives in the phytoplankton of Monterey Bay, off California, tripled between 1955 and 1969 as shown by James Cox of the Hopkins Marine Station, Pacific Grove, Calif. Various species of eagles, falcons, pelicans, and gulls were all failing to reproduce adequately because of DDT by-products. It was feared that the brown pelican might die out along the California coast, although colonies in other parts of the world were not in trouble.

DDT derivatives also tended to become concentrated in the yolk sacs of fish eggs. The eggs hatch into small fry with the yolk sacs attached, and the fry begin to swim about. At this stage, the yolk sacs are absorbed by the baby fish, and the DDT derivatives then kill them.

Organic mercury compounds were presenting a similar problem. Metallic salts of mercury had been discharged wantonly into streams and lakes with the belief that they were insoluble and simply settled into the mud or on the bottom. However, it

Electron micrograph shows siliceous marine rock dredged from the Atlantic continental slope off Long Island. The rock contains zeolite in a form that indicates the occurrence of volcanic action in the western North Atlantic during the Eocene epoch.

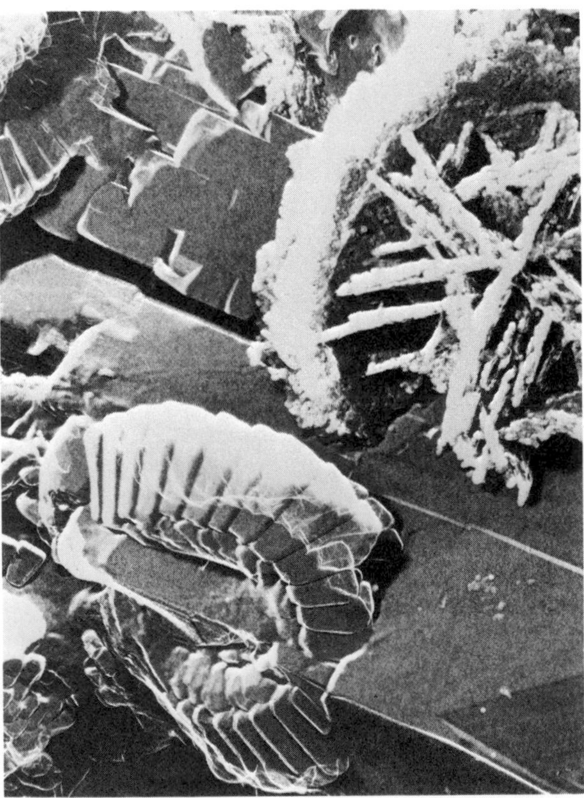

Courtesy, Dr. Kenneth M. Towe, Smithsonian Institution

was recently learned that instead of remaining an insoluble salt, the mercury was changed by bacterial action into organic mercury compounds such as methyl mercury, which is soluble in oils and in plant and animal fatty tissues. Once in the zooplankton and phytoplankton, these organic mercury compounds move up the food chain just as the DDT derivatives do. Excessive concentrations of mercury were found in a wide variety of food fish, and, as a result, swordfish sales were for a time prohibited. In May a woman in New York who had eaten swordfish twice a day for many months was found to be suffering from mercury poisoning.

Both the use of DDT and the release of mercury into the rivers and lakes were substantially curtailed during the past year. However, the amounts already in the environment were large, and the length of time required for the rivers, lakes, and oceans, particularly the coastal waters, to cleanse themselves was not known.

Oil pollution. The intentional, though often unlawful, discharge of oil containing sludge from ships as their tanks were being cleaned was a source of oceanic pollution of increasing concern. The Coast Guard, which was charged with halting this kind of pollution in coastal U.S. waters, had a great deal of difficulty in proving that a particular ship was responsible for a particular oil discharge. Court cases often were lost for lack of adequate proof.

The days of oil pollution might be numbered, however. Norman Guinard and his associates at the Naval Research Laboratory, Washington, D.C., demonstrated that a side-looking imaging radar could easily detect oil slicks both day and night, even through clouds. Moreover, the nations of the North Atlantic Treaty Organization resolved "to achieve by 1975, if possible, and by the end of the decade at the latest the elimination of intentional discharges of oily wastes into the oceans." Success in this endeavor by 1975 or 1980 is essential. If the amount spilled remains in the same proportion as the amount transported, and projections through 1980 hold up, the result would be catastrophic if this goal were not achieved.

Are the oceans an infinite sink? One of the concepts involved in dumping sewage and other wastes in the oceans is that mixing and dispersion processes dilute the material throughout such a large volume that the effects of the discharge soon become unimportant. The analogy is that of a finite amount of material being diluted by a practically infinite ocean. The fallacy in this concept is that the pollutants are being introduced along the coastlines of all the populated areas of the world in amounts that are not negligible compared to the total capacity of the world ocean to disperse them.

The concentration of polluting material tends to vary inversely with distance from populated coastlines, as shown by John Ryther of the University of Connecticut. Pollutants tend to stay in the surface waters, attaining high concentrations, and they are not mixed through the entire volume of the ocean. As Philip Abelson wrote in *Science,* "The recent measurements of mercury in tuna and of DDT in other fishes should warn us that we cannot count on the oceans as an almost infinite sink. . . . Having been warned, we would be imprudent not to take action against heavy metal pollution."

For additional information about pollution *see* Feature Article: WASHING OUR DIRTY WATER; Year in Review: ENVIRONMENTAL SCIENCES.

Physical oceanography. An exciting area of theoretical physical oceanography was that of double diffusion. When gradients of salt content and heat content are both present in the oceans, the diffusion in the combined system is markedly different from the diffusion of either substance alone. The liquid becomes highly stratified with thin horizontal zones of rapid molecular diffusion separating thick horizontal bands where convection transports the salt or the heat. The study of this effect in the laboratory by Richard B. Lambert of the University of Rhode Island, using salt and sugar instead of temperature and salt, resulted in patterns quite similar to the structures observed in the ocean.

Advances were also made in the measurement and study of ocean waves. Philip Rudnick and Raymond W. Hasse of Scripps Institution of Oceanography, La Jolla, Calif., and of the U.S. Naval Underwater Sound Laboratory at New London, Conn., respectively, described motion pictures of extreme waves passing "Flip," the floating instrument platform in the Pacific operated by Scripps, on Dec. 1, 1969. In these motion pictures, the crest of one wave was 40 ft above the mean water level at "Flip" and one of the troughs was 45 ft below mean water level, resulting in a crest-to-trough range of more than 80 ft. These high waves occurred during rather light winds, and they were actually generated many hundreds of miles to the northwest in a large area of 70-knot winds. In the storm area, although no measurements were made, some of the individual waves must have been more than 100 ft high.

Willard J. Pierson, Jr., and Y-Y Chao completed an interesting series of studies of wave refraction. Chao succeeded in solving the theoretical problem of how high waves on water get when they pass through regions of partial focusing, called caustics. Waves on water have many properties similar to light waves. For example, they travel more slowly in shallow water and are refracted when the depth changes, just as the speed of light changes on passing through different media. Pierson and Chao made a small square tank for water and generated waves in shallow water that traveled toward a sloping inclined edge where the water became deeper. The waves turned back into the shallow water at the caustic, unable to penetrate into the deep water. The same effect has long been known for light. For a prism with two perpendicular faces and one 45° face, light will be totally internally reflected at the 45° face.

In the study of ocean currents, the past year saw much work on the Gulf Stream and Florida Current complex. Several papers presented at the annual meeting of the American Geophysical Union were concerned with meanders as well as with warm and cold eddies breaking off from the Gulf Stream. These phenomena were detected by space satellite, aircraft, and ship observations, both separately and in various combinations, to show a complete picture of many features of the Gulf Stream.

Marine biology. New evidence appeared showing that biological activity in the deep sea proceeds at a very slow rate. According to H. W. Jannasch of the Woods Hole (Mass.) Oceanographic Institution and his co-authors in an article in *Science,* "Food materials from the sunken and recovered research submarine 'Alvin' were found to be in a strikingly well-preserved state after exposure for more than 10 months to deep-sea conditions. Subsequent experiments substantiated this observation and indicated that rates of microbial degradation were 10 to 100 times slower in the deep sea than in controls under comparable temperatures." (*See* Year in Review: MICROBIOLOGY.)

The area of mariculture, which parallels agriculture in the same way that fishing parallels hunting, was expanding rapidly. The National Marine Fisheries Service reported that: (1) shrimp hatchery facilities had been improved to the extent that one million juvenile shrimp were produced between April 1970 and April 1971; (2) the reproductive behavior of pompano was studied in efforts to improve the production and harvesting of this fish; (3) a source of phytoplankton that could be easily grown as food for the very tiny anchovy larvae was developed; (4) a desirable marine bait, the lugworm, was grown in a running seawater tank for 12 months and there was considerable interest in growing this animal commercially; and (5) scientists had identified a number of factors limiting oyster production that, when eliminated, increased such production more than tenfold.

Four women aquanauts leave their Tektite habitat to begin a scientific experiment. The purpose of their mission was to study marine life and record the reactions of marine creatures to various stimuli.

UPI

O. A. Roels and his colleagues at Columbia University studied the possible uses of cold, nutrient-rich water pumped to the surface from the deep sea. Two areas under investigation were its use for air conditioning, where its cold temperature produces the heat sink needed, and its use in ponds and closed bays as the nutrient-rich water for the start of the food chain needed for mariculture. In mid-1971 both uses appeared to be promising.

—Willard J. Pierson, Jr.

See also Year in Review: EARTH SCIENCES, *Geology and Geochemistry,* Marine Geology.

Mathematics

The mathematical production of the last half century rivals that of the preceding 2,000 years. Pure mathematics came modestly to life in the United States about 50 years ago, at a time when it was already flourishing in Germany, France, Great Britain, Italy, and Poland. The U.S. then profited from the exodus of European scholars caused by World War II and since that war has been the major contributor to an amazingly vigorous and creative mathematical epoch. The Soviet Union and Japan have also recently been in the forefront.

It is remarkable, therefore, that mathematics remains relatively untreated in books and periodicals dealing with science. This is particularly true in the U.S. and reflects something about America and something about modern mathematics.

On the one hand, many Americans publicly rationalize mathematics merely as part of a technological arsenal; it cannot be properly understood or appreciated in such terms. On the other hand, mathematics has evolved historically so far from its empirical roots that a vast part of modern research is addressed to problems having virtually no direct contact with ordinary experience. This makes it difficult to communicate in any convincing way to the nonpractitioner the extraordinary depth and brilliance of some of the most important recent mathematical achievements.

It thus seemed futile and inappropriate to offer an unavoidably technical survey of the past year in mathematics. Instead, this article deals with only a few of the most important areas (excluding number theory, discussed in the 1971 yearbook) and over a more extended time period. In two cases, group theory and topology, there are informal descriptions of some current themes that, perhaps, suggest some of the flavor of modern mathematics.

Classification, isomorphisms, and groups. These terms are often employed by mathemati-

"The study of mathematics is apt to commence in disappointment . . . We are told that by its aid the stars are weighed and the billions of molecules in a drop of water are counted. Yet, like the ghost of Hamlet's father, this great science eludes the efforts of our mental weapons to grasp it."
Alfred North Whitehead

cians in formulating problems and conclusions (theorems). Since they will be useful in the discussion that follows, it is necessary first to explain them.

When a biologist classified animals, he is simply giving names to large aggregates of them (families, genera, species) that share certain specific characteristics. Some of these classifications make finer distinctions than others. A classification therefore involves both a designation of the objects to be classified and a criterion determining when two such objects are not to be distinguished. When these two conditions are present, the classification is then a description of the totality of classes of mutually indistinguished objects (for example, a complete list of names of all animal genera).

A simple mathematical example of this procedure is as follows: The objects to be classified are regular polygons, such as an equilateral triangle, square, and regular pentagon. The classification is indeed simple: there is one object for each integer $n \geq$ (is equal to or greater than) 3, that object being the one with n sides. This classification is meaningless, however, until a criterion is established that determines when two polygons P and Q are not to be distinguished. Two such criteria from elementary geometry come to mind, "congruence" and "similarity." A congruence is a "one-to-one correspondence" which mathematicians denote by a symbol, $s\colon P \rightarrow Q$, assigning to each point x of P a point denoted $s(x)$ of Q so as to preserve distances:

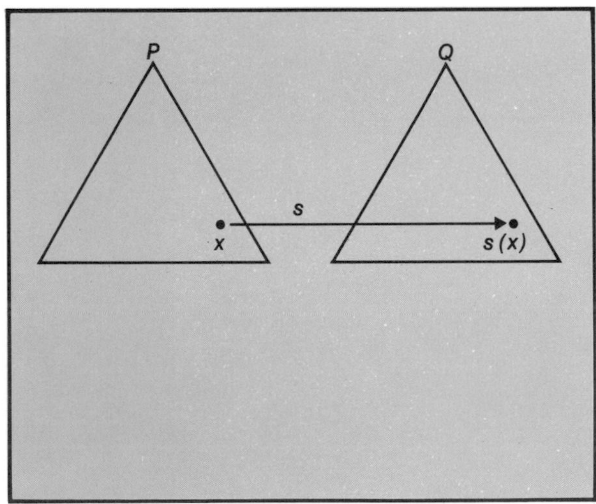

A similarity, on the other hand, is required to preserve only angles. Clearly, similarity is the appropriate notion for the polygon classification, since congruence would continue to distinguish one equilateral triangle from another one twice its size.

Both congruences and similarities are examples of what mathematicians call "isomorphisms." This concept applies also to objects quite unlike the geometric ones described above. For example, suppose P and Q are not polygons but, say, some collections of numbers or vectors or other kinds of algebraic quantities on which operations like addition and multiplication can be performed. In such a setting, the appropriate notion of isomorphism would be a one-to-one correspondence, $s\colon P \rightarrow Q$, which preserves addition $[s(x + y) = s(x) + s(y)]$ and multiplication. When such an s exists, P and Q are isomorphic. Decreeing that isomorphic objects should not be distinguished, mathematicians could then classify these algebraic objects just as they did the polygons above.

Isomorphisms are interesting not only for comparing different objects but also for studying the internal symmetry of a fixed one. The symmetry of an object P is expressed by the isomorphisms from P to itself; these are called "automorphisms." For example, if P is a regular hexagon, there are 12 automorphisms, the rotations by $\frac{h}{6} \cdot 360° \; (h = 0, 1, 2, \ldots, 5)$ and the reflections in six diameters.

The set G of all automorphisms of an object P is called the "automorphism group" of P. If s and t are automorphisms (that is, elements of G), t can be followed by s to obtain another automorphism, denoted st: $st(x) = s(t(x))$ for each point x in P. For this "multiplication of automorphisms," the automorphism e which does nothing, $e(x) = x$ for all x, multiplies like the number 1: $es = s$ for all s in G. Moreover, each s has an inverse, s^{-1}: $ss^{-1} = e$, and so it is also possible to "divide." All the usual rules for multiplying numbers hold *except* that there is no need to have $st = ts$.

Automorphism groups turn out to be interesting objects in their own right. In the course of studying them, a mathematician finds it useful to abstract what is essential, and so he defines a "group" to be *any* collection G of elements given with *any* kind of multiplication table having properties like those of the examples above. He further calls two groups G and G' isomorphic when there is a one-to-one correspondence $s\colon G \rightarrow G'$ preserving multiplication.

Finite groups. Group theory, with historical roots in number theory, crystallography, geometry, differential equations, and quantum physics, has deeply penetrated virtually all of modern mathematics. The classification problem posed above is therefore not so frivolous as it may have appeared. On the other hand, the question is hopelessly out of reach in the generality stated.

If we consider only finite groups (those with a finite number of elements), the problem remains untouchable at present. However, there are certain natural constructions known that allow one to build from two given groups some new and larger ones. Groups which cannot be obtained from smaller ones in this way are called "simple." They are like the "elementary particles" of finite group theory, and the now-vigorous search for all of them shares some characteristics with particle physics. Lists of finite simple groups were initiated in the 19th century. Efforts to verify the completeness of these lists unveiled in recent years a fascinating series of so-called "sporadic" groups.

This vast enterprise—the classification of finite simple groups—owes its basic methodology to the work of Richard Brauer at Harvard University. The fact that one can now seriously contemplate its completion, perhaps during the 1970s, is attributable to the prodigious contributions of John Thompson, Walter Feit, Michio Suzuki, Daniel Gorenstein, and their many collaborators and students, as well as to the increasing application of high-speed computers.

Topology. Numbers can be thought of as points on a line, pairs of them (x,y) as points in a plane, triples (x,y,z) as points in space, and "n-tuples" $(x_1, \ldots, x_n)$ as points in (by definition) n-dimensional (affine) space. Formulas for distance and angle in dimensions $n \leq 3$ extend naturally to all n and permit one to do n-dimensional Euclidean geometry. For example, one can speak of n-dimensional spheres or polyhedra, compute n-dimensional volumes, etc.

Many "geometric objects" or "spaces" that mathematicians study are of the following general type: One starts with certain kinds of sets of points in an affine space. These are then "glued together," using an appropriate kind of isomorphism as "glue" to attach a portion of one set to a matching portion of another. For example, we might start with polyhedra and glue them using congruences or similarities between matching faces:

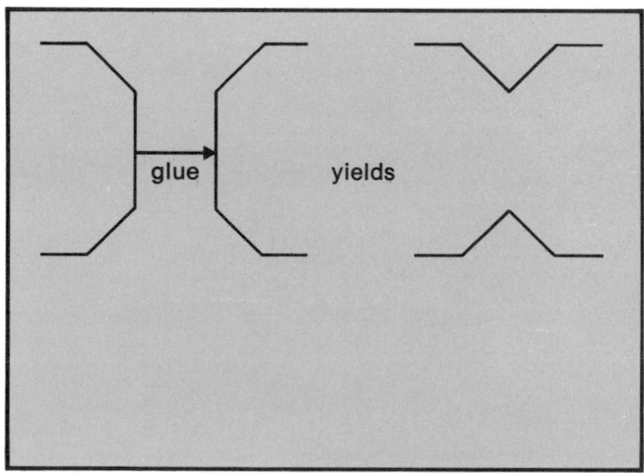

The "spaces" obtained in this way are called "piecewise linear" (PL), and there is a corresponding notion of PL-isomorphism.

Another process starts with solid balls in affine space and glues portions of them together to form what are called "manifolds." For example, a suitable gluing of two elastic discs will produce a sphere:

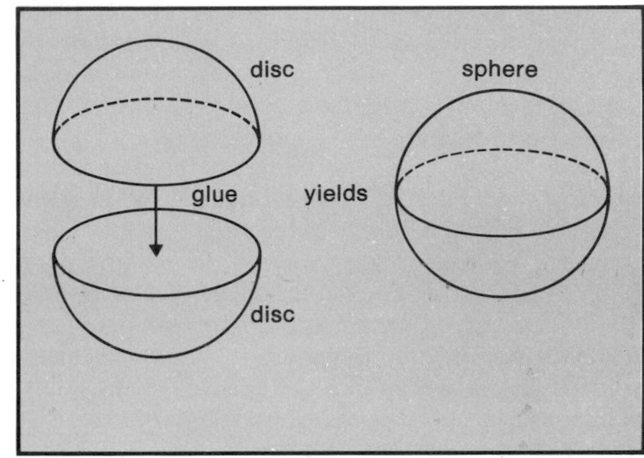

A manifold is said to be topological or differentiable according as the gluing functions are continuous or differentiable. Continuous functions are those that "carry nearby points to nearby points." These furnish the isomorphisms of topology.

PL-spaces are attractive because they can be described combinatorially by a finite amount of data, easily susceptible to computations. However, a space in general can be cut into polyhedra in many distinct ways, and it is not clear whether any two of these yield PL-isomorphic results. This uniqueness question goes back to the origins of topology. It was solved negatively about 10 years ago by John Milnor.

The spaces of classical interest are manifolds, which Milnor's example was not. For a manifold X, one therefore asks: (1) Does there exist a PL-structure on X (that is, a topological isomorphism with a PL-space)? (2) Is such a structure unique to within PL-isomorphism?

These long-standing problems were recently solved in collaboration by Robion Kirby and Larry Siebenmann using a brilliant piece of earlier work by Kirby. The solution describes precisely when a PL-structure exists, and then classifies all of them. Several other important problems were felled as a consequence. Partial contributions to these problems were made also by Dennis Sullivan and Richard Lashof.

Much of modern research in topology uses a technique called surgery theory. This was brought into sharp form through the work of C. T. C. Wall as well as that of Sergei Novikov and William Browder. Among its spectacular applications was the recent construction by Ted Petrie and Ronnie Lee of certain types of groups of automorphisms of spheres whose existence was not previously known.

Lie theory. Mixing the notion of group with that of differentiable manifold or of algebraic variety (a set of solutions of several polynomial equations in several variables), one arrives at the notion of a Lie group or of an algebraic group, respectively. The classification of simple Lie groups is one of the towering achievements of our grandfathers. One owes mainly to E. Borel and C. Chevalley the extension of this theory to algebraic groups.

These groups, their "representations on various kinds of linear spaces and varieties," and properties of certain "discrete subgroups" of them became a domain of convergence for an extraordinary diversity of mathematical phenomena, from number theory through algebraic and differential geometry and analysis to quantum theory. This exhilarating kind of synthesis made Lie theory, in its many facets, the target of some of the deepest and most ambitious mathematical research.

—Hyman Bass

Medicine

This review of events in medicine in the past 12 months begins with a survey of the most noteworthy developments that occurred throughout the range of the medical sciences under the heading of *General Medicine*. It then focuses on four specialized areas of medicine. Of these, *Child Abuse* and *Drug Addiction* discuss recent advances in understanding and treating medical problems that have social implications, and *Biomedical Engineering* and *Malignant Disease* look at current research efforts that hold out promise for dramatic changes in health care.

General medicine

More than one sage contemplating the strife and progress that highlighted medicine in recent months came up with the phrase, "the year of the patient." How else could any observer tie together the trends that tangled politics with the clinical laboratory, disease with "the quality of life," and every aspect of medicine with virtually every other issue of modern living? For the physician, the patient always has been the center of attention; but the sick person was now known by other names: "the health consumer" and "the impatient taxpayer," to name two.

It is impossible to separate out all the factors that combined to bring on "the year of the patient." A few prominent forces, however, were the increased public and political pressure for better, more evenly distributed medical care, diminishing funds for basic and clinical research, and increased public awareness of how medicine does and should function. No clinical advance, no new therapy could avoid being examined by patient and physician alike in this spotlight of social reality.

One example was the seemingly basic discovery, by investigators in both Europe and the U.S., that chromosomes could be made to fluoresce (emit radiation) by staining them with quinacrine mustard, a chemical derivative of the compound usually used against malaria. The technique was quickly found to be valuable in identifying abnormalities of the sex-determining chromosomes (particularly the "Y," or maleness, chromo-

Top, three of four red blood cells from a patient in sickle cell crisis are grossly abnormal. The fourth still retains its typical doughnut configuration. Bottom, red blood cells from the same patient after intravenous treatment with urea show essentially normal shapes, which persisted for eight hours after the treatment period.

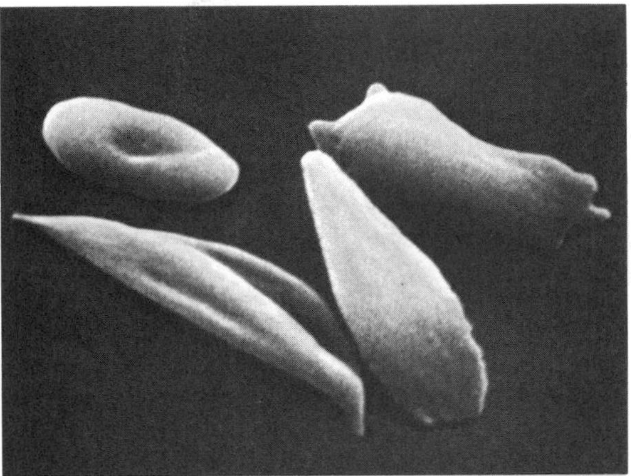

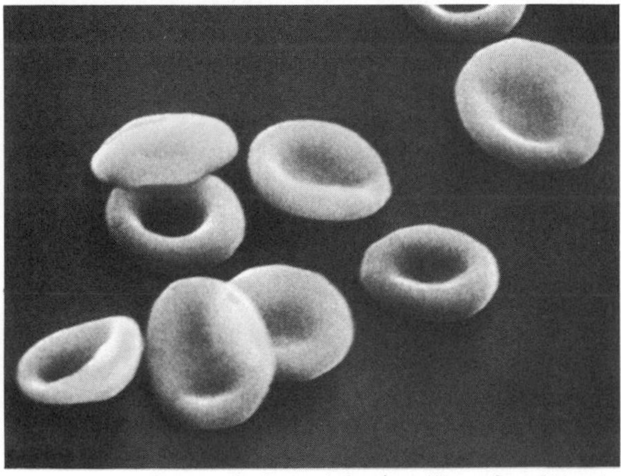

Courtesy, Dr. Marion I. Barnhart, Wayne State University

Sidney Harris, MEDICAL WORLD NEWS

"It cures it in chickens, it causes it in mice."

some) and in picking out breaks in the autosomes (non-sex-determining chromosomes) that were not visible under older, much slower tissue culture methods. In many instances, determinations could be made from sloughed-off fetal cells found in the amniotic fluid taken from the mother weeks prior to birth.

The question that immediately arose was both social and scientific: Could and should the fluorescent staining technique be used to determine if a woman should undergo an abortion? An increasing number of states allowed pregnancies to be interrupted if there was a reasonable chance that the newborn would have a serious genetic defect. Medical scientists were still debating whether this laboratory advance, which would teach us so much about the basic nature of the chromosome, should be adapted to this task.

At the other extreme of the reproductive process, investigators mounted more intensive searches for better ways to control population growth. Clinicians reported that the people-overgrowth concepts of such ecologists as Stanford (Calif.) University's Paul Ehrlich (author of *The Population Bomb*) were causing couples to inquire about vasectomy or tubal ligation (operations producing sterility in male or female, respectively) and other methods for permanently forswearing procreation.

At the same time, many investigators believed that a safe and reliable contraceptive that is reversible might soon be available. This substance, which one expert called "the penicillin of contraceptives," is prostaglandin, a substance everyone already carries about in his body. Prostaglandin in women was known to control key segments of menstruation and labor as well as general uterine function. Research underway indicated that the right prostaglandin (there are several forms) at the right time would prevent conception without the need to resort to the more artificial means that are now employed.

The quality of health care. Meanwhile, the emphasis continued to be on improving the "quality of life" for the average American, sick or well. No less than seven plans for national health insurance were introduced before Congress, extending coverage to rich and poor by various blanket plans. The key difference for most of the plans was whether the insurance system resulting should be privately or federally controlled. The exact profile of the system had not yet emerged, but both Republicans and Democrats had decided that such a system was in the nation's future interests.

But once Americans had the means to pay for adequate health care, how would they obtain that right? In part, the national plan hoped to correct the inequity of rural versus urban, rich versus poor that finds many far from help when they are ill. New methods of health care delivery were needed, and some hopeful experiments were underway. In one small New England town, for example, residents needing psychiatric help were interviewed by a local physician as psychiatrists at Dartmouth College in Hanover, N.H., watched on closed-circuit television. Thus, the patient could be interviewed and provided with highly skilled counsel even when deep snowdrifts or heavy hospital schedules prevented specialists from reaching his bedside. In a similar application, a doctor in the Ozarks had a computer terminal at his desk that linked him

Gerald Brimacombe, LIFE Magazine © Time Inc.

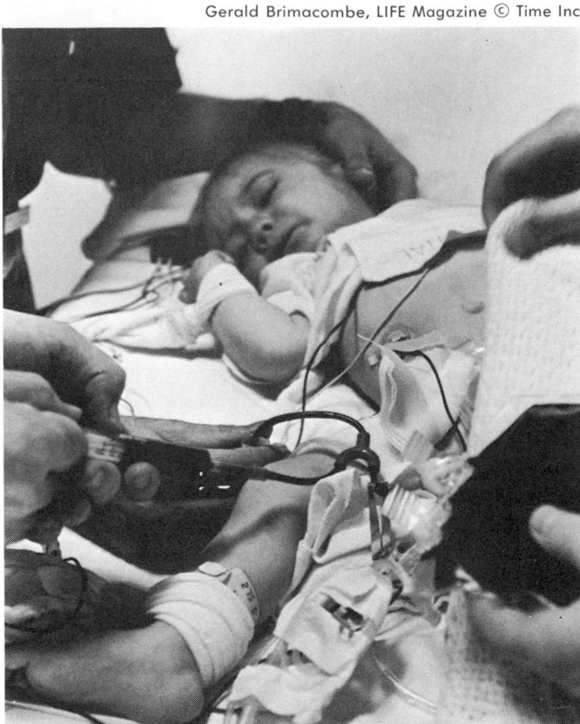

Doctors inject bone marrow donated by an older sister into the abdominal cavity of an infant born with an inadequate immunological defense system. The transplanted cells would be carried by the child's bloodstream to his bones, where they were expected to begin manufacturing mature white blood cells.

with the University of Missouri and its expertise in interpreting test results, recalling patient records, and placing disease problems in perspective. The electronic hookup enabled the physician to get answers as fast as he could punch questions onto the keyboard. (See *Biomedical Engineering*, below.)

The harassed, isolated physician also needed more personal help. One answer being tried was the "physician's assistant" or "P.A." The idea was pioneered at Duke University (Durham, N.C.), where returning Army medics were retrained to take over such routine duties as immunizations and examinations of healthy babies. The P.A. was also someone for the patient to talk with while the doctor and his nurse proceeded with more urgent problems. Several states passed legislation that enabled a P.A. to work beside a physician without legal restraints. In rural areas of Florida, California, and elsewhere, senior medical students were assigned to work with routine patient problems in areas where the full-time physicians were few and overburdened.

If a patient is entitled to equally good medical care wherever he lives, then it also becomes necessary to ensure that each doctor is fully qualified. This logic was getting increasing acknowledgment from governmental and medical groups. The American Medical Association (AMA) began giving a special award to members who undertake postgraduate education. The American Academy of Family Practice, which includes most general practitioners, began to ask its members to submit proof periodically that they continued to be qualified. In July 1970, New Mexico became the first state to add a provision for regular postgraduate study to its medical licensure law. New Mexico and Hawaii were among several states whose physicians were threatened with loss of insurance coverage because of the size of malpractice judgments against them. Most physicians freely admitted that this legal pressure from the "health consumer" was proving a major impetus to efforts at periodic recertification and relicensure.

The tenet that because medicine serves the public it is directly responsible to it was not so firmly held by the practitioner, however. A lively debate raged as to how far statutes should infringe on professional ethics and responsiveness. Increasingly, doctors and patients clashed in the highly visible arena of the Congressional hearing.

Disease eradication efforts. During the winter of 1970–71 U.S. Pres. Richard M. Nixon, Sen. Edward M. Kennedy (Dem., Mass.), and numerous prominent laymen proclaimed that the time was right to forge a research force that would do for cancer what the Manhattan Project had accomplished for the atomic bomb, or what the National Aeronautics and Space Administration (NASA) had done for the race to the moon. Cancer researchers replied eagerly that the extra attention and money was most welcome, but they also stressed that there was no way to set a deadline on the cure for this killer because it is really a hundred or more diseases that must be assaulted simultaneously, from all sides. (See *Malignant Disease*, below.)

Public health officials announced that two "conquered" diseases, measles (rubeola) and diphtheria, were back. The number of cases of both began to rebound in 1970–71 because vaccination campaigns had not reached young children in rural and urban poverty areas. The medical achievement of an effective vaccine is fruitless without public and governmental effort to make it work, said J. D. Millar of the Center for Disease Control in Atlanta, Ga.

But the news from the disease fighters was not all bad. Rubella (German measles) infections declined abruptly with the wider use of a vaccine authorized in 1969. Smallpox continued its worldwide decline, causing World Health Organization experts to predict, cautiously, virtual eradication within five years. In addition, experimental vaccine against serum hepatitis had encouraging results in early human trials conducted by Saul Krugman in New York. The vaccine, prepared from the serum of hepatitis patients was unusual in that the actual virus that presumably makes it work had never been isolated. Workers presumed many years ago that at least this type of hepatitis has a viral origin and proceeded to devise their experiments accordingly. Recent immunologic studies, originating with Baruch Blumberg of Philadelphia, showed that the Australia antigen (so-called because it was first found in an Australian aborigine) is often seen in the blood of hepatitis victims. The antigen, the foreign substance that stimulates the production of antibodies, may prove to be the virus itself, many investigators thought.

Advances in immunology. The physician-immunologist provided many of the most exciting clinical advances during the past year. In Indianapolis, Ind., a rotund man named Louis B. Russell, Jr., continued to live actively in mid-1971 with a transplanted heart he received in August 1968. The second longest surviving heart transplant beat in the chest of Donald Kaminski of Michigan. He received the transplant in December 1968 and survived a subsequent automobile accident. Many transplant surgeons believed both men survived so long because their donors matched them closely in terms of immunologic characteristics identified on the white blood cells. Cardiac transplants came to a virtual halt in recent months while immunologists tried to find out how to prevent transplant rejection and create techniques to render such operations better long-term risks.

Another type of transplant, while considerably less dramatic, was having good results for the same reasons that a heart transplant like Louis Russell's finds success. Bone marrow transfers from close relatives to patients with certain types of leukemia and immunologic deficiency diseases found success at the University of Wisconsin and University of Minnesota medical centers. In effect, the "factory" in the human body that manufactures white cells is replaced completely, giving the patient all the immunologic capability of the bone marrow's original owner. Only a relatively small amount of healthy marrow is needed to overpower and supplant the deficient cells.

Another new form of immunologic therapy was not yet fully understood by those using it. "Transfer factor" does not fit into any concept of how the body defends itself, but it is a substance that is real and valuable, according to investigators at Stanford and New York universities. A single injection of the factor, which is extracted from a normal donor's sensitized lymphocytes (one type of white blood cell), confers on the recipient the donor's immunologic memory for so-called cellular immunity. This is the same sort of immune skin reaction that takes place in response to a tuberculin vaccination. Patients with such skin diseases as the Wiskott-Aldrich syndrome and sarcoidosis and with advanced malignancy lose their ability to respond to such stimulation and become easy prey for normally innocuous infections. For such persons, transfer factor confers protection for up to two years.

The University of Minnesota's Robert A. Good estimated that, at any given moment, 10 million–15 million people in the U.S. are fighting disease centered in the immune system. Such techniques as bone marrow transplantation and transfer factor therapy are, he believed, only the first of a whole generation of therapies to combat illness with the body's own weapons.

Cost-cutting efforts. But such techniques were still experimental and applicable to a relatively small number of cases. Meanwhile, what was being done to assist those with the more common, chronic, and often costly types of disease? This question was raised with increased frequency in Congress, state and local governments, and all the new-found forums of the "health consumer." Hospital costs had risen approximately 150% in recent years; medical care of all types was costing the federal Medicare and Medicaid programs more than double the dollars originally estimated. The patient who reached the limits of his personal insurance might be on the threshold of hopeless debt. This dilemma was part of the debates surrounding national health insurance programs, but systems for cheaper, more efficient care were needed in the meantime.

Under the stimulus of the AMA, local hospitals and medical centers were stressing the process of Peer Review. At regular intervals, committees of physicians were meeting to review the work of their colleagues in the perspective of cost, hospital facilities, and efficiency. Any physician who tended to keep his patients in the hospital longer than usual, for instance, would be asked for an explanation by the Peer Review Committee. Clinical problems that normally required lengthy hospital stays, such as the use of hemodialysis for renal failure, were being done in the home by trained relatives of the patient. At the Albert Einstein College of Medicine Bronx Municipal Hospital Center, poor persons with moderate or advanced tuberculosis were treated at home, after their infectivity had been controlled. Over an average of 3.5 months, the estimated savings per individual was $2,991;

The level of lead absorption in the general population may prove to be a subtle health risk. An electron micrograph (below) shows that excess lead not sloughed by a lead-poisoned rat complexes with protein to form an inclusion body, the large, dense structure, within the nucleus of a renal cell. On X-ray plates of an 18-month-old infant, bits of lead-containing paint eaten by the child appear as dark specks in the large intestine (bottom, left), and dark "lead lines" indicate that excess lead is stored at the ends of the long bones of the child's legs (bottom, right).

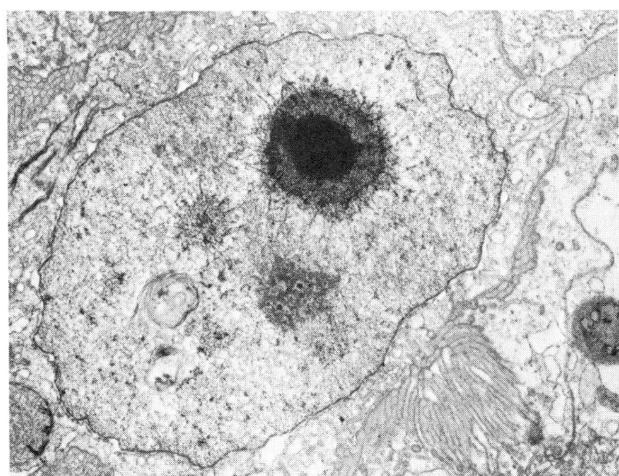

over two years, a group of 69 home-care patients saved some $288,000 in hospital costs.

Public health problems. If the average American took better care of himself, his health care problems would be pared considerably. Much of the hepatitis rampant in younger age groups was transmitted by needles used for injecting drugs. In New York City, 447 teenagers died in 1970 as the result of complications traced to drugs, primarily heroin. Many succumbed so rapidly that they were found with the needles still in their arms. The University of Chicago's Daniel X. Freedman said that about 1.5 million persons in the U.S. were now *dependent* on marijuana. The case for marijuana addiction had not been made, he pointed out, but dependence leads to a personal neglect in many persons that is an invitation to illness of many types.

An estimated 1.8 million cases of the venereal disease gonorrhea occurred in the U.S. in 1970, according to William J. Brown of the Center for Disease Control. Only about 675,000 of these were actually reported, leaving two out of three victims outside the control of public health officials. Gonococci bacteria themselves had become so resistant to the antibiotics used that a dose double that

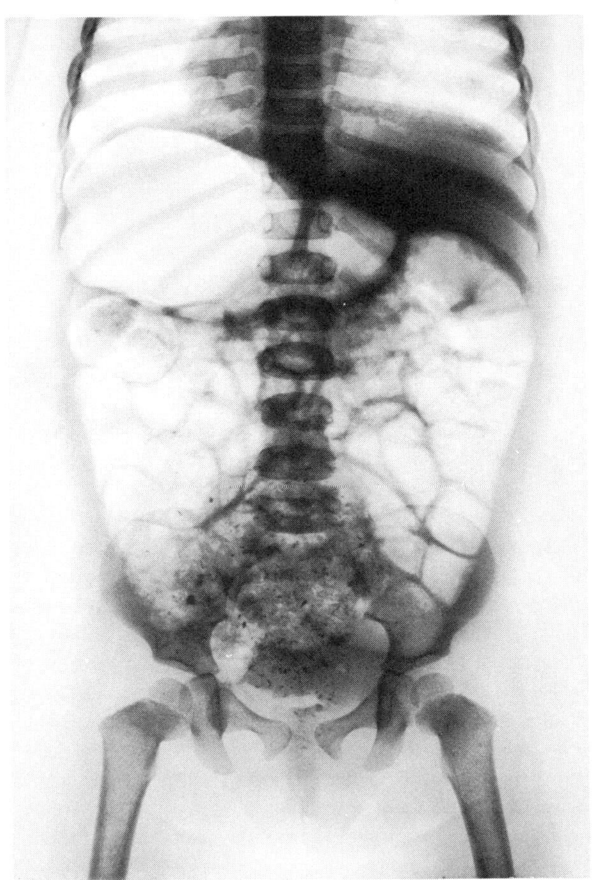

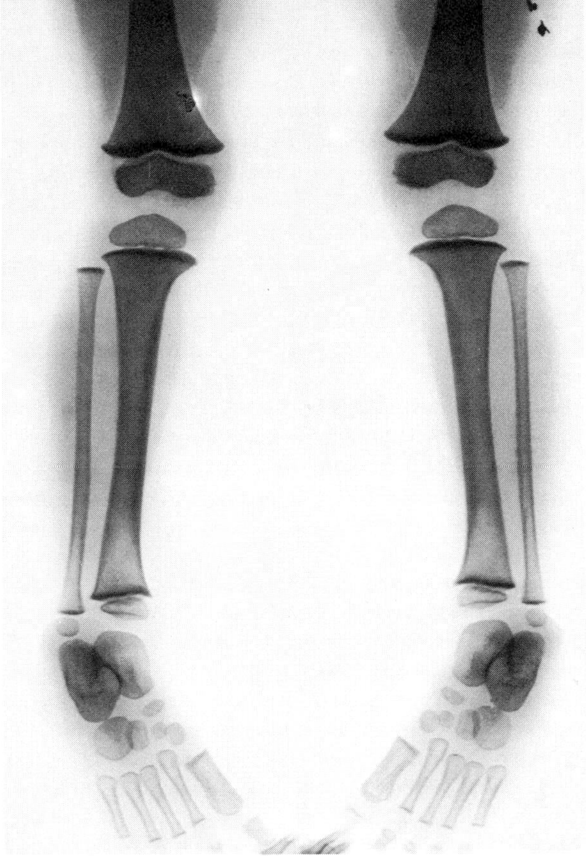

(Top) courtesy, Dr. Robert A. Goyer, University of North Carolina; (bottom) courtesy, Dr. J. Julian Chisolm, Jr.

formerly recommended must be used. About a half-million cases of infectious syphilis, much of it untreated at its susceptible early stages, also afflicted Americans. Public ignorance of venereal diseases and their damaging long-term effects was blamed for a seemingly casual attitude toward detection and control.

Of course, many public health problems of modern times are tied to the conditions of everyday living. An estimated 2–5% of children living in urban ghettos suffered lead intoxication from the ingestion of chipping paint. A number of cities, including Chicago and New York, were mounting extensive detection campaigns, but actual prevention was a much larger problem.

"Internal pollution." Among the ecology-conscious, there was also a growing awareness of "internal pollution." High traces of mercury in swordfish led the Food and Drug Administration to advise against its use. Acute mercury poisoning caused the serious illness of several children in New Mexico. They had eaten pork from animals that had inadvertently received mercury-contaminated feed. Several studies showed that motorists ingested measurable quantities of lead and other potentially dangerous substances in the course of their daily commuting. A New York study found significant deposits of asbestos in the lungs of lifelong city dwellers. Several construction projects in New York and Chicago were halted until contractors built shields to prevent sprayed asbestos, applied for fireproofing, from contaminating the air. Public pressure caused many builders to begin using other substances for this purpose.

Indeed, public pressure was central to the Congressional votes to discontinue funding of the supersonic transport (SST) plane. In addition to noise and air pollution, some scientists feared that the aircraft would cause a depletion of atmospheric ozone and a concomitant rise in skin cancers. "The environment issue may elect a president someday," commented Sen. William Proxmire (Dem., Wis.), who led the SST opponents.

Even such familiar household items as sugar and detergents came under fire. Washday products containing certain "dirt-cutting" enzymes were ordered withdrawn from the market because they contributed to the eutrophication (aging) of the streams and rivers into which they ultimately flowed. A spate of enzyme-free detergents then appeared; but at least two of these had to be withdrawn when they caused skin reactions on some housewives' hands and arms.

In Great Britain investigators noted the close relationship between the increased incidence of coronary heart disease and rising consumption of sugar. Americans were using twice as much refined sugar as they had 70 years before, the average being some two pounds per person per week. The relationship of this dietary habit to heart disease stirred considerable controversy among bioscientists.

Easily the most controversial observation of this type, however, also was shared with the public. Two-time Nobel Prize winner Linus Pauling published a small book titled *Vitamin C and the Common Cold,* maintaining that large doses of the vitamin would prevent colds. It made an immediate best-seller of both the book and every product containing a high concentration of "the sunshine vitamin." Even though many persons have felt intuitively that a glass of orange juice in the morning wards off those winter colds, scientists were quick to point out that Pauling's theory was not based on coherent scientific studies.

The physician-authors. The year was a banner one for books on medical topics. No less than six physician-written volumes appeared on the best-seller list at one time or another—and their authors became familiar faces on late night television talk shows. There was *MASH,* the campy classic by Maine surgeon Richard Hornberger, written under the pseudonym Richard Hooker ("It refers to my golf game"); the movie adaptation of life in a Korean War surgical hospital made paperback sales soar. A rural Minnesota surgeon, William Nolen, penned *The Making of a Surgeon,* which told of his misadventures involved in becoming a medical professional. Several psychiatrists found the best-seller range, including David Reuben with his frank and casual work, *Everything You Always Wanted to Know About Sex—But Were Afraid to Ask.*

If one common thread ran through all these books, it was that physicians are just as human as their patients. This realization was a central one in "the year of the patient."

—Byron T. Scott

Biomedical engineering

Biomedical engineers in academic environments traditionally involve themselves in applications of mathematics and physics to the study and analysis of cellular, organ, and bodily functions and their communication and control mechanisms. In industry, bioengineers generally develop and build improved instrumentation to assist in the medical diagnosis and treatment of patients. In the past year, as a direct result of decreasing federal financial support for basic research and increased funding for applied research in medical care delivery systems, impressive accomplishments in biomedical engineering were visible in direct patient care,

Courtesy, Downstate Medical Center

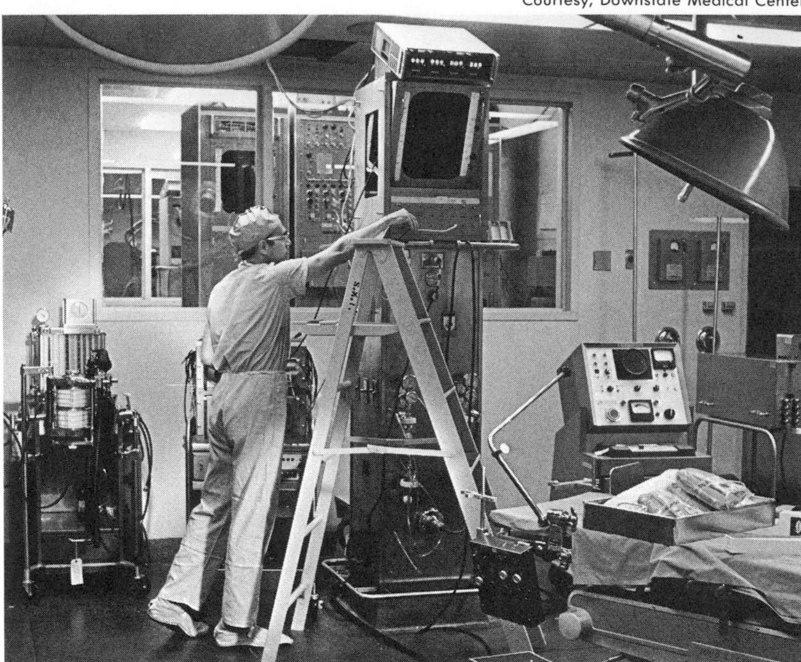

A technician at the Downstate Medical Center, Brooklyn, N.Y., examines operating room equipment for efficiency and safety prior to its use in an operation. The U.S. Food and Drug Administration expressed concern that hospitals may be receiving faulty equipment from manufacturers and that its use by unqualified personnel may result in patient deaths.

especially in instrumentation, cardiac monitoring and support systems, automation of laboratories, and computer applications.

Instrumentation. Significant progress was made in improving such instruments as endoscopes for visualizing interiors of hollow organs and color thermographs to measure skin temperature. Use of these instruments eliminates the need for an incision or to penetrate the skin or bodily membranes for diagnostic purposes. Fiberglass flexible endoscopes with external light sources and cameras were used to examine the uterus and even to explore for abnormalities inside the fallopian tubes, which was previously impossible without abdominal surgery.

Color thermography was being used to measure small differences in skin temperature that might indicate cancer. The presence of a breast cancer impairs return circulation and dilates the surrounding veins. This delays the cooling rate of the overlying skin of a person exposed for 10–15 minutes to a cold room. A thermograph can measure the degree of temperature elevation of the overlying skin and detect any warm spot that would suggest an underlying cancer. The recent addition of color to this method increased resolution and quantification of temperature differences.

Sensory aids, such as reading devices for the blind, were improved by new transistorized equipment that enhanced the efficiency of scanning machines so that printed characters could be identified under computer control and reproduced in visual, tactile (Braille), or audible forms. A reading service utilizing equipment of this nature could be implemented on a large centralized computer and could make current newspapers and magazines available on a time-shared basis to a number of blind persons at a reasonable cost. Computer programs were already converting printed English texts to Braille at high speeds by means of electric typewriters or moving-belt Braille embossers.

Laser beams had been used for several years by ophthalmologists to weld back detached retinas of the eye. Their more recent use for extirpation of cancer tissue indicated an even greater potential, especially in neurosurgery, where they could incise, bore, excise, or vaporize as needed. When used with an operating microscope, they permit high-precision surgery.

An artificial lung made of silicone rubber and polyurethane was reported to have performed successfully in laboratory dogs for as long as eight days. When perfected, the prosthetic lung would be clinically tested in humans for implantation in the chest cavity following the surgical removal of a lung, or as a lung supplement for patients with severe respiratory insufficiency. The device, which was modeled closely after the human lung, is encased in a thin silicone membrane. Clusters of capillary tubes to carry the blood are formed of white silicone rubber. The tubes are spaced with air sacs (alveoli) made of polyurethane foam, through which oxygen enters the capillaries. The capillary tubes are connected to the subject's pulmonary blood supply by Dacron grafts.

Cardiac monitoring and support systems. The

most frequently employed form of instrumental surveillance of human organ systems is cardiac monitoring. It is used in hospital surgeries and postoperative recovery rooms as well as in coronary and other intensive care units. It differs from the usual diagnostic electrocardiography in that it continuously registers heart function and is programmed to alert the nurse and doctor when distress occurs.

Monitoring of other vital signs of the critically ill within intensive care units was advanced considerably recently when the feasibility of computerized on-line monitoring was demonstrated. The new system provided for continuous graphic displays of the arterial and venous blood pressure; cardiac rate, rhythm, and output; and the concentrations of oxygen and carbon dioxide in the blood. In addition, programmed alarms alerted hospital personnel when a parameter exceeded safe limits. Similarly, for patients with advanced respiratory failure, continuous measurements could now be made of lung capacity, respiratory rate, the exchange of gases between arteries and alveoli, and the oxygen and carbon dioxide tensions of the blood as well as its pH (acidity or alkalinity). Computerized calculations of these findings could also aid in diagnostic and treatment decisions.

Mobile monitors. Extensive experience with special-purpose intensive care units demonstrated that, despite their high cost, they are indeed lifesaving. Two-thirds of the deaths attributed to heart attacks, however, occur before the patient can reach a hospital's coronary care unit. Therefore, considerable work was in progress to develop ways of monitoring the patient during this prehospital period.

Recently, a portable wireless monitoring system was improved by using low-cost, solid-state circuit devices that function in high-frequency bandwidth ranges. A patient could now be kept under surveillance by radio telemetry while in an ambulance, emergency room, or small peripheral community hospital. Pocket-sized radio transmitters could transmit several channels of vital information, including the electrocardiogram, temperature, respiration, and blood pressure, to a central hospital's monitoring nursing station that was equipped with a radio receiver and a small computer to store the data and to provide visual displays and printed output for the nurses and doctor. The major disadvantage of radio telemetry was loss of reception —and, hence, a loss of the data—caused by "nulls" and reflections of radio waves from steel structures or from the earth. To provide the most reliable monitoring service, therefore, a dual antenna receiver system was required, and transmitting and receiving antennae needed to be oriented parallel to each other.

Cardiac pacemakers. Until this past year standard cardiac pacemakers about the size of a package of cigarettes were inserted under the skin of patients with heart block and operated by chemical batteries that had to be replaced every one to two years. Extensive studies to insure a greater reliability of pacemakers, however, had now made feasible miniature, long-life, intracardiac instruments that were completely self-contained and about the size of a pencil-tip eraser. Under local anesthetic, these devices could be attached to the inner wall of the right ventricle of the heart by insertion through the jugular vein. An atomic battery that could operate the pacemaker for 5–10 years was also available.

Two major categories of pacemakers were presently in use: fixed-rate (asynchronous) pacers, which had low-frequency oscillators designed to deliver pulses at preset stimulation rates; and noncompetitive pacers, which had, in addition to the oscillator, an electronic circuit designed to sense inherent electrical cardiac activity. The latter is capable of modifying the output of the pacer should a natural heart contraction occur within a preset time interval.

External interference by high-frequency electric fields remained one of the hazards associated with the use of pacemakers. However, because of the shielding characteristics of the skin and body, the

An artificial lung modeled closely after the human lung was successfully tested for up to eight days after implantation in laboratory animals. Developers reported that they expected the device would be ready for clinical testing in humans within three years.

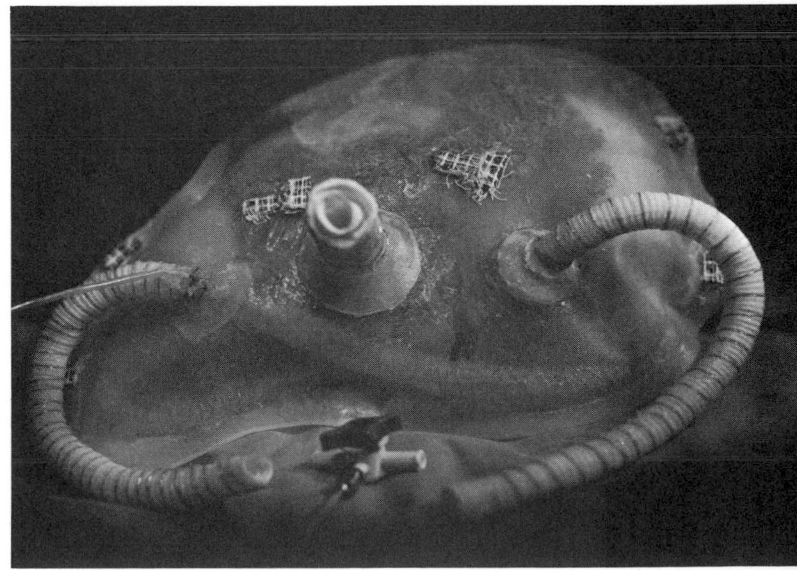

Courtesy, Chicago Wesley Memorial Hospital

probability was slight that external electrical fields would interfere with the function of these implanted devices. Nevertheless, fixed-rate pacers could be affected by electric cautery and radiofrequency diathermy applied within a few inches of or directly over the pacemaker site. Non-competitive pacers were more susceptible to high-frequency radiation; they might be in some danger when within a distance up to 10 ft from such medical equipment as electrocautery, diathermy, ultrasound, ultraviolet light emitters, and betatron units.

All patients with pacemakers were also subject to internal instrument failure caused either by battery exhaustion or by a break or an unstable connection between battery and myocardium. To forestall such failures, regular checkup stations were being established so that the pacemaker pulse could be checked electronically every six months.

Counterpulsion. A newly developed method of treating severe shock of heart attack resulting from myocardial infarction (a withering of a portion of the heart muscle) was intra-aortic balloon counterpulsion. In this method, a cannula (tube) is inserted in the neck, usually through the left subclavian artery. A balloon is then passed through the cannula into the aorta (the main artery leading from the heart), where it lies just above the heart and the openings of those arteries that supply blood to the heart muscle itself (the coronary arteries). When the balloon is inflated and deflated, it not only helps draw blood from the heart during each cardiac contraction (systole), thereby decreasing the pressure work of the heart, it also increases the blood flow through the coronary arteries to nourish the heart muscle during relaxation (diastole).

In experimental animals counterpulsion was also shown to revitalize the damaged area of the heart muscle by expanding previously dormant coronary collateral blood vessels. Problems with the technique that still had to be overcome included safeguarding the balloons from rupture and the precise synchronizing of their action with the cardiac cycle. (See *1971 Britannica Yearbook of Science and the Future,* Feature Article: THE HEART: A PUMP THAT MUST NOT FAIL.)

Automated laboratories. Various manufacturers throughout the world provided a wide selection of automated chemical analyzers for hospitals and for automated multiphasic screening programs. These analyzers could handle up to 360 blood samples an hour and could perform as many as 40 different clinical chemistry tests simultaneously on one sample at processing speeds of from 10 to 120 min for each sample. Most were available interfaced with computer processing systems.

The several commercial automated multitest laboratories that appeared in the last year constituted the first major systems engineering application in the practice of medicine. Approximately 100 automated multiphasic health testing programs were identified as being in operation in the U.S. in 1971, and another 20 were planned for 1972. One concept introduced in 1970 was the use of multiphasic testing to separate patients who were in need of physician care from those relatively healthy ones who would derive more benefit from health education and preventive health maintenance programs. Such services could be staffed by nurses

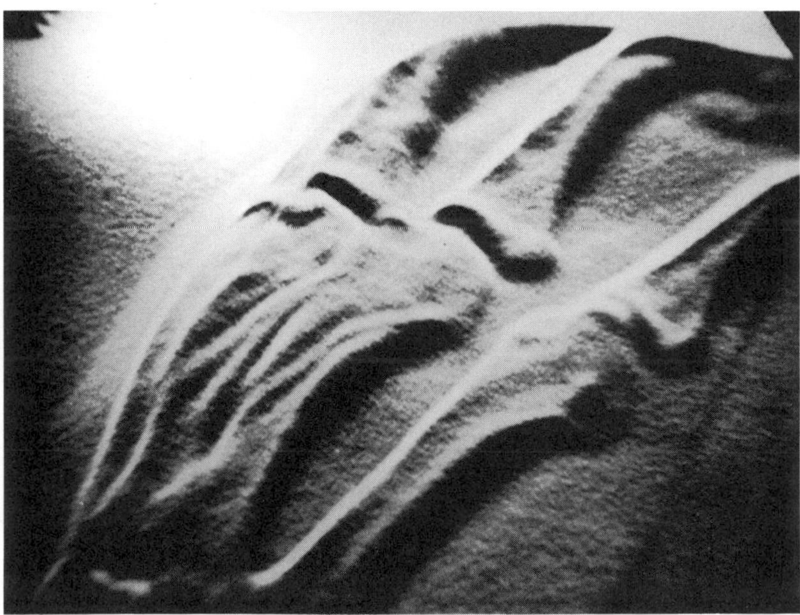

An ordinary X-ray image of a fracture of the metacarpal bones has been translated into an isometric display on a television screen. A camera scans the X-ray plate and bends signals according to the density of the tissue so that the fracture is delineated.

Courtesy, Dr. Wilfred Roth, University of Vermont

and health educators working under physician supervision.

Computer applications in medicine. Although the computer was generally agreed to be the greatest help to medicine since the invention of the microscope, it was only beginning to have a noticeable impact on direct patient care. Various computer programs were being developed to assist hospitals and practicing physicians with patient medical records, and particularly with the reorganization of the physician's documentation of his management of a patient in terms of a problem-oriented medical record.

Automated medical histories developed for multiphasic health testing programs were an important engineering innovation in direct patient care. By permitting a patient to select his choice of responses (usually "yes," "no," "don't know," or "don't understand"), a computer program could lead the patient through a series of branching questions to provide a medical history that was more complete and readable than that usually obtained by the busy physician. The responses are stored in a computer, which, at the completion of the procedure, provides an organized printout of the patient's medical history for the doctor. The extent to which routine medical histories for preventive health maintenance could be facilitated by such relatively inexpensive and tireless interviewing techniques was still being evaluated. It was shown, however, that patients enjoy being interviewed by a computer.

Reports of physical examinations, diagnoses, and patient progress were also being entered directly into the computer in some hospitals. Most recently, the use of television-type display terminals has made it possible for physicians to enter on displayed pictures the anatomical locations of abnormalities.

Computerized electrocardiograms that accurately separate normal from abnormal cardiograms and that identify the majority of cardiac abnormalities were becoming available commercially. Such electrocardiograms can be performed in the on-line mode, where the signals are read directly into the computer and responded to immediately, or they can be recorded on magnetic tape for more economical batch processing at a later time.

Future computer applications. Recent estimates that the volume of medical laboratory test data is doubling every five years, coupled with studies showing that one-third of hospital operating costs are expended in handling information, increased the expectations that computerized hospital information systems not only would be able to meet this accelerating load but would also slow down the rate of increase of costs of hospital care. Many hospitals already had computerized administrative and business functions, and several had pilot programs for the computerization of various medical subsystems. No hospital, however, had all of its medical functions computerized, even though many, notably in the U.S., Sweden, U.K., West Germany, and Japan, utilized the computer for a variety of functions that assist physicians in direct patient care.

Among the difficulties that continued to delay the development of acceptable hospital computer systems were the large capital investments required over several years, a need for experienced teams of physicians and engineers, and the lack of terminals to communicate with the computer in both computer and natural language.

It was anticipated that in the near future a significant increase in the number of physician-engineers, (graduates with both medical and engineering degrees) would decrease the serious manpower handicaps now plaguing interdisciplinary teams. Moreover, continued emphasis on applied rather than basic biomedical engineering research should result in developments in computer technology that could have a greater beneficial impact on medical care. In addition, the establishment of health-care centers, rather than sick-care centers, should enable health professionals to exploit more fully the capabilities of computer-programmed decisions and of bioengineering instrumentation to monitor early abnormalities of patients.

—Morris F. Collen

Child abuse

Child abuse, or child neglect, is one of the most serious health problems of young children and is only now becoming properly understood. A physician has to consider seriously the possibility of a diagnosis of child abuse whenever a parent's explanation of the child's injury does not adequately explain or is inconsistent with the child's history and the medical findings. All injuries to small babies must be suspect because true accidents are rare in infancy.

Child abuse is being reported in the U.S. at a level of 300 cases per one million population per year; 10% of all accidents seen in emergency rooms and 25% of all fractures in children under three years of age are found to be inflicted. The frequency of reinjury in children under two is as high as 60%, and the incidence of permanent brain damage is 30% when the first injury occurs before the child is one year old. Although physicians report many medical reasons for "failure to thrive" in the first two years of life, the reason listed in up to 40% of

UPI

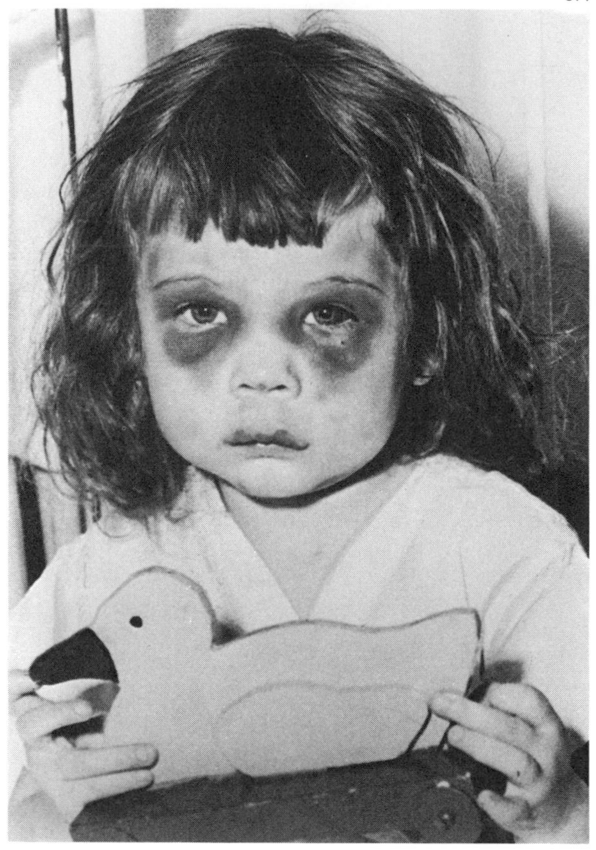

A three-year-old child displays severe head injuries believed to be the result of a beating in her home. A neighbor had heard the child's screams and called the police. Battered child syndrome is viewed as only one extreme of the entire syndrome of child abuse.

all cases in children's hospitals is insufficient love, food, or both.

In the mid-1950s researchers at the University of Denver, Colo., first described graphically the "battered child syndrome," a condition that was seen by 1971 as only one extreme of the entire syndrome of child abuse. Currently, the definition of child abuse syndrome has been expanded to include a child's failure to thrive because of pathological or insufficient mothering, mild repeated injuries, more serious injuries including fractures, and to massive brain damage with or without skull fracture. Research in recent years has provided physicians, courts, and welfare personnel with the beginnings of a comprehensive view of the child abuse syndrome, and means of detecting it early and treating it successfully.

The child abuse syndrome. Child battering of an episodic kind generally occurs as an explosive outburst of unpremeditated violence. Generally, four conditions have to coexist for it to occur. (1) Almost invariably the two marriage partners have similar personal backgrounds; one gives the child the beating and the other condones, covers up, and sometimes arranges for the abuse. (2) One child is often singled out because of some pathological feeling of one or both the parents about the child. He is regarded as slow, spoiled, demanding, and generally undesirable. (3) There is almost invariably a family crisis (persistent crying, pregnancy, family strife). (4) There is generally no helping person available to "rescue" the parents.

Parents who physically abuse their babies and children, while coming from all walks of life and socioeconomic levels, share a pattern of parent-child relationships characterized by a high demand for the child to perform so as to gratify them, and by the use of severe physical punishment to ensure the child's proper behavior. Abusive parents also show an unusually high vulnerability to any criticism, lack of interest, or abandonment by a spouse or other person important to them, or to anything that lowers their already inadequate self-esteem. Such events create crises of unmet needs in which the parent turns to the child with exaggerated demands for gratification.

Both this pattern of behavior toward the child and the crises that trigger it stem from the parents' own childhood experience and learning. Abusive parents were raised in a similar system; they were expected to perform well and to gratify parental needs very early in life, and were criticized, punished, and often abused for failing to do so. They felt their own needs were neither met nor adequately considered and, as adults, have no firm cushion of self-esteem or awareness of being loved. In a crisis of insecurity, they repeat what they learned in childhood about how parents behave and turn to their own child for the nurturance and reassurance they need to restore their self-esteem.

Two elements in the family crisis are sometimes detected. First, the child did something that taxed the patience and tolerance of the parents to the breaking point. He did not eat right, soiled his bed, broke something, was disobedient, or in some way responded improperly to the parental wish and need. More subtle, more important, and more difficult for the parents to describe is the second element of the crisis—what caused their unbearably low sense of value and lack of self-esteem that resulted in a desperate need of reassurance. A common precipitant of such feelings is the withdrawal or alienation of a spouse, ranging from simple emotional coolness or lack of conversation to actual separation. Verbal criticism can also produce the same desperate emotional state. Some crises may appear quite trivial, such as a dented fender or a breakdown in the automatic washing machine, but for a person with a precarious sense of stability, the slightest untoward event may topple the balance and produce a feeling of worthlessness and anger.

Usually each child has a special meaning to the parents. One child may be perceived as entirely comforting, gratifying, and satisfactory. This child usually learned very early in life just what to do to please the parents and how to avoid displeasing them. He may appear overly controlled, submissive, cooperative, and without individual initiative, but, even though he is deprived of what might be considered an optimum amount of loving consideration, he is rarely physically abused. The child likely to receive abuse is perceived as a complete failure, rarely performing those things that the parents want for their own satisfaction. Even worse, he may do things that thwart the parents' desires.

Abusive parents view infants and children as if they were much older than their chronological age and mistakenly endow the child with an inappropriate ability to understand and meet their expectations. When the child fails to perform properly, the parents attribute this behavior to deliberate stubbornness, willful disobedience, or a malicious desire to thwart their wishes, and, in their style of child rearing, such behavior calls for severe punishment. The child who has "too many accidents," who shows unusual bruising, who is thought to be slow or clumsy, or who "fails to thrive" eventually comes to medical attention and is ready to be rescued by interested health professionals.

Prediction of child abuse. Most experienced clinicians rule out abusing by a parent as the cause of problems normally encountered with children, or at least they assign it a very low probability of occurrence. Nevertheless, general practitioners or obstetricians and their nurses and receptionists can often recognize mothers who are likely to have difficulty with their babies. Among these are women who for one reason or another have strong desires not to bear the child and may have tried to induce or obtain an abortion. Others are extremely young and have unrealistic, blissful views of motherhood, or are unduly concerned that the baby's sex and appearance be "just right," or express considerable concern about the baby's being disciplined at an early age.

At her first contact with a new baby, a mother may spontaneously express pity, but no love or tenderness. Others may make very disparaging remarks and express extreme distaste or profound rejection. While in the hospital some mothers either fail to name the baby or else they select a highly impossible name. It is common practice to assume that the mother will come to love the baby. In fact, a particularly rewarding "good" baby who gives little trouble to the mother or a very warm and giving husband may enhance minimal mothering capabilities.

The routine six-week or ten-week well-child visit to a pediatrician or clinic often allows both nurse and physician to see mother and baby interaction at a time when the mother has come to know the baby and to form definite attachments. At this time some mothers may feel quite depressed. In the past, physicians assumed that this was a physiological and probably universal event that rarely had serious impact. It is now recognized that while some mothers cannot love their babies because they are depressed, many more are depressed because they cannot love their babies.

The opportunity to express negative or aggressive feelings toward a new baby rarely exists unless the pediatric nurse or physician is able to allow the mother to express them. When the baby has been expected with too much hope of repairing the mother's past emotional deprivation, the mother may describe it almost from the start as "spoiled," "bad," "slow," or "miserable." The physician's impression that the mother is describing a baby quite different from the one he sees is his most useful subjective point of warning. Not infrequently, the child fails to thrive and is not following his predicted growth curve. This, to the mother, simply proves that the child is "not right." If early encouragement followed by diagnostic hospital evaluation reveals that the baby thrives in the hospital, where normally a baby should not thrive, the diagnosis of maternal deprivation as the cause can then be made. The child then can be placed in foster care, and therapy of the parents can often be initiated.

Treatment through "mothering." Treatment consists of placing a mother substitute meaningfully into the life of the abusing parent. This can be done by professionals, such as psychiatrists, social workers, public health nurses, and physicians. The worker assumes the "good mother" role that the parents so desperately need to reverse the effects of many years of unmet needs and the feelings of worthlessness and failure. This requires skill and much more flexibility, openness, and time than most workers are used to giving. Many social workers feel that it is unprofessional to permit this degree of dependency, but without it their efforts almost certainly will be limited, if not useless. Recently, however, experience with family aides (lay therapists) has been excellent. These paid individuals are available 24 hours a day by phone, and, in time, provide the deprived parent with an experience of love and support.

The response of parents to therapeutic intervention may be rapid or slow. Marked improvement can occur within three months in those parents who had the least amount of difficulty in their own childhood and whose reality situation is reasonably good. In some cases, however, the parents will use their life-long ability to be sensitive to what is expected of them and perform accordingly. Such pseudoimprovement does not hold up under stress.

A clear indicator of real growth and improvement is the way parents report and handle crises. Early in therapy, parents tend to report numerous crises with a sense of being overwhelmed and helplessly unable to cope. Eventually, however, they are able to report with a sense of pride and achievement that a crisis has been handled reasonably well without the therapist's help, or that it was managed with help from some other source.

The child can return home safely (1) when the battering parents develop better appreciation and love of themselves, as shown by their ability to make a few friendships and join into some outside activity; (2) when they see the child with some pleasure as interesting and lovable; (3) when they demonstrate the use of a number of individuals in moments of crisis and have shown that they can get help at this time; and (4) when, through brief weekend reunions, they are showing an ability to provide basic parental care. Parents will often pressure the court to return the child because its removal exposes them to public or family disapproval. It is rare for a child to do well after returning to a home where there is this pressure.

This drowsy girl was one of Cleveland's two "Sleeping Beauty" sisters, victims in a classic child abuse case. Doctors spent 10 months seeking the cause of their comas before it was found that their mother fed them barbiturates, even while they were hospitalized.

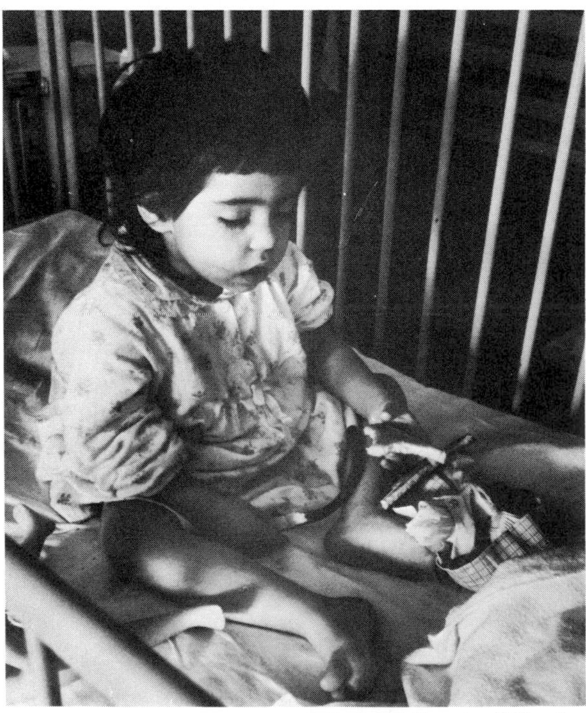

J. Clark, LIFE Magazine © Time Inc.

Experience in Denver indicates that 80% of all battered children can be returned to their parents within eight months after foster home placement and that the continuation of treatment is crucially important at this time. It is necessary not only to consolidate the gains already made but also to see that the inevitable shifts in emotions and living patterns caused by the return of the child can be managed without regression to previous unhealthy patterns. The child may have received a great deal of attention and love in the foster home and have responded well to it. On being returned home, he will have a reservoir of affection, therefore, and his parents will be delighted with him. Unless, however, the parents can continue this empathic giving to the child, the child will run out of his store of affection, and the parents may see him again as unrewarding and unsatisfying. Such a happening indicates that the child was probably returned prematurely and that the parents will need an extra amount of help to tide them over this period.

Future developments. A number of studies that involve interviews of parents prenatally and postnatally to determine the potential of child abuse are under way in Scotland and the U.S. If successful they will provide help in the early detection and successful intervention on behalf of a troubled family. Other studies are attempting to determine the possible relation of child abuse to later aggressive delinquency, to evaluate new therapeutic approaches involving the use of crisis nurseries and day-care centers, and to follow up on children known to have been abused or deprived in order to assess the long-term damage of their abuse and deprivation.

In Europe and the U.S., efforts are being made to improve ways in which the child can have a professional advocate from his earliest infancy. Such a person, probably a nurse health visitor, would be able to intervene in behalf of the child before injury or deprivation progressed too far. Increase in reporting and the obligatory involvement of the juvenile court in all reported cases of suspected child abuse are likely to develop. A number of states are seeking a U.S. Supreme Court ruling on the possible future requirement of having independent counsel and expert witnesses represent a baby in a dependency petition, or are considering laws to bring all reported cases of suspected child abuse to the attention of the juvenile court rather than have them handled administratively by welfare departments.

—C. Henry Kempe

Drug addiction

In common parlance, drug addiction refers to the frequent compulsive use of psychoactive drugs in such a manner as to be harmful to the individual or to society. Several years ago, the Expert Committee of the World Health Organization abandoned the term "addiction" as being too imprecise, and substituted "drug dependence," embracing both physical (physiological) dependence and psychological dependence (intense craving leading to compulsive drug use). It seems doubtful, however, that semantic clarifications of this sort can have much real effect upon the problem. There is a certain elementary clarity about the term addiction, for it is correctly understood to imply self-administration of drugs.

If every patient receiving a drug to which dependence develops were regarded as an addict, the issue would be hopelessly confused. Are patients receiving opiates for pain under a physician's direction, and who may indeed become dependent, to be called addicts? What about epileptics treated with anticonvulsants or diabetics treated with insulin? In all these cases the patients are dependent upon the medication in that they would not function normally without it; but it is not helpful to say they are "addicted." This controversy is not purely academic. When a narcotic such as methadone (*see Treatment of heroin addiction,* below) is administered regularly to heroin addicts as part of a medical treatment program, the recipients are certainly dependent upon it, but they are regarded as ex-addicts because they have broken away from the world of illicit drug self-administration.

As the table accompanying this article indicates, there are several distinct classes of addicting drugs. Their actions range from virtually harmless to extremely dangerous; their legal status from freely available to totally banned. Curiously, there is no definite relationship between harmfulness and illegality.

Basic research. Basic research into drug addiction asks: "What is the psychological, physiological, and biochemical basis of addiction?" This question cannot easily be answered until we know how the various addicting drugs alter brain function. Such research is in its infancy, since so little is yet known about how the brain works. (See *1970 Britannica Yearbook of Science and the Future,* Feature Article: THE BRAIN, pages 34–53.) Much interest has centered during the past few years upon the amine neurotransmitters, chemicals that serve in the communication network between neurons (nerve cells) in the brain. Some of the addicting drugs are closely related to certain of these neurotransmitters.

TYPES OF ADDICTING DRUGS

Drug or Drug Class	Legal Status	Physical Dependence
Caffeine	U	Mild
Tobacco (nicotine)	U	Moderate
Marihuana (THC)	B	None
Hallucinogens (LSD, mescaline)	B	None
Cocaine	B	None
Opiate narcotics (heroin, morphine)	B, R	Severe
Barbiturates and other "downers"	R	Severe
Amphetamines and other "uppers" ("speed")	R	Moderate (?)
Alcohol	U	Severe

Legal status: U = unrestricted, or subject only to age limitation; R = regulated for medical use, illicit except on prescription; B = totally banned; importation, sale, possession, and use illegal.

Important progress was made during the past year in identifying and isolating receptors, the components in nerve cell membranes upon which the neurotransmitters act. Two groups, one led by E. De Robertis in Argentina and the other by R. Miledi in England, succeeded in obtaining what appears to be the acetylcholine receptor in a moderately pure state. The Argentine team was even able to construct a synthetic membrane, into which they could introduce the material extracted from brain that contained the acetylcholine receptor. Remarkably, the artificial membrane then responded to acetylcholine much in the way sensitive nerve cells do, by undergoing a sudden increase of electrical conductivity. The importance of these advances for understanding the actions of addicting drugs is immense, since they open the way for developing model systems in which such drugs could demonstrate their actions in the test tube.

Marihuana. Research on marihuana was stimulated by the recent isolation of its active principle, tetrahydrocannabinol (THC). It is impossible to study the pharmacology of a drug intelligently unless one knows what that drug is and has it available in pure form. Thus, the availability of active THC permitted quantitative experimental investigations of marihuana for the first time. These experiments are delineating the main characteristics of marihuana intoxication (the high), the duration of action, and the differences among people with respect to sensitivity, metabolism, and so on. One of the main issues to be settled is whether THC itself is the psychoactive substance, or whether

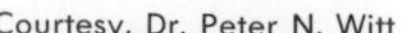
Courtesy, Dr. Peter N. Witt

The effect of a drug on the ability of a spider to make a web may give some indication of the drug's effect on human behavior. This typical web of an adult female Araneus diadematus *spider took 20 minutes to complete.*

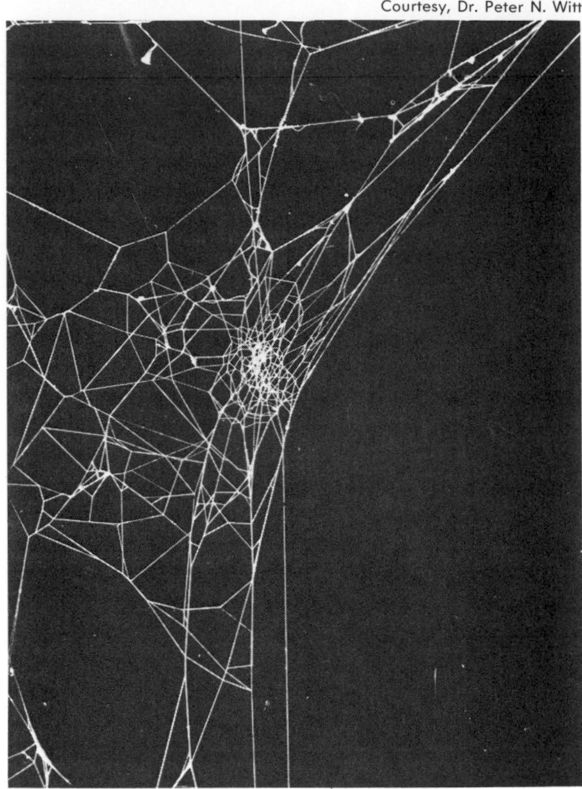

The same spider attempted this irregular and incomplete web 12 hours after drinking sugar water containing dextro-amphetamine. It was several days before she could again build a web similar to the first one.

THC has to be converted in the body to a derivative that is the real psychoactive compound.

Research with animals. Using a method introduced by Chilean investigators a few years ago to produce physiological dependence upon narcotics in mice, the author and his co-workers at Stanford (Calif.) University found that a small amount of dependence is produced by even a single injection of a narcotic. This wears off in about half a day, but if repeated injections are given at shorter intervals, the dependence builds up. The process is probably the same in man, where it is well known that occasional injections of heroin do not lead to dependence; only if heroin is used at least a few times weekly does the addict become "hooked."

Studies of this kind shed light on a question that is currently very controversial: When a narcotic is withdrawn from an addicted animal or person, is there a complete return to the normal state, or are there long-lasting (perhaps even permanent) changes in brain and body function? The evidence is contradictory. The Stanford results show clearly that dependence and tolerance, as we measure them in mice, disappear entirely. Moreover, an animal that was addicted before does not become more easily dependent and tolerant the next time. On the other hand, several investigators have found persistent abnormalities after a cycle of addiction. Abraham Wikler, at Lexington, Ky., has shown that after once being addicted, a rat will more readily choose to drink an opiate solution than will normal rats, and this preference lasts most or all of the animal's life.

Narcotic antagonists such as naloxone precipitate an acute and full-blown withdrawal sickness in narcotic-dependent animals or man but have no effect in normal individuals. A group at the University of Michigan found that for many months after withdrawal of opiates from dependent monkeys, naloxone caused physiological disturbances very reminiscent of withdrawal effects. This implies again that there is some long-lasting aftermath of the addicted state. If we could understand exactly what these long-lasting changes were, we might be in a good position to understand and perhaps deal with the tendency of narcotic addicts to relapse.

Alcohol dependence in animals. Most animals will reject alcohol, probably because of its taste. If forced to take it in their drinking water, they will drink as little as possible. Recently, however, a useful method for producing intoxication and eventual dependence in animals was developed at Stanford by Dora B. Goldstein. Mice were housed in a vapor chamber, thereby forcing them to breathe a constant concentration of alcohol. Because this was absorbed steadily through the lungs, a constant blood level of alcohol was established and controlled at will.

When the intoxicated mice were removed from the chamber, the alcohol was metabolized rapidly and left the body. As that happened, withdrawal effects developed, very much as they do in human alcoholics. Using this system, it has already been possible to show that the intensity of dependence, as measured by the severity of withdrawal symptoms, is directly related to the total alcohol exposure (amount and duration).

Drug detection. A significant development in the detection and assay of addicting drugs was the perfection of antibody-based laboratory procedures. During the year Sydney Spector, at the Roche Institute of Molecular Biology, Nutley, N.J., succeeded in obtaining an antibody to morphine, using the classical procedure of inoculating rabbits with morphine coupled to a protein. The antibodies, isolated from the rabbit blood serum, were then allowed to combine with radioactive morphine. If a solution such as urine, containing an unknown amount of morphine, is added to this mixture, any morphine present will compete for the antibody sites, thus displacing the radioactive morphine. The amount of radioactivity in the solution after chemical removal of the antibody (with any radioactive morphine still bound to it) is a measure of the amount of morphine that was added.

Subsequently, a research team in Palo Alto, Calif., introduced a method called spin immunoassay, using morphine antibody but employing free-radical technology instead of radioactivity. A free radical is a molecule with an unpaired electron, giving it the capacity to absorb energy in a magnetic field pulsed at microwave frequencies. One can determine not only the presence of a free radical but whether or not it is bound to an antibody. The starting material of the spin immunoassay method is a mixture of morphine antibody with spin-labeled morphine. Adding a small amount of urine or other body fluid containing morphine displaces a certain amount of the spin-labeled morphine, which is immediately detected by the change in its energy absorption. In practice a tiny drop of the fluid to be assayed is sufficient, and the entire procedure is complete in 30 seconds without any chemical manipulation.

Treatment of heroin addiction. The most important contribution of recent years to the treatment of heroin addiction has been methadone maintenance. First employed in New York by Vincent Dole and Marie Nyswander, it has gained widespread acceptance as a means of bringing narcotic addiction under control. The past year has seen state after state authorize methadone programs, and some very large ones have begun to operate. This rapid expansion of methadone treatment was based on a careful independent evaluation of the New York program by Frances Gearing of Columbia University, which showed that about three-quarters of all hard-core addicts admitted to methadone maintenance were still in the program three years later, having substantially altered their life styles.

Action of methadone. If an addict in daily contact with a heroin-using culture tries to stop using heroin without help, he almost invariably fails. Methadone, a synthetic narcotic that is taken by mouth and has a long duration of action in the body, facilitates giving up heroin if the addict desires to do so. By maintaining the state of narcotic dependence, it prevents withdrawal sickness. Because of its long action, methadone maintains the dependence in a stable manner, without the ups and downs characteristic of heroin. For the same reason, methadone itself produces no high once stabilization has been achieved (after a week or two). Thus addicts stabilized on methadone can function normally.

Youths huddle around another suffering the effects of drug use during a peace march in Washington, D.C. Area hospitals anticipate drug-related problems during mass demonstrations and establish extra emergency facilities to cope with them.

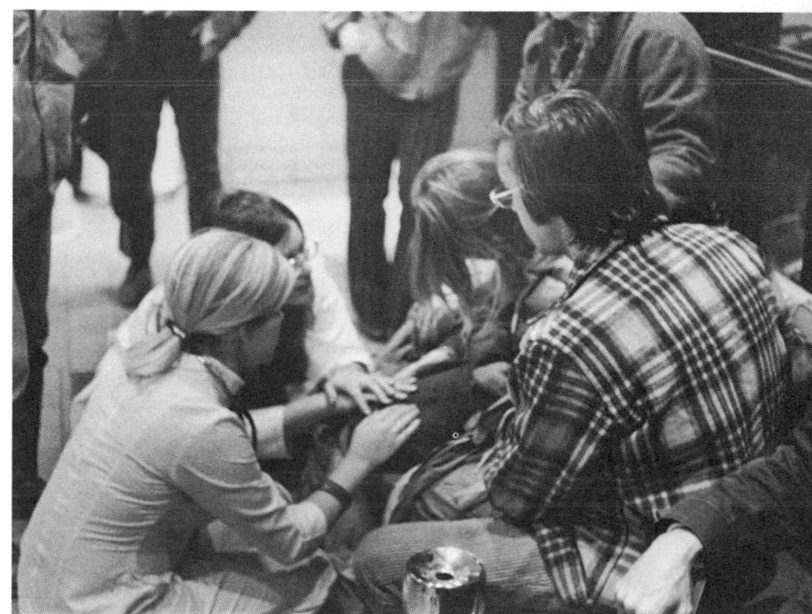

Jay Nelson Tuck, MEDICAL WORLD NEWS

Culver Pictures, Inc.

A woodcut made in 1876 entitled "Interior of a hasheesh hell on Fifth Avenue, while in full blast," depicts a then current view of drug abuse in New York City.

Another important action of methadone is the establishment of a cross-tolerance, sometimes known as a "blockade" of heroin effects. Since methadone is a narcotic, as its dose is increased to the stabilization level, the body becomes tolerant to a higher narcotic level. Consequently, if the patient does inject heroin, he finds it ineffectual. If he used a sufficiently large amount, he could exceed the tolerance level of his body, but this behavior is rarely seen. Clearly, if the addict were determined to get high on heroin, the logical course would be to drop out of the methadone program; yet, as we have seen, a remarkably high percentage do not.

Opposition to methadone has come principally from those who see any narcotic as bad by its very nature. But pharmacologists (to cite Vincent Dole) do not classify molecules as moral and immoral. If a particular synthetic narcotic can help addicts reenter the mainstream of society and lead satisfying and productive lives, it is difficult to deny its medical value. A major question for future research is whether or not addicts can be withdrawn gradually from methadone, once they have substantially changed their life styles, have removed themselves from the drug subculture, have become self-sufficient, and have found alternative satisfactions in life.

Naloxone treatment. Another approach to the treatment of addicts now under study is the use of narcotic antagonists, substances that do not produce any high themselves but block the high produced by heroin. One such compound, naloxone, appears promising. If a person is given a small dose of naloxone by injection and then takes heroin, the heroin has no effect on him whatsoever. Theoretically, then, if the addict took naloxone regularly, he would soon stop using heroin, simply because he would derive no benefit from it. Unfortunately, the effects of naloxone last only a short time, and an addict wishing to obtain a high from heroin would only have to interrupt the naloxone schedule for a day or so. What is needed is a modified naloxone that could be taken by mouth and would last for at least several days. Alternatively, an implant lasting a month or so might be developed.

Another problem is encountered with naloxone. Since antagonists precipitate withdrawal illness, in order to institute naloxone treatment it would be necessary first to withdraw the addict completely from narcotics. Experience has shown that this cannot be accomplished reliably "on the street," so expensive in-patient hospital withdrawal programs would presumably be needed. Finally, if the addict is eventually to function without naloxone, the same problem presents itself as with methadone—how to prepare him to live in the real world without relapsing to narcotic use.

Future trends. The past year has seen a convergence of pharmacologic and psychotherapeutic approaches to heroin addiction. Abstinence-oriented movements such as Synanon, Daytop, and various "halfway house" programs have had only limited success as measured by the numbers of addicts achieving long-term abstinence after they return to their communities. Nevertheless, the methods pioneered in such programs have had a very widespread influence. The trend, as exemplified by Jerome Jaffe's work in Chicago, is to combine the best available methods, including methadone. If an addict is to function in the real world without dependence on narcotics, he will have to be self-motivated—not because it is illegal to use heroin, but because he sees the value to himself of being free of the narcotic habit.

A major concern during the next year will be how to extend the existing treatment modalities, especially methadone, to the estimated quarter of a million addicts in the United States. What no one really is sure about is how to treat the increasing numbers of adolescent heroin addicts, many of them white, middle-class youths. Experimental treatment programs are urgently needed to see if methadone has a proper role in normalizing this group. Here the question of motivation looms large—it is not evident how young people whose drug use may be primarily a symptom of alienation from society can be weaned away from drugs until they truly wish to establish a new life style for themselves.

—Avram Goldstein

Malignant disease

Twice since World War II science has been convinced it was on the verge of conquering cancer. Although the current campaign, intensified early in 1971, seemed likely to be successful because of the revolutionary developments discussed in the articles MICROBIOLOGY and MOLECULAR BIOLOGY, it had its roots in the earlier, abortive attempt.

The search for a chemical solution. Because of the great advances that developed during World War II in electronic instrumentation, computational machinery, fundamental physics, and the chemotherapy of infections (the use of drugs or other chemicals to treat or control diseases), the U.S. entered the 1950s with confidence in the power of science to solve medical problems. The stage was well set for a vast expansion of cancer research work: The American Cancer Society had matured into a powerful mobilizer of public and professional interest in cancer research that could ensure Congressional support for the funding of a national research program. Wartime experience with the antimalarial program had established a pattern of organization for involving pharmaceutical laboratories, government facilities, and university and clinic medical scientists in large-scale efforts toward the development of chemical methods of disease control. Chemical warfare experimentation had brought to light the intracellular damage responsible for the toxicity (poisonousness) of certain chemicals.

In the light of its origins in that particular time and place, it is not surprising that most cancer research work was undertaken on the assumption that cancer could be adequately described in terms based upon an analogy with infectious disease. The cancer cell was accepted as a given; there was relatively little interest or support for studies into its origin. It was thought of as an invader, a destructive parasite similar to the microbes of tuberculosis, cholera, or malaria. The problem of research was assumed to be that of finding a selective poison for the cancer cell.

The optimism of the public for a chemical solution to cancer was stimulated by each new development of successful chemotherapies for tuberculosis and virtually the entire range of bacterial diseases. Scientific confidence in the chemotherapeutic approach also continued to expand despite the fact that a growing minority of cancer research specialists were taking a sober view of the empiricism of the national research effort.

By 1956, more than 200,000 chemicals had been tested against experimental cancer in animals. The criterion for this screening program was the

B-type virus particles from the milk of a mouse with mammary tumors (top), with distinctive antigenic spikes, and spikeless monkey mammary tumor virus particles (bottom) are two of the animal-infecting viruses being studied extensively in the hope they will shed light on, or even serve as a vaccine against, human breast cancer.

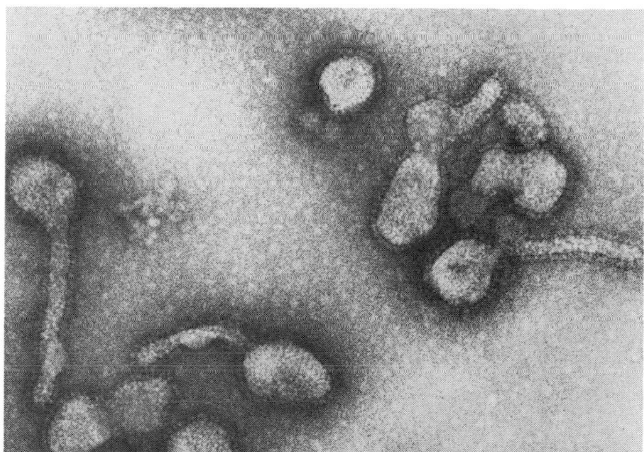

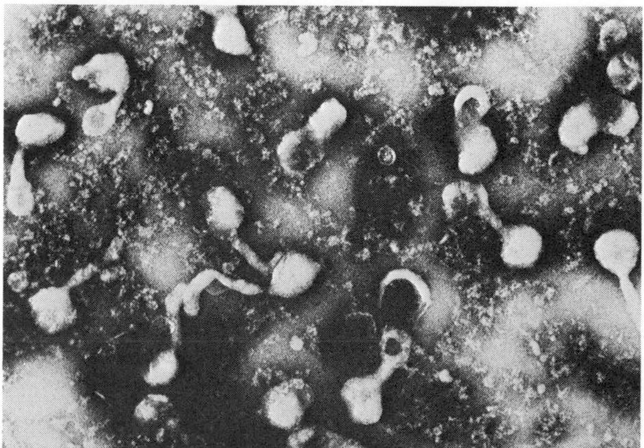

Courtesy, Dr. Harish C. Chopra, National Cancer Institute

ability of the agent to retard or abolish the growth of a transplantable animal cancer without seriously injuring the host animal. But compounds successful in this screen were only rarely useful in clinical medicine; the toxic effects produced during their use in man were severe while the benefits they gave were transient.

Federal support for cancer research began to decline in the early 1960s. But the impetus provided by the manpower training, laboratory building, and scientific organizational efforts of the years of prosperity kept a large complement of workers active in the field and inevitably gave rise to new discoveries of fundamental significance.

A rekindling of interest. The demonstration in the early 1950s by Ludvik Gross, Sarah Stewart, and others that DNA (deoxyribonucleic acid) viruses could cause cancer in rodents rekindled interest in cancer virology. Also in the 1950s, great advances in the understanding of the internal anatomy of cells were made and techniques for cell culture were brought up to an industrial level of control and productivity. Side by side with these gains, methods for isolating cellular components and studying their physiology and biochemistry were developed with the aid of cancer research funds. It became increasingly clear that the fundamental disorder that gives rise to cancer is a disorder of structure and function within the cell and that the understanding of cancer causation and the ability to design chemical treatments effective against the disease would depend upon a deeper comprehension of intracellular biochemical processes.

The critical discovery of the structure of DNA by James D. Watson and Francis H. C. Crick in 1953 had been made possible by intense and sustained efforts devoted to the understanding of chromosomes and the chemistry of cell heredity. Led by Max Delbrück in the 1930s and 1940s, this work, fundamental to modern biology, had found its subject in the seemingly remote fields of bacterial genetics and bacteriophage (bacteria-infecting virus) transformation of bacteria. Out of it grew the new discipline of molecular biology, which united fundamental physics and mathematics through physical and organic chemistry to cell physiology. It is today the predominant discipline of cancer research and of modern biological study generally. In the early 1960s, François Jacob and

The physical similarities of three tumor virions, or virus particles, are evident in micrographs. The top row shows positively stained thin sections of (from left to right) mouse virions, monkey virions, and the human milk particle. The bottom row shows negatively stained whole virions from mouse, human, monkey, and, again, human samples.

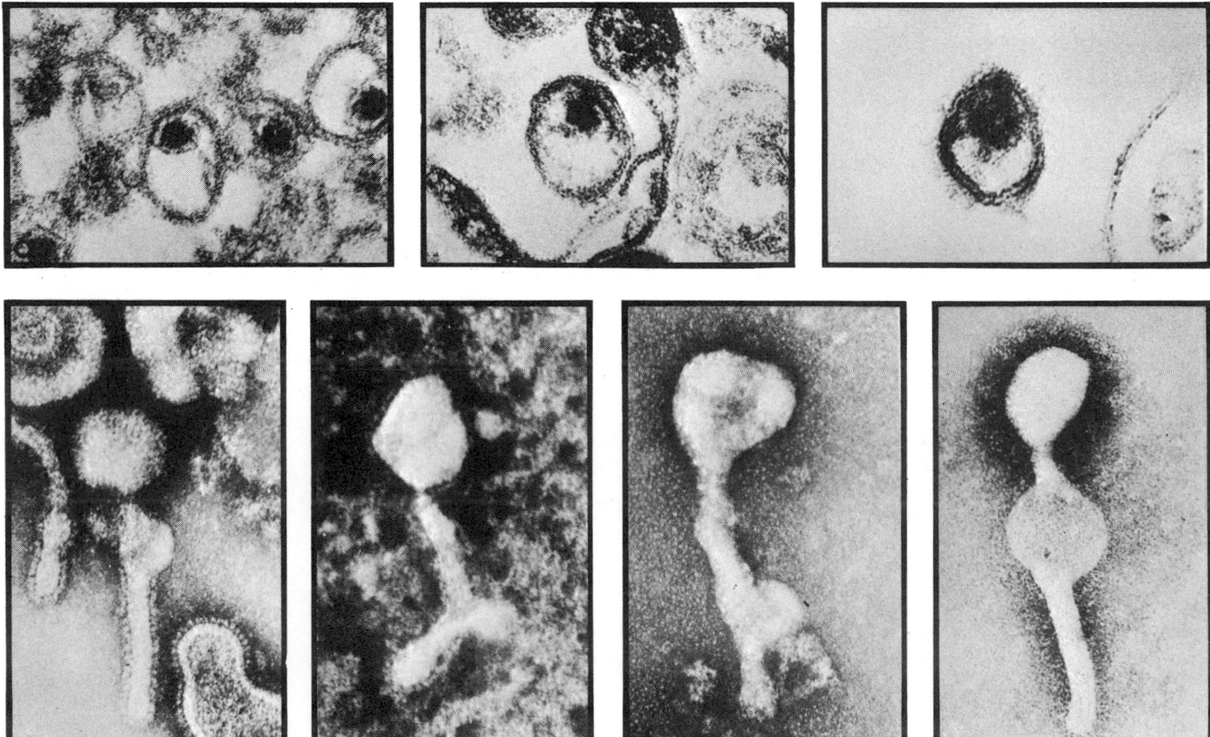

Courtesy, Dr. Dan Moore, Institute for Medical Research

Jacques Monod at the Pasteur Institute in Paris developed a mechanistic theory of gene function that laid open a whole new set of hypotheses for the origin of the cell characteristics produced by the transformation of normal cells to cancer cells.

As a consequence of these and many other major advances, skepticism was dispelled by 1970. Hopefully, the return of excitement and optimism in the attitude of scientists toward cancer research, and of responsiveness toward requests for support by the public and in the Congress, is better justified than it was in the post-war years. At least it is now possible, scientifically and medically, to speak of the cancer process in direct operational descriptive terms and to develop and test specific hypotheses rather than draw analogies from another kind of disease.

The current state of cancer research. The resurgence of confidence for the achievement of a radical and general solution to the problems of cancer was well symbolized by the success of the 10th International Cancer Congress at Houston, Tex., in May 1970. More than 5,000 cancer research and treatment specialists attended this quadrennial international assembly, coming from every advanced country with the exception of China. The proceedings of this congress detailed cancer progress to that date in every area of study and action.

Precise and powerful studies at the level of internal cell structure and physiology were reported in the field of cell transformation from the normal to the malignant state by both chemical and viral carcinogens (cancer-causing substances). Advances in the knowledge of hormonal action within cells, critical to the control of breast, prostate, uterine, and other major types of cancer, were detailed. Information on the special characteristics of cancer cell membranes, energy-producing mechanisms, and replication (duplication) cycles was abundant. The worldwide experience in organ-transplant work in the previous decade had vastly expanded knowledge of the immunological system of mammals and had shown the relevance of immune responses to the problems of cell-growth control, genetic coding, and the ecology of cell populations, all of which are central to the cancer-cell problem.

At the same time, experience with protracted immunoparalysis maintained with the aid of immunosuppressant chemicals over periods of years to prevent destruction of transplanted organs was observed to be associated with a definitive increase in the frequency of cancer in these patients. These developments and earlier classical experimental studies on cancer immunity led to a proliferation of new experimental and clinical efforts to control cancer by immunological means. Further impetus to this work was given by evidence that virus-transformed cells possess special antigens (foreign substances) not detectable in their normal progenitors and by the increasing strength of evidence that viruses are causes of human cancer.

Current chemotherapy. But practical and regularly efficacious ways to utilize immunological methods in human cancer treatment remained wanting, even though definite treatment successes in experimental cancer were reported. Consequently, it was in chemotherapy that rational methods were best able to show their great potential for practical medical treatment.

The chemotherapeutic agents provided by the earlier empirical effort, though few in number and limited in their efficacy, were fundamental tools for the development of the new cell-biological knowledge. Detailed study of their action led to more precise understanding of the chemical mechanisms by which they interfere with cancer cell growth. Since each of the several families of cancer chemotherapeutic drugs has been found to possess unique modes of interference with cell growth processes, the rational design of combination treatments has become possible.

In the early 1960s, work began with combination chemotherapy in childhood leukemia. An intensive initial treatment course was aimed at the destruction of the largest possible fraction of cancer cells. This was followed by carefully timed maintenance treatments designed to eradicate by attrition the entire cancer cell population. In 1971 the probable curability of this form of cancer was announced; a small number of children with this uniformly fatal disease had remained free of it more than five years after discontinuance of treatment.

Acute leukemia of children is a rare and atypical kind of malignant disease. Therefore, perhaps even greater significance to the general problem of cancer chemotherapy was to be found in the success reported in the treatment of Hodgkin's disease, a condition of the lymph nodes resembling more closely in its frequency, chronicity, and pattern of development the major kinds of cancer afflicting mankind. Advanced Hodgkin's disease, only minimally responsive to single chemotherapeutic agents, gave a response rate of 81% to combination chemotherapy.

The thrust of combination chemotherapy based on intensive initial treatment followed by scheduled maintenance therapy had as of 1971 not yet been brought to bear on the most common varieties of cancer. The reason for this was that susceptibility to presently available agents is much less marked in the cells of breast, lung, colon, and other common cancer types than it is in acute childhood leukemias and Hodgkin's disease. Furthermore, these conditions appear predominantly in older populations, rather than in children or young adults. Consequently, the capacity of patients with the more frequent kinds of cancer to withstand the inevitable stress of chemotherapeutic toxicity reduces the degree and extent of drug treatment to which they may reasonably be subjected. Nevertheless, it is highly probable that successful and curative interventions with combination chemotherapy supplementary to radiation and surgery will be accomplished in some of the common cancers in the near future.

An avenue of chemotherapy not yet explored was opened by the new evidence for viral causation of human cancer uncovered during 1970. An international cooperative research effort demonstrated for the first time the presence of an RNA (ribonucleic acid) virus of the B-particle type in the milk of young women with family histories of breast cancer. A virus of this kind had been known since prior to World War II to be the causative agent of mouse mammary tumor. Its demonstration in breast cancer families studied in Bombay, India, Camden, N.J., and Detroit, Mich., and differing so widely in location, genetics, and environment, was strong argument for its significance in breast cancer causation. A similar virus was also found in a breast cancer appearing in monkeys.

RNA tumor viruses. The discovery in 1970 that RNA tumor viruses contain certain enzyme (polymerase) activities capable of reversing the familiar direction of genetic transcription—that is, using RNA as a template (model) for the formation of DNA, the genetic coding material—promised an extraordinarily rich harvest for cancer researchers. (*See* Year in Review: MOLECULAR BIOLOGY, *Genetics* and *Breakthrough in Molecular Biology.*) Thus, later in 1970 a group of researchers at the National Cancer Institute, Bethesda, Md., found that extracts of leukemia (lymphoblast) cells of three patients with acute lymphoblastic leukemia contained this reverse-transcriptase system and also RNA-dependent DNA polymerase. There followed almost immediately reports of the detection of apparently the same polymerase in nine leukemia patients, and the tally has since increased considerably.

These findings raised the hope that a sensitive assay for reverse transcriptase may prove to be, at the very least, a useful tool for the diagnosis of leukemia and perhaps other forms of cancer. Moreover, should it be found that virion-specific reverse transcriptase occurs only in tumor cells, it may be possible to design a specific inhibitor for its activity.

The epoch-making discovery of reverse transcriptase in RNA tumor viruses provided insight

Courtesy, National Cancer Institute, National Institutes of Health

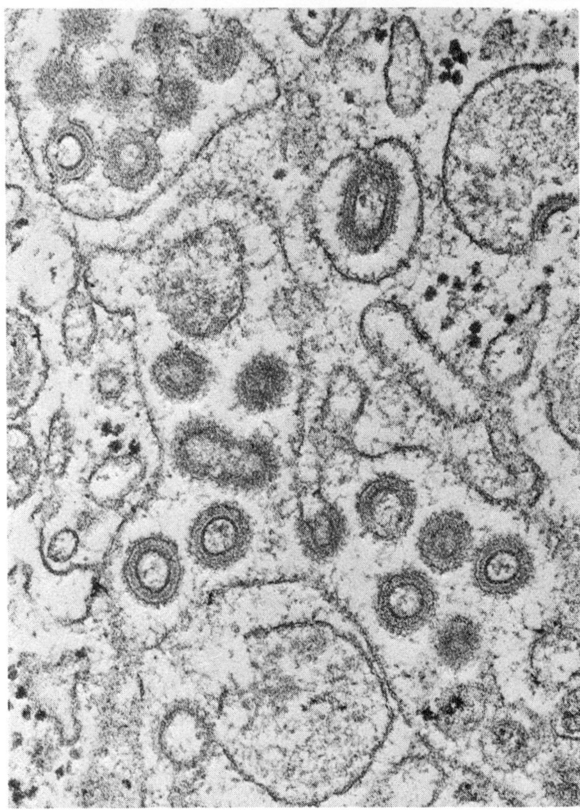

The formation of type C virus particles is visible in this photomicrograph of a bone marrow cell of a mouse with Moloney leukemia. In 1970 the reverse-transcriptase system by which such viruses are thought to infect normal cells was detected in human leukemia cells.

into a possible mechanism of malignant transformation, which had previously been a riddle for molecular biologists attempting to explain the heritable character of the cancerous transformation of cells in a way consistent with the established laws of molecular genetics. Thus, the reverse transcriptase can use the single-stranded RNA of the virus as a template for the synthesis, first, of a complementary DNA strand, and then a double-helical DNA. The latter, containing all the genes of the infecting virus, including those responsible for transformation, can be integrated into the chromosomal DNA of the infected cell and inherited by each daughter cell at the time of cell division (mitosis).

Howard M. Temin and his co-workers at the University of Wisconsin provided evidence that the virus carries with it all the enzymatic machinery necessary for the integration of tumor (virus) DNA into a host-cell chromosome. They found that virus particles contain, in addition to reverse transcriptase, an enzyme (DNA-endonuclease), which can cut long DNA chains into shorter species, and still another enzyme (DNA-exonuclease), which can clip the nucleotide constituents of DNA one by one from the end of a chain. Furthermore, these viruses also contain an enzyme (ligase) that can join together the free ends of two DNA chains and thereby incorporate the viral template permanently into the cell's central genetic machinery, and ensure its participation in the cell's governance and in that of its progeny cells.

The presence of these enzymes in the virus particles suggests the following sequence of events in transformation to malignancy: After infection, reverse transcriptase makes a double-stranded viral DNA. The two nucleases then cut some part of the DNA of the host chromosome and trim away the gap. Finally, the viral DNA is inserted in the gap and sealed into the host DNA molecule via bonds generated by the DNA ligase.

The integration of tumor-virus DNA into a host-cell chromosome could also conceivably be brought about by enzymes existing in the cell before it is infected. In fact, the hypothesis has been advanced that the genetic message of an RNA tumor virus, which is mostly switched off for infectious expression, may be transmitted (vertically) through the operations of the natural genetic apparatus of a normal mammalian cell in which it lies dormant. The transmitted genetic message can, however, be turned on by physical, chemical, hormonal, or other viral agents, leading to cell transformation and cancer development.

While it was accepted as probable, therefore, that RNA viruses possess in themselves the necessary functional elements for the conversion of a normal cell to a cancer cell, it was necessary to recall that, at least in breast cancer formation, the role of hormones secreted by the pituitary and ovary remains critical to the development of a progressive neoplastic (tumorous) growth in the organism. Thus, RNA cancer virus, while quite probably a necessary causative element in breast cancer, is not of itself sufficient to produce the disease. The interaction of viruses with hormones, or even with environmental chemicals, may be required in order to trigger the cancer-producing process in a host organism. In 1970 groundwork was laid at a fundamental molecular level for identifying the chemical properties that chemical reactions must confer upon an organic compound to enable it to interact with nucleic acids for mutagenesis (inducing a mutation) and cancer induction.

Implications for patient care. The likelihood that viruses are fundamental causative agents in many mammalian cancers has implications for diagnosis, for immunological treatment, and for cancer prevention. Cancers caused by a given virus contain specific antigens capable of evoking an immune response in the host. In many experimental virus-induced cancers, more than one class of "neoantigens" (immunity-stimulating proteins not demonstrable normally in the cells from which the cancer has been derived by virus conversion) can be identified. Antiserums directed against these antigens can selectively destroy the cancer cells containing them when these cells are transplanted into an animal protected by prior administration of these antiserums.

The question of immunological treatment of human cancer, therefore, received increasing attention as the probability of virus causation grew. For example, the supplementation of chemotherapy with immunological stimulation in acute childhood leukemia was reported to produce an encouraging increase in the percentage of patients with long periods of disease-free remission. In addition, strong evidence was found for the existence of an immunological response on the part

High-voltage generators at the Los Alamos (N.M.) Meson Physics Facility inject particles into an accelerator designed to produce pions and muons. Among the device's practical applications will be the use of pions to destroy certain kinds of deep-seated cancers.

Courtesy, Los Alamos Scientific Laboratory

of cancer patients, which creates antibodies directed against their own cancer cells.

Cancer antigens circulating in the blood stream can provide a delicate index of the presence of cancer in an organism. While not all cancer-associated antigens need be viral in origin, the presence of virus in a cancer can be expected to make its serological detection more feasible and specific. The antigen found in human bowel cancers and the presence in bowel-cancer patients of an antibody against this protein are not related to a virus causation of these cancers. Yet their usefulness in the early detection of bowel cancer is a testimony to the potential benefits that immunological studies of virus-related cancers may bring.

Finally, there is at least some hope that vaccination programs against human cancer viruses may prove effective in the primary prevention of the disease they cause. Reports citing the protection by vaccination of experimental animals against virus-induced cancers have been published. For this reason major efforts to prepare large quantities of primate cancer viruses by cell culture methods are under way.

The task ahead. This necessarily limited review of a small range of the cancer research accomplishments in 1970–71 may serve to illustrate the new vigor of this field of work and its growth beyond the empiricism that formerly limited its prospects. The initiatives of the U.S. Congress, and the purpose announced by Pres. Richard M. Nixon in January 1971, indicate that a second, rationally founded phase of the effort to achieve a cure for cancer through chemical means has begun to merit massive support.

The vast efforts now in the offing will have the advantages not only of the great scientific accomplishments of fundamental cellular biology that occurred in the third quarter of the century, but of automated techniques and instrumental advances never before available. Progress in planning and administration gained from the fields of operations analysis and simulation methodology, which helped to achieve the rapid successes of the space program, are available for the organization of a worldwide cancer research program commensurate with the goal of achieving, in the last quarter of the 20th century, the deliverance of mankind from malignant neoplastic disease.

The peace of the world and the salvation of our environment depend upon the formation of a worldwide community, one that is above all the product of the sharing of work and its yield. With this realization in mind, the late French Pres. Charles de Gaulle was among the first to call for a great common effort against cancer. Hopefully, the initiatives now advocated for the support of cancer research in the U.S. signal a national intention to devote science and technology to tasks that will serve the genuine needs of all men in a framework of world brotherhood. The cure of cancer is a noble purpose with which to begin this redirection of our capabilities.

Cancer research is a field ripe for harvest in the 1970s. If the effort is fully supported and rightly managed, it should be reasonable to hope that ours may be the last generation open to the ravages of cancer.

—Michael J. Brennan

Microbiology

The subject that dominated scientific news during the past year was cancer—its causes, its prevention, and the probability of a cure. It overshadowed news of such recently popular subjects as molecular biology and environmental/ecological issues. But the study of cancer is not a subject that is foreign to those other areas. Much of our recent understanding of the disease was made possible through such basic studies in the field of molecular biology as the interrelationships of deoxyribonucleic acid (DNA), ribonucleic acid (RNA), proteins, and genes. Similarly, certain environmental pollutants have been identified as contributing causes of cancer.

The American Cancer Society estimated that 350,000 cancer deaths and 650,000 new cancer cases would occur during 1971 alone. Several recent breakthroughs in cancer research, however, raised expectations that a "cure" for cancer might be imminently possible. (*See* Year in Review: MEDICINE, *Malignant Disease.*)

Viruses as causative agents of cancer. The evidence continued to mount that most forms of cancer are caused by viruses. More than 100 viruses are known to induce cancer in vertebrates, and 20 of these can grow in human cells and transform them into cell types showing similarities to cancer cells. Recently, scientists reported that cells from 12 human sarcoma-type cancers were grown in tissue cultures (mammalian cells cultured "in the test tube" outside of the body of the animal or human from which they were derived). These cultures exhibited cells having characteristics similar to the cell types produced by cancer-inducing viruses in cultures of animal cells. Normal human cells were also transformed into cancer-type cells when exposed to cell-free material obtained from these tissue cultures. This suggests the presence of a virus capable of causing transformation of normal cells into cancerous cells. Since sarcomas are caused by viruses in animals, scientists have suspected for a long time that human sarcomas are also viral-induced.

Scientists from three laboratories reported further evidence linking human breast cancer with a virus. Particles were observed in the milk of women breast cancer patients that were almost identical in appearance to viruses known to cause breast cancer in mice. Data were also reported that blood serums from human breast cancer patients could neutralize the mouse breast cancer virus. The presence of neutralizing antibodies in serums from human breast cancer patients provided strong evidence not only that a virus is the causative agent of the human disease, but also that the human and mouse viruses are closely related.

More than 100 distinct diseases are called by the common name of cancer. All are characterized by rapid cell growth and a tendency to spread from one part of the body to another. While most researchers agreed that viruses cause cancer, in most instances there appeared to be a less direct cause and effect relationship than in the usual virus/disease relationship. An exception is Marek's disease of chickens, which was shown to be caused by a DNA-containing virus "horizontally" transmitted from one chick to another.

Whether there is horizontal transfer of human viruses that cause cancer remained to be determined. An alternative line of reasoning holds that a piece of genetic material of the RNA cancer virus is transmitted "vertically" from generation to generation as part of the natural human genetic apparatus. The viral genetic material is "switched off," or repressed, during most of life. Cancer results when the RNA viral material is "turned on" by chemical or physical irritants.

A micrograph shows particles of Australia antigen found in a blood sample of a patient with serum hepatitis. In 1971 researchers at New York University Medical Center announced that they had successfully immunized a small group of children against serum hepatitis using serums containing Australia antigen.

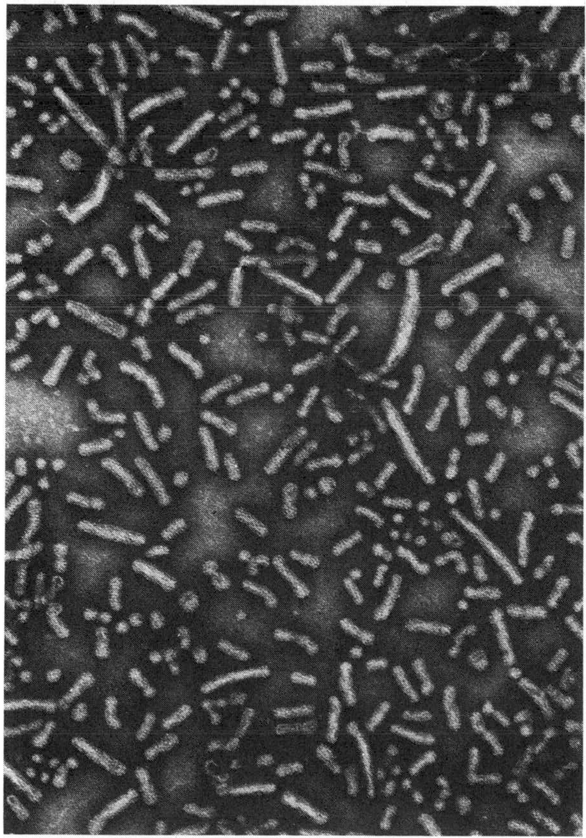

Courtesy, Dr. William T. Hall, Electro-Nucleonics Laboratories, Inc.

Cure and prevention of cancer. The long accepted "central dogma" of molecular biology was that, through the actions of enzymes, only DNA could make RNA, which in turn made protein. It recently came as a shock to the biological world that two scientists independently discovered a polymerase enzyme in RNA-containing cancer viruses that formed DNA from RNA (*see* Year in Review: MOLECULAR BIOLOGY, *Breakthrough in Molecular Biology*).

A drug, rifampicin, was developed that can interrupt the actions of this polymerase enzyme found in cancer-linked viruses and in cancer cells. It was used with a high degree of success in the treatment of leukemia. But because the polymerase enzyme was now thought to be present even in normal cells, the usefulness of rifampicin may be limited. At any rate, the development of drugs directed against cancer viruses appeared to be a promising approach to a cure for cancer.

A new bacterial enzyme was discovered that inhibited the growth of cancer cells in tissue culture. Thus it offered promise of being a useful treatment of cancer. This enzyme causes the depletion of the B vitamin (folic acid), which results in the inhibition of the proliferation of cancer cells.

There were two encouraging reports in recent months on the use of benign viruses that in some way prevented or cured cancer. In one case, an innocuous virus was found in turkeys, which when injected into chickens with Marek's disease, prevented the formation of tumors. The preventive effects of the innocuous virus were unknown, but it apparently kept the tumor cells from switching on. In the other case, a virus was discovered whose effects appeared to be limited to cancer cells. Administration of this virus to animals with developing tumors resulted in the inhibition of the tumors.

Another promising avenue of approach to the

Courtesy, State University of New York at Buffalo

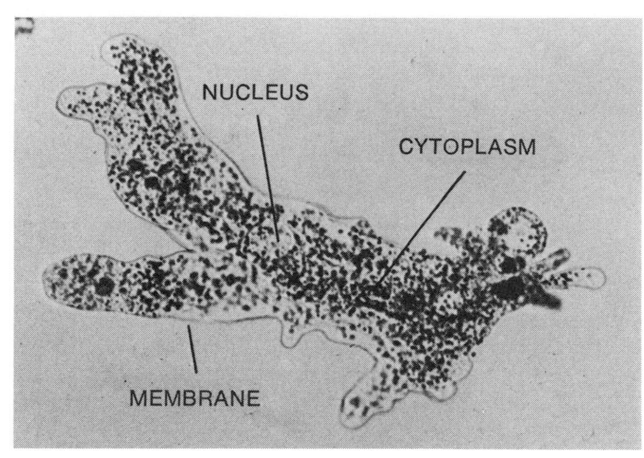

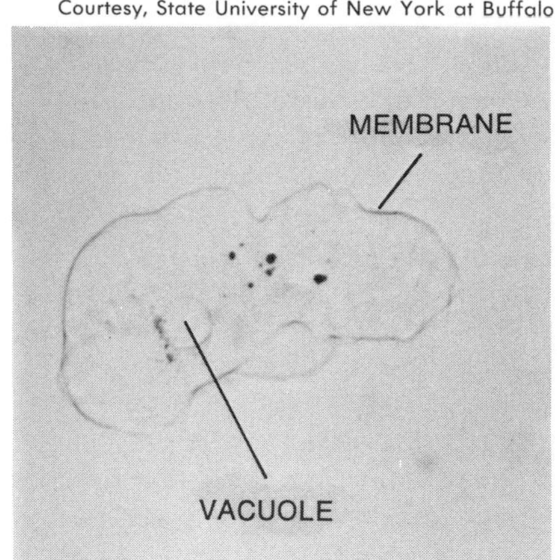

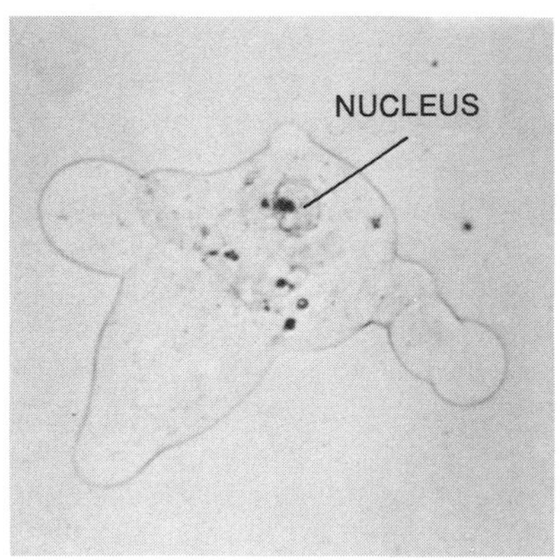

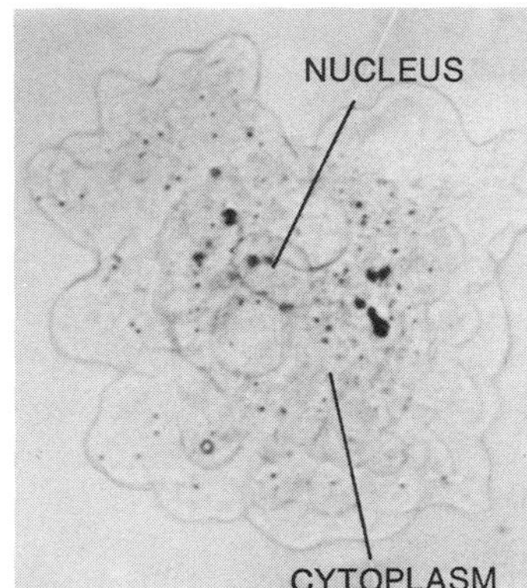

The breakup and reassembly of an amoeba as carried out by James Danielli of the State University of New York at Buffalo is illustrated. An amoeba (top, left) is centrifuged until little more than the membrane remains (top, right); a new nucleus is inserted (bottom, left) and new cytoplasm injected (bottom, right), forming a new amoeba.

prevention and cure of cancer was through immunization. Statistical studies showed that patients who received drugs to suppress the immune response of the body had an increased tendency to develop cancers. This suggested that susceptibility to cancer may be due to an inadequate immune defense system and, conversely, that development of cancer may be suppressed by immune defenses of the body.

An unlicensed vaccine that protects chickens against Marek's disease was being produced. Its existence clearly demonstrated that a cancer known to be caused by a virus can be prevented by a conventional vaccine.

Findings on other viruses. Within the past year, an accurate and simple test became available for the detection of hepatitis virus in donated blood. The hepatitis virus is frequently transmitted through blood transfusions, with the result that a serious liver infection occurs in the person receiving the blood. With the new test, it was possible to determine if the donated blood was free of hepatitis virus before use. There was also a report that a group of scientists succeeded in making antibodies that protect against serum hepatitis (*see* Year in Review: MEDICINE, *General Medicine*).

"Slow virus" infections are those in which the viruses are present in the infected individual for most, or all, of his life. Several slow virus infections in animals are known and may serve as models for counterpart infections in humans. Current thought was that many chronic and degenerative diseases of humans may be caused by viruses of this type. Evidence recently came to light that implicated slow viruses in such neurological disorders of humans as multiple sclerosis and polyneuritis as well as in rheumatoid arthritis. The slow viruses also display activities similar to the cancer-causing viruses, and some had been implicated in cancer.

There was also increasing emphasis on developing drugs that act directly on viruses. Two were described recently: rifampicin, which, as mentioned above, acts on an enzyme found in

RNA-containing viruses; and isoprinosine, which, although still being tested, appears to work against a number of viruses, such as influenza viruses and the herpes viruses. The latter cause a variety of diseases, including chicken pox and shingles. A herpes-like virus was also found to be associated with sarcoidosis, a chronic disease of the lungs that resembles tuberculosis. At one time, pollen from pine trees was suspected of being the contributing agent.

Environmental and applied microbiology. Concern over the biological quality of the environment continued to make news. Several items emphasized the role of microorganisms in marine ecology; others were concerned with bacteria and the environment.

When the research submarine "Alvin" sank in the Atlantic in 1968, the three crew members escaped safely, but their lunches sank with the ship to a depth of 5,000 ft. At that depth, the temperature is 37–39° F and the pressure is 150 atmospheres (1 atm = 14.7 pounds per square inch). Food materials recovered from the sunken submarine were found to be in a well-preserved state after exposure for more than 10 months to these deep-sea conditions. The rates of microbial degradation were as much as 100 times slower in the deep sea than in controlled environments under comparable temperatures. In fact, the preservation of the fruit in the lunches equaled that of careful storage, while the preservation of starch and proteinaceous material surpassed that of normal refrigeration. The implications of these findings concerned the value of using the deep sea as a dumping site for organic wastes. Such dumped materials may be effectively removed from the natural recycling process because of their slow degradation.

A U.S. Army laboratory technician handles cans of frozen biological warfare agents destined for destruction. In the spring of 1971 the U.S.S.R. announced its willingness to negotiate a treaty barring the development, manufacture, and stockpiling of such weapons.

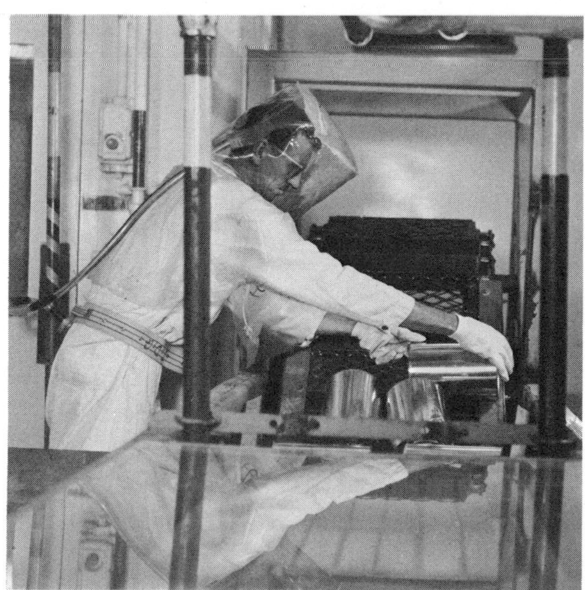

Courtesy, U.S. Army

U.S. scientists developed a mixture of 20 different species of microorganisms to be used to clean up oil spills by microbial biodegradation. Similarly, Soviet scientists combined marine bacteria, seaweed, and mollusks into an artificial ecosystem designed to degrade and to remove oil from water.

In another attempt to curb marine pollution, nitrilotriacetic acid (NTA) was being considered as a replacement for phosphates in detergents. Although the use of NTA was banned by the U.S. government, it was expected that the ban would soon be lifted. Unfortunately, few studies had been conducted to establish NTA's long-range environmental effects. Its release without such assessments might produce a situation similar to that of the release of DDT 25 years ago. Recent evidence indicated that the removal of phosphate from detergents may not slow the eutrophication (aging) of coastal marine waters since nitrogen, not phosphate, appears to be the limiting factor in the growth of algae, which are implicated in the eutrophication process. In fact, the evidence suggested that the degradation products of NTA in sufficient quantity might even promote undesirable algal blooms and that NTA may not be as readily degraded by microorganisms as had been assumed.

Biological pest control continued to receive emphasis as an alternative to chemical pest control. Recently, scientists used two bacterial species that specifically infect Japanese beetle grubs to kill this insect at major airports. It was thought that the beetle invades new territory by "hitchhiking" with man and his goods.

Many bacteria contain an R factor, so-called because it confers on them simultaneous resistance to several antibiotics. The R factor can also be transmitted to other bacteria. It has long been suspected that the presence of antibiotics in animal feeds leads to the selection of populations of bacteria that become resistant to antibiotics because of long-term low-level exposure to them. With the subsequent release of antibiotic-resistant bacteria into the environment, a potential health hazard to man exists. The British government recently banned the use of antibiotics in animal feeds, and in the U.S. a task force was formed to study the problem. (*See* VETERINARY MEDICINE.)

—Robert G. Eagon

Molecular biology

Developments in the molecular sciences during the past year—in biochemistry, biophysics, and genetics—were spectacular. As in previous years, work done in these areas had implications throughout the life sciences.

Biochemistry

Inspection of research literature of recent months revealed that biochemists in increasing numbers were studying the nature and function of the constituents of biological membranes. Traditionally, biochemistry had been concerned with characterization of molecular components of living things, but attention was given primarily to those molecular species that could be studied in pure form after isolation from aqueous (watery) extracts of disrupted cells. This approach provided an invaluable amount of information about the structure and function of many important molecules, such as deoxyribonucleic acid (DNA) and ribonucleic acid (RNA) or proteins of specialized function. But it suffered from the fact that little was being learned about the molecular species in complex structures, such as membranes, that are not readily soluble in water and exist in cells as aggregates with proteins or lipids (fatty substances).

It is essential to understand membranes in greater detail because they coordinate a variety of cellular activities. They permit selective entry of nutrients and exit of wastes; they aid in organization of subcellular organelles (specialized bodies), such as the mitochondrion and the chloroplast; they bind complex enzyme systems involved in energy production, protein synthesis, and DNA replication (duplication); they have specialized structures that permit cells to recognize one another and to form tissues; and they interact specifically with hormones or drugs, which in turn can alter cellular processes. In other words, an understanding of cell membranes will give considerable insight into many biological processes, including transport, energy production and utilization, cell division, cell recognition, cellular differentiation, transmission of the nerve impulse, hormone action, and drug activity.

Unit membrane hypothesis. Early investigations into membrane structure, based especially on electron microscope studies, suggested that all membranes in higher animals have the same basic molecular structure. This suggestion led to the unit membrane hypothesis, which proposes that a biological membrane consists of a bimolecular layer (two molecules thick) of polar lipids coated with a single layer of protein molecules. Most of the lipids found in membranes are rod shaped molecules with hydrophilic (having affinity to water) groups occupying one end of the molecule; the other part of the molecule consists of long hydrophobic (lacking affinity for water) hydrocarbon chains. Within the lipid bilayer the individual lipid molecules are oriented so that their hydrophilic groups face outward to form two hydrophilic surfaces, which are assumed to be covered with protein layers. The hydrocarbon chains fill the interior space, thereby creating a continuous hydrophobic phase through the center of the membrane. The hydrophobic phase usually appears in electron micrographs as a band of relatively low density between two dense bands.

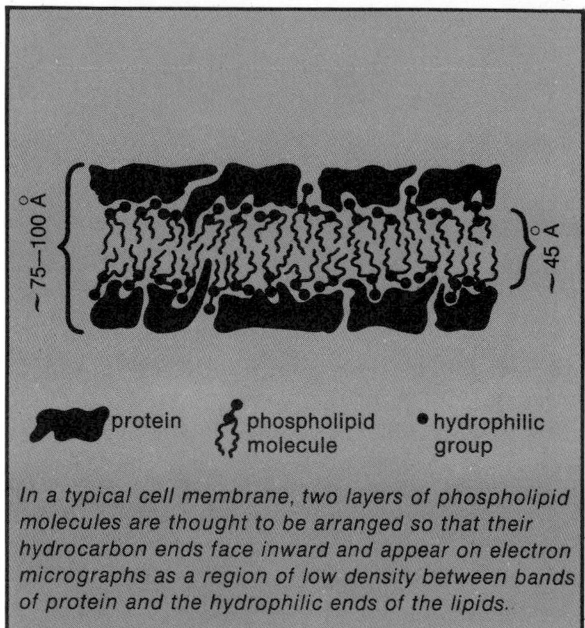

In a typical cell membrane, two layers of phospholipid molecules are thought to be arranged so that their hydrocarbon ends face inward and appear on electron micrographs as a region of low density between bands of protein and the hydrophilic ends of the lipids.

The unit membrane hypothesis accounts for several known properties of membranes and has served as a focal point for interpretation of many recent investigations. For example, W. L. Hubbell and H. M. McConnell at Stanford (Calif.) University used a technique called electron paramagnetic resonance spectroscopy to examine both biological membranes and synthetic bilayers of phospholipids (one type of lipid), which serve as membrane models. These workers synthesized lipid derivatives that form stable organic free radicals (spin labels), and then examined the orientation of the labels in membranes and synthetic bilayers.

The spin labels appeared to behave very similarly in the membranes and in the synthetic bilayers. Their behavior suggested that the long hydrocarbon chains of lipids in the membrane and the bilayers are locally fluid; the chains become more and more fluid near their hydrophobic ends. This conclusion was consistent with a widely held view that the membrane is not a static structure but a solid solution that can allow considerable movement of special constituents throughout it.

M. H. F. Wilkins and his co-workers at King's College, London, used another physical technique, X-ray diffraction, to examine biological membranes and lipid bilayers. They showed that membranes from the microorganism *Mycoplasma,* myelin (the sheath surrounding nerve cell axons), and other tissues give X-ray diffraction patterns of a type that strongly support the view that a lipid bilayer is a major structural component of membranes from widely different cells. Although the unit membrane hypothesis had received support from these studies, specialized functions of different cells must be explained by structural differences in membranes rather than by structural similarities.

Membrane proteins. There has been a keen interest for many years in the exact molecular composition of a biological membrane. Although the principal lipid components had been identified, little was known until recently of the nature of membrane proteins. Membrane proteins exist in intimate contact with lipids and are difficult to study because of their insolubility in aqueous solutions. During recent months several workers demonstrated that membranes are soluble in dilute solutions of detergents, such as sodium dodecyl sulfate, and that the dissolved membranes can be submitted to gel electrophoresis. This technique uses electric charges to cause the proteins in the solution to move on short columns of gels of polyacrylamide, an acrylic acid compound. The proteins remain in solution and migrate on the gels in proportion to their molecular weight; the smaller the protein, the faster its migration. Different membranes examined this way display considerable dissimilarity in their protein constituents.

Recent reports indicated that the proteins of membranes may have molecular weights ranging from 15,000 to 200,000. (Molecular weight is the sum of the atomic weights of all atoms in a molecule.) In the red blood cell membrane, the proteins in greatest amount are two very large ones with molecular weights of about 200,000, one of about 100,000, and six with weights between 40,000 and 80,000. There are also many proteins in minor amounts with different molecular weights. It was unknown whether these proteins act strictly as structural components of the membrane or have specific biological functions. Whatever their role, the red blood cell proteins from five mammalian species including man are very similar.

J. Gwynne and C. Tanford at Duke University, Durham, N.C., also dispersed red cell membranes. By using chromatographic techniques which separate the various substances in a solution, they obtained a molecular weight range for the red cell proteins that was the same as that determined by the electrophoretic methods.

The fact that membranes contain proteins with polypeptide chains as large as 200,000 is particularly intriguing, since only a few polypeptide chains of that size have been found in proteins that are not membrane-bound. One group of workers put forward the view that 50% of the protein in red cells is a protein containing polypeptide chains with a molecular weight of about 5,000. Others had not

continued on page 301

Courtesy, Dr. Arnold L. Shapiro and Dr. John G. Birch

Biological membrane proteins can be studied economically and efficiently because of the development of this apparatus for polyacrylamide gel electrophoresis. Electric charges cause proteins to migrate on columns of gels in proportion to their molecular weight. Any polypeptide chain can be examined in this manner.

Breakthrough in molecular biology: proving the provirus hypothesis

In 1958 Francis Crick enunciated what was to become the cardinal tenet of molecular biology. He asserted that the flow of information in living systems invariably follows the pattern DNA→RNA→protein. Crick's hypothesis held up in all kinds of applications and had the advantage of appearing to be logical. Of the three molecules involved in it, DNA was known to be very stable. The only way known to alter its structure was by genetic mutation, whereas RNA and proteins were subject to many influences, such as enzyme or antigen activity or directions from DNA. It was unlikely, therefore, that any message initiated by RNA or protein would be able to affect DNA.

The one area in which the theory did not seem to apply was that of the RNA tumor viruses, which cause leukemias and sarcomas in animals. Most viruses can infect a cell and use its genetic material to produce more agents that spread and infect other cells. But these agents are recognized as foreign by the host, which moves to reject them. RNA tumor viruses, however, manage to induce a permanent infection that does not destroy the host cell and that is not rejected as foreign. Because they transform a normal cell into a genetically stable cancer cell, it could be assumed that they act somehow on the host cell's most stable element, its DNA.

While many workers felt that the explanation for this could be found within the framework of Crick's theory, Howard M. Temin at the McArdle Laboratory for Cancer Research of the University of Wisconsin questioned the applicability of the hypothesis itself. In 1964, based on an accumulation of circumstantial evidence (see *Genetics*, below), Temin proposed the provirus hypothesis to explain the action of the RNA tumor viruses. He suggested that, following infection, the viral RNA is transcribed into a DNA "provirus," which is then integrated into the DNA of the chromosomes of the host cell and maintained as such in subsequent cell generations.

If this were true, an RNA→DNA information transfer would have to occur, in contradiction to Crick's theory, and an enzyme, a polymerase, would have to exist to cause the transfer. Since no such enzyme had been found in any type of cell, it was assumed that the virus carries the enzyme into the host cell as part of the virion, the complete virus particle. In fact, several virions of animal viruses had been shown to contain enzymes (RNA polymerases) that permit the synthesis of RNA. Temin and Satoshi Mizutani, and David Baltimore, who was working independently at the Massachusetts Institute of Technology, examined the virions of RNA tumor viruses and found a specific enzyme that would synthesize DNA from an RNA template, that is, an RNA-dependent DNA polymerase.

The tests used to prove the enzyme was a DNA polymerase included placing purified viruses in solution with the four chemical building blocks of DNA. The product of the reaction of these molecules was tested and found to be soluble by deoxyribonuclease, an enzyme that breaks down DNA, but to be unaffected by ribonuclease, which breaks down RNA. It, therefore, had the properties of DNA. Further tests using centrifugation techniques established that most of the DNA-polymerase activity was associated with the virions, and that, therefore, the enzyme and the template (RNA) were constituents of the virion.

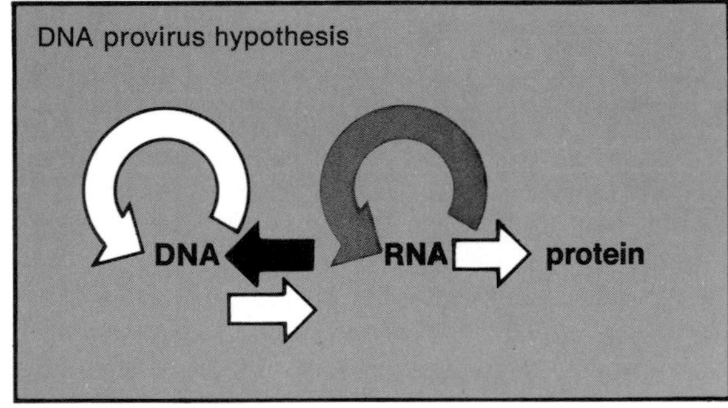

Biological information can be transferred by DNA, which duplicates itself (circular white arrow) or produces RNA, which directs protein synthesis (white arrows). RNA in viruses can replicate itself (circular gray arrow) and, in some cases, direct DNA synthesis (black arrow).

Since the announcement of this success in May 1970, the provirus hypothesis has been confirmed for a variety of RNA tumor viruses. Four separate enzymes in the virion are presumed to be involved in completing the process, called Teminism, by which the RNA-coded DNA is integrated into the DNA of the host chromosome (*see* Year in Review: MEDICINE: *Malignant Disease*, RNA tumor viruses).

But proving one hypothesis only raises many others. Temin since has looked at possible evolutionary origins of viruses with this unique information transfer system and suggested that they all arose from a "protovirus." This viral form could evolve if RNA copies from the protovirus region of

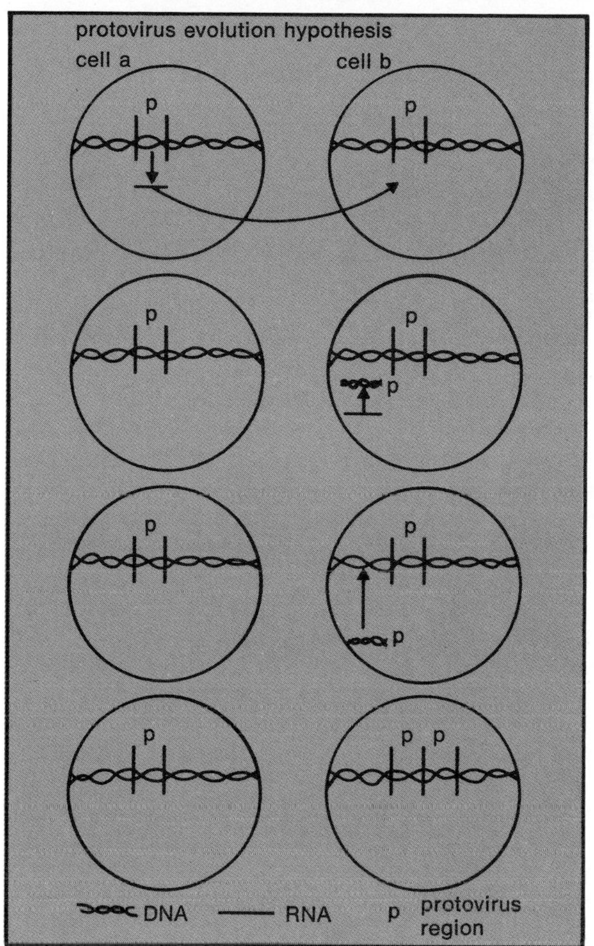

RNA tumor viruses may have evolved when RNA coded for by one cell was transferred to a second cell, where it produced new DNA that the host DNA incorporated. In time, a DNA "protovirus" region might be established and might code for a complete RNA tumor virus particle.

DNA in one cell were carried to another, transcribed into new DNA, and inserted into the host DNA in accordance with the provirus mechanism. By repetition of this process, possibly enhanced by continued evolution of altered cells, a region of chromosomes could arise with the information necessary to form a virion. Depending on where the new DNA was inserted, the result would be slightly different and would give rise to different RNA tumor viruses.

Other ideas about the origin of cancer-causing viruses have been suggested and may prove more plausible than the protovirus hypothesis. Temin's second proposal, however, has the same advantage his first one had: it provides for a genetically stable medium at all steps along the way.

continued from page 299
confirmed this view, and it was agreed generally that proteins with such small chains are unlikely to contribute to membrane structure.

Gel-electrophoresis technique. Although there was much to be learned about proteins in biological membranes, the ability to begin studying them was attributable in large measure to development of the technique of gel electrophoresis in dodecyl sulfate, which was accomplished by A. L. Shapiro, E. Viñuela, and J. V. Maizel at the Albert Einstein College of Medicine of Yeshiva University, New York, N.Y. This technique, described only four years earlier, can be used to examine any kind of protein, whether or not it is bound in a membrane. Previously, determination of the molecular weight of a polypeptide chain (a compound of amino acids of the type that forms proteins) required many hours of analysis with expensive apparatus.

Gel electrophoresis in dodecyl sulfate had, perhaps, more impact in biochemical research in the past year or two than any other technique, and was being applied to great advantage in a variety of biochemical problems. In addition to the studies of cell membranes, the protein components in the membranes of several subcellular structures were examined by this method.

For example, the inner membrane of rat liver mitochondria (energy-producing organelles) appears to contain a minimum of 23 polypeptide species ranging in molecular weight from 15,000–80,000. Very likely each of these chains is derived from a protein with a very specific enzymatic or transport activity, but just as with the red cell proteins, it is difficult to ascribe a specific biological function to proteins observed in polyacrylamide gels. Nonetheless, many specific permeases (a type of enzyme) are known to be associated with the inner mitochondrial membrane and it should be possible to reveal their identity on acrylamide gels in the future.

The fact that clear structure-function relationships were unknown for mitochondrial proteins did not prevent some workers from offering rather detailed models of mitochondrial membranes that propose to give insight into the mechanisms involved in oxidative phosphorylation, the process by which mitochondria burn food to release energy in the form of an energy-storing molecule called adenosine triphosphate (ATP). These models perhaps serve to stimulate further studies, but they required considerably more experimental support to be accepted widely.

The proteins in membranes of other subcellular bodies, such as the Golgi apparatus and the smooth and rough microsomal fractions of rat liver

Courtesy, Dr. Robert L. Hamilton, University of California School of Medicine

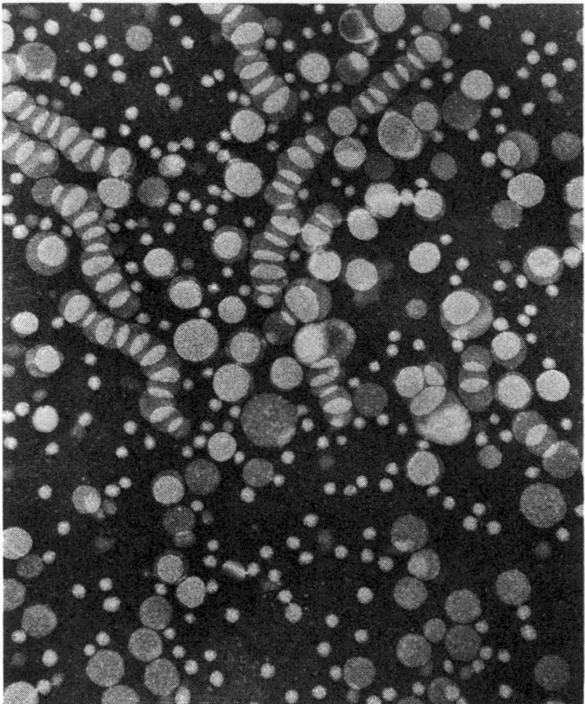

In an electron micrograph of negatively stained whole human serum, abnormal lipoproteins appear as large overlapping disks, in contrast to smaller, spherical normal ones. Analysis by various methods, including the new, highly useful gel electrophoresis technique, indicated that abnormal lipoproteins resemble cell membranes in that they are lipid bilayers enclosing water spaces.

cells, also were examined. Each of the membranes from these structures can be distinguished by its unique set of proteins; the protein patterns also differ from those of the plasma membrane and the mitochondrion membrane.

Identifying protein functions. One major goal of membrane research is to identify specific functions for each of the membrane proteins. Only modest success has been achieved in this regard because methods for isolating and identifying membrane proteins destroy their biological activity. Nevertheless, researchers at Hebrew University (Israel) reported studies that specifically allowed identification of an enzyme (adenosine triphosphatase, or ATPase, which acts on ATP) in sacroplasmic reticulum, a type of muscle membrane. They succeeded in labeling the enzyme radioactively so that it could be distinguished easily by electrophoresis of the membrane proteins. Other workers specifically labeled the enzyme acetyl cholinesterase and an ATPase in red blood cells that is associated with ion transport across the membrane. Clearly this type of approach should provide valuable information in the future about the functions of specialized membrane proteins.

Another membrane protein with a very specific function was described recently in nerve tissue. The conduction and transmission of the nervous impulse has been known for many years to be closely associated with local changes in the permeability of a nerve-fiber membrane to sodium and potassium ions. Thus, analysis of membranes in the nervous system in molecular terms is essential for further insight into nerve conduction.

One event that changes the permeability of a membrane is initiated by acetylcholine, the compound that transmits nerve impulses. Acetylcholine is released at nerve endings and binds specific receptor sites, called cholinergic sites, on the nerve membrane. Many attempts made in the past to identify the molecules associated with the receptor sites failed because of the difficulty of distinguishing the receptor sites from other substances, such as the enzyme acetylcholinesterase, which also can bind the compound. Recent reports, however, indicated that work done by J-P. Changeux and M. Kasai in Paris, C-Y. Lee in Taiwan, and R. Melidi, P. Molinoff, and L. T. Potter at University College, London, allowed specific identification of the receptor sites as protein molecules that are readily distinguished from the enzyme acetylcholinesterase.

The key to the success of these studies came from earlier work by Lee and his co-workers, who in the course of studies on the components of the venom of a snake from Taiwan (*Bungarus mueticinctus*), isolated a polypeptide toxin (poison) of molecular weight 8,000, which was named α-bungarotoxin. Examination of the action of the toxin suggested that it bound irreversibly with the cholinergic receptors. Changeux and his co-workers confirmed this suggestion and showed that the toxin was without effect on the activity of the enzyme acetylcholinesterase. Thus, the action of the toxin is highly specific, and, because of its tight binding to receptor protein, it could be used to identify receptor macro-molecules in solubilized membranes from the electric organ of the eel, *Electrophorus electricus*.

Melidi and his co-workers performed similar studies and were able to isolate and purify what seems to be the receptor-site protein. They prepared radioactively labeled bungarotoxin and allowed it to react with disrupted electric tissue of *Torpedo marmorata*, the electric ray. The labeled toxin-electric tissue preparation then was treated with dilute solutions of a detergent, which rendered the toxin-receptor site complex soluble. When the complex was purified and characterized, it was found that the toxin was bound to a protein of molecular weight of about 80,000. This receptor protein has one binding site for the toxin or acetylcholine and is readily distinguished from acetylcholinesterase.

Cell division. Membranes also serve a vital role in cell division. Cell division involves the regulation of many cellular activities, including DNA replication, segregation of the replicated DNA, and formation of new cell membranes. The role of the bacterial membrane in cell division was examined recently by several investigators.

Y. Hirota and his co-workers at the Pasteur Institute in Paris isolated several mutants of *Escherichia coli* (*E. coli*) that were defective in different aspects of cell division, particularly in DNA synthesis or replication. From study of these mutants it was established that during cell division DNA becomes attached to the bacterial membrane at the point at which the membrane divides and that it is oriented in such a fashion that orderly assortment of the replicated DNA to the daughter cells occurs.

M. Inouye and A. B. Pardee at Princeton University also studied cell division in *E. coli* and were able to show that two changes occurred in cell membranes during division. When DNA synthesis was inhibited, less protein appeared in the membrane corresponding to a species of 34,000 molecular weight. In contrast, when cell division was inhibited, extra protein components that change in the membranes were found not to be precursors or products of a specific reaction accompanying cell division. The exact function of these components remained unclear, but it would appear that they serve a vital role in cell division.

Trends in biochemistry. It is clear from these examples of recent biochemical research that new and important insights into cellular structure and function at the molecular level should be forthcoming. In addition, research on membranes is vital to understanding human disease. For example, many malignant diseases appear to result in large part because of defective membrane functions. It is evident that without continuing efforts in basic research, the hope for managing cancer successfully will be diminished. (*See* Year in Review: MEDICINE, *Malignant Disease*.)

It also seems clear that the increasing efforts of biochemists in the membrane field reflected a major change in the direction of biochemistry. Now that most of the major molecular constituents of cells were known in some detail, the field of biochemistry seemed to be entering a new phase in which greater emphasis would be placed on studies designed to define the interactions of molecular constituents in cellular ultrastructures of specialized function.

—Robert L. Hill

Biophysics

The scientific resources embodied in physics, chemistry, and biology continued to be directed toward a clearer understanding of molecular processes and composition. The methodology used included electron microscopy, nuclear magnetic resonance, electron spin resonance, radiotracers, and an increasing use of lasers. To a large degree the past year was an interval of summarization, concerned with where molecular biology was in more than usual detail.

With this overview, it was not surprising that considerable speculation on the future trends in molecular studies and their implications should be manifest, culminating in the tantalizing possibility that man may soon be able to effect molecular repair and construction in living systems. The implications for the repair of defects and actual manipulation were most significant in the case of genetic material. Biophysical techniques, however, were being used with equal effectiveness to unravel the past and probe the future.

Applications in plant studies. Biophysical techniques were put to good use in recent months to establish new information about plant systems. One system of particular interest was transpiration, the emission of watery vapor from the plant parts exposed to the air, especially the stomata, the pore-like openings on leaves. The studies undertaken established at least two new methods for determining a plant's transpiration rate.

Ozone uptake and transpiration. The gas ozone is usually thought of as a major air pollutant causing damage to plants. This faintly blue-colored, irritating, and pungent form of oxygen causes extensive damage to leaf crops, including beans and tobacco, during weather conditions that permit its concentration in the atmosphere. At the Connecticut Agricultural Experiment Station in New Haven, researchers studying the damage to bean leaves caused by ozone found that there is a close relationship between the factors in both the plant system and the environment that regulate a plant's ozone uptake and its transpiration.

Bean plants were subjected to light intensity and ozone concentration studies in controlled-growth chambers. After ozone was pumped into the chambers, its rates of depletion from the atmosphere and of uptake by the leaves were watched and found to increase as light intensity was increased toward the level known to be needed to open leaf stomata and enhance transpiration. The transpiration rates of the plants were also measured and used to confirm this data. It was also found that if the humidity in the atmosphere was increased, rates of transpiration and ozone uptake slowed appreciably.

These studies established that both transpiration and ozone uptake are controlled primarily by the stomatal opening and that, therefore, a measure of a plant's ozone uptake can be used to determine its transpiration rate. One application of this work to the protection of leaf crops was the suggestion contained in it that if the humidity around a crop being subjected to an air-pollution incident involving ozone was increased by spraying, the plant's ozone uptake and the damage it causes should be reduced.

Transpiration measured by tritiated water. In another study, scientists at Argonne (Ill.) National Laboratory established a more direct way to measure transpiration rates under a variety of circumstances. Tritiated water (water in which one atom of hydrogen is tritium, or ^3H) was injected into trees near the base and samples of the trunk, leaves, and even whole small trees, were later analyzed for the presence of tritium. A pattern of water movement throughout the plant was established. In other tests tritiated water was sprayed on the soil and injected into it at various depths. This permitted measurement of water balance and water use throughout the soil-tree-atmosphere system.

Studies on trees injected before leaf-sprouting had occurred revealed that soil temperature, rather than air temperature or sunlight duration and intensity, was the principal factor in the movement of water up through the tree. In fact, both transpiration and plant growth were found to be limited by available ground moisture during the dry but warm late summer season in the midwestern U.S., where the tests were conducted.

Botanical applications of SEM. Until recent months it was necessary to section, chemically fix, stain, or even freeze-dry plant material before it could be examined microscopically. These conditions restricted the type of studies that could be performed. For example, entire structures, which might have revealed information about plant growth and development, could not be viewed and most treated specimens were subject to drying or other distortion-causing events.

Recently, however, workers at the University of California at Davis applied scanning electron microscopy (SEM), the most recently developed of the microscopic tools and one that permits three-dimensional images, in a way that permitted the examination of fresh plant material in its natural state. Specimens of such young leaf parts as primordia (earliest growth), floral appendages, and even the shoot's pointed end, or apex, were simply attached to a specimen holder and examined in detail for as long as 10–15 minutes before appreciable drying effects were noted. It was expected that this SEM technique would be particularly valuable in the examination of the responses of such whole formative tissues as these to various chemicals, particularly growth-regulating hormones.

Biogeochemical applications. The use of biophysical techniques to study geological formations containing remnants of life forms had several interesting results. Various marine sediments, for example, had been found to contain small amounts of the amino acids characteristic of organic proteins. The amino acids were presumed to have been derived from organisms that lived in the top layers of the sediment and from the settling of the material of rock disintegration in the water.

Scientists at Woods Hole (Mass.) Oceanographic Institution reported that they can establish the age of marine sediments by testing the amino acids, and specifically isoleucine, for changes in optical activity, that is, in their ability to cause plane-polarized light to rotate. The effects of temperature on amino acids cause them to undergo a slow change from an optically active to an optically inactive state. Hence, in areas where a temperature history can be established sufficiently, the rate of change in optical activity can be a useful measure of sediment age up to approximately 20 million years ago. Optical inactivity is known to be essentially complete in amino acids of Miocene Age (26 million years ago). This method was one of those used to determine the rate of spreading of the ocean's floor from the mid-Atlantic ridge (*see* Year in Review: EARTH SCIENCES, *Geochemistry and Geology*).

In a somewhat similar approach, chloroform extracts from coal seams from the Middle Eocene (55 million years ago) obtained from the Geisel Valley, East Germany, were separated chromatographically at Indiana University. One of the pigments exhibited readings that indicated it was methyl pheophorbide a, a derivative of chlorophyll, the coloring matter in plants. This finding was the oldest documentation of fossil chlorophyll derivatives reported to date.

Cellular studies. The mammalian cell may be thought of simply as a solution of enzymes surrounded by a membrane. While the structure of the membrane itself was the subject of many recent scientific efforts (see *Biochemistry,* above), the passage of substances across the membrane was the subject of others. Any extracellular substance that is to be metabolized by enzymes within the cell must first get through the cell membrane. For most naturally occurring compounds this penetration is accomplished by proteins called membrane carriers. Without the presence of these membrane-penetrating proteins many extracellular substances would not be accessible to enzymes and would be functionally useless.

One example of a specific transport mechanism that was studied thoroughly was the passage of glucose into intestinal cells where its conversion into an organic phosphate (phosphorylation) by the enzyme hexokinase occurs. By examining the capacity of isolated intestinal sacs to accumulate sugars against a concentration gradient, it was established that the glucose-protein complex, that is, the transport system, must have very specific structural features for effective transport to occur. Examination of the hexokinase showed that in several respects its structure was the inverse of that of the sugar transport system. The fact that the two molecules were essentially complementary was regarded as the key to effective transport across a membrane and successful utilization within a cell.

Laser methodology had progressed to the state where individual molecular bonds may be severed. Some molecular biologists envisaged using laser beams to slice through DNA molecules at desired points and "burning out" faulty genes. The defective DNA could be replaced by segments of DNA tailored in the laboratory to provide a properly functioning gene and then introduced into the body as artificial, but beneficial, viruses. The concept was sound; viruses are merely segments of DNA or RNA surrounded by protein sheaths which, penetrating the cell nucleus, take over the cellular DNA. In fact, one virus, the Shope papilloma, which contains the DNA that triggers synthesis of the enzyme arginase, was used to treat children having an hereditary inability to produce the enzyme.

Molecular or cellular repair is important, but the same methodology might also be applied to building in certain features and functions at the molecular or cellular level, creating, in a sense, an organism better adapted to survival and well-being. If man, for example, could be given the genes to produce a two-compartment stomach that could digest cellulose, he might "acquire" the ability to survive at a time when the production of

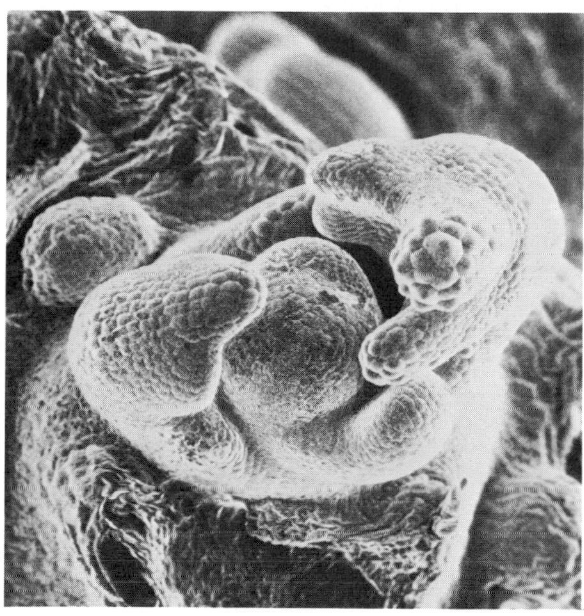

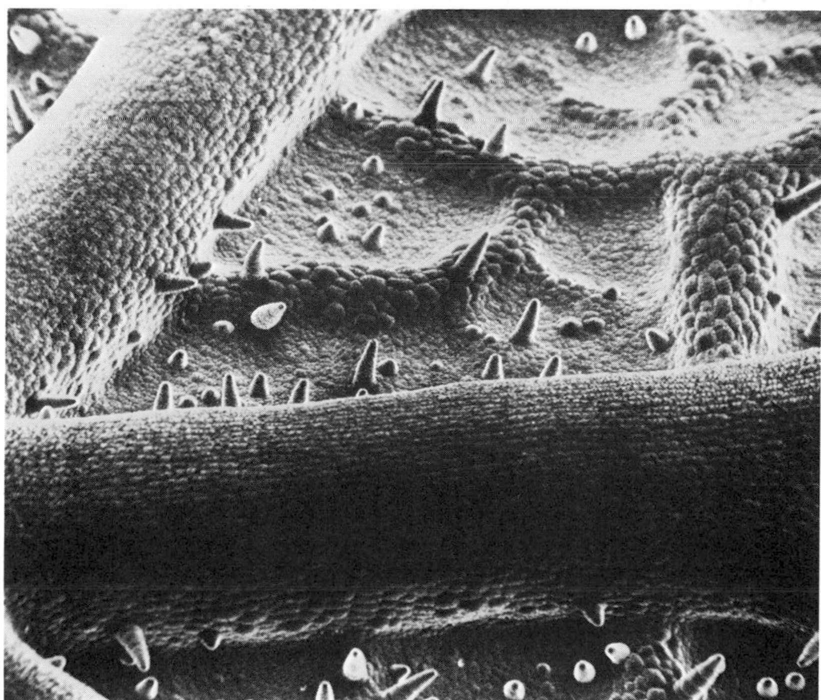

The formative leaf structures of young nasturtium plants can be observed in these scanning electron micrographs in more detail than was previously noted. A pointed tip, or apex, and four leaf forms in the earliest stage of development (above) and the prominent veins on a surface of a young leaf (right) became clearly visible with the aid of new techniques.

Courtesy, Dr. Richard H. Falk and Dr. Ernest M. Gifford, Jr.

food is not keeping pace with increasing population.

Future trends. Molecular studies, along with virtually all other areas of scientific research, had received decreasing financial support in the U.S. and abroad in the past several years. There were positive indications that this trend was ending and that support would increase in the near future. Funds, however, would doubtless be awarded more strictly than in the past on the basis of the scientific merit and relevance of the research contemplated to program goals of the funding agencies. As a result, advances in molecular biology were likely to be occurring only in certain discrete areas, such as those likely to benefit the national effort to eradicate cancer, rather than across the broad spectrum of the molecular sciences.

—Philip F. Gustafson

Genetics

Reverse transcription. According to the central dogma of molecular genetics, DNA is the hereditary blueprint; it can be transcribed into RNA messengers, which then are available for translation into polypeptide sequences. These sequences fold into the proteins with characteristic shapes that determine the working functions of the parts of the cell. An essential tenet of the dogma is that genetic information cannot flow from protein back to RNA or DNA. No plausible mechanism for such a backflow has been found. Were it to be, it would provide a theoretical basis for the long-discredited Lamarckian view of evolution (based on theories proposed in 1815 by French naturalist Jean Baptiste Lamarck), that is, adaptive changes in the hereditary content of an organism result directly from its own life experience.

The transcription processes, and the replication of DNA and of viral RNA, could be understood in terms of three distinctive enzymes: DNA-replicase, transcriptase, and RNA-replicase. In June 1970, Howard M. Temin and Satoshi Mizutani of the University of Wisconsin and David Baltimore of the Massachusetts Institute of Technology independently reported that particles of Rous sarcoma virus contained another enzymatic activity, a "reverse transcriptase" (see *Breakthrough in Molecular Biology*). This process produces DNA sequences corresponding to the information in RNA templates. A continued flood of reports reflected intense interest in "reverse transcription" as a potential key to studies of cancer and of normal developmental processes. (*See* Year in Review: MEDICINE, *Malignant Disease,* RNA Tumor Viruses.)

The choice of the Rous sarcoma virus for these studies was based on earlier evidence that the growth of this RNA virus (in contrast to, say, the influenza virus) depended on the integrity of DNA synthesis in the host cell. Rous virus RNA also shows some homology (similarity) with the DNA of normal cells and, according to some workers, even more with cells previously transformed by the Rous virus. Temin then had hypothesized that the life cycle of the Rous sarcoma RNA virus included the elaboration of a DNA copy that might be incorporated in the host cell chromosomes, in a fashion similar to that established for some DNA viruses. His speculation that RNA could be a template for DNA synthesis motivated his search for the reverse-transcriptase enzyme activity in the virus particles.

Many questions, however, still remained. The postulated DNA equivalent of the Rous RNA remained to be isolated and identified as a biologically active unit. Furthermore, the purified virus enzyme was found to have a broader range of activity than was at first appreciated: it also will produce DNA copies of certain DNA or DNA-RNA hybrid templates. Under certain conditions, the bacterial DNA polymerase previously studied by Arthur Kornberg of Stanford (Calif.) University also will accept RNA templates. Reverse-transcriptase activity also was reported in extracts of normal, supposedly uninfected human cells, as well as in several other RNA tumor viruses and "slow" viruses. The biological significance of reverse transcription was, therefore, not yet firmly established. It seemed unlikely that it is merely an artifactual byproduct of DNA synthesis; even if this should prove to be so, molecular biologists will have a more powerful tool for transferring information experimentally from RNA back to DNA.

Repeated sequence DNA. When DNA solutions are heated, the strands are separated (or "denatured") and dispersed randomly. Upon slow cooling, homologous strands may find one another, and reassociate to reform the double-stranded structure of native DNA. The rate and efficiency with which any strand reassociates will obviously depend on its finding a homologous partner. This is then a way to measure the concentration of a given DNA species or, conversely, the variability of the strands in a solution with a given total concentration of DNA. Early studies on the reassociation of vertebrate DNA showed that a substantial part of the DNA must be highly redundant, the reassociation occurring much more rapidly than would be expected if each segment of DNA in the chromosomes were unrelated to any other. Some of this redundant DNA is manifest as a "satellite band" in ultracentrifuge experiments, the satellite

Courtesy, Dr. H. M. Golomb, Armed Forces Institute of Pathology

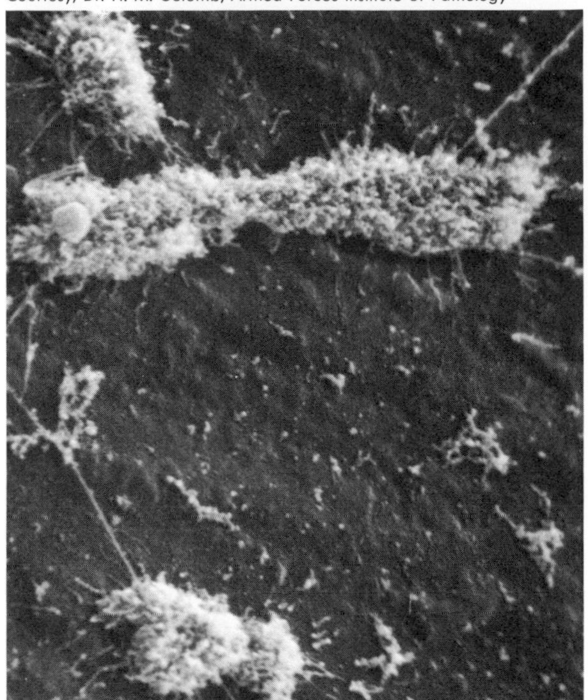

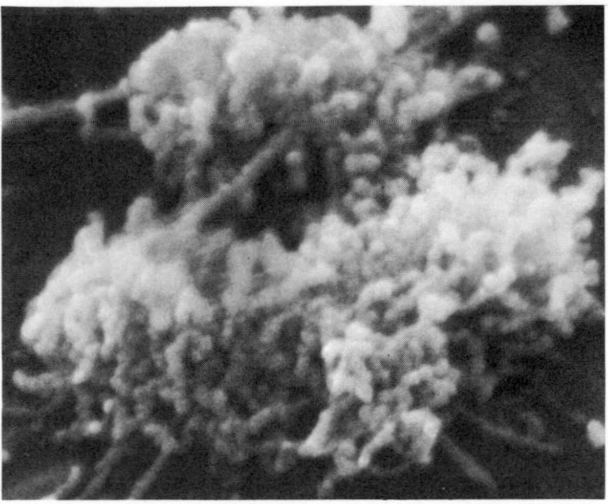

Three human chromosomes separating during cell division (left) and a single chromosome (above) look like skeins of yarn in scanning electron micrographs. Critical-point drying removed water without disturbing the chromosomal structure and permitted micrographs that confirm concepts of chromosome configuration.

having a base composition deviating from that of the average DNA.

The so-called satellite band of the guinea pig accounts for about 5% of the DNA in this species. Kinetic studies of reassociation had suggested that the satellite DNA consisted of repeated segments about 100,000 units long. (Units of DNA consist of two of the chemical side groups, or bases, either adenine (A) and thymine (T) or guanine (G) and cytosine (C), which are linked by hydrogen bonds to form the rungs of the double helical structure.) E. M. Southern, of the University of Edinburgh, reported in 1970 on direct chemical analyses of the satellite that amplify this picture. The basic repeating sequence is only six units long: $\frac{\text{CCCTAA}}{\text{GGGATT}}$ represented in millions of copies per cell. However, about one-fifth of the bases have mutated, superimposing on the sequences a second-order variety that had been interpreted as a much longer repeat length.

The prevalence of this redundant DNA poses many questions about the evolution of the chromosome that are difficult to understand in terms of a string of genes subject to independent mutation and selection. W. Hennig and P. M. B. Walker, also working at Edinburgh, pointed out that closely related species of rodents show little homology among their DNA satellites, although, by definition, these are rather homogeneous within individual samples. Satellite DNA evidently is not transcribed.

Hennig, however, previously had reported that the Y chromosome of *Drosophila* fruit flies is transcribed only in a single tissue, Y-containing male germ cells (spermatozoa), the homologous RNA being absent from other tissues of male or female cells. This gene product evidently is required for the proper maturation of the spermatozoon.

As to the evolutionary origin of the repeated sequence, one suggestion was that some unspecified event, concurrent with the separation of different species approximately 50 million years ago, established many copies of a new basic sequence in a given species. These sequences will have undergone random variation since then to account for a 20% frequency of changes.

An alternative is suggested by recent speculations about reverse transcription—namely that a single sequence is amplified every generation, perhaps during oogenesis (formation of the female germ cell), and many copies (often faulty ones) reincorporated into the chromosomes of the developing zygote (fertilized cell). If compositional differences in satellite DNA can be found between hybridizable varieties, for example, the horse versus donkey, bison versus domestic cattle, or among inbred mouse strains, a genetic test of these ideas will soon be available.

—Joshua Lederberg

See in ENCYCLOPÆDIA BRITANNICA (1971): CELL; GENE, *Genes and Development;* NUCLEIC ACIDS; PROTEINS; VIRUSES.

Obituaries

The following persons, all of whom died between July 1, 1970, and June 30, 1971, were noted for distinguished accomplishments in one or more scientific endeavors. Biographies of those whose names are preceded by an asterisk (*) appear in *Encyclopædia Britannica.*

Beberman, Max (Aug. 20, 1925—Jan. 24, 1971). One of the creators of the "new math," Beberman was educational director of the Computer-Based Education Research Laboratory at the University of Illinois. He graduated from City College of New York in 1944 with a bachelor of arts degree in mathematics. After two years' teaching (1946–48) in Nome, Alaska, he obtained his master's degree in 1949 from Teachers College of Columbia University, and in 1953 received a doctorate in education from Columbia. Beberman joined the University of Illinois faculty in 1950 as a math teacher in the experimental high school. A year later, as director of an experimental teaching curriculum for young children, he became involved in what was to be known as new math, which sought to develop the reasoning powers of children rather than the stressing of rules and formulas.

Boyd-Orr, Lord John (Sept. 23, 1880—June 25, 1971). Scottish scientist and nutritionist, Lord John Boyd-Orr was a founder (1945) and the first director-general (1945–48) of the United Nations Food and Agriculture Organization (FAO). He was awarded the 1949 Nobel Peace Prize for his work with the FAO. Boyd-Orr received a degree in medicine from Glasgow University, and in 1914 became director of the Institute of Animal Nutrition at Aberdeen University; in 1929 he founded the Imperial Bureau of Animal Nutrition at Rowett. In 1945 he was appointed rector of Glasgow University and a member of Parliament for the Scottish universities.

*__Carnap, Rudolf__ (May 18, 1891—Sept. 14, 1970). A U.S. philosopher, Carnap was a member of the so-called Vienna Circle, a group of philosophers, mathematicians, and semanticists of the 1920s and 1930s. He was a founder of logical positivism, a major school of modern philosophy that rejected the concepts of older philosophers and preferred to believe that philosophy should describe, clarify, and criticize science and language. In 1931 Carnap joined the German University in Prague, Czech., but, with the expansion of Naziism, he emigrated to the United States in 1935. He was professor of philosophy at the University of Chicago (1936–52) and at the University of California at Los Angeles (1954–62).

de Kruif, Paul Henry (March 2, 1890—Feb. 28, 1971). A U.S. bacteriologist, de Kruif was best known as a writer on medical and other scientific topics. Author of more than a dozen books, his contributions to the best-seller lists included *Microbe Hunters* in 1926, *Hunger Fighters* in 1928, and *Men Against Death* in 1932. He was co-author (with Sidney Howard) of *Yellow jack!*, a 1934 Broadway play that dramatized Walter Reed's fight against yellow fever. De Kruif's own investigations in bacteriological research were conducted at the Rockefeller Institute in New York (1920–22) and at the Pasteur Institute in Paris. He helped develop an early treatment for syphilis, and worked on an antitoxin for gas gangrene.

Dobrovolsky, Georgi Timofeyevich (June 1, 1928—June 30, 1971). Soviet cosmonaut Lieutenant Colonel Dobrovolsky, commander of the Soyuz 11 spacecraft, died with his two fellow crewmen at the controls of their craft near or during the time of its reentry into the earth's atmosphere. (*See* Year in Review: ASTRONAUTICS, *Manned Space Exploration.*) Dobrovolsky graduated from Odessa's special secondary school for training boys to serve in the Soviet Air Force, then completed a four-year fighter-pilot course at the flying school at Chuguyev. He was with the Soviet space program eight years before making his first flight with the Soyuz 11. He was awarded posthumously his nation's highest title, Hero of the Soviet Union (see *Patsayev, Viktor I.*, and *Volkov, Vladislav Nikolayevich,* below.)

Fairchild, Sherman Mills (1896—March 28, 1971). U.S. industrialist and inventor, Fairchild numbered among his devices the Fairchild aerial camera used in mapmaking and aerial surveying. He attended Harvard University and studied engineering at Columbia University before starting work on his aerial camera for the U.S. War Department in 1918. In 1920 he established his own company, Fairchild Camera and Equipment Co., and, as well as developing and marketing the camera, he also went into aircraft manufacturing, having invented several improvements for planes to better handle the work in mapping and surveying.

Farnsworth, Philo Taylor (Aug. 19, 1906—March 11, 1971). U.S. pioneer in electronics, Farnsworth developed the techniques that made television possible. At the age of 15, Farnsworth worked out the basic features of his system; six years later his first patented television film was filed; and by 1935 he gave a 10-day public demonstration of his new entertainment device, on which he held more than 165 patents.

Karrer, Paul (April 21, 1889—June 18, 1971). A Swiss chemist and pioneer in vitamin research, Karrer was awarded the 1937 Nobel Prize for Chemistry (with W. N. Haworth) for his studies concerning the constitution of the carotinoids, flavins, and vitamins A and B_2. Karrer received his doctorate from the University of Zürich in 1911, was

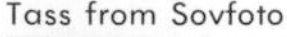

Tass from Sovfoto

Cosmonauts Georgi Dobrovolsky, Viktor Patsayev, and Vladislav Volkov

professor of chemistry there from 1918 until 1959, and also served as director of the chemical Institute during the same period.

Kronberger, Hans (July 28, 1920—Sept. 29, 1970). An Austrian-born physicist, Kronberger was member for reactor development at the United Kingdom Atomic Energy Authority. With the spread of Naziism in his native land, Kronberger went to England at the age of 18. He studied mechanical engineering at King's College, University of Durham, Newcastle upon Tyne. He was interned during World War II and his associates at that time persuaded him to switch his interest to physics. Released in 1942, he completed his degree and then obtained a doctorate at Birmingham University. He went to Harwell in 1946 and worked in the industrial group of peaceful applications of nuclear energy. In 1956 he became chief physicist at Risley and in 1958 director of research and development there.

Patsayev, Viktor I. (June 19, 1933—June 30, 1971). Soviet cosmonaut Patsayev, civilian engineer aboard the Soyuz 11 spacecraft, died with his two companions at the controls of their craft as it neared a soft landing on the earth. (*See* Year in Review: ASTRONAUTICS, *Manned Space Exploration.*) Patsayev graduated from the Industrial Institute in Penza and worked as a design engineer in precision instruments before joining the Soviet space program in 1968. He received posthumously the U.S.S.R.'s highest title, Hero of the Soviet Union (see *Dobrovolsky, Georgi Timofeyevich,* above; and *Volkov, Vladislav Nikolayevich,* below).

***Raman, Sir Chandrasekhara Venkata** (Nov. 7, 1888—Nov. 21, 1970). An Indian physicist, Raman was awarded the Nobel Prize for Physics in 1930. He graduated from the Presidency College in Madras in 1904, at the age of 16, and three years later earned his master's degree. Raman became professor of physics at the University of Calcutta in 1917, then in 1933 went to the Indian Institute of Science at Bangalore. In 1947 he became director of the Raman Research Institute, also in Bangalore. He was awarded the Nobel Prize for his discovery that light scattered by any medium is emitted in frequencies equal to the infrared frequencies of that medium, a phenomenon which came to be known as the Raman effect. The use of this effect in determining fine molecular structure was considered a forerunner in the making of laser spectrometers.

Stanley, Wendell Meredith (Aug. 16, 1904—June 15, 1971). A U.S. virologist and Nobel Prize winner, Stanley, whose pioneering work into the nature of viruses laid the foundation for many later medical advances, was professor of molecular biology and biochemistry at the University of California at Berkeley. In 1946 Stanley, with John H. Northrop, also a virologist, shared one-half of the Nobel Prize for Chemistry for their preparation of enzymes and virus proteins in a pure form (the other half went to James B. Sumner, researcher in enzymes and viruses). Stanley's work was on the tobacco mosaic virus, which was destroying vast tobacco crops. He found that the virus, although acting like an inanimate chemical, presented evidence that it was a living and growing organism.

***Svedberg, Theodor** (Aug. 30, 1884—Feb. 27,

1971). Swedish nuclear scientist Svedberg was awarded the Nobel Prize for Chemistry in 1926. After graduation from Uppsala University in 1907, he developed a method for studying the movements of very small particles. He also introduced a number of techniques for solving difficult experimental problems. One important achievement was the development of the ultracentrifuge, by which he could measure accurately the size of very large molecules such as the proteins found in living cells.

*__Tamm, Igor Evgenyevich__ (July 8, 1895—April 12, 1971). A Soviet physicist, Tamm was the recipient of a Nobel Prize for Physics in 1958. He was able to combine Einstein's theory of relativity with the theory of quantum mechanics at a time (Stalinist period of the 1930s and '40s) when both were discredited officially in the U.S.S.R. as anti-Marxist. Tamm, one of three scientists who shared the 1958 Nobel Prize for Physics, was honored for his work in explaining the Cerenkov effect in nuclear physics (the emission of light waves by electrons or other electrically charged atomic particles moving in a medium at any speed greater than the velocity of light in that medium). Tamm held the title of Hero of Socialist Labor (1950) and was the winner of two Stalin Prizes and three Orders of Lenin.

Van Slyke, Donald D. (1883—May 4, 1971). A U.S. biological chemist, emeritus professor at Rockefeller University, New York, Van Slyke was internationally known for his work in the fields of physiology and medicine, as well as in chemistry. His classic studies on acidosis revolutionized the treatment of diabetes; his measuring techniques were widely used in determining blood carbon dioxide and carbon-14; and his discovery that hydroxylysine was an amino acid found in collagen was helpful in studying such diseases as arthritis.

Volkov, Vladislav Nikolayevich (1936?—June 30, 1971). Soviet cosmonaut Volkov, civilian engineer of the Soyuz 11 spacecraft, died with his two companions at the controls of their craft as it was about to land on the earth. (*See* Year in Review: ASTRONAUTICS, *Manned Space Exploration.*) Volkov graduated from Moscow Aviation Institute in 1959, joined the Soviet space program in 1966, and made his first flight with the three-man crew of the Soyuz 7 in October 1969. He received the title Hero of the Soviet Union following that successful mission, and was awarded posthumously the nation's second highest medal, the Gold Star (see *Dobrovolsky, Georgi Timofeyevich;* and *Patsayev, Viktor I.*, above).

Warburg, Otto Heinrich (Oct. 8, 1883—Aug. 1, 1970). A German physiologist and biochemist, and a leading cancer researcher, Warburg was the recipient of the Nobel Prize for Physiology or Medicine in 1931. After gaining doctorates in chemistry at the University of Berlin (1906) and in medicine at the University of Heidelberg (1911), he became a prominent figure in the institutes of Berlin-Dahlem. He first was recognized for his work on the metabolism of various types of ova at the marine biological station in Naples, Italy. His Nobel Prize was awarded for his research into respiratory enzymes. In 1944 he was nominated for a second Nobel Prize, but was prevented from accepting it by a ban imposed by the Hitler regime. In 1934 he became a foreign member of the Royal Society and in 1962 was awarded the highest West German civilian decoration, the Pour Le Mérite order. He headed the Max Planck Institute for Cell Physiology in Berlin.

Photography

In recognition of the increasing value of photography in entertainment, science and technology, education, the graphic arts, business, and communications, the major photographic companies of the world were investing increasing amounts of money in research to provide the basic knowledge for further advance in these fields, and to discover new ones. It was probable that the amount spent on research alone during the past year was between $50 and $100 million.

Practical photography. The great photographic event of 1970 was Photokina, the photographic fair in Cologne, W.Ger., which featured the products of more than 700 firms from around the world. The fair revealed no significant trends or startling innovations, at least in the public showings. Noteworthy, however, were the vast number and variety of new lenses. Their proliferation was the result of the development of lens computation by computer, which allowed for a great savings of time in creating new lenses. Most prominent were the extra-wide-angle lenses of increased aperture for single-lens reflex cameras (especially a Nikkor 220°, 6.3-mm focal length, f/2.8 fisheye), the new extra-long (up to 2,000-mm) focal length telephoto and mirror lenses, and zoom lenses of improved optical quality, with the Japanese manufacturers' offerings surpassing those of the West Germans in variety, if not in quality.

Automation was again the main trend in the 35-mm camera field, with nothing startling in the way of new designs. Electronic exposure automation appeared to be on the way for 35-mm single-lens reflex cameras with focal plane shutters.

Color. Rumors of a new concept for producing Polaroid color pictures continued to sweep the market. A recent patent issued to the Polaroid Corp. related to the combination of the Polaroid image transfer system and the additive screen-film process of making color transparencies. At the Photokina fair Agfa-Gevaert Inc. introduced the new Agfachrome CU 410 dye-silver-bleach color print process (positive-to-positive) and announced that it would be in use in the United States by 1972 for a direct color printing service. The Eastman Kodak Co. introduced a new Ektachrome Color Duplicating Film for duplicating sheet and aerial color positive transparencies.

Pre-press color proofing for the printing industry continued to assume great significance. Its growing popularity was due to the fact that proofing from the separation color negatives or positives rather than from printing plates made from them provided a cheaper, faster approach. There were more than 20 such methods in 1970, but it was expected that not more than five would eventually survive. The latest was a system using pigmented photopolymers that changed their melting point on exposure. The softer images could then be transferred in succession and in register to paper.

In-camera processing. Polaroid expanded its line of cameras during the year to include low-priced color cameras with transistorized electronic shutters, photocell exposure control, and coupled rangefinders. Although an agreement between Kodak and Polaroid ensured that Kodak would continue to supply color negative film to Polaroid for at least five years, Kodak expressed its interest in entering that market on its own. The Kodak company stated that it was "increasingly predictable that Kodak black-and-white and color films would be available for in-camera processing prior to 1975."

A normal human sweat pore is shown greatly magnified in a micrograph obtained using the technique known as scanning electron microscopy (SEM). SEM's advantage over other microscopic techniques is the ability to present a three-dimensional image more realistically.

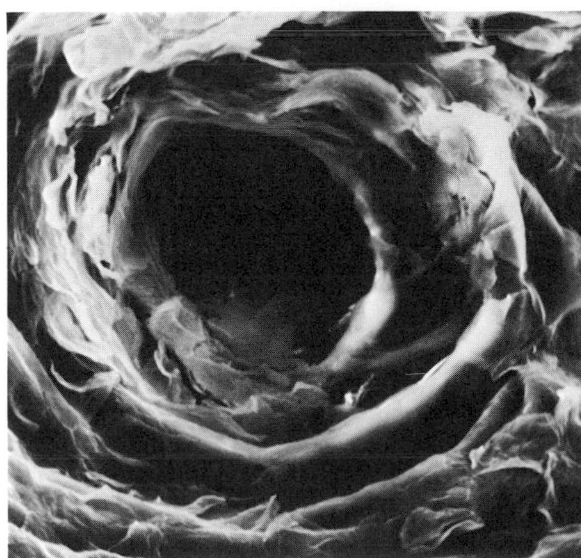

Courtesy, Dr. C. M. Papa and B. Farber, Johnson & Johnson

Light sources. An improved flash cube, the "Magicube," was introduced this past year. Instead of using batteries, it works mechanically, depending on the operation of the camera shutter to release a torsion spring that hits the base of the flash lamp and activates an impact-sensitive material that causes the lamp to flash. A series of five new Kodak Instamatic X cameras was specially designed to use the Magicube.

Polaroid introduced its "400 Series Color Pack" cameras using "Focused Flash," a flash-cube system coupled to the focusing system in such a manner that louvers move in front of the flash cube in relation to the distance of the subject focused on, thereby adjusting the amount of light thrown on the subject. A new General Electric "Hi-Power Cube" with increased light output was incorporated in the "Focused Flash" system. The Hi-Power Cube emits 2.4 times as much light as regular cubes, relying on hafnium instead of zirconium foil for the flash material.

More than 20 auto-flash units (often called "computerized" electronic flash units) have become available. These units automatically adjust their light output depending on the distance of the flash (not necessarily the camera) from the subject. General Electric also announced the development of a "molecular arc lamp" that depends on an electric arc passing through a vapor of tin chloride to produce light. The molecular arc lamp gives off a uniform spectrum approaching that of noonday sun and has a far higher efficiency than normal incandescent lamps. When generally available, this lamp should be of great importance for studio lighting for color photography.

Special applications. Photography continued to play a major role in a variety of scientific and technical fields. In medicine, for example, a new family of rapid-process films was recently introduced for a wide range of diagnostic uses. These include high-speed films that reduce the X-ray dosage to the patient and that permit radiography where there is movement (as in the case of the heart and with children), special film for close examination of small areas, duplicating film to give copies for consultation and teaching, and a new dental X-ray system using film in a special jacket that can be inserted directly into a compact processor and be processed in four minutes while still in the jacket.

Photography from the air or from above the atmosphere continued to play an increasingly significant role not only in military reconnaissance and map-making, but also in weather forecasting and in evaluation of the world's natural resources.

Courtesy, U.S. Geological Survey

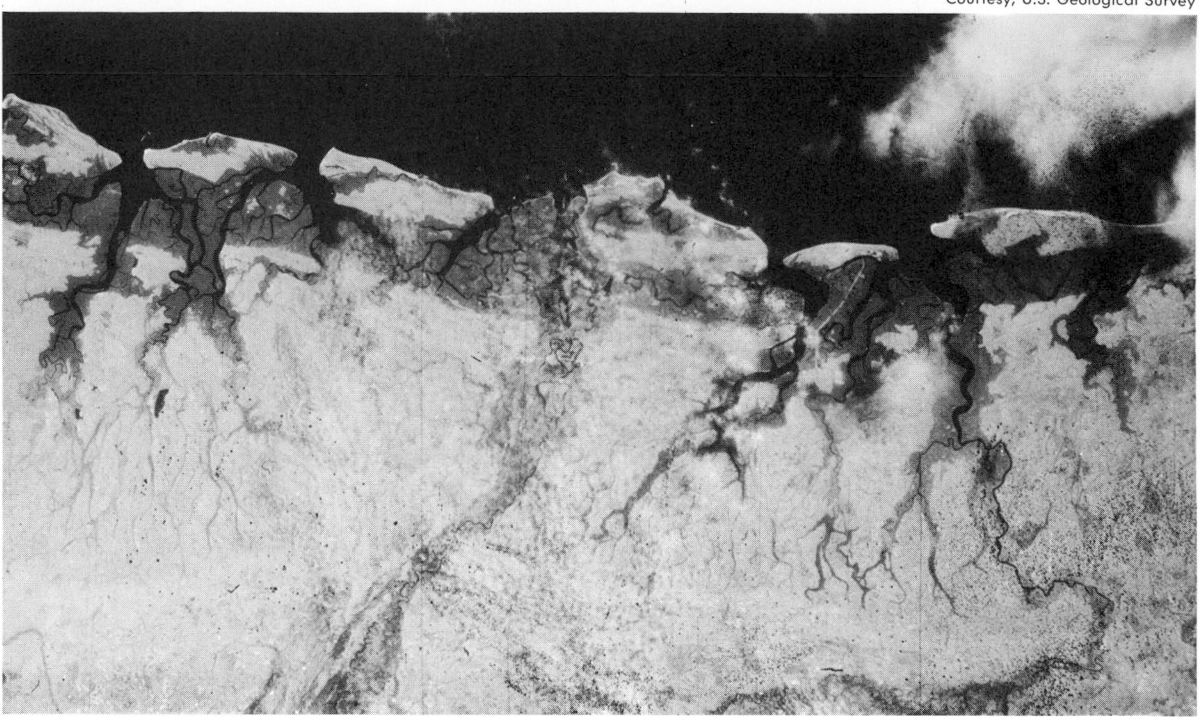

An infrared photograph taken during the Apollo 9 space flight reveals the geographical features of a coastal region, with the areas covered by water appearing dark in contrast to the lighter land areas. Such photographs were proving particularly valuable in the evaluation of the earth's resources.

"Multiband" photography was particularly useful in that it allowed the earth to be photographed simultaneously in several separate wavelength regions, in color, and in infrared. Agricultural, forest, geological, and oceanographic information could thus be obtained in far less time and at far less expense than by normal survey methods. International cooperation in aerial photography of the disastrous earthquake that took place in Andean and coastal areas of Peru in May 1970, for example, helped in short time to assess the damage, plan reconstruction, and even determine property lines. It was estimated that the Earth Resources Technology Satellite program would use up to 100 million color prints annually to take inventory of the earth's resources.

Photography remains one of the major tools of scientific study in many other areas. A special stereo camera has photographed the surface of the moon, and spectrographs have been carried outside the earth's atmosphere to record radiations on film without interference from the air. High-speed photography can record events in nanoseconds. In 1970, for example, scientists at the Bell Telephone Laboratories succeeded in taking the first stop-motion photograph of the world's fastest moving object—a pulse of laser light. Scanning electron microscopy, neutron radiography, holography, and a myriad of new scientific techniques all rely on photography.

Current trends in photography. The photographic industry based on conventional silver systems and the current basic design of cameras was expected to continue to expand for many years. Although significant progress was made in continuous-tone electrophotography, it did not appear at the moment to present a serious challenge to normal silver photography in its major applications.

There appeared no doubt that, in the home and in the educational fields, video players using prerecorded cartridges and cassettes that could be "played" through conventional television sets would have an increasingly strong market in the near future. Eight-mm film would probably be used in a fair percentage of the players, but it remained to be seen how the field would be divided between photographic and nonphotographic systems once standardization and market stability were attained.

The uses of microfilming were expected to increase in variety and to develop greatly enhanced markets. Especially strong growth for microfilm was forecast in business applications, although three other main areas of its use—security, engineering, and publishing—were expected to grow at an average annual rate of 15%. In the field of electrophotography, the main commercial application of which was document copying, new developments toward the achievement of color reproduction were expected, as were applications of the xerographic process to radiography.

—Walter Clark

Physics

New developments in particle accelerators, investigations that might lead to the production of new heavy elements, and a sensitive technique for measuring the roughness of a metal surface were among the highlights of the past year in physics. Casting a shadow over the field in the United States were budgetary restrictions that made it difficult for young physicists to find jobs and also caused the cancellation or postponement of some experiments.

High-energy physics

The past year was a relatively slow one for high-energy physics. Many theorists were concerned with inventing new models to analyze and systematize the massive amount of experimental data on properties of the hadrons (strongly interacting particles, such as protons and neutrons) that had been obtained at various laboratories. The experimentalists, while continuing the routine collection of such data, were busy preparing new experiments for two new high-energy machines, the European Organization for Nuclear Research (CERN) intersecting storage rings (ISR), which began operation in January 1971, in Geneva, and the U.S. National Accelerator Laboratory near Batavia, Ill., scheduled to begin operating in late 1971. Physicists hoped that the great increase in energy range available at these accelerators would enable them to settle several of the outstanding problems in the field, such as the existence or nonexistence of "quarks," a set of hypothetical particles that are supposed to be the constituents of all the known hadrons.

In the United States budgetary restrictions forced the premature shutdown of a high-energy accelerator at Princeton, N.J. The shortage of funds was also a major factor in causing an extremely tight job market for young high-energy physicists, forcing many of them to leave the field entirely and others to remain unemployed. The financial prospects for high-energy physics seemed somewhat brighter in Western Europe, where the CERN council of European nations agreed after several years of delay to begin construction of a 300-GeV (billion electron volt) proton accelerator near the present CERN accelerators. Construction of this machine was expected to take about eight years, and the cost was forecast at about $300 million. It was designed to be comparable in performance with the Batavia accelerator.

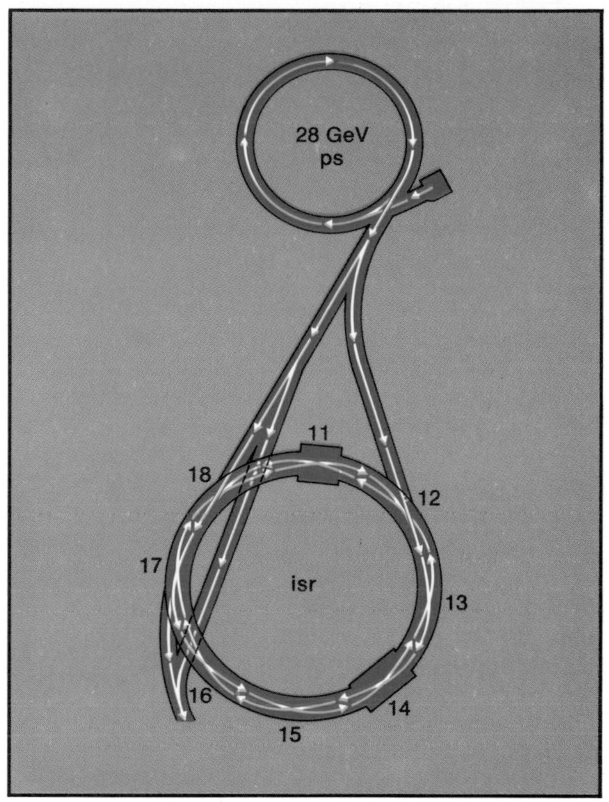

At CERN beams of protons move from the proton synchrotron (ps) to the intersecting storage rings (isr). At four of the eight intersections (12, 14, 16, 18) the center of mass of the proton collisions moves inward, and in the others, outward.

Colliding-beams experiments. One recent development in high-energy accelerators, the "colliding beams" machine, reached an important milestone in 1971 with the first operation of the CERN ISR. The basic difference between ordinary accelerator experiments and colliding-beams experiments is that in the former the rapidly moving particles produced by an accelerator are directed into a fixed target, where they interact with particles that are either at rest or traveling very slowly. On the other hand, in a colliding-beams experiment, two beams of particles, both of very high energy and traveling in nearly opposite directions, are allowed to cross each other's path and interact while crossing. This arrangement cannot be obtained by building two adjacent accelerators, because the beam of an accelerator has too few particles. Instead, all of the particles involved are accelerated by a single machine. The output of the machine is magnetically deflected into two rings of magnets, which may be either adjacent or overlapping, as they are at CERN. The beams circle the rings in opposite directions, so that where the rings intersect, the particles meet almost head on. The key feature is that the beams may be stored for periods of several hours in the rings, allowing hundreds of pulses from the accelerator, each containing a trillion protons, to be stored in each ring. These protons each have the energy imparted to them by the accelerator, which at CERN is 28 GeV.

The major advantage of colliding-beams experiments as compared to those involving fixed targets is that, for a given accelerator energy, the former allow a much higher range of energies and, correspondingly, a much smaller scale of distances to be probed. This somewhat surprising fact results from a consequence of the special theory of relativity, which is that the behavior of physical processes, such as the collision of particles, can depend only on quantities that have the same value for observers in motion relative to one another. It follows from this principle that a head-on collision of two particles, each of energy E and mass m, probes the same region as a collision of a particle at rest with a particle of energy E_L, where E_L and E are related by $4E^2 = 2mE_L + 2m^2$. To illustrate, a collision of two electrons, each of energy 500 MeV (million electron volts), measures the same things as would a collision between a fixed electron and an electron with an energy of 1,000,000 MeV, an energy far beyond that of any existing accelerator.

Spurred by this fact, in the past few years several storage rings for electrons have been operated with electrons of energy in the range of 500 MeV. Experiments with colliding electron beams have given detailed information about the production of such particles as pi-mesons. One such experiment, reported by a team of Italian physicists, indicates that in head-on electron-positron collisions with about 2,000 MeV total energy the production of more than two pi-mesons at a time is the most prevalent process. No simple explanation for this unexpected observation is yet available.

For a head-on collision of two protons, each of energy 28 GeV, the equivalent laboratory energy would be about 1,700 GeV, or several times more than the highest energy attainable by any present proton accelerator. Thus, the CERN ISR will probe an unknown region in proton-proton collisions.

Disadvantages of colliding-beams experiments. Colliding-beams experiments do have several serious limitations in comparison with those using a fixed target. One is that the expected number of individual collisions per second is much smaller. This is a consequence of the fact that the density of particles in the storage ring is much less, by a factor of many trillions, than the density of particles in a target. A proton beam that hits a solid target produces about one collision for each particle in the beam, or about one trillion collisions per second. In a colliding-beams experiment, on the other hand, the particles in one beam rarely hit a particle in the other beam; hence, the number of collisions is less than one million per second. This reduced number of collisions makes it much more difficult to measure rare events than in fixed-target experiments.

A hydrogen bubble chamber at CERN, seen partially dismantled in this view, is used to observe the collisions of high-speed charged particles. The beam of particles enters the chamber from the left. On the upper platform is the control system.

Another disadvantage of colliding-beams experiments is that only a single primary process, which at CERN will be proton-proton collisions, can be studied. This is again in contrast to fixed-target experiments, where, in addition to the primary beam produced by the accelerator, secondary beams of other particles that are produced when the primary beam hits a target can be used. Pi-mesons are among those particles. Such secondary beams are not available for colliding-beams experiments, as the particles are not produced by the experiments in sufficient numbers and are too short-lived to store. The one exception to this is

Courtesy, CERN

Star-shaped counters (center) surround both storage rings at one of their points of intersection in the CERN laboratory. These counters recorded the first head-on collisions between the intersecting beams of protons in January 1971.

that it might someday be possible to store antiprotons in the same way as protons, and, hence, perform proton-antiproton colliding-beams experiments.

Particles from proton-proton collisions. One of the most important areas being explored with the CERN ISR is the production of new particles in proton-proton collisions. One search is for the so-called intermediate boson (W particle), a hypothetical particle that is supposed to mediate weak interactions, such as neutrino-neutron collisions, in the same way that pi-mesons mediate strong interactions, such as electron-neutron collisions. According to this theory, the scattering of a neutrino by a neutron occurs in two steps. First, the neutrino emits a W particle and changes into a muon. Then the W particle is absorbed by the neutron, which is thereby converted into a proton. In such a process, the W particle is said to occur *virtually*, because it exists for such a fleeting time interval as to be unobservable by any direct measurement.

Although most physicists find this hypothesis of a W particle to be the most satisfactory way of describing the weak interactions, the crucial test would be the actual production and detection of the particle. In order to perform this test, it is necessary to have sufficient energy to create at least one W particle. But because the rest energy of the W particle, which determines the minimum energy needed to create one, is unknown, it has not been possible to tell whether the previously unsuccessful searches for it have failed because of a lack of sufficient energy or because there is nothing to find. It is known that previous experiments would not have been able to produce W particles if their rest energy is greater than about 5 GeV.

At the ISR it should be possible to produce W particles with rest energy as high as 27 GeV, if they are produced as a pair with opposite electric charge. If a single W particle can be produced at a time, then the ISR would be capable of making one with a rest energy as high as 54 GeV. This last possibility may be important because one hypothesis puts the mass of the W particle at about 39 GeV. The W particle is unstable and does not live long enough to leave a track in the same way as such longer-lived particles as the pi-meson. Instead, it has to be detected through its decay products. Two characteristic decay modes of the W particle useful for this purpose would be into a high-energy electron and neutrino, or into a muon and neutrino. Consequently, one experiment being performed by a group from Columbia University has sought high-energy electrons emitted perpendicular to the direction of the colliding proton beams. If such electrons were detected in large quantities, it would be strong evidence in favor of the existence of the W particle. Physicists working on weak interactions have awaited eagerly the outcome of this and other W particle searches, which should be known within the next year or so.

Inelastic electron collisions. One of the interesting discoveries in high-energy physics in 1969 was the behavior of those collisions of high-energy electrons with protons in which additional hadrons are produced. These are called inelastic collisions. It was found then that in such collisions the proton behaves very much as if it were made up of small, pointlike, charged constituents, resembling somewhat the structure of the proton as suggested by the hypothetical quark model. During the past year these results were extended to the inelastic collisions of electrons with neutrons, and it was found that the neutrons also behave in these collisions as if they were made up of pointlike charges. The main difference between a neutron and a proton is that in the neutron the charges add up to zero, while in the proton they add up to one.

The results on neutrons, like those on protons, were obtained at the Stanford (University) Linear Accelerator Center. The neutron experiments are less precise than those with protons, mainly because the neutron is an unstable particle; free neutron targets are, therefore, unavailable. Instead, the electrons collide with deuterons, nuclei of heavy hydrogen consisting of a loosely bound neutron and proton, and the result is compared with the electron-proton collisions. The difference between the deuteron and the proton collisions then measures the scattering by neutrons.

The successful extension of the pointlike quark model to inelastic electron-neutron collisions does not necessarily prove the truth of this model, and other models that fit the data are available. A new set of experiments to test various hypotheses will involve collisions of high-energy neutrinos with neutrons and protons. Such collisions involve weak interactions rather than the electromagnetic ones involved in collisions of high-energy electrons with neutrons and protons. Physicists generally believe that, although there are many similarities between these two classes of interactions, enough differences exist so that the neutrino collisions will give important new information which may allow for a decision among the various hypotheses.

Experiments on neutrino-proton and neutrino-neutron collisions should be carried out in 1972 when the proton accelerator near Batavia begins operation. Since neutrinos are neutral, they cannot be accelerated; instead, high-energy neutrinos are obtained from decays of charged particles. To obtain high-energy neutrinos in sufficient numbers to study collisions in detail, it is necessary for large numbers of very-high-energy charged particles to be produced and decay, which the Batavia accelerator will, for the first time, be able to accomplish. Hence, the study of high-energy neutrino reactions was established as one of the first priorities at Batavia.

—Gerald Feinberg

Nuclear physics

The past year was one of many surprises and new discoveries in nuclear physics as well as one of consolidation and much more fundamental understanding of old phenomena. In a field as old as radioactivity, a totally new type of reaction—one involving spontaneous emission of protons rather than the more familiar alpha or beta particles—was discovered. As more and more energy was added to a nucleus, totally new and unexpected nuclear configurations, which bear marked resemblance to the molecules of atomic physics, were also discovered.

In other ground-breaking work, uncertainties concerning elements 104 and 105 were resolved, and preliminary evidence for element 107 was announced; even more important, the very-high-energy proton bombardment of tungsten provided evidence for the possible production of element 112, with a lifetime measured in months. New insight was also gained into the microscopic motions of the nucleons (neutrons and protons) in both light and heavy nuclei, and significant new technological developments were reported in the areas of nuclear detection and instrumentation as well as particle acceleration.

Although many of these activities took place in the United States, the U.S. nuclear science community, which had long enjoyed international preeminence, was in grave danger of losing its position of leadership as a result of dwindling financial support and of rapidly increasing activity abroad.

The anti-omega-minus baryon, a rare and elusive nuclear particle, was first detected in 1971 as a short line (arrow) in this bubble chamber photograph. The baryon decays into a neutrally charged antilambda particle and a positive K meson.

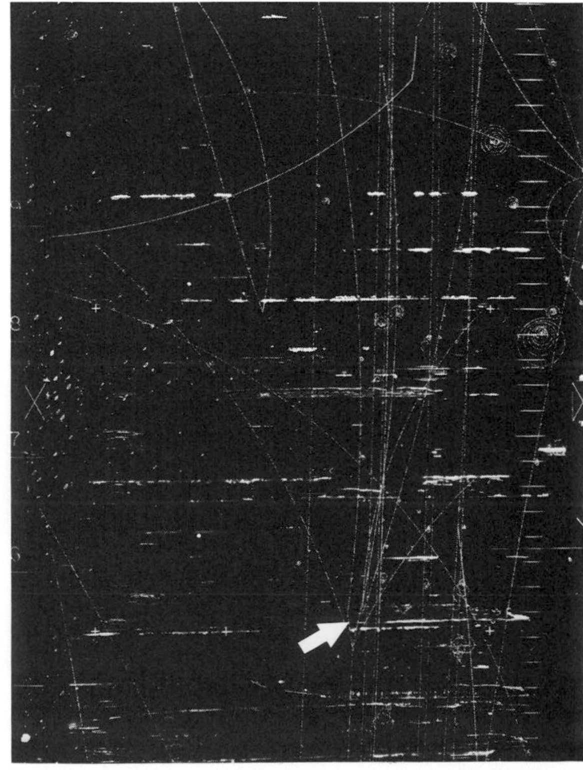

Courtesy, Lawrence Berkeley Laboratory, University of California-Berkeley

Courtesy, Oak Ridge National Laboratory

Highly radioactive reactor fuel is buried in underground salt deposits during Project Salt Vault, an experiment sponsored by Oak Ridge National Laboratory. The success of the project demonstrated the feasibility of initiating a proposed radioactive waste repository project near Lyons, Kan.

Proton radioactivity. Since the earliest days of nuclear physics, when attention was focused on the alpha particles (helium nuclei) that were emitted spontaneously from radioactive nuclei, nuclear physicists have searched for evidence for the direct emission of a single proton (hydrogen nucleus) as well. They found that if a nuclear state emitted a proton at all, it did so instantaneously instead of as a radioactive process with a characteristically much longer lifetime. In the late 1960s, and almost simultaneously, Georgi N. Flerov's group at Dubna in the Soviet Union and Robert Bell's group at McGill University discovered protons with longer lifetimes emerging from nuclear systems; however, although interesting in its own right, this phenomenon was soon shown not to be direct proton radioactivity. Rather, an initial nuclear state was undergoing beta decay (electron or positron emission) with the observed lifetime, and the resultant nuclear state instantaneously emitted a proton.

During the past year, however, Joseph Cerny and his collaborators, first at Oxford University and subsequently at the University of California at Berkeley, finally found unambiguous evidence for direct proton radioactivity. In their experiment an excited state of the cobalt isotope of mass 53 emitted protons with a half life of 243 msec, a long lifetime by nuclear standards. Reflecting the current sophistication of microscopic nuclear theory, L. K. Peker of Moscow University has already demonstrated that this is the only nuclear state with a mass less than 80 that could possibly show this behavior. Further studies of this new type of radioactivity will provide a very delicate probe of our understanding of the structures involved.

Extensions of the nuclear shell model. In recent years it has become clear that almost all nuclear phenomena can be understood within the framework of a general shell model wherein the individual nucleons move in rather well-defined orbits, similar to those in an atom or indeed in a planetary system. Moreover, it appeared possible to work a vast simplification in these models by assuming that the observed nuclear characteristics were largely attributable to the valence nucleons—those outside of closed shells—again in analogy with atomic physics or chemistry. In explaining and correlating the wide range of available experimental data, it was possible to assume that these valence nucleons remained in the orbits of lowest energy available to them.

Within the past year, however, there has been a growing suspicion that this picture is much too simple; in effect, the experiments that have been done to date have not been sensitive to possible part-time adventuring of the valence nucleons into much higher orbits. Groups at the University of California, Yale University, and at Copenhagen demonstrated that such adventuring does indeed take place, opening up whole new areas of both experimental and theoretical study. This was accomplished by searching for, and finding, states of very high spin which would not occur if the valence nucleons spent all their time in the orbitals of lowest energy.

Spin dependence of the nuclear forces. It is an inherent property of nucleons as they move in their orbits that they spin about an internal axis, just as the planets do. An important question has long been whether the direction of the spin is correlated with the orbital direction; that is, if one is clockwise, does that determine the other? New measurements by Willy Haeberli and his group at the University of Wisconsin, using a beam of deuterons having all spins aligned in a given direction, demonstrated that the correlation is different in different nuclei, depending sensitively on details of the fundamental nuclear forces. They found that if the incoming neutron component of the deuteron is captured by a target nucleus into a valence orbit, then the emerging protons are deflected to the right or to the left, depending upon the spin-orbit correlation involved.

The phenomena involved are reminiscent of the way in which a billiard cue ball caroms off a cushion after the ball has been struck left or right of center, causing it to spin on its vertical axis (have "english"). The resulting preference for deflection to right or to left off the cushion depends upon the ball's initial spin direction. Such nuclear measurements will play an increasingly important role in elucidating the many remaining puzzles concerning the nature of the strong nuclear force that binds nucleons together into nuclei.

Molecular phenomena in nuclei. A number of years ago Allan Bromley and his collaborators at the Chalk River Nuclear Laboratories in Ontario discovered certain states in the magnesium nucleus of mass 24 that could be considered as "nuclear molecules." Within such molecules the 24 nucleons present were separated into two groups of 12, each having the structure of the carbon nucleus of mass 12 and each moving relative to one another in characteristic molecular fashion—that is, rotating end over end like a dumbbell and at the same time vibrating along the axis of the dumbbell. Apart from representing a new and interesting mode of nuclear behavior, these molecular configurations can, according to David Arnett and James Truran at the California Institute of Technology, play a vital role in the twilight of a star when the burning of carbon is of particular importance.

Within the past year, groups at Yale University, the University of Pennsylvania, the University of Washington, the University of Heidelberg, and the Saclay Nuclear Research Center in France undertook extensive searches for additional molecular phenomena. While searching, again in magnesium, the first two of these groups discovered what appears to be a new and more general molecular configuration. Using heavy-ion reactions in which larger clusters of nucleons can be transferred between projectile and targets, new states were found that appear to have the structure of two helium nuclei moving relative to an oxygen nucleus and of three helium nuclei moving relative to a carbon nucleus. These would be the nuclear analogs of the chemical molecular complexes H_2O and CH_3, with the tightly bound helium nucleus playing the same role as the hydrogen atom in the chemical molecules. Techniques were developed for isolating and studying these new configurations more generally, and extensive theoretical activity was being directed toward their understanding.

Scientists have long questioned whether, as more and more energy is added to a nucleus, its nucleons move more or less independently in orbits of ever higher energy, or whether smaller aggregations of nucleons form, which, in turn, move relative to one another. The latter now seems established, at least in some cases. These new configurations can have important consequences in nuclear astrophysics as well as in nuclear physics, and the techniques evolved for their study may have application in the new fields of accelerator chemistry, where the individual molecules are assembled and studied in a manner completely analogous to that traditional in nuclear physics.

Transuranic elements. During the past three decades, Glenn Seaborg, Albert Ghiorso, and their many collaborators at the University of California at Berkeley produced and identified the transuranic elements 93 through 103. These are all heavier than uranium, the heaviest naturally occurring element, with an atomic number of 92. In 1964 and in 1967, respectively, Flerov's group at Dubna announced the discovery of elements 104 and 105. This led to a long dispute involving the Berkeley and Dubna groups that was only recently resolved, when the production and identification of several isotopes of both elements were confirmed. Part of the dispute arose because the Soviets used the spontaneous fission of those heavy nuclei as their primary means of identification whereas the Americans used alpha radioactivity. Because each group used a different technique for identification purposes, until recently there was uncertainty regarding the exact isotope under study.

In February 1971 Flerov's group announced preliminary results—some 10–12 events—which are believed to indicate production of element 107. Such discoveries by the Dubna group, with equipment unmatched either at present or about to be matched in the near future, resulted from an intensive attack on the study of the transuranics using intense beams of heavy ions between neon (atomic number 10) and zinc (atomic number 30) on all available heavy targets. What is already clear, however, is that in moving from element 100 to 107 there is a systematic decrease in the production probability and in the lifetime of the species once they have been produced. Consequently, although these elements are of great scientific importance in permitting testing of a wide variety of nuclear models, it is improbable that they will enjoy practical applications.

Supertransuranic elements. A lack of practical applications is not necessarily the case, however, for the supertransuranic elements. Detailed knowledge of the nuclear shell model orbitals in heavy nuclei suggests that elements 112, 126, and 164 could, if ever produced, have very long lifetimes. Because these elements represent closed proton shells, they could be combined with the next closed neutron shell, expected at neutron number 184, to achieve considerable stability. None of the natural production mechanisms in our region of the universe would have been able to reach these species, but their absence in nature does not imply that they could not be stable for indefinite periods.

Extensive searches for supertransuranics in various sources were conducted by several teams: in primary cosmic radiation by Peter Fowler and his collaborators at the University of Bristol, Eng.; in natural minerals by Stanley Thompson and his collaborators at the University of California; and in a wide variety of natural minerals, sea-floor manganese nodules, and ancient lead samples by Flerov and his group at Dubna. These studies yielded suggestive data but no definite evidence. The heavy-ion accelerator facilities at Dubna were utilized during the past year in an intensive program aimed at direct production of the supertransuranics. Again, no definite evidence has been obtained.

Of great interest was a report by an Israeli, British, and U.S. team led by John Batty of the Rutherford High Energy Laboratory, in the U.K., on the possible production of long-lived element 112 through extended bombardment of a massive sample of tungsten with protons accelerated to 30 GeV (billion electron volts). After a cooling period of several months to reduce its intense radioactivity, the sample was dissolved and subjected to radiochemical elemental separation. Because it involves the simplest chemistry, the mercury fraction was extracted with the anticipation that element 112, if present, would have sufficiently similar chemistry to be extracted with it. When the mercury fraction was analyzed, a spontaneous fission activity from it was indeed observed; if it is element 112, a lifetime for the element of at least many weeks is indicated. With newer, more carefully controlled sample irradiations and

Courtesy, Oak Ridge National Laboratory

Composite photo reveals both the exterior and interior of the ORMAK, a plasma-containment device developed at Oak Ridge National Laboratory to help achieve controlled nuclear fusion. The doughnut-shaped core is made of gold-plated steel.

chemical separation, it will also be possible to search for a whole range of supertransuranics from each sample.

When one recognizes that theoretical calculations indicate that neutron-induced fission of the supertransuranic elements may produce as many as 10.5 neutrons per fission as compared with 2.5–3.0 for currently utilized fissile nuclear fuels, the practical applications become obvious. The concentrated, portable energy sources which these supertransuranics would represent, if of very long life, could revolutionize society and provide a uniquely powerful means of combating hunger and poverty throughout the world.

New detectors. It must be emphasized that a large fraction of all the new work in nuclear physics would be totally impossible were it not for impressive developments in the field of nuclear detectors. In X-ray work dealing with mesons, for example, lithium-doped germanium detectors provide an energy resolution that is 100 times better than was previously attainable. In dealing with transuranic elements, Curt Bemis and his collaborators at Oak Ridge (Tenn.) National Laboratory demonstrated that it is now possible to identify elements in quantities as small as 2,500 atoms through measurement of the energy of the characteristic K X rays with such detectors.

In both transuranic and supertransuranic work, the only possible detector that could isolate the spontaneous fission events of interest from the enormous radioactive backgrounds present was the Lexan plastic film developed by the General Electric Co. This detector demonstrated that a counting of fission fragments could be obtained by preferentially etching out of the track region many background events in the amorphous materials that were damaged by the fission fragment.

Combinations of traditional detection systems, such as magnetic analysis, time of flight, energy-sensitive detectors, and the like, have become possible under computer control and have had a revolutionary impact. An excellent example is found in the work of Vladimir Volkov and his collaborators at Dubna, who used such techniques to separate out such exotic isotopes as ^{8}He, $^{17-19}$C, $^{19-20}$N, $^{21-24}$O, and $^{24-25}$F from the enormous flux of nuclear species resulting from high-energy ^{22}Ne bombardment of ^{232}Th.

New accelerators. Within the next year the meson factory at Los Alamos, N.M., an 800-MeV, high-intensity proton accelerator, is scheduled to come into operation as a unique tool for both nuclear and elementary particle research. It will be designed to use the proton beam itself and also the secondary neutron and meson beams that the proton beam will make available.

For several years the nuclear community in the U.S. has been blocked in its desire to obtain the best possible accelerator capability for heavy-ion physics. The only approved U.S. project is the upgrading of the Berkeley heavy-ion linear accelerator (HILAC) to a super HILAC capable of producing nucleon beams with a particle energy of somewhat over 8 MeV per nucleon. All other proposals have been deferred because of other competition for the federal funding required.

At the same time, at Dubna, three new approaches were being actively pursued. A cyclotron producing beams of 18-MeV per nucleon came into operation; the unique Soviet heavy-ion sources are currently being installed in the old 10-GeV Dubna proton synchrotron ring with the stated hope of attaining several hundred MeV per nucleon during 1971. Moreover, and perhaps most important, with the group under V. P. Sarantsev at Dubna having already demonstrated excellent performance of a model electron ring accelerator (ERA) on alpha particles, Flerov's group at Dubna began moving vigorously to construct a very large heavy-ion ERA. These ERAs are uniquely powerful for heavy ions since the energy produced depends directly on the mass of the ion involved; in essence, the ions are carried along in a rapidly moving smoke ring of electrons.

Future prospects. Reflecting present funding strictures, the fear of even more serious ones to come, and the rejection of nuclear energy by many citizens, the morale of the U.S. nuclear community was in 1971 lower than at any time in the last several decades. If current funding trends continue, a substantial fraction of the existing U.S. nuclear activity will be dismantled with losses, both short- and long-range, to the nation. Such losses are difficult to assess with accuracy but will certainly be great. This comes at a time of marked expansion of nuclear activities abroad and will lead inevitably to a much weakened U.S. competitive position.

—D. Allan Bromley

Solid-state physics

The past year saw a marked reduction in the emphasis on basic solid-state physics and an increased effort to develop research projects that had a strong connection to practical applications. This appeared to be a worldwide trend. A detailed discussion of its causes and effects in the United States appears at the end of this article; at the beginning is a report of the developments in one area, that of superconductors, where the shift from fundamental to applied work was clearly visible.

Superconductivity. In previous years two aspects of superconductivity research were emphasized. The first was research aimed at providing a better fundamental understanding of the physics of superconductivity; the second was theoretical and experimental work to develop a room-temperature superconductor. Although work continued during 1970–71 in these two areas, one could see impressive efforts in a different direction: that of making practical applications of the superconducting technology presently available. Among the leading nations in these attempts were Japan, Great Britain, and the United States. (See *1971 Britannica Yearbook of Science and the Future,* Feature Article: CRYOGENICS: THE WORLD OF SUPER-COLD, pages 318–323.)

In Japan work was under way on a 400 km/hr train which would not touch the ground; rather, it would be floated several feet above the surface by the repulsion between two magnetic fields, one on the train and the other on the ground. The magnetic fields could be supplied by ordinary electromagnets; however, there was strong optimism that a suitable superconducting magnet would be developed for this. The superconducting materials being used were niobium-based alloys that become

Courtesy, Dr. M. Lubin, University of Rochester

Disk amplifier is used to amplify a low-energy laser pulse to one with an energy of more than 1,000 joules and a diameter of only 15 cm. Neodymium-glass laser disks are mounted at an angle to the incoming laser beam to minimize reflection and are excited efficiently by flash lamps on each side.

superconductive at temperatures of about 9.5° K. The superconducting magnet would be mounted on the moving train so that only a relatively small amount of material would have to be kept at such low temperatures.

Other applications of superconductivity were being investigated in Japan. Work was underway on superconducting magnets designed for thermonuclear power generators. For this purpose a 10-kw superconducting motor was built and was in operation at the Central Research Laboratory of the Toshiba Electric Company in Tokyo. In Great Britain efforts were being made to develop still larger superconducting motors and/or generators.

In the United States a large superconducting linear particle accelerator was under construction at Stanford (Calif.) University. By using superconducting material to make the chambers through which the particles were accelerated, much larger accelerating fields could be built up and particles (normally electrons) could be accelerated to high energies in much shorter distances. The work at Stanford was motivated by the hope of producing meson beams for use in the treatment of cancer. (A meson is a nuclear particle with a mass typically between that of a proton and an electron.) Meson beams would be preferable over the normal X-ray or electron beams because much more of their energy can be expended in the cancerous cells. This would allow for heavier therapeutic dosages with less damage to healthy cells.

Synchrotron radiation sources. From the beginning of solid-state physics, infrared, ultraviolet, and X rays provided important methods of probing the properties in solids. In the infrared and visible spectral regions, black-body radiation sources such as the sun or incandescent lamps provide continuous, strong sources of radiation that can be used by experimental solid-state physicists; however, no such sources were available in the past for the more energetic ultraviolet and X-ray regions. In these spectral ranges only relatively weak lines at discrete wavelengths were available. This severely limited the work that could be carried on in these regions. As a result much more became known about optical transitions in solids for the visible and infrared spectral ranges than for those in the ultraviolet and X-ray regions. This situation was being vastly changed, however, by making use of the radiation produced in high-energy particle accelerators. Called synchrotron radiation, it is produced when beams of relativistic electrons (electrons moving at speeds near that of light) are bent around the circular paths characteristic of many accelerators.

A problem arose with using accelerators for that purpose, however. This was that the beam is only in operation for a relatively short period at any one time. More satisfactory are storage rings into which high-energy electrons are injected and kept in a closed circular orbit for periods as long as several hours. A storage ring at Stoughton, Wis., works at an energy of 240 MeV (million electron volts) with a current of 10 milliamperes. Another such ring was under construction in California at the Stanford Linear Accelerator Center. The electrons there will have an energy of 2.2 GeV (billion electron volts) and a current of one-fourth of an ampere.

Although much of the pioneer work on synchrotron radiation was carried out at the National Bureau of Standards in Washington, D.C., the strongest group in recent years has been one in the Federal Republic of Germany. The West Germans have concentrated on studying transitions of electrons from the tightly bound, atomic-like, levels of ion cores to the extended, bandlike conduction states characteristic of solids. Before the advent of synchrotron sources, such studies could be done at only a few wavelengths, but now they can be made over a continuous range of wavelengths. On the basis of recent findings it appears that, in some cases, many-body theory must be invoked for an explanation of the data rather than the one-electron theory traditionally used to explain optical data in the visible and infrared spectral ranges. A discussion of these two theories follows.

Surface plasmon excitations. In principle, scientists should be able to solve all problems in solid-state physics through the application of the quantum or wave mechanical theories developed in the 1920s and 1930s. Practically, however, this is impossible because only problems involving a few particles can be solved exactly and, of course, solids are made up of an immense number of particles. A typical solid contains about 10^{23} electrons per cc.

Thus, one cannot normally find exact solutions to theoretical problems involving solids. One way out of this dilemma is to identify a single particle whose properties are essential for the problem under consideration and to lump the effect of all the other particles into a potential which stays constant independently of what happens to the single particle under consideration.

Because of its basic simplicity, this one-electron approximation has been used extensively and successfully in many materials. For example, the optical properties of germanium in the visible spectral range were explained by means of such theory. The theory of the transistor and many other semiconductor devices is based on a modification of one-electron theory. There are, however, other areas of solid-state physics where such theories are not applicable. In such cases the "many-body" theory must be applied. It attempts to take into account simultaneously the interactions among a large group of particles.

In an increasing number of cases, it is becoming apparent that many-body theory must be applied in solids. A case in point concerns the optical reflectivity of aluminum films. According to the one-electron theory, the reflectivity should be about 92% in the energy range 3 to 11 eV. However, it was found experimentally that the reflection was usually less than that predicted by the one-electron theory. The reflectance was also found to vary considerably from sample to sample.

Work at the Oak Ridge (Tenn.) National Laboratory of the U.S. Atomic Energy Commission, at the Michelson Laboratory of the U.S. Navy, and at Stanford University showed that this variation in reflection was due to a many-bodied excitation called a surface plasmon. The surface plasmon is produced by an absorption of energy by all of the conduction electrons near the metal's surface. (A conduction electron is one that is free to move under the influence of an electric field.) Ordinarily, light cannot excite such an absorption in thick films; however, it was established that the optical effect was made possible by surface roughness. As the surface roughness becomes larger, the surface plasmon absorption becomes larger.

The remarkable feature of this phenomenon is that the surface roughness necessary to induce considerable absorption is so small. An average surface roughness of only 18 Å (1.8×10^{-7} cm) produces a change in absorption of as much as 27%. The work at Stanford University also showed that the collective oscillation dies by giving up its energy to a single electron, which often escapes into a vacuum where it can be measured. The measurement of this current due to the surface plasmon provides perhaps the most sensitive method developed to date for measuring the roughness of a metal surface. In fact, changes on the scale of atomic dimensions can be measured.

Lasers. The first lasers were developed during the early 1960s. Because lasers produced extremely narrow beams of light in almost the same frequency, many fabulous usages were envisioned for them. However, a number of difficulties had to be overcome before any extensive practical use could be made of lasers. By 1971 many of these difficulties had been overcome, and a number of interesting practical applications were under way.

One area of application was communications systems. The higher the frequency at which such a system operates, the more the information that can be transmitted over a single communication link. Because it operates at frequencies millions of times higher than conventional telephone, microwave, and radio systems, the laser has tremendous potential for communications systems. The main problems are those associated with impressing a useful signal on the laser beam and those associated with absorption of the laser radiation by the atmosphere. The most powerful laser sources fall in the infrared spectral range, where atmospheric absorption is particularly severe.

Work was under way on at least two types of systems to overcome these obstacles. The first utilizes a very-high-powered neodymium laser at a wavelength of 1.06μ. Because this system was designed to communicate with satellites in space, only a few kilometers of atmosphere must be traversed. The second system was being developed in Japan for telephone communication in highly congested areas, such as that around Tokyo. In this case Japanese scientists and engineers were working on a system that used relatively low-powered gallium arsenide lasers. In this system repeater stations would be placed about every 1.5 km; at each of them the laser beam would be received, amplified, and redirected. By using such a system, the capacity of a municipal telephone system could be greatly increased without the necessity of stringing more wires.

Another development was that of "laser radar" to monitor pollution. In this application, a laser beam is directed at air suspected of being polluted and the reflected beam is then measured. By determining the change in wavelength from the original to the reflected beam, scientists can identify the foreign gas in the atmosphere.

New directions in research and applications. It would be unrealistic to discuss the year's developments in solid-state physics without commenting upon the changing atmosphere in which the work is being done. It is likely that the future direction of solid-state physics (as well as that of many other areas of research) will be greatly affected by external social forces. What are these forces and how did they develop? Two of the most obvious are reduced funding for both fundamental and applied research, and reduced employment opportunities for scientists and engineers. Another is strong pressure from both government funding agencies and industry to channel efforts into applied areas. Even those government agencies that in the past have emphasized fundamental research now often speak in terms of research projects that will return practical results within a year or two.

Many of the employment and funding problems facing solid-state physicists in 1970 arose from the fact that the electronics industry, which had provided most of the non-academic jobs, was changing from a fast-growing industry into a mature one with restricted growth. Thus, workers were only being replaced rather than added on to staffs.

In addition to employment by the electronics industry, many solid-state physicists had been hired by the expanding universities and government laboratories; many of those hired were committed to fundamental research with little concern as to applications. However, university expansion has slowed considerably, and government spending is moving away from fundamental research; thus, this source of employment was also drying up.

—William E. Spicer

Science, General

The American scientific establishment has grown and the capacities of its researchers have developed to the point that our capability in basic research has made us preeminent in the world. Having achieved that position, basic natural science finds itself in a crisis of both financial and social support. Historically, Federal funding, the main source of basic scientific research, has been large relative to the scientific resources available to do the work. In the recent past, as the scientific establishment continued to grow, the supply of funds leveled off so that the previous relationship has in effect been reversed. There is too little money relative to the number of scientists involved. At the same time, in the past half decade, scientists and their works began to come under fire as a result of the association of scientists with the military, and with industrial technology which has produced environmental pollution. In concert with these two developments, our national priorities have shifted to the solution of social problems, and basic scientists are being asked to shift their focus of work from the development of knowledge for its own sake to working on basic problems which have relevance for today's social issues.

The result is serious strain on an institution which furnishes us with our most fundamental understanding of ourselves and of our world, and which has been the source from which technology has evolved in recent times to serve economic growth. In the past few decades, we have been very successful in making basic science useful, but now we find ourselves in a crisis as to how to ensure its future usefulness, and of how to balance the long-range utility of basic knowledge with present urgent needs.

One of the major decisions with which we are faced is that of the level of support we will furnish basic science in the future. This is clouded by the problem of making basic research "useful" in the short run. It is in the nature of basic research that answers to practical problems may be found in unsuspected areas of inquiry. Some problem areas, at a given time, have a greater potential for exploitation than others. Setting research priorities on the grounds of probable utility is often a choice of possible short-term benefits against the longer-term ones which might result from a more rapid expansion of the basic pool of knowledge by permitting science to pursue the internal logic of its own development.

What is needed, and may in fact be developing, is a forum in which the partially conflicting needs for maintaining the integrity of the core of basic research and the practical needs of the society are resolved.

In conjunction with the need to work out an appropriate level and distribution of funding, we must face the fact that an articulate minority are attacking the very rationale and spirit of science and of rational inquiry itself—the most elementary tools man has for the orderly guidance of his affairs.

—*Toward Balanced Growth: Quantity with Quality* (July 4, 1970)

This remarkably concise description of the state of U.S. science at the beginning of the year under review is all the more notable in that it was prepared by the National Goals Research Staff of the White House. Staff director was Leonard Garment, who served Pres. Richard Nixon in a similar capacity during the 1968 presidential campaign. At that time, the president-to-be had said:

> If science and technology were to founder or stagnate many of our hopes would collapse. To the extent that we neglect this source of our greatness and to the extent that we fail to preserve the conditions of openness and order that made our progress possible, we are living off the land of civilization without refertilizing it. Instead we must bring about a new dawn of scientific freedom and progress.

Comparison of the two statements reveals how much more difficult it is to state goals for the national scientific endeavor when one is occupying the seat of power. And without a clear statement of attainable goals, it is even more difficult to enunciate a national science policy. Indeed, the search for a national science policy had preoccupied the leadership of both the federal government and the scientific community for a decade or more without result.

The search for a science policy

Leading the search was a unique committee of Congress—unique in that it was the only locus on Capitol Hill for concern over the strength of the national science effort, in contrast to those committees responsible primarily for dealing with specific federal agencies or specific national problems. Under the chairmanship of Emilio Q. Daddario (Dem., Conn.), the House Subcommittee on Science, Research, and Development conducted a series of hearings on national science policy during three months of the summer and fall of 1970. The most cursory examination, the staff noted, disclosed the lack of any formal or structured policy with regard to the use, support, or management of science and technology. Further investigation, however, showed that, in the absence of such a policy, a pattern of generally accepted premises had developed to guide short-term policy-making. They were (1) continuing development of science and technology at an optimum rate is vital to the nation; (2) federal support goes to science and technology where and when they appear promising; and (3) control of the support for science and technology should not be centralized.

"I hate to say this, but, even under our brutally slashed research budget, I've made an important discovery."

Ed Fisher © 1971 Saturday Review Inc.

The testimony of the distinguished scientists, engineers, educators, and government officials who paraded through the hearing room filled 963 pages of proceedings. It led the subcommittee to declare that the need for a national science policy had been established without question and to issue a series of recommendations, the purpose of which was to bring such a policy into being. First, it called on the administration to form a special task force whose sole assignment would be to draft a national science policy for submission to the Congress no later than Dec. 31, 1971. Second, it recommended that the Office of Science and Technology be strengthened and separated from its present advisory relationship to the White House. Third, it reiterated a previous proposal for the creation of a National Institute of Research and Advanced Studies. Fourth, it recommended that a greater portion of federal funding of basic research be moved from such mission agencies as the National Institutes of Health and the Department of Defense to the National Science Foundation (NSF). Finally, it urged the Office of Management and Budget to develop a "stable funding" procedure with regard to basic research in order to avoid seriously disruptive fluctuations in funding levels.

Whether the earnestness of the subcommittee report and the prestige of the witnesses on whose testimony it was based would actually produce a national science policy remained to be seen. The outlook was chancy when the report was issued, and the odds lengthened still further when Representative Daddario gave up his House seat to run for the governorship of Connecticut. His place was taken by the next ranking Democrat on the subcommittee, John W. Davis of Georgia. Although Davis had served on the subcommittee since its inception in 1963, considered himself as having an "honest-to-gosh interest in science" (he was an amateur astronomer), and professed many of the views of his predecessor, he had yet to earn the grudging respect that Daddario had won from the scientific community over seven years of attentive concern.

A change in the White House. Another change in the political leadership of science took place in the late summer of 1970—one that had, in the short run, at least, a far more telling effect on the national science effort.

Lee A. DuBridge, who had enjoyed an illustrious career as a distinguished contributor to scientific research and one of its staunchest defenders during the bleak period of the early 1950s, had been chosen by President Nixon to serve as his special assistant for science and technology and director of the White House Office of Science and Technology. The appointment had been greeted by the scientific community with considerable enthusiasm, but enthusiasm gave way to doubt as the budgets for science continued to lose ground to inflation. As the scientific leadership looked to their man in the White House, it became all too clear that he was not often to be found there. A prominent science historian observed bitterly that "science has apparently dropped out of the high councils of the Nixon Administration." Later it was revealed that officials in the Bureau of the Budget no longer invited staff members of the Office of Science and Technology to participate in the last—and hence most crucial—round of budget cuts preparatory to the annual presentation of the budget to the Congress.

DuBridge withstood the criticism quietly. Nor did he respond when major changes in the White House staff were announced and an accompanying organization chart did not even include him. But when the highly respected international science journal *Nature*, published in Great Britain, traced the muddled state of the U.S. science effort to inadequacies in the Office of Science and Technology, and *its* problems, in turn, to the president's science adviser, DuBridge thrust aside his natural shyness and called in the press. In an almost unprecedented session with reporters, he justified his inability to restore the lost prosperity of the scientific enterprise by blaming a budget squeeze that had forced the president to make hard decisions. Would there be a change? Not soon, DuBridge admitted: "It is evident that fiscal '72 is going to be a very tight budget situation."

When the fiscal 1972 budget was issued by the president in early 1971, however, it turned out to be not so tight after all. It also turned out that the man who received much of the credit for the turnaround was the president's new science adviser, a 45-year-old communications researcher named Edward E. David, Jr., formerly of Bell Telephone Laboratories.

Twenty-three years younger than DuBridge, David represented a break with the past in several respects. Chief among these was that, unlike his five predecessors, he had not made his reputation in the hierarchy of scientific leadership through service in the scientific advisory councils of World War II. He was also the first to come from an industrial laboratory rather than an academic position. Except for his colleagues at Bell Laboratories and a relatively small group familiar with his work in the revision of engineering curricula, no one had a form chart on the new man.

It was not long in coming. When the '72 budget was made public on January 29, David announced that there would be a significant increase in federal funding of research and development (R and D), with particular emphasis on academic research. The scientific community, he said, "can look to the future with confidence."

The budget. The total R and D budget called for expenditures of $16.7 billion, an increase of 7.6% over fiscal 1971 and close to the peak of $16.8 billion in 1967. David reported that support for research and development in colleges and universities would rise almost 15%, with the entire increase provided by the civilian agencies of the government. The NSF budget itself was up 22% to $622 million; its research budget alone was up an astonishing 44.5%—from $324 million to $468 million.

The president's budget does not provide money to spend, however. That has to be appropriated by the Congress, and there is often a long and bloody trail from the budget announcement to the final appropriations act. Nevertheless, the budget does serve to characterize the attitude of the administration toward national priorities. On the face of it, said *Nature*, "it seems as if the United States has once more begun to regard scientific activities as valuable investments of public funds." *Science*, the leading U.S. magazine in the field, was somewhat less sanguine, observing that "although significant funding increases are in the budget, they are, so to speak, far from in the bag."

Nor were all scientific leaders jubilant over the budget itself. Although the NSF budget had nominally been increased by $116 million, $74 million represented research that had been funded previously by other government agencies, and $47 million represented expansion of the NSF's experimental program for support of applied research on national problems. Meanwhile, support for training of science students through the postgraduate phase was cut by $16.2 million and support for training of teachers was cut by $18.8 million. It appeared to many that the origin of these deep cuts lay in the Office of Management and Budget, several of whose officials had been talking openly of their dissatisfaction over outright grants to science students. They preferred, instead, the system of preferential student loans customary in other fields.

Opponents of the science education cutbacks—notably Philip Handler, president of the National Academy of Sciences and former chairman of the National Science Board—argued that the production of highly trained talent was a national necessity and an appropriate responsibility of the federal government. Instead of cutting support of postgraduate science education down to the level that obtained in other fields, he said, the federal government should accept the responsibility for support of all fields essential to a rich national life—including music, history, education, and other branches of the arts and humanities.

In response to the pleas of Handler and others, the House Subcommittee on Science, Research, and Development, responsible under recent legislation for preparing the House version of the bill authorizing the NSF to spend the funds proposed in the president's budget, took the unprecedented step of shifting some of the budgeted funds from one category to another—principally from applied-research back to educational support. As of mid-1971 the bill had yet to pass, and the responsible committees on appropriations still had to provide the actual funds. Nevertheless, a critical committee of the Congress had signaled that it was not ready to let the federal government abandon its role in the production of scientists in the United States.

An exchange of opinions. At the end of March 1971, a New York woman named Marilyn Harbater addressed the following poem to George P. Shultz, director of the Office of Management and Budget.

Sir:
The National Science Foundation
  is just about reaching its prime,
This year marks its twenty-first birthday,
  which should be a most happy time.
But now in its new-found adulthood,
  NSF leaves us no cause to sing,
For in making its new yearly budget,
  NSF has omitted one thing!

The Secondary Science Training Program,
  also known as SSTP,
Has just been dropped from its roster—
  this is now a fait accompli!
And all because someone decided
  that science is now over-rated,
That scientists in the future
  will certainly be outdated!

But how can America function
  in this technological age
If our scientific community
  is stifled and put in a cage?

Should our youngsters stop taking courses
  in bio and math and in chem
Because they are told in the future
  they'll have no need of them?

These high-ability students
  haven't wasted their summers in play.
They've worked hard in NSF's program
  every moment of the day.
I know, from personal experience,
  the good that this program has done,
I saw the exciting awakening
  in the person of my own son.

Please restore this aspect of study
  only SSTP can provide,
Allow youngsters the opportunity
  to do work in which they take pride,
To study with fine professors
  who encourage inquiring minds—
America will be the winner,
  with future scientific finds.

Respectfully yours,
Marilyn Harbater

On April 22, Shultz replied:

Dear Mrs. Harbater:
We thank you for your poem
  about the NSF
and the cut in training
  to which you think we're deaf

Alas, this type of program
  has had a small effect
on changing students' choices
  and the courses they select

And further we don't see now
  the need for a reliance
on special Federal efforts
  to add more men in science

While research is important,
  our problems to surmount,
it's the quality of people
  and not numbers that will count

With more demands arising
  to draw on our tax take
priority decisions
  are something we must make

But funding is increasing
  in fiscal seventy-two
in many fields of science
  where there is work to do

For many men in research
  this will increase the chance
to utilize their talents
  —our nation to advance

We praise your son's high purpose
  in using his bright mind
and think that still in science
  a future he can find.

Sincerely,
George P. Shultz
Director

But it appeared that Mrs. Harbater had had the last word. Her poem had been read into the record at the NSF authorization hearings before the Subcommittee on Science, Research, and Development on April 7, and on April 29 the parent committee reported out the authorization bill restoring the cuts in educational support.

The clouded employment picture

Although the administration's budget proposals heartened the scientific community, the overall picture was still bleak in mid-1971. The Battelle Memorial Institute predicted that, although the total national investment in R and D would increase some 3.6% in 1971 to approximately $28.5 billion, the effects of inflation would result in a real decrease of 2.3%.

Scientist unemployment. Hardest hit by R and D cutbacks were the aerospace and electronics industries and the larger physics laboratories; in consequence, the engineers and physicists appeared to absorb the brunt of the blow. News stories told of engineers in the tens of thousands desperately looking for work. The June 1971 *Bulletin of the American Physical Society* reported that 4% of new Ph.D.s in physics were unemployed. In 1970, according to the chairman of the society's Economic Concerns Committee, 1,500 new Ph.D.s and 1,700 experienced physicists were looking for positions, and more than 30% of those who were looking for jobs in the traditional areas of physics in the United States failed to find them. Of the 1,700 experienced workers, only about 900 found new jobs in their field.

Discrimination. The increasing competition for faculty positions among scientists, especially in those institutions that permitted or encouraged faculty research, heightened the concern of those who felt discriminated against in employment opportunities. During 1970–71 the most persistent voices of complaint were those of women. In a survey of several hundred campuses, the American Association of University Women found that only 9.4% of the female faculty members were full professors, while 24.5% of the males held that position. It noted, too, that whereas 41% of all students were female, 21% of all the institutions surveyed had no women trustees.

An article in *Science*, calling on universities to improve the status of women on campuses before that question reached the crisis stage, cited a study disclosing that women lagged two to five years behind men in reaching full professorship in the biological sciences and as much as a decade in the social sciences. The study also found that salaries received by married women were 70–75%

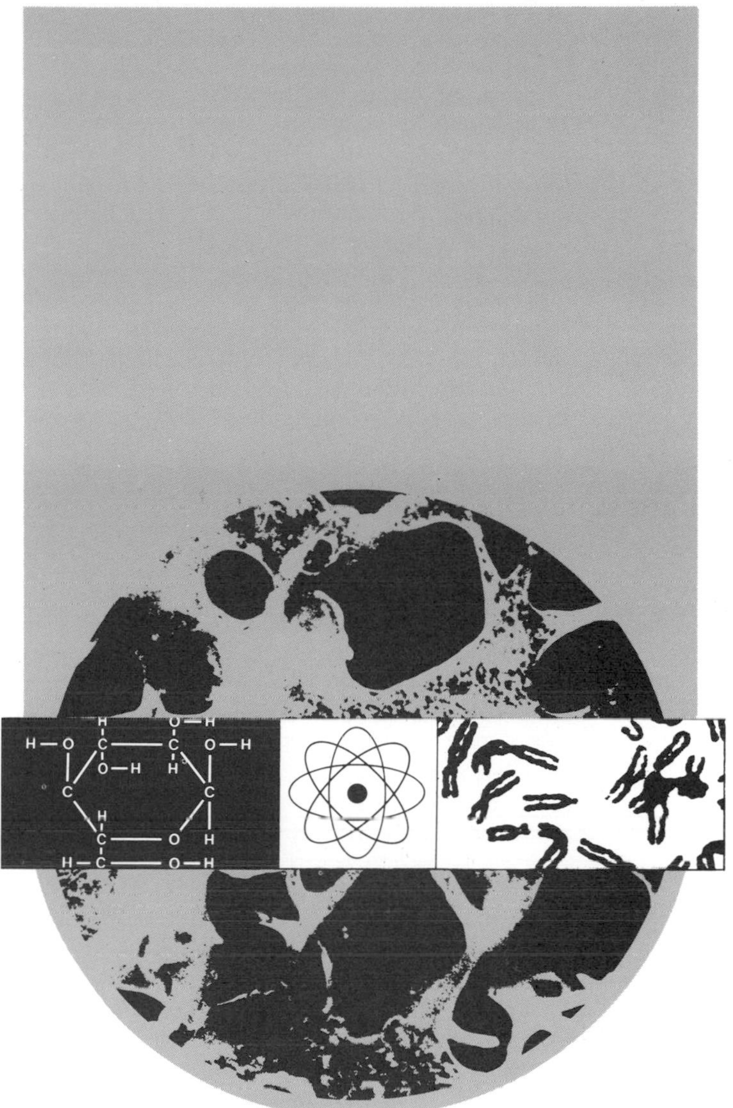

"The most beautiful thing we can experience is the mysterious. It is the source of all true art and science."
Albert Einstein

of those received by men who had obtained their doctorates at the same time—despite the fact that married women Ph.D.s publish slightly more than either men Ph.D.s or unmarried women Ph.D.s.

The aggrieved ladies found an ally in the U.S. Department of Health, Education, and Welfare (HEW). In November 1970 it was revealed that HEW had not only demanded that all universities and colleges cease discriminatory treatment of women students and employees under the considerable threat of losing all federal contracts, but had also insisted that all female employees be compensated for financial loss due to discrimination since October 1968. Action was taken to withhold new contracts from the University of Michigan as well as a number of other, unspecified institutions, pending compliance with the HEW directives.

Dissension

The academic year 1970–71, like the preceding one, was marred by political dissension within the scientific community and occasionally by mindless violence. On August 24 terrorists blew up the Army Mathematics Research Center at the University of Wisconsin at Madison, killing one graduate student, injuring four others, and causing an estimated $6 million damage. Four months earlier to the day, arsonists had set three fires in the serene study and meeting rooms of the Center for Advanced Study in the Behavioral Sciences at Stanford, Calif. There was no loss of life in that case, but the fire came close to destroying the life work of a visiting scientist from India.

The changing face of protest. There was, however, a detectable difference in the nature of the dissension over the two-year period. The change can perhaps best be characterized by comparing the annual year-end meetings of the American Association for the Advancement of Science (AAAS) in 1969 and 1970. At the Boston meeting in 1969, a substantial number of students and younger faculty participated in protest and disruption involving a variety of issues. Their principal concern seemed to be that the "Establishment" had gotten its funding priorities all wrong. A symposium on the space program was disrupted by individuals calling attention to the plight of the inner city. A number of groups concerned with the deterioration of the environment issued handbills. There were outbreaks of protest over discrimination, and there were angry accusations from the floor against participation by the scientific leadership in weapons development.

December 1970 at Chicago was a different scene. There were fewer protesters but they were far more disruptive. All other causes were subordinate to the general charge that the scientific community —partly through its fear of politicization—was serving as handmaiden to a power elite, which was chiefly characterized as a warmongering oligarchy, indifferent to the needs of the underprivileged. The statement of protest, handed out to all visitors and emblazoned with a startling obscenity, said in part:

> We are scientific workers who see science being used for militarism, exploitation, and oppression. We believe that the scientific community must organize itself together with other groups to struggle for a society in which science can be used for human welfare.

A second striking difference was that there seemed to be less student participation, although this might have been due to a smaller population of campus radicals in the Chicago area. In any event, the average age of those who engaged in disruptive tactics was estimated at 27 or 28, about the age of assistant professors.

Whether the abusive tactics of the disrupters—the use of bullhorns, holding placards in front of speakers' faces, and disabling amplifiers—served to advance their cause remained an open question. It must be noted, however, that when the wife of a distinguished biology professor speared the arm of an especially energetic dissenter with a knitting needle, there were few cries of outrage.

The radicals themselves were split on the advisability of the tactics of outrage, but, as a series of letters to *Science* revealed, even those radicals who were discomfited by the tactics of physical force argued that they were made necessary by the denial of the platform to "radically different perspectives." The *Washington Post* may have answered for the majority in an editorial that also mildly reproved the AAAS for not managing its affairs more tidily. It declared:

> It is a scientific fact, we believe, that only a single speaker can be heard at a particular place. Those who want to hear him should be free to do so; those who do not should be free to go away. This is not alone the basis of science; it is also the essence of freedom.

The question of responsibility. These postmortems were concerned primarily with tactics. The larger question—the social responsibility of the scientist—was not dealt with so easily. The traditional response of the scientific community to charges concerning its relationship to such depressing possibilities as thermonuclear war, test-tube babies, and global pollution was that science merely produces knowledge; the people, through their political leaders and through the marketplace, use it for good or evil.

To the revolutionary philosopher Herbert Marcuse this answer was no longer acceptable. He declared:

The traditional distinction between science and technology becomes questionable. When the most abstract achievements of mathematics and theoretical physics satisfy so adequately the needs of IBM and the Atomic Energy Commission, it is true to ask whether such applicability is not inherent in the concepts of science itself. The question cannot be pushed aside by separating pure science from its applications and putting the blame on the latter only; the specific purity of science facilitated the union of construction and destruction, humanity and inhumanity in the progressive mastery of nature.

The revolutionary thought of Marcuse was echoed by others closer to the scientific community. Hilary Rose of the London School of Economics and Political Science wrote in *Nature*: "Science is no longer to be assumed to happen in a social vacuum; the traditional belief that it was driven solely by its own inner logic is challenged." Philip Siekevitz, a professor at Rockefeller University in New York City, told a meeting of the Biophysical Society that scientists can no longer claim to be immune from responsibility for the ultimate product of their labors. At the very least, he said, they should devote 10% of their time to acquainting the public with the potential consequences of their work.

The ecology of ecology

Even the relatively immature field of ecology, which embraces the study of the interrelationship between living things and their natural environment, found itself the source of contention. One of its problems was that concern over the quality of the environment had burst so suddenly upon the lay public that some of its most vocal protagonists had no clear understanding of what all the words meant. "Ecology" and "environment" were used interchangeably, and both had reassuringly positive connotations. Articles of commerce were advertised as "good for the ecology," although it turned out that at least one of them was not very good for the people who used it. But if the public was prepared to believe that ecology was good for you, there was something less than unanimity within the scientific community on the condition of the environment, the principal causes of pollution, or what priorities should be given to various pollution-control programs.

One of the most frightening bogeys with which environmentalists had become concerned was the "greenhouse effect." It was conjectured that the increased use of such fossil fuels as coal, oil, and gas would lead inexorably to an increase in the amount of carbon dioxide (CO_2) in the upper atmosphere. As the CO_2 content rose, there would be a commensurate increase in the tendency of the upper atmosphere to retain the radiant energy from the sun. The consequence would be a steady and irreversible rise in the average mean temperature within the global envelope. This, in turn, would gradually melt the polar icecaps; the sea level would rise; and New York and other coastal cities would be inundated.

This particular fear appeared to have been laid to rest by a study held at Williamstown, Mass., in the summer of 1970. According to papers presented at that conference, the exchange of atmospheric carbon dioxide with other global reservoirs, such as natural plant life and the surface of the oceans, not only slowed down the accumulation of carbon dioxide in the upper atmosphere, but also provided a means to repair the damage should it become apparent that a greenhouse effect was beginning to take place.

An even more frightening spectre was raised with regard to the oxygen supply in the atmosphere, on which almost all life depends. Two threats were cited: one was that the burning of fossil fuels itself would deplete the oxygen supply to dangerous levels; the other was that the amount of DDT already in the oceans would so reduce the amount of microscopic plant life there—and consequently its capacity to produce oxygen via the process of photosynthesis—that within two decades available oxygen might fall below the amount necessary to sustain life.

Both theories have been disproved, according to S. Fred Singer, chairman of the Committee on Environmental Quality of the American Geophysical Union. But, he hastened to add in a *Science* editorial, this does not mean that we can forget about ecological disasters:

On the contrary, it is absolutely necessary to investigate each and every one of the side effects of our modern technology to its final conclusion and examine their possible influence on the global climate and on the ocean . . . The history of the Earth gives abundant evidence of cataclysmic happenings . . . The possibility that we might inadvertently set off an irreversible reaction must constantly be kept in mind.

Science in the spotlight

There were few who would disagree with the thesis that in order to protect the delicate ecological balances that sustain human life on this planet, some means must be found to predict the environmental consequences of new technologies. Yet the first effort to deal with risk-versus-benefit considerations in the political arena did not augur well for the developing process of "technological assessment."

Focus for the debate was congressional action on an appropriations bill containing funds for the production of a prototype model of a U.S. supersonic transport plane. Claims and counterclaims concerning the economic and social values of the SST surged across the political stage, but as the crucial Senate vote drew closer, a new factor emerged. On March 2, 1971, in testimony before the House Appropriations Committee, James E. McDonald, an atmospheric scientist of some reputation, voiced concern that engine emissions from the SST could initiate a series of photochemical reactions resulting in depletion of the ozone layer of the atmosphere. Such an effect, he warned, might allow more of the solar radiation that appears to produce skin cancer in humans to reach the surface of the earth.

Although the chairman of a panel on weather and climate modification, of which McDonald was a member, assured the leading Senate opponent of the SST that McDonald's tentative conclusions needed further examination, the skin cancer scare found its way into almost every news story concerning the debate. It played a significant, if unmeasurable, role in the eventual negative vote.

Even those who had pioneered in the argument for technological assessment were unhappy—not so much with the outcome of the SST debate as with the failure to resolve the central issues. Raymond Bowers, a Cornell University physicist who was one of the authors of a report on technological assessment prepared for the House Subcommittee on Science, Research, and Development, commented:

> Do we have no better way of assessing and deciding technological priorities than to convert the process into a national prizefight replete with posturing, misrepresentation, and abuse? Must these decisions wait until millions of dollars have been invested and thousands of people are already employed on the project?

Whether or not the techniques for cost-benefit judgments had matured, however, they were being called into play at an increasingly rapid rate. During the year, the leading scientific institutions of the country found themselves suddenly confronted with an almost bewildering array of public-policy problems involving science and technology. The National Academy of Sciences alone was called upon by acts of Congress or requests from federal agencies to: advise the Environmental Protection Agency on factors affecting determination of permissible levels of radiation from atomic power plants; advise on the technological feasibility of developing, by 1975, automobile engines with emissions 90% below those of current models; determine the efficacy of a variety of prescription and over-the-counter drugs; evaluate the risks to users of detergents containing enzymes; establish guidelines on nutrition for manufacturers of food products; and estimate the long-term effects of the military use of herbicides on the population and ecology of Vietnam.

So heavily involved was the academy in advising federal agencies on critical matters of public policy that consumer advocate Ralph Nader announced in May 1971 that his Center for the Study of Responsive Law was launching a nine-month study of that 108-year-old institution.

Word of the study was carried by the wire services to every newspaper in the nation. If there had been any doubt that the decade of the '70s had thrust the scientific community, unaccustomed and blinking, into the glare of the spotlight, it ended then.

—Howard J. Lewis

Transportation

The two-time rejection by the U.S. Congress of funds for further development of a 1,800-mph supersonic air transport, the collapse of Rolls-Royce, Ltd., due to underestimating costs of its engines for the L-1011 wide-bodied jet air transport, and severe financial problems of Lockheed Aircraft Corp. because of similar cost overruns for the huge C-5A military air transport are examples of why the aircraft manufacturing industry experienced a bad year. On the plus side, Congress passed a new Airways/Airport Development Act, with a user-charge financing package, that should assure the modernization of facilities needed to eliminate congestion and costly delays.

In the railroad field, the creation of a National Railroad Passenger Corp. marked a sharp departure from the predominantly privately owned U.S. transportation system, with the federal government assuming responsibility for maintaining a basic network of intercity passenger trains. Other significant congressional action set a 1975 deadline for new automobiles sold in the United States, including imports, to be virtually pollution-free; approved a 10-year, 300-ship program to modernize the U.S.-flag merchant fleet; approved a 12-year $10 billion program to modernize urban mass transit in U.S. cities; and set strict financial liability standards on all large vessels that discharge oil and waste in U.S. waters. All these new acts were expected to have a major impact on transport technology.

Maritime transportation entered a new era with the start of regular commercial service with lighter aboard (LASH) ships that carry on their decks barges that have been loaded with cargo at inland river ports for overseas movements to similar inland ports. Oil interests discontinued tests with a huge icebreaker tanker that had opened the

Courtesy, McDonnell Douglas Corporation

Assembled rear fuselage and aft engine/bulkhead sections of the Douglas DC-10 tri-jetliners are viewed in area where they are connected. Interior installations are completed before the three main parts of the aircraft are joined together to allow workers direct entrance through the open end of the sections.

Northwest Passage to the new oil fields of Alaska's North Slope, after deciding that pipelines would be more economical.

Air transportation. The recession in the U.S. caused a sharp reduction in expected payloads for the long-range 300–375 passenger Boeing-747 jumbo air transports and a cutback or stretch-out of orders for additional deliveries of this aircraft. Also experiencing a cutback or stretch-out of orders were the medium/long-range, 250–300 passenger wide-bodied Douglas DC-10 and Lockheed L-1011 jet transports, scheduled for entry into commercial service in late 1971. In addition, many options for the latter aircraft were dropped.

The collapse into receivership of Rolls-Royce, Ltd. threatened the discontinuance of the L-1011 program, although the British government agreed to give financial support to the company provided the U.S. government did likewise to the financially hard-pressed Lockheed Aircraft Corp., builder of the L-1011. Citing commitments to other aviation projects, however, the British government decided not to help fund other wide-bodied jet programs, such as the BAC 311 and the A-300B. The French and West German governments would thus have to assume about 92% of the development cost of the short–medium range, 260-seat A-300B Airbus, with a British firm contributing the remainder.

Reentering the commercial aircraft market for the first time since 1938, the West German aircraft industry developed the twin-engine VFW-614. With its wing-top engines less susceptible to sucking up pebbles and other debris on low-grade runways at smaller airports, this aircraft is uniquely designed to handle feeder-type service.

Following the same course of action on two separate occasions, the U.S. Congress voted down further federal financial participation in the American supersonic air transport (SST) program despite the fact that more than $1 billion (about 85% of which came from the government) had been spent over a four-year period to develop the 280-passenger, 1,800-mph plane. This apparently left the market open largely to the smaller, 128-passenger, 1,400-mph British-French Concorde SST, which was test-flown successfully but which continued to have such serious problems as excessive noise, limited range, and small capacity. The Concorde was not expected to be in commercial service until late 1974 or early 1975. Meanwhile, the SST of the Soviet Union, the Tu-144, continued extensive flight tests, and the aircraft—which has about the same size and speed as the Concorde—was reportedly near the production stage. If put into service late in 1972 as predicted, the Tu-144 was expected to be flown mostly internally over the long distances between the Soviet Union's eastern and western cities.

Airports just completed or under construction were utilizing new technology to help move people and baggage more rapidly within the terminal areas. The new Tampa (Fla.) International Airport, for example, featured a fixed track shuttle system that moves passengers rapidly between the main terminal building and four satellite loading terminals. It used fully automated bus-type vehicles on rubber-tired wheels. Automobile parking at the main terminal allows passengers to be no more than five minutes away from the aircraft departure point. Similar systems, some underground, were being constructed or planned for other major airports.

The Canadian government, which already had promoted the development of commercial STOL (short-takeoff-and-landing) aircraft, was asked by a special government-industry advisory group to step up this effort in order to enable that country to gain a commanding lead in this aviation field. If approved, such a program would cost the government a minimum of $150 million, although a potential market for such aircraft of $500 million to $1 billion was forecast. The program, which would be given national priority, would seek to develop progressively larger STOLs; it would embrace all aspects of a STOL network: aircraft, navigation aids, and the stolports themselves.

Highway transportation. Regular use of turbine engines for large over-the-road trucks and buses came closer to reality, with both Ford Motor Co. and General Motors Corp. announcing plans to start building them by the end of 1971. General Motors also announced the availability of a new, four-speed automatic transmission for use in such engines, which it claims "is the first automatic developed for a vehicular gas turbine powerplant." Designed for turbines of from 250 to 400 hp, the transmission, according to General Motors, will make large tractor trailers and buses as easy to drive as passenger cars.

A vehicle guided by a magnetic field makes a test run in May near Munich, W.Ger. A pollution-free linear-induction motor provided the power for the vehicle, which, on the 656-yd track, reached a speed of 62 mph.

UPI

Major advantages of the turbine engine over the reciprocal engine are that it is smaller per given horsepower and much lighter. Turbine engines also are virtually odorless, have a clean exhaust, and a lower noise level. On the negative side, their higher fuel consumption and initial cost were expected to limit their use to the larger, over-the-road vehicles that have a minimum of stop-and-go driving.

General Motors agreed to purchase production rights to the German Wankel rotary piston engine for $50 million. Compared to like-rated conventional engines, the Wankel is 30–50% lighter and smaller, has 40% fewer parts, operates on nonleaded fuel, and is very responsive to antipollution devices. Considerably more development work must be completed, however, before any full-scale production begins.

The Rohr Corp., which claimed a breakthrough in its efforts to meet 1975 motor vehicle emission standards for bus engines at small additional cost, offered to supply 20 buses for tests in the San Francisco, Calif., area. The engines would operate on liquefied natural gas fuel carried in special cryogenic (low-temperature) tanks. Tests by the U.S. General Services Administration on 344 motor vehicles operating on a dual-fuel system indicated such a potentially "dramatic reduction in exhaust pollution" that an additional 550 vehicles were converted. The dual-fuel system used high-pressure tanks containing compressed natural gas for a driving range of about 70 mi, with easy conversion to regular gasoline by throwing a switch and pulling a mechanical device on the dashboard. More advanced tests were being made on vehicles that operate solely on liquefied natural gas.

While widespread use of battery-powered motor vehicles still appeared to be a long way off in the U.S., West Germany's largest power company, RWE, announced the formation of two subsidiaries to develop such vehicles and a supporting network of service stations. Concentrating initially on small buses and city delivery trucks—hopefully on passenger cars by 1975—the specially built vehicles would be able to drive into the service stations for a one-minute replacement of batteries, which would be rented. Batteries now available, RWE claimed, could be charged up to 2,000 times and used to power a vehicle up to 62,000 mi. RWE also said that the added power load could be handled by doing most recharging during the night, when demand is at its lowest.

The revival of the steam engine continued to be pushed, and industrialist-inventor William Lear claimed development of a revolutionary low-pollution model that could meet the 1975 clean-air standards for new auto engines. He offered General Motors first opportunity to put it on the market to meet the deadline, but it would cost the company from $75 million to $300 million just to retool for mass production.

In another highway transport area, the use of exclusive lanes for city buses during commuter rush hours continued to be tested as a means of getting people out of their private cars and thus reducing traffic congestion. Two major tests took place in Washington, D.C., and New York City. The former was saving about 4,300 passengers approximately 30 min a day. The latter, which included a newly built approach to the Lincoln Tunnel, handled 35,000 daily commuters traveling on 800 buses and reportedly was saving each one 15 min each way.

Pipelines. Clearance from the U.S. Department of the Interior for construction of the 800-mi, 48-in. trans-Alaska crude oil pipeline continued to be withheld pending reports from various government agencies and interested organizations, plus an environment impact report required by law. Secretary of the Interior Rogers C. B. Morton said that he planned to make a final decision by October 1971, but expected further delays because of court suits. Canadian government officials expressed a strong interest in having the line built across their country, although strong opposition to this was expressed by several major U.S. agencies because of the diversion of economic benefits. Humble Oil and Refining Co., major backer of the successful voyage through the Northwest Passage by the specially built icebreaker tanker SS "Manhattan," suspended further tests after determining that pipelines had an economic edge for moving North Slope crude oil to refineries.

Renewed operation in the U.S. of a long-distance slurry pipeline came about with the start of commercial movements of slurry coal through an 18-in., 273-mi pipeline from strip mines in Arizona to a power plant in Nevada. The Black Mesa Pipeline, a subsidiary of the Southern Pacific Transportation Co., moved the coal in a 50% coal–50% water mixture at the rate of 14,000 tons a day in around-the-clock operations. The cost of the pipeline, including pumping stations, slurry preparation plant, and other facilities, was $35 million.

The potential of such lines in the U.S., a heavy consumer of coal, was brought out by a researcher in this field, an official of the Bechtel Corp., who predicted construction of from 8,000–10,000 mi of coal slurry pipelines during the next 10 years. This same official predicted another 5,000 mi of slurry lines hauling iron ore, sulfur, potassium chloride, and other similar products and a substantial growth of waste commodity pipelines from cities to outlying dump areas during the same 10-year period. Major interest in such slurry lines was in Canada, a heavy producer of raw materials for one-way movements to industrial consumers. Plans were announced to build 1,492 mi of such pipelines in three large projects that would cost $230 million. They included a 242-mi, $70 million iron ore line between Labrador and Quebec to a pelletizing plant on the St. Lawrence River for overseas shipment; a 500-mi, $100 million, 24-in. coal line from western British Columbia to the Roberts Bank Superport near Vancouver, B.C.; and a 750-mi, $60 million, 12-in. sulfur line from Alberta to Vancouver.

Canada was the major promoter of the solids pipeline concept, which would move cylindrical containers filled with cargo through pipelines. The containers would be suspended in and propelled by a liquid such as water or petroleum. Tests were conducted by the Alberta Research Council on a 3,500-ft layout of 4-in. pipe that included both grades and curves; capsules 15 in. long and 3⅝ in. in diameter would be used in this pipeline. The Canadian Ministry of Transport awarded a new three-year, $535,000 contract to the council for continuation of these tests and for use of more sophisticated instrumentation.

Rail transportation. *Freight.* The U.S. railroads continued to add unit-trains to their freight service. Such trains usually consisted of approximately 100 freight cars, many with 100-ton capacity, that haul a single bulk-type commodity in a shuttle-type, point-to-point service. The cars are usually loaded and unloaded in motion, using built-in devices and special terminal facilities.

One unique development in this area was a new service by the Illinois Central Railroad Co. that provided a 36-car unit train that carried 3,000 tons of stabilized organic residues from Chicago to a storage facility 160 mi south. The residues were prepared in slurry form and moved in 20,000-gal tank cars. The slurry itself is odorless, black, and 10% solid and can be used by farmers as an organic soil conditioner without need for drying.

A modification of the unit-train concept, the all-piggyback train, became increasingly popular with the railroads as a means of capturing general freight business. About 85 such trains, which haul both truck trailers and straight containers on special flat cars, were operating in 1971 on daily schedules in the U.S. and were providing a level of service comparable to the best truck schedules. These trains made nonstop trips of about 400 to 500 mi in 10 to 12 hours, and total daily use of the

"We do not ride on the railroad; it rides upon us."
Henry David Thoreau

flat cars averaged 500 to 600 mi, or about 10 times the average for all rail freight cars.

The U.S. railroads were also pressing to get even a greater share of the new automobile shipments through use of specialized freight cars. Adopting such innovations as triple-decker, auto-rack cars, they were able to increase their share of this lucrative traffic from only 8% in 1959 to nearly 55% in 1971. Two innovations were the result of joint engineering efforts by General Motors and the Southern Pacific Railroad, Co. One, the Vert-A-Pac car, hauled the new subcompact Vega in a vertical position, front bumper down, 30 autos per car. About 183 Vert-A-Pacs were in service in 1971, and they were said to have delivered the Vegas "promptly, smoothly, and without damage."

The second General Motors-Southern Pacific innovation involved the use of cocoonlike containers, each with space for three horizontally stacked autos. The containers are open only at one end, and the autos fit so snugly that they must be pushed into position and pulled out. Special cranes load the containers onto a flat car in an end-to-end configuration, and remove them at destination. The pilot program called for movement of 200 Cadillacs to California over a six-month period, and the first shipment of 12 arrived at its destination without damage. The purpose, as in the case of the Vert-A-Pac car, is to protect new automobiles from inclement weather, dirt, vandalism, pilferage, and other types of damage en route.

U.S. railroads, financially hard-pressed to expand their freight car fleet, joined together to maximize the fleet's utilization through computerized and centralized information about the availability, location, and status of all cars used in interchange service. An integral part of this program was the labeling of all such freight cars for automatic car identification by roadside scanners. After five delays, the deadline for labeling finally passed, with more than 90% of the freight cars properly labeled. The next big step was to install the scanners; by mid-1971 only a small fraction of the total had been put in place. Once the program gets underway, the railroads hoped to increase freight car utilization by about 10%, the equivalent of adding 180,000 cars.

Passenger service. U.S. railroads, largely freight-oriented as compared to European and Japanese railroads, were relieved of further responsibility—and a $200 million-a-year financial burden—for providing intercity passenger service when the newly formed National Railroad Passenger Corporation (called Amtrak) took over in 1971 as a quasi-government enterprise. Part of Amtrak's basic network included the demonstration of high-speed service between Washington and New York with electric Metroliners, and between New York and Boston with Turbotrains.

While the Metroliners proved to be attractive from the standpoint of high load factors, with passenger count averaging 65% of capacity, they still were not paying their way. Electrical and mechanical problems remained, and $3 million was allocated to solve them, with both General Electric Co. and Westinghouse Electric Corp. making special studies. Because of poor roadbed, the New York-Boston Turbotrains were held down to speeds averaging 60 mph. Improvements were expected, however, as a result of a new contract to continue this experimental service for at least two more years.

For the second time the Canadian National Railways removed Turbotrains from service between Montreal and Toronto because of "an accumulation of operating" and mechanical problems. The French apparently were able to overcome many of these problems, because by 1971 they had nine Turbotrains in service that operated at speeds of 100 mph on both short and intermediate runs.

Although common in Europe, the hauling of private automobiles by train along with their drivers failed to catch on in the U.S. An innovation may indicate changes in the future, however. The Interstate Commerce Commission (ICC) approved a petition by Auto-Train Corporation to operate a daily auto-on-train service between points near Washington, D.C., and the new Disneyland in Florida. Family cars would be driven onto and transported in special rail cars, and their drivers and passengers would ride in special, two-deck coaches that would be air-conditioned and offer stereophonic music, movies, and food service at the passengers' seats.

The cost was expected to be less than $200 per carload for the 14-hour trip. Present plans do not call for use of sleepers, although they would be added if the demand warranted them. The new service, scheduled to begin in late 1971, was called "highly speculative" by the ICC, which said it is requiring Auto-Train to make "a full disclosure to the investing public of the risks involved."

Water transportation. A major change in the maritime field was predicted by ocean shipping officials as a result of the entry into commercial service of LASH (lighter aboard) ships. These ships were designed to carry preloaded barges that could be transshipped to both U.S. and European inland river ports without further handling. Because they carry their own cranes to load and unload the barges, LASH ships can serve ports not able to handle container ships, which require costly terminal facilities.

Two LASH ships were built in Japan and were in operation in 1971, along with several built in U.S. yards. As more were built, they were being placed into service on a regular, scheduled basis; a total of 11 U.S.-built LASH ships were scheduled to be operating by the end of 1972. A new contract for three additional LASH ships for Delta Steamship Lines Inc.—the first ship construction contract under the new 10-year, 300-ship U.S. maritime program—called for only 44% of the total cost being covered by government subsidy (as compared with 50% normally) and for a 32-man crew (as compared with 38 normally).

Work progressed on schedule on the SEABEE vessels, which are larger than LASH ships and haul bigger barges that are floated on and off the ship's stern, elevated to one of three decks, and positioned by special hydraulic transporters. The first of three such ships was scheduled for delivery at the end of 1971, with the other two in 1972.

The potential of gas-turbine power for large merchant ships was demonstrated by the G.T.S. "Euroliner," a gas-turbine container ship built in West Germany for Seatrain Lines Inc. It set a new average speed record of 26.5 knots for transatlantic cargo ships, using 80% of full power. Seatrain cited several advantages of the gas-turbine engines over conventional steam-turbine engines: gas-turbine engines can reach full service speed from dead start in under five minutes as compared with up to four hours for conventional engines; a complete gas-turbine powerplant costs 10–20% less than a conventional one; the lighter weight of the gas turbine makes it possible to carry 10% more cargo.

Another innovation with possible major potential was the launching of an 11,000-hp oceangoing tug near New Orleans, La., that will link up with a special 35,000-ton tank barge as a single unit for over-the-seas movements at speeds of up to 14 knots. Aided by the U.S. Maritime Administration, Ingram Corp. was scheduled to test the new concept, which utilizes a rigid link-up system using a notched stern of the barge for a push-type operation. At the front of the barge a remote-control propeller, called a bow thruster, helps the tug operator maintain full control. Advantages over conventional tankers include the much lower initial cost and a 67% reduction in crew. The main disadvantage is the much higher insurance costs for such ships, but this could be reduced sharply if the linkage system works satisfactorily.

Considerable differences of opinion were expressed about the size of U.S.-flag tankers that should be built under the new maritime program. The Navy preferred smaller tankers for its varied needs, while commercial interests wanted very large tankers for maximum economy. While a number of foreign ports were able to handle the

Courtesy, Bay Area Rapid Transit District

The interior of an underwater tunnel through which trains will travel is part of the 75-mile Bay Area Rapid Transit system under construction in San Francisco. The new urban transport system proposes to reduce air pollution and traffic congestion, and provide inner-city residents access to jobs in outlying areas.

200 tankers of 100,000 tons or more that were in the world fleet, all were barred from using U.S. ports because of channel depths and width limitations. Studies to determine the feasibility of building multipurpose offshore terminals to handle such large ships were being funded by the U.S. Maritime Administration.

Nuclear-powered merchant ships took a double blow from both the U.S. and British governments. The NS "Savannah" was finally beached and awaited lay-up at Galveston, Tex. The British government accepted the findings of a special study group that nuclear merchant ships would not become economically viable in the foreseeable future and that further financial support was, therefore, not warranted. Problems cited included the lack of adequate shore facilities and qualified technicians to man such ships, and their unacceptability in some foreign ports.

Urban mass transit. A new $40-million grant from the U.S. was expected to enable San Francisco's fully automated, electric-powered, 75-mi rapid transit system to be completed, with operation beginning in early 1972. Final clearance for building its 250 commuter cars was, however, being held up pending tests on the last of 10 prototypes that must incorporate all final design features. A huge tube under New York City's East River for two-level commuter and subway operations was under construction; it was to be part of a $70-million tunnel that was about 40% completed.

The U.S. Department of Transportation awarded a number of contracts to promote various methods of improving urban mass transit. They included $1.5 million to start a personalized door-to-door system, called Dial-A-Ride, using 15-passenger buses linked to a central communication control; two studies totaling $373,787 to determine the feasibility of electronically locating and tracking city buses for better utilization; and two contracts totaling nearly $4 million to design and build prototypes of a fully automated system using small, rubber-tired, buslike cars on a guideway.

Air-cushion and LIM vehicles. The U.S. Department of Transportation started full-scale construction of its 45-sq mi test center near Pueblo, Colo., where emphasis would be placed on the development of high-speed ground transport vehicles. The department's plans included vehicles that operate over a fixed track, some suspended by air pressure (air cushion) and more advanced ones (LIMs) suspended magnetically. Propulsion would be supplied by a linear induction motor in which electromagnetic forces act between a coil in the vehicle and a fixed secondary conductor in the guideway. The first LIM vehicle, one with rail wheels, passed low-speed tests of up to 35 mph and was being shipped to the test center for high-speed trials.

—Frank A. Smith

Veterinary medicine

Ecology became a dominant theme in veterinary medicine during the past year, as it had in many other fields of endeavor. In most instances, however, this was less a matter of joining a popular cause than acknowledging that veterinarians had always been concerned with conservation of animals as a major national resource. After World War II livestock practitioners had turned increasingly from treatment of individual animals to development of comprehensive herd-health programs designed to improve efficiency of meat and milk production. Even when dealing with one sick cow or horse, however, veterinarians had been aware of environmental problems in disease, perhaps to a greater extent than physicians since animals live closer to nature and in a largely uncontrolled environment.

An epidemic of Venezuelan equine encephalomyelitis (VEE), a viral disease carried by mosquitoes, killed more than 1,000 horses in Texas in July 1971. The disease, which causes flulike symptoms in humans, swept up Mexico's Gulf coast in June and entered the U.S. at Brownsville, Tex. Approximately 70% of the horses infected with VEE died. The U.S. government made efforts to contain the disease by quarantining all horses in Texas, inoculating horses there and in adjacent states with a vaccine, and spraying mosquito breeding grounds.

Large-scale animal production. With increasing demands for more efficient meat, milk, and egg production, during the past two or three decades animals were moved into larger and larger producing units, thus creating new problems. One of these was the increased hazard of communicable disease occasioned by the stress of confinement and close contact with many other animals. Thus, a susceptible animal was more likely to contract any of the various common infections from another animal who otherwise might have been only an unaffected carrier.

For some years the more prevalent disease problems arising under such intensive production methods were largely controlled, and some were eliminated, by closer attention to sanitation, quarantine of animals entering the premises, widespread use of medicated feed, and development of specific vaccines or other disease preventives. Veterinarians engaged in herd-health practice, therefore, were concerned with such environmental matters as housing, adaptation to confinement stresses, disinfection, and waste disposal, together with feeding for optimal meat or milk production; this was in addition to the more conventional aspects of disease prevention and treatment. But even the latter underwent considerable change; animals that in a small herd would have been segregated and treated individually when ill became a menace to an entire large herd or feedlot and often were disposed of without any attempt at treatment. If a group of animals became ill, they might be moved to a hospital pen and treated en masse through medicated feed or drinking water.

From the outset, these methods made it necessary to ship animals in from widely scattered areas where breeding herds were located. The rigors of shipping led to a great increase in stress-related diseases, especially of the respiratory system. These were combated in part by minimizing handling and time in transit, providing adequate rest and water en route, and by vaccinating the animals for such diseases as shipping fever (hemorrhagic septicemia), parainfluenza, and rhinotracheitis (red nose).

These measures were not adequate, however, and a new concept—that of preconditioning—began to take hold about 1970. This involved assembling the animals a month or longer before they were to be shipped, treating them for internal parasitism (deworming), castrating the males, and giving whatever vaccines were considered necessary. Meanwhile, they were well fed to condition them for the stress of transportation and becoming acclimated to new surroundings.

Antibiotics in feeds. For a decade or so, most of the disease problems related to close confinement of animals had been minimized by feeding antibiotics and other antibacterial agents. There was a tendency to use these drugs as a substitute for good management, and because they kept even subclinical disease at a low level, they also acted to promote the rate of growth and efficiency of weight gain.

The almost universal use of antibiotics, however, created new problems, as certain strains of disease-producing bacteria became resistant to one agent after another. Then it became apparent that resistance to various antibiotics could be transferred among and between species of bacteria. There was concern in both the United States and Great Britain as to whether such drug resistance might be transferable to human pathogens, and when it became evident that such transfer may, in fact, have occurred in a few instances, the broader implications of this problem were widely studied.

In 1969 the Swann Committee had recommended that feed use of antibiotics in Britain be almost completely curtailed, and implementation of its report in 1971 ended the use of penicillin and two of the common tetracyclines in animal feeds. Attention was turned toward producing a few "safe" feed antibiotics that would not be used in human medicine. The U.S. Food and Drug Administration, through its Bureau of Veterinary Medicine, promulgated similar but less stringent regulations designed to ensure that meat of slaughtered animals would not contain antibiotic tissue residues in excess of established tolerances. Both veterinarians and owners were made responsible for the proper use of such agents, and 1971 saw the first prosecution for violation of these regulations.

Waste disposal and pesticide poisoning. With increasing public awareness of ecology, other problems, partly of a veterinary nature, became apparent. One was the disposal of animal wastes in ways to avoid their becoming a nuisance. Domestic animals in the United States produced some 1,700,000,000 tons of solids and about the same amount of liquid waste last year, 10 times the amount produced by man, and half of this was concentrated in areas near large human populations.

Veterinarians had recognized pesticide poisoning of animals long before Rachel Carson's book *Silent Spring* alarmed the public. Thus, in addition to treating such outbreaks when they occurred, veterinarians found it necessary to become experts on the safe use of pesticides as a preventive medicine procedure. Some of the large-scale deaths of wildlife attributed to pesticides, however, proved to be natural outbreaks of disease, and in this regard veterinarians had the responsibility of seeing that blame was properly fixed.

World food problems. Another aspect of the environmental crisis of increasing concern to veterinary medicine was that of the world food supply. Over the past several years, expanded programs designed to increase the world supply of animal protein were undertaken by veterinarians working at all levels. A Florida veterinarian engaged in a mariculture (sea farming) project developed foot-long shrimp, and by combining hormone treatment with artificial methods of rearing increased the productiveness of pompano several hundred times. Researchers in many nations worked to produce improved strains of domestic animals and to seek better ways of reducing the toll of animal disease and of improving reproduction rates.

Despite the great strides made over the past two decades, UN Food and Agriculture Organization veterinarians recently estimated that a further 50% reduction of animal disease losses was a realistic goal for the less developed nations they serve. They believed that this reduction would result in a 25% increase in available animal protein. Through the urging of two young veterinarians at the University of California at Davis, the UN voted to establish, as of June 1971, a volunteer Veterinary Corps to work with the FAO.

Unwanted pets. Increased awareness of environmental problems led veterinarians and governmental officials in the United States and Great Britain to recognize the excess population of unwanted dogs and cats, many of which had become strays, as a form of pollution. Attention was therefore turned to means for reducing their numbers, one approach by the city of Los Angeles being to establish, in the spring of 1971, the world's first municipally operated spaying and neutering clinic. This proved so successful that other clinics were planned, and the project attracted the attention of animal control officials in many other cities. Following the lead of various humane groups, the states of California and Pennsylvania passed legislation requiring that animals adopted from shelters be neutered.

—J. F. Smithcors

Zoology

There was activity in a great number of zoological and related fields during the year. In protozoology (the study of single-celled animals), work on classification, taxonomy, and the description of new species was decreasing, while work on electron microscopy, biochemistry, physiology, and ecology of protozoa was increasing. Moreover, because many protozoa, which are an excellent and often neglected tool for the biologist, are easily cultivated and because their enzyme systems are similar to those of other animals, numerous tests can be carried out on them which are difficult to do in higher animals.

Working with insects, one group of investigators learned that members of the family Dytiscidae, the diving beetles which are generally thought to be extremely voracious predators in both the larval and adult stages of their life cycle, are actually scavengers, preferring dead food to live prey. How do aquatic larvae overcome surface tension and break away from the surface when necessary? An investigator in Scotland found, in the pupae of the mosquitoes *Aedes togoi* and *A. aegypti*, a mechanism at the base of the spiracles (external openings of the breathing system) which produces, under muscular control, a backward movement of the spiracles and aids in the escape from the surface.

There had been increasing skepticism on the part of certain investigators who felt that Karl von Frisch's experiments with bees 25 years earlier lacked adequate controls to discount the influence of a multitude of variables that might affect the bees' foraging behavior, dancing, and recruitment. However, their own experiments during the past year confirmed Frisch's contention that the directional information contained in the forager's waggle dance does, indeed, include additional information that can be utilized by recruits.

Further research on bees showed how and why queen honeybees lay fertilized eggs in worker cells but unfertilized eggs in the slightly larger drone cells. N. Koeniger of the Bee Research Institute at the Johann Wolfgang Goethe University, in Frankfurt am Main, W.Ger., reported noticing that when the queen inspects a cell before laying in it, she inserts not only her head into the cell, but also her front legs. By attaching "spurs" of tape to a queen's forelegs, Koeniger made it impossible for her to get these legs into a cell during inspection. All the eggs the queen then laid were fertilized, whatever the type of cell, but when the tape "spurs" were removed, the normal pattern of laying was resumed. These experiments show that the queen uses her forelegs in some way to differentiate between the two types of cell. It was also reported that worker bees seem to have some control over the queen's egg laying. When a colony is preparing to swarm, egg laying is reduced and the workers stop feeding and grooming the queen and shake her constantly and quite violently, sometimes several hundred times in an hour. The result is that the queen is slim and muscular enough to swarm.

X-ray studies of Lower and Middle Devonian slates from West Germany revealed a surprising amount of pyritized soft parts of fossil cephalopods (members of a large family of mollusks, modern representatives of which include the squid and cuttlefish) and trilobites (a group of extinct invertebrates). These structures are so delicate that they are normally destroyed by the usual mechanical preparation of the specimens. The most surprising discovery was a system of light guides leading from the facets of the eyes of the trilobite *Phacops* nearly to the center of the head. There also seems to be a significant difference between the facet eye structure of the horseshoe crab *Limulus polyphemus* and the eye of *Phacops* which earlier work on the lens structure of trilobites did not reveal.

A group at the Ocean Research Institute in Tokyo discovered that in the starry flounder, a bony fish, the urinary bladder may play an important role in the maintenance of body chemistry. Formerly, the maintenance of the internal environment through the regulation of osmotic pressure (the difference in pressure between two solutions separated by a membrane that is permeable to only one of them) was ascribed to the gills, kidneys, and intestine of the fish, and the urinary bladder's function was thought to be only that of storage.

Response to sound as opposed to vibration is generally considered insignificant in salamanders, although some aquatic species are known to produce underwater sounds. *Siren intermedia* makes and responds to underwater sounds, and the discovery of the function of the acoustic behavior in this genus is new. During the past year it was learned that *Sirens* are attracted to recordings of their natural and simulated natural sounds, make yelps in distress, and make vocal sounds only when other *Sirens* are in the vicinity.

The remarkable imitative powers of the Indian hill myna (*Gracula religiosa*), which make this bird such a popular pet, were the subject of an investigation carried out in Assam in northeastern India. Although adult mynas are seen in pairs for much of the year, a bird does not imitate its mate. Birds will, however, tend to make at least some of the same calls as those made by neighboring mynas of the same sex. There are distinct local dialects confined to one sex. Captive mynas can imitate human noises probably because low-pitched human voices are similar to wild myna calls.

UPI

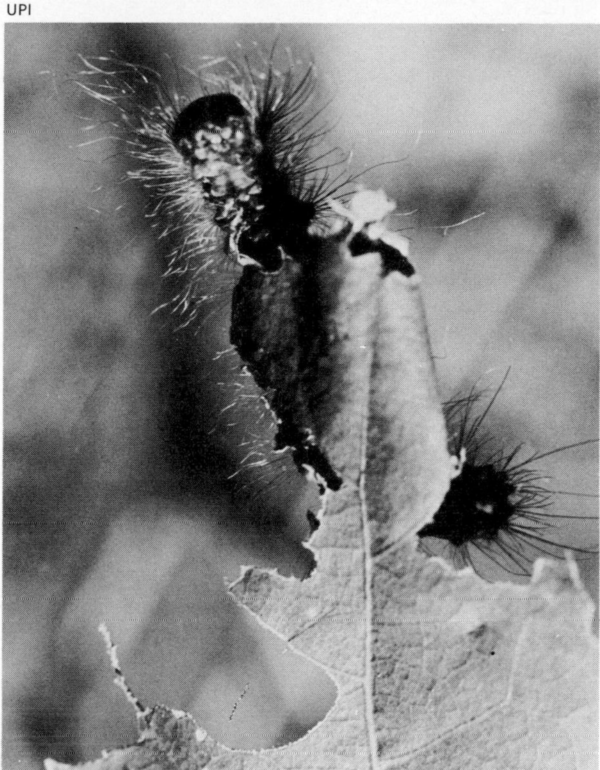

An immature gypsy moth, Porthetria dispar, *rapidly devours a leaf. In an attempt to control these pests, which have been responsible for denuding millions of trees, synthesized female sex attractant has been used to lure and trap the males.*

Developmental zoology. Developmental zoologists, often called embryologists, work with unborn animals. In the past their work has been mainly on the description of the processes of embryological development. More recently, however, their investigations have been directed toward answering questions about how various developmental processes take place within the embryo. Work on cellular differentiation, the process by which cells become unlike each other, is very much in the limelight. It relates to those processes that produce different tissues from the single fertilized cell or zygote.

Cells reject other, foreign cells through immune

responses. To date, relatively little is known about the cell's immunological function—or, again, about its genetic control, which is really the problem investigators would like to solve. Rapidly increasing knowledge of immunological function in the fetuses of many animals emphasizes the need for some means of comparison between the species, since they pass through similar phases during fetal development. Students of fetal immunology would like to know, for instance, if a newborn guinea pig is immunologically as mature as a rabbit at birth and if any direct comparisons of immunological status can be made with a newborn human baby. By applying age equivalence to available information, a Scottish group showed that the onset of the ability to make immune responses occurs at the same stage of physiological development in all mammals and birds.

In September 1970, a British group presented data on fertilization and cleavage to the 16-celled stage in vitro (in tissue culture in a test tube) of human eggs recovered just before ovulation. In the following January the group reported two human embryos that reached the stage of becoming fully developed blastocysts (the hollow sphere of cells formed at an early stage of fetal development) in vitro because of improved handling of the cultures. Another British group reported another first: a technique for growing human digits in a synthetic medium. These kinds of studies have been performed with lower animals for many years. Since much of the work done by zoologists relates ultimately to the work of physicians (who are zoologists), these studies should provide valuable material for investigating the anatomical and biochemical effects of various physiological and pathological conditions.

Cellular and genetic zoology. Many attempts were in progress to extract and characterize the proteins of cell membranes. These studies attempted to localize and isolate a particular biological activity, or to discover some new features of general structural importance. Both these lines were generously represented in the current literature. For instance, recent studies suggested that most animal cell membranes contain many different proteins instead of a single predominant protein, as had been proposed earlier.

The transmission of nerve pulses, heart function, and kidney processes are coupled to a process known as active transport—the movement of sodium and potassium ions across cell membranes. "One of the central unsolved problems of biology," says David E. Green, co-director of the Institute for Enzyme Research at the University of Wisconsin, "is the mechanism by which ions are actively transported across cell membranes and concentrated in the interior spaces." Green advanced a theory suggesting that ions, instead of being pumped out, flow smoothly through membranes because of cyclic changes in the structure of the membranes themselves. The key remaining question is whether his model, established from experiments with mitochondria (small bodies in the cytoplasm, the protoplasm surrounding a cell nucleus), will hold for active transport in all types of cell membranes.

A series of experiments at Princeton University led to the identification of a chemical that harnesses the runaway multiplication of cells common to cancer. The chemical, called Con A, repairs the surface damage found on cancerous cells, thereby causing them to return to normal growth behavior. This reversibility of the phenomenon seems to indicate that the cells' genetic machinery was not altered during the experiments. For the first time, investigators can stop the wild multiplication of cells without completely killing them.

A research team in Baltimore, Md., reported a way of introducing mouse embryo deoxyribonucleic acid (DNA) into mouse cells in vitro. The research has implications for the eventual development of the use of DNA for gene therapy in man, especially with regard to inherited diseases. Until these experiments, no evidence existed that a fragment of host DNA encapsulated in a protein coat could be uncoated by an animal cell. Now that this uncoating has been demonstrated, the question arises as to whether the genetic information in this DNA is expressed.

In other work it was found in Stockholm that the X chromosomes in female cattle start to replicate later than the other chromosomes of the complement, and the Y chromosome is the last chromosome to initiate replication in the normal male. (The X and Y chromosomes carry sex characteristics; the union of two X chromosomes produces a female while an X and a Y produce a male.) Other researchers postulated what they called a phase change in cells, in which mitochondria undergo a major slowing of activity when threshold low temperatures are reached. It was discovered that hibernating animals convert their mitochondria to the type possessed by cold-blooded animals when the hibernating season arrives.

A simple method of cultivation and fixation of the adrenal gland was developed at Yale University. Another group at Yale found that cells in crustacean blood maintain large quantities of polysaccharide (a complex carbohydrate that can be dissolved into two or more simple sugars), in the form of glycogen. The possibility that these cells act like liver tissue for storage and metabolism of polysaccharides was under investigation. A method was described for inducing cell division without mitotic nuclear division (in which division of the nucleus precedes division of the cytoplasm) and nuclear division without cell division in the soil amoeba *Acanthamoeba rhysodes.*

Ecology. Incorporation of carbon-14 from atmospheric nuclear tests in 1961–62 into the microscopic or near microscopic marine plant life known as phytoplankton through photosynthesis, and so into zooplankton and higher levels in the food chain, was used during the past year to determine the flux of organic carbon from the euphotic zone (0–100 m) into the deep sea and into the bottom sediments of the ocean.

One of the prime ecological liabilities of DDT and other chlorinated hydrocarbon pesticides is that they are not biologically degraded except over long periods. Entomologist James W. Butcher of Michigan State University reported that a team led by him demonstrated that two insects—a kind of Collembola called a springtail and a soil mite—are capable of breaking DDT down into the less toxic compound DDE. Other species might also have this ability in varying degrees. The intestinal microscopic plant life of the northern anchovy *Engraulis mordax* was reported as being capable of dechlorinating DDT to the less toxic form called DDD.

The soil mite Folsomia candida, *shown here in both adult and egg form, is one of two insects found to break down DDT into less toxic DDE. This ability, presumably related to resistance to DDT, may also prove to exist in many related species from different habitats.*

Courtesy, Dr. J. W. Butcher, Michigan State University

If a pesticide should increase activity (restlessness) in an organism in the wild, it could work to the disadvantage of that organism by increasing the risk of predation. An investigator studied whether warty newts will preferentially prey on hyperactive, DDT-treated tadpoles of the common frog rather than on untreated tadpoles. Hyperactivity of the treated tadpole apparently did cause newts to locate tadpoles more readily and then stimulated attempts to kill them.

Biological pest controls fall into three major categories: (1) pathogens—various microscopic plants or viruses that are natural enemies of insects but that can be cultivated and applied in far larger numbers than in nature; (2) interference with metabolism or reproduction in insects, as through the sterile male approach; and (3) introduction of natural predators. The U.S. Department of Agriculture hoped to begin an integrated attack on the boll weevil, using chemical pesticides first, then sex attractant (pheromone) traps, and, finally, sterilized males. Ultimately, results might reduce total hard pesticide use in the United States by one-third, since about 80 to 90% of all hard pesticides are used against 100 major insect pests. Investigators predicted that there would be specific biological controls for half of these pests within 10 years.

Scientists at the University of Notre Dame, Ind., developed a new mosquito that they hoped could reduce the world's mosquito population. The researchers irradiated the yellow-fever mosquito, *Aedes aegypti*, to produce mutations. Surviving male mosquitoes with the new chromosome appear normal and strong, but produce sperm that are 75% incapable of fertilizing eggs. Of those offspring that are produced, approximately 80% of the males inherit the sterility factor and thus pass the lethal characteristic to succeeding generations. The new breed of mosquito appears capable of competing with normal males. The World Health Organization planned to test the new strain in India.

Sex attractant has been used to combat gypsy moths. First collected from females and then synthesized in the laboratory, the attractant was used to seduce males into congregating in a limited area, where insecticides were used to kill them. Sex attractants are especially promising for use against bark beetles, which spend much of their life under the bark of trees where most pesticides cannot reach them. A group of researchers in New Zealand, attempting to locate possible sites of sex-attractant production in the female grass grub beetle *Costelytra zealandica (White)*, the principal insect pest of pastures in New Zealand, found that symbiotic bacteria situated in glands in the female produce the substance that attracts the males.

Florida was described during the past year as a biological cesspool of introduced life. As many as 57 species of exotic vertebrates were found in Florida, 10 of which were well-established fish species. Several species from both South American and African stocks had become widespread and were thriving, it was feared, at the expense of native fish. Some of these fish had been released to improve sport fishing; others were the result of accidental escapes from stock ponds.

A new study of deadly Pacific sea snakes reinforced ecologists' concern about the proposed canal across the isthmus connecting North and South America. There are many possibilities of the mixing of similar species from the Atlantic and Pacific sides, and the results are unpredictable. The proliferation of the venomous sea snakes in the Caribbean is one adverse possibility. It was urged, therefore, that when the canal is dug, a foolproof biological lock—preferably of fresh water—be incorporated, and that the most thorough studies possible be done prior to construction.

As of the early 1970s, about one vertebrate species per year was becoming extinct, and probably more than 100 would disappear from the earth in the next 30 to 50 years. The International Union for Conservation of Nature and Natural Resources (IUCN) listed 275 species of mammals and 300 species of birds as rare and endangered. There were 49 species and subspecies of primates on the IUCN list of endangered mammals, constituting more than 10% of all living primate species.

Evolution. A 10-in. skeleton of a cynodont, a mammal-like reptile that was a contemporary and distant relative of *Lystrosaurus*, was found in Antarctica. A *Lystrosaurus* skull had been found in Antarctica in 1969. Both cynodonts and *Lystrosaurus* are members of the order Therapsida—the reptilian ancestors of the mammals. These fossils make a significant addition to the rapidly accumulating data in support of a view that Africa, South America, Australia, India, and Antarctica were once parts of a supercontinent called Gondwanaland, which began to split up 200 million years ago.

Research on dinosaurs during the past year showed that, although pictures depicting sauropods—which reached a length of 100 ft and a weight of 30–40 tons—always show them shoulder-deep in swamps, the anatomy of these dinosaurs points to fully terrestrial, elephant-like habits rather than to more aquatic hippo-like habits. It was found that brontosaurs may not have needed molar-like teeth for grinding tough food; they may have had a powerful gastric mill like that of crocodilians in which ingested animals are ripped and crushed by the stones embedded in the muscular stomach wall. Isolated dinosaur skeletons have been found with concentrated masses of small stones among the ribs. Among the most prominent dinosaurs of the Upper Cretaceous, about 75 million years ago, were the duckbills, so named because of their broad elongated jaws. Their average length was about 30 ft, but in California a team of paleontologists unearthed a duckbill whose fossilized bones measured a full 100 ft in length. It appeared to be of a previously unknown species.

Was Neanderthal man's simian shape the result of the vitamin D deficiency disease known as rickets, as Rudolf Virchow claimed nearly 100 years ago? During the year Francis Ivanhoe of London suggested that Neanderthal man as a race suffered from vitamin D deficiency, and that his diagnostic characteristics are due to the disease, but Ernst Mayr of Harvard University and Bernard Campbell of the University of Cambridge retorted that skulls of the characteristic Neanderthal shape occur in low latitudes and periods of warm climate, when a vitamin D shortage would have been unlikely. Ivanhoe had argued that the incidence of rickets fell off with the passing of a cold stage. Mayr and Campbell, however, pointed out that Neanderthal was replaced by modern man at the peak of this cold stage, and that modern man passed through several subsequent cold phases without reverting to Neanderthal shape. Another Englishman suggested that bone changes in Neanderthal remains are not unlike those seen in such treponemal diseases as congenital syphilis.

The way evolution is taught came under attack in a number of states. The emphasis was on opposition to exclusive presentation of current theories of the origin of life and the diversity of species. The challenge came not from theologians but from qualified scientists belonging to the Creation Research Society, who were pressing to give the theory of divine creation equal time with other explanations of the orgin of man. The society's major success was the inclusion of language it favored in new guidelines adopted by the California State Board of Education. The proposed guidelines had indicated that life probably arose from "a soup of amino-acid-like molecules" some 3,000,000,000 years ago, and that the diversity among present species is the result of evolution through natural selection.

Several members of the society testified against the original guideline language. George F. Howe of Los Angeles Baptist College and Theological Seminary, in Newhall, Calif., held that creationism is no less "scientific" and no more "religious" than the general theory of evolution. Indeed, creationism solves what he called a problem for evolutionists: the large odds that a living protocell—or even a specific protein molecule—could have been formed merely through the accretion of various substances in an ancient ocean. The evolutionist's reliance on mutation as the mechanism for evolution, he said, is contradicted by evidence that the vast majority of mutations are harmful; creationists maintain that cells of each living "kind" were formed rapidly by the Creator. Moreover, the manner in which species appear in the fossil record in a "now you see them, now you don't" manner fits the creationist—not the evolutionist—view.

The board removed the two disputed paragraphs. It substituted a compromise statement that "all scientific evidence to date concerning the origin of life implies at least a dualism or the necessity to use several theories. . . . Creation in scientific terms is not a religious or philosophic belief. . . . Creation and evolutionary theories are not necessarily mutual exclusives. . . . Aristotle proposed a theory of spontaneous generation. In the nineteenth century, a concept of natural selection was proposed. . . . More recently efforts have been made to explain the origin of life in biochemical terms."

Brain and neurobiology. The sympathetic nervous system regulates a vast array of involuntary body processes, including blood pressure and the rapid mobilization of energy. For work elucidating the chemicals that carry messages between nerve cells of this system, three scientists shared the 1970 Nobel Prize for Physiology or Medicine. They were Julius Axelrod of the National Institute of Mental Health in Bethesda, Md., Ulf von Euler of the Karolinska Institutet in Stockholm, and Sir Bernard Katz of University College in London. Axelrod pointed out that von Euler also discovered prostaglandins, hormone-like substances with a score of physiological effects, and predicted that there was another Nobel waiting for someone in that field.

A group in Australia showed that the tympanic membrane of crickets is mechanically tuned to a rather narrow spectrum of frequencies between four and six kilohertz. A group at Berkeley, Calif., found that one central auditory neurone, called the ϕ neurone, of an East African field cricket responds to recorded songs in terms of both rhythm and intensity. Unlike other Orthopteran central auditory neurones, the ϕ neurone shows little or no response decrement during prolonged stimulation. There are obvious advantages to this: orientation and location of a desired individual by another cricket involves continuous listening for minutes at a time, while singing between alternate males can go on for hours.

At the University of Southampton, Eng., cockroaches (*Periplaneta americana*) were mounted in such a way that when one animal placed its leg in a solution, it received shocks at the rate of one per second. A second, control animal got a shock each time the experimental animal was shocked, regardless of the position of its leg. The experimental animal learned within 35 minutes to keep its legs out of the solution, whereas the control animal did not learn to raise its leg.

Courtesy, Dr. George B. Craig, Jr., University of Notre Dame

Bronze-colored eggs of a mutant of Aedes aegypti, *the yellow fever mosquito, contrast with normal black eggs. Knowledge about the mosquito's breeding habits that may help in eradicating this pest can be obtained by observing the easily spotted mutants in field tests.*

This system was used by a British group to study some of the chemical changes that take place in the metathoracic ganglion (the mass of nerve tissue in the posterior part of the thorax or middle division of the insect) during the learning process. The metathoracic ganglia of the experimental animals exhibited approximately one-third as much activity of the enzyme cholinesterase as the ganglia of control or resting animals (which received no shocks). The results indicate that cholinesterase activity decreases markedly in the experimental ganglia on learning. The demonstration of a greater turnover of ribonucleic acid, enhanced protein synthesis, a rapid decrease in cholinesterase activity, and other changes in experimental animals led this group to suggest that these changes constitute the basis of short-term memory storage (one to three days).

—John G. Lepp

Our advanced industrial civilization of the 20th century has generated a problem never before encountered on such a scale—how to dispose of huge and ever growing quantities of refuse and waste. Man always has been confronted with the need to dispose of his wastes, but rapid population growth and strong emphasis on consumption in the last 50 years have made the task increasingly formidable. As a result, our senses are assaulted daily by littered roadsides, fouled rivers, junked automobiles, and other residues of our consumer culture. On the following pages the *Britannica Yearbook of Science and the Future* reviews this situation in a cluster of three special articles. The primary emphasis in these reports is on the individual—how man as a private citizen has contributed to the waste problem and what steps he can take to help resolve it. The next seven pages illustrate in dramatic pictures how our senses are besieged by refuse. Following this photo essay are articles dealing with solid wastes and with sewage.

The Senses Besieged

(Below) Tor Eigeland from Black Star; (opposite page) Stuart Finley, Inc.

John Launois from Black Star

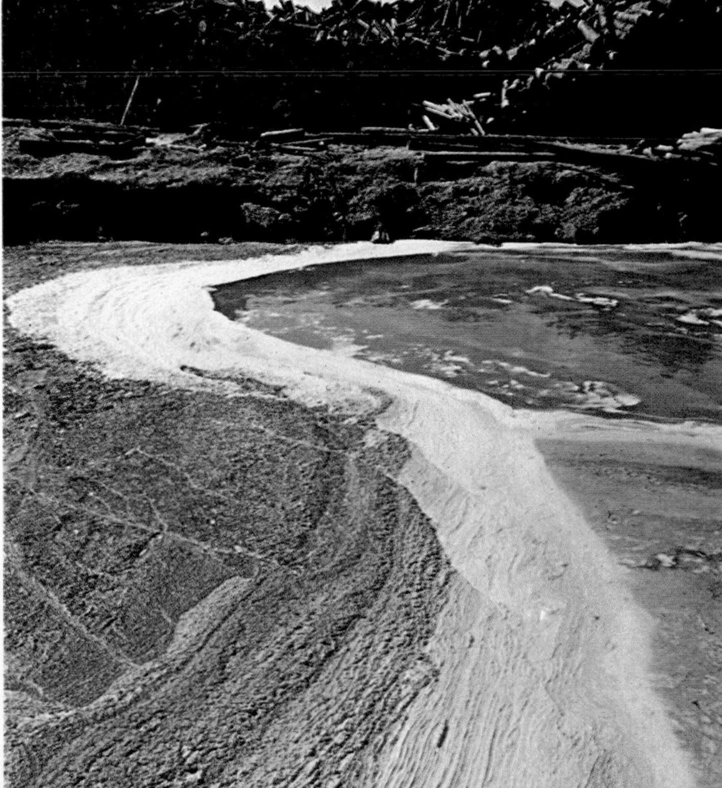

Burning sulfides from flared natural gas hang over a goat pasture in Iran (above), while industrial wastes foul the Rouge River near Detroit (left). Metal cans and paper are major constituents of an unsorted municipal refuse pile (opposite page).

(Below) Jack Fields from Photo Researchers; (opposite page) Jack Rosen from Pix

Industrial smoke darkens the skies in many parts of the world, as evidenced by the control burning of oil that has seeped from an offshore well in Indonesia (above) and the discharge from the stacks of a steel mill in the United States (opposite).

Burk Uzzle from Magnum

Larry Dale Gordon

Dan Morrill

Urban congestion produces such assaults on the senses as the demolition of an old multistory building (above), rush-hour jams on streets and sidewalks (opposite, top), and a horizon cluttered with television antennas and filled with industrial haze (opposite, bottom).

"The great question of the '70s is: Shall we surrender to our surroundings or shall we make our peace with nature and begin to make reparations for the damage we have done to our air, to our land and to our water?" So spoke Pres. Richard M. Nixon in his 1970 State of the Union message. In the following pages "the damage we have done... to our land" is discussed. The author describes how man is becoming buried in the residues of his affluence. He outlines the advantages and drawbacks of the proposed solutions—which range from sanitary landfills to the recycling of refuse, as pictured here. Recycling, he cautions, may not be the all-embracing answer to the problem that many now envision it to be. Rather, a combination of methods may be the most effective way of dealing with our mountains of refuse.

Buried in Affluence

by Percy H. McGauhey

Stuart Finley, Inc.

A compacter moves across a sanitary landfill, spreading and crushing a refuse pile. In an eight-hour day such a machine often can compact 1,500 tons of refuse to a density of approximately 1,100 pounds per cubic yard. Each compacted layer is covered with dirt.

PERCY H. McGAUHEY is director emeritus of the Sanitary Engineering Research Laboratory and special faculty adviser to the chancellor at the University of California at Berkeley.

Our wealthy industrial civilization has produced a problem of growing dimensions—it has created an ever larger quantity of refuse and ever fewer acceptable ways to dispose of it.

Nowhere is the problem more acute than in large cities. A case in point is San Francisco. On the west, that city fronts on the Pacific Ocean, and oceans long have been considered as possible repositories for solid wastes. At San Francisco, however, waters near the shore are shallow, rough, and shrouded seasonally in cold fog that extends many miles out to sea. Thus distance, weather, and physical danger make waste disposal in the Pacific off San Francisco an unattractive alternative. Economics, logistics, water-quality objectives, and the propensity of sunken plastics and other materials to float ashore make it downright unacceptable.

Moreover, north of San Francisco the bay is physically unsuited to any sort of underwater refuse fill. To the east, however, where shallow water and mudflats have been irresistible fill sites for all manner of purposes, one estuary of San Francisco Bay served the city's refuse disposal needs for many years. This area had not yet reached its capacity when, in the mid-1960s, the use of the bay as a sink for refuse began to be questioned. A growing public concern for the environment forced the enactment of restrictive legislation and the creation of special commissions dedicated to "saving the bay." Then, in 1968, crisis struck suddenly. The town of Brisbane, south of San Francisco, reneged on its long-term contract with the city and refused to permit further filling within its boundaries.

Inside the city of San Francisco itself there was no open land except for the public parks and a few hilltops. Beyond the city limits, to the south, other communities were searching desperately for answers to their own problems of refuse disposal. A strict air-pollution authority discouraged incineration by any economically acceptable current technology. Moreover, in San Francisco as elsewhere the only incinerator site acceptable to the public was "somewhere else." The attitude of the general public made matters worse. Refuse collection and disposal had been contracted to two efficient scavenger companies for years. Citizens merely assumed that their wastes were the responsibility of these contractors and increased their demands for an aesthetic environment in which there was no place for wastes.

In this unhappy situation, when all solutions seemed equally impossible, San Francisco was deluged with proposals of every sort, many of which were simplistic and opportunistic. An imaginative and feasible system of long-distance transport via railway eventually was worked out, and an area willing to accept the fill material was found. Preparations were made to contest in court a series of hastily enacted ordinances prohibiting the passage of refuse trains through towns along the route. Then, at the last moment, the plan fell through. The railway company involved revised its contractual terms to a degree unacceptable to the city.

Fortunately for San Francisco, it was then able to find an alternate solution involving the use of refuse filling to increase the land resource of the city of Mountain View, about 30 miles south of its original fill site at Brisbane. This bought a few years of time, which San Francisco and its contractors are using to construct facilities that will rapidly expand present salvage operations and also to install new systems to recover and recycle materials that traditionally have been wasted.

Not every city has found as satisfactory an interim answer to its refuse disposal problems as has San Francisco. In the San Francisco Bay area, for example, 84 agencies still are trying to sequester wastes in each other's backyards while seeking a more suitable solution.

What can be done to deal effectively with mounting piles of refuse? Before discussing possible solutions, an understanding of the way in which the problem arose may be helpful. Although the following circumstances deal solely with the United States, they can be applied with little alteration to most other industrialized countries.

Our heritage of waste

Three distinct periods mark the history of our husbandry of the North American continent: (1) conquering the wilderness environment; (2) exploiting natural resources; and (3) reconquering the environment. In the initial period our potential for affluence lay in the land. To realize this potential, men had to strip away hardwood forests and break up sod with all possible speed. Therein lay the beginnings of a habit of wastefulness that unconsciously became a part of our national and cultural heritage.

Only a few of us living today participated in that pioneer phase of American life. Ours has been, instead, the age of exploitation of resources through the ingenuity and inventiveness of technology. But while applying our genius to the mass production of material things, we have clung to our heritage of wastefulness. The residues of this era of exploitation have been discarded with the same abandon as were the hardwood forests of another era.

The third period of our history only now is beginning to emerge. It is characterized by such concepts as "total environment," "ecology," and "quality of life." In a sense it returns us to our pioneer beginnings, but this time the wilderness is more subtle and complex than trees that stand in the way of the plow or hard rock that overlays the hidden gold. It is a wilderness of polluted waters, fouled air, and automobile graveyards. To many the prospect is foreboding. We have an uneasy feeling that, in order to conquer this wilderness, drastic changes in our concepts of the world are overdue—changes that may force us to surrender much of the affluence associated with our heritage of waste.

Each of the three periods of our history has special significance. From the first we inherited the attitudes that encourage waste production. In the second we raised waste production to the level of a national culture. The third offers promise of solutions to the problems created by the second—only the cost in terms of affluence is unknown and, therefore, frightening.

An important part of the recycling process takes place when crushed glass from used bottles is mixed in a furnace with the raw material used to make new glass.

Courtesy, Adolph Coors Company

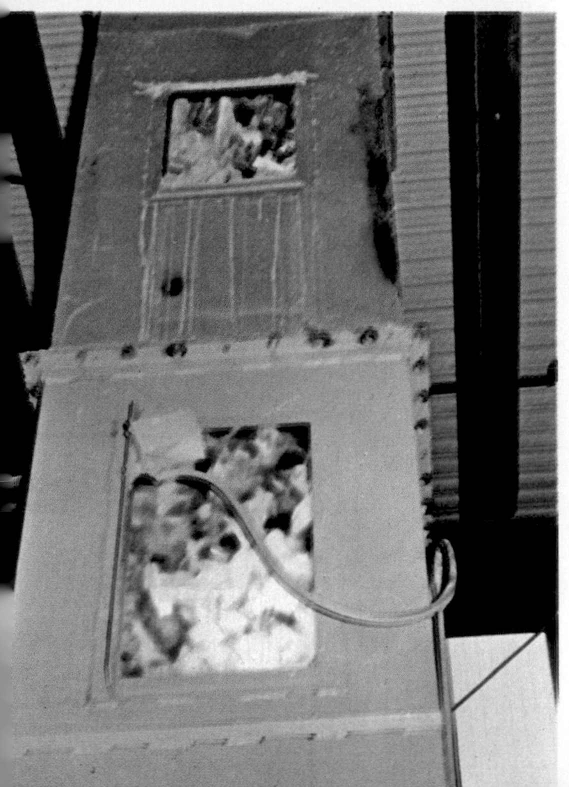

Affluence and waste

The relationship between economic wealth and the wastage of resource materials is a close one. In our affluent society, economic growth inescapably depends upon continual acceptance of goods by consumers. To make room for this new merchandise, the consumer often must discard some of the old, even though it may still be innately useful. Automobiles, household appliances, furniture, and many other hard goods often are discarded long before they are worn out. Thus, affluence is characterized by a rejection of material possessions as "wastes" simply because someone has lost interest in owning them.

In the case of some products this discarding is almost instantaneous. The daily newspaper, for example, is converted from a necessity to a waste merely by looking at it for a few minutes. Bottles, cans, and packaging of all types are converted to wastes by the simple act of removing the product they contain. This discarding is, of course, magnified by the increasing number of people to be accommodated each year at an increasing standard of living. The result was expressed succinctly by Richard J. Sullivan, commissioner of the New Jersey State Department of Environmental Protection, when he stated in late 1970, "growth has outdistanced control over the waste this growth generates."

Magnitude of the waste load

Hollis Dole, assistant secretary of the interior, testified before Congress in 1970 that "in the past 40 years the United States has consumed more minerals, mineral products, and fossil fuels than the entire world in its recorded history." In the long run, wastage of the residues and products of such exploitation will exhaust the nonrenewable resources and could spell the end of affluence, but in the meantime such wastage threatens to overwhelm urban man by its sheer volume. Omitting an estimated 1,700,000,000 tons of mining wastes, 1,500,000,000 tons of agricultural and animal wastes left in fields, and a backlog of 12 million junked automobiles representing about 22 million tons of metal, Table 1 summarizes estimates of the annual load of solid waste on the land environment.

Of all the components of the total waste load cited in Table 1, municipal refuse is by far the most pervasive and heterogeneous. It has, therefore, attracted the most attention. Enough such refuse is produced in the United States each year to make a compacted fill 10 feet deep covering an area of 50 square miles. Much of it originates in the individual household, where a family of four throws away about four tons per year.

Table 2 indicates the types of materials aside from bulk products that are discarded by the householder. It reports the amount and changes in composition of refuse collected from a particular cross-section of predominantly residential areas in Berkeley, Calif., on two occasions 15 years apart. Of particular note is the increase in plastics over the years at the expense of metal cans and metals in general. The "decom-

Courtesy, National Association of Secondary Material Industries, Inc.

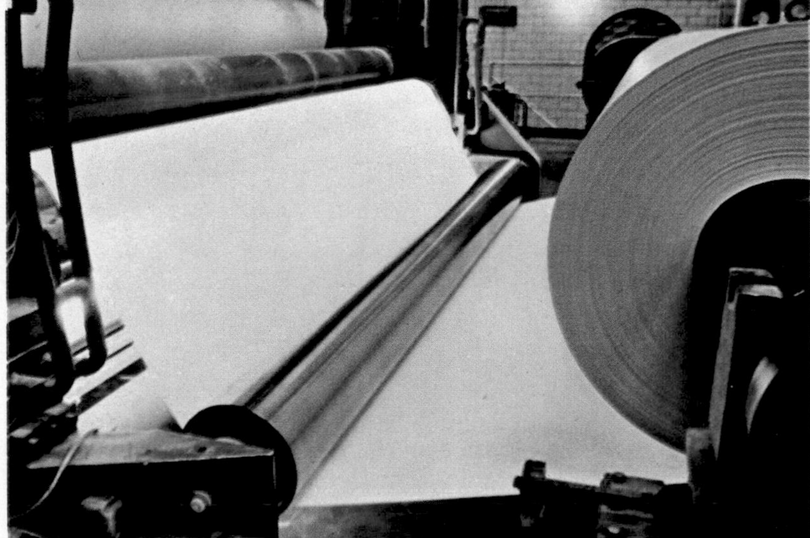

Over 11 million tons of paper stock are recycled each year, providing a means for solid-waste disposal and for conserving our natural resources. Following the delivery of waste paper to a recycling plant (opposite page, top), it is processed and then conveyed by a flow of air to a baler (center), after which it is converted into new fibers in a hydropulper (bottom) and then quality tested in a laboratory (below). At the left, finished paper made with recycled fibers is seen coming off a paper-making machine in rolls.

posable organic matter" cited in Table 2 as a fraction of household refuse is only about one-sixth garbage. From 60 to 75% is paper. In fact, approximately 75% of the total municipal refuse of the nation is paper.

Bottles, cans, and plastics, which represent a large fraction of the packaging wastes category in Table 1, also appear in the municipal refuse along with rubber tires and miscellaneous debris. They originate as wastes throughout the entire community, whereas junked automobiles, industrial wastes, agricultural residues, and animal manures, although of vast amounts, tend to be concentrated at certain points in the environment.

The dilemma of solid-waste disposal

The life of a city depends upon the continual inflow of goods and products from many points of origin. Eventually, these materials either are used and become wastes within the community or are shipped away to be used and become the refuse of another community. Those wastes that remain at home can be managed in one of two ways: by stockpiling within the community, or by export to some other location. The first of these two is limited by the level of human tolerance of wastes or by the availability of land surface. Consequently, export has traditionally been the preferred choice.

To export wastes, there must be both a transport system and a destination. Moreover, the wastes must be loaded upon the transport system, which generally requires that they first be assembled. Three types of transport systems are used throughout the world: the atmosphere, flowing water, and man-made vehicles. The atmosphere is particularly convenient because it is accessible to everyone and is assumed to be free. Burning refuse on the waste producer's own premises or in the town incinerator loads the atmosphere with particles that the winds then scatter elsewhere. Water has been similarly utilized

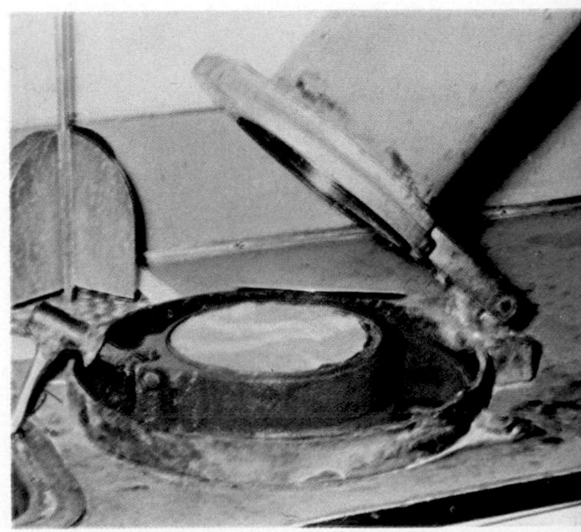

TABLE 1:
SOLID-WASTE LOAD ON U.S. LAND

Nature and amount of waste	Source
Total solid wastes, all types	
(but not including mining and field wastes of agriculture)	
30 pounds per capita per day (national average)	(1)
1,200,000,000 tons per year	(1, 6)
Some components of the waste load	
municipal refuse	
250 million tons per year	(3)
900 million pounds per day	(1)
3 to 4.5 pounds per capita per day (national average, 1965)	
6 to 8 pounds per capita per day (some localities, 1968)	(1, 2, 4)
paper and paper products	
50 million tons per year (1967)	(5)
glass	
14 million tons per year	(1)
metals, mostly junked automobiles	
10 million tons per year	(6)
rubber	
2.2 million tons per year	(1)
packaging wastes	
29,000,000,000 + glass containers per year (1967)	(7)
54,000,000,000 + metal containers per year (1967)	(7)
600 million plastic containers per year (1967)	(7)
total tonnage per year: 83 million tons (1969)	(8)
industrial residues	
13 - 14 pounds per capita per day	(9)
agricultural residues	
total, including portion left in fields: 2,280,000,000 tons per year	(3)
demolition and construction debris	
variable and seasonal; no estimate of total	
cannery wastes	
Varies regionally and seasonally; for example, five canneries in the San Francisco Bay area produce some 4,000 tons per week during canning season.	(11)
animal manures	
national problem of unestimated dimensions; for example, 20 million cubic yards per year produced at margins of major California cities by dairies, animal feed lots, and egg- and poultry-producing installations.	(10)

Sources: cited

(1) Various reports and papers of the U.S. Bureau of Solid Waste Management.

(2) **California Waste Management Study,** a report to the State of California Department of Public Health. Report No. 3056, Aerojet General Corp., Azusa, Calif. August 1965.

(3) **Environmental News Digest,** November-December 1970.

(4) **Comprehensive Studies of Solid Waste Management,** Public Health Service Publication No. 2039, Bureau of Solid Waste Management, 1970.

(5) **Policies For Solid Waste Management,** National Academy of Engineering, 1970.

(6) Value computed from Source (3) on the basis of 6 million automobiles junked per year.

(7) **The Role of Packaging in Solid Waste Management, 1966 to 1976,** Public Health Service Publication No. 1855, Bureau of Solid Waste Management, 1969.

(8) **Proceedings of the First National Conference on Packaging Wastes,** Bureau of Solid Waste Management and University of California at Davis, September 22 - 24, 1969.

(9) Basis of this value uncertain. Source (3) shows only 3 pounds per person per day. Quoted value may include residues disposed of by industry on own property, including chemical residues.

(10) Hart, S. A., and McGauhey, P. H., "The Management of Wastes," Food Technology, April 1964, page 30.

(11) Data from National Canners Association, Berkeley, Calif.

TABLE 2: AVERAGE PERCENTAGE COMPOSITION OF HOUSEHOLD REFUSE; AND CHANGES OVER A 15-YEAR PERIOD (BERKELEY, CALIF.)

Type of material (values in percent of total)*								
Year	Metal cans	Bottles, broken glass	Rags	Metals	Shoes	Plas-tics	Decom-posable organic matter†	Misc. wastes (no value)
1952	10.0	11.4	1.5	0.6	0.2	trace	69.0	7.4
1967	8.4	11.3	1.1	0.3	0.3	1.9	69.7	7.1

*Estimated 2 pounds per person per day, 1952; 3 - 4 pounds, 1967
†Paper, yard trimmings, grass
Source:
Comprehensive Studies of Solid Waste Management, Public Health Service Publication No. 2039, Bureau of Solid Waste Management, 1970.

TABLE 3: SUMMARY AND EVALUATION OF DISPOSAL PROCESSES

Process		Evaluation
Open dumps	75%	Generally unsatisfactory
Sanitary landfills	8%	Satisfactory, although land limited
Incineration	12%	75% of 300 incinerators substandard because of incomplete burning and air pollution
Compost Discharge to sewer Salvage	5%	Generally satisfactory

Sources:
Environmental News Digest, November - December 1970;
Solid Waste Report, November 30, 1970.

by the citizen and the community, especially to carry away bodily wastes and the garbage from the household (*see* Feature Article: WASHING OUR DIRTY WATER). The wastes that man himself has had to transport traditionally have been loaded on vehicles and hauled to a dump site beyond the confines of his own city and there abandoned in as simple a manner as society will tolerate.

A threat of burial by our own affluence results from a breakdown of all three of these systems at a time when we are still adhering to the notion that both the residues and the products of resource use are to be treated only as wastes. The use of the atmosphere as a transport system is each year less permissible because it could eventually become so polluted that man could not breathe it. The flowing water method is being limited by increasingly stringent laws requiring industries and municipalities to remove dissolved and suspended solids from the waste water before releasing it to the environment.

But more significant than the loss of air and water transport systems is the loss of destination for the refuse truck. Economic considerations make refuse disposal noncompetitive with other urban land-use alternatives, and it long has been overlooked in land-use planning. Thus, as one incorporated city has impinged on another, the freedom of any

Metal refuse, separated from other municipal waste, awaits further treatment at a U.S. government recycling center in Madison, Wis. The metal will be shipped to a federal laboratory in Maryland for experiments aimed at finding effective and economical ways to reprocess it for reuse by manufacturers.

Ron Sherman from Nancy Palmer Photo Agency

Courtesy, Heil Co.

The load of a refuse collection truck (left) is transferred at a central station to a specially designed compaction trailer (right). The trailer is equipped to haul the loads of many collection trucks to the disposal area. Thus, the transportation of refuse is made more efficient.

one to export its solid wastes beyond its own boundaries in traditional fashion has drastically declined. By 1971 only limited land area suitable for refuse fills existed within most urban areas. The U.S. Bureau of Solid Waste Management reported in November 1970 that of 34 cities relying on sanitary landfills the time remaining before the fills reached capacity varied from 0 to 25 years. Of these cities 14 had less than 2 years, 18 less than 5 years, and 28 less than 10 years of remaining fill capacity.

Possible solutions

Dedicating land area to the disposal of the entire solid-waste load and not permitting materials to become wastes at all represent the two theoretical extremes of possible solutions to the refuse problem. In the practical sense, however, man must resort to a combination of methods. Only about 50% of the waste load of Table 1 is combustible or biodegradable (capable of being broken down by bacterial action), and there is little prospect that such materials as broken concrete, brickbats, and ashes are going to be treated as resource materials. This was recognized by the authors of the federal Resource Recovery Act of 1970, when they included both new methods of disposal and recovery of resource materials in the objectives to be encouraged through government funding.

As a basis for evaluating the present and future of waste disposal and recovery, Table 3 summarizes what is now being done with most materials other than junked automobiles, animal manures, and agricultural residues.

Land disposal

Because of the scarcity of land in the vicinity of urban centers, long-distance transport to land of lower economic value will be needed in most communities. To this end considerable research has been supported in the United States by the Bureau of Solid Waste Management on the technology and economics of transporting shredded refuse by pipeline. Both San Francisco and Philadelphia are among those cities that have explored another alternative—long-distance rail hauling of compacted refuse. High costs and the lack of regional jurisdictions, however, have delayed development of practical systems, but there seems to be little doubt that the future will see landfills at distant locations that serve a number of communities simultaneously. Such solutions, which will be entirely compatible with the national objective of recycling resource materials, involve three conceptual possibilities for landfills as replacements for open dumps:

1. Landfills should be limited to the minimum number and acreage necessary to dispose of the nonbiodegradable materials that have no resource value, such as broken concrete, earth materials, and similar rubble.
2. Unsorted refuse, except for obviously reclaimable fractions, should be considered as cheap fill material for reclaiming a land resource for such beneficial uses as parks and playgrounds.
3. Landfills of unsorted refuse should be constructed in areas where competition for land use is at a minimum, and should be maintained as stockpiles of miscellaneous resource materials for reclamation at some appropriate time in the future. In the meantime, such stockpiles, in the form of man-made hills, might be used to enhance the environment or for sledding, tobogganing, or other recreational use, as in the much-publicized "Mount Trashmore" concept.

Incineration

Incineration as an alternative to open dumping and landfills has been given serious consideration in recent years. Various new types of furnaces have been proposed, and high-temperature burning capable of reducing mixed refuse to a slag has been demonstrated. A 100-ton-per-day pilot plant of such a nature was built at Whitman, Mass. There are, however, deterrents to high-temperature incineration. These include maintenance costs, generation of nitrogen oxide air pollutants, and production of hydrochloric acid from the burning of polyvinyl chloride plastics.

Municipal refuse burns in a Milwaukee incinerator. Unable to meet new air-pollution standards, the incinerator was closed and the area converted to a waste-removal transfer station.

Courtesy, Waste Management, Inc.

How a national resource recovery policy will affect incineration is not yet clear. However, preliminary signs indicate that conventional systems are losing favor. For example, a citizen's commission in New York City has recommended that the city drop plans for conventional incinerators in favor of new technologies. One such technology is pyrolysis, the fractionating (breaking up into constituent parts) of organic matter by heat. A number of research groups are investigating this process, and the Monsanto Co. has developed at St. Louis a plant that will fractionate 35 tons of matter per day. Such a system could create materials that might become useful for some chemical industry, or be burned efficiently to carbon dioxide and water without polluting the atmosphere.

Such prospects notwithstanding, there were in 1971 about 300 in-

cinerators in use in the United States. From the ash of one ton of municipal refuse in such installations the U.S. Bureau of Mines has recovered from 2 to 9 ounces of silver and from 0.02 to 0.05 ounces of gold. The bureau also discovered that incinerator residues contain as much as 30% iron, 1.5% nonferrous metals (aluminum, copper, lead, zinc, and tin), and 44% glass.

Composting

The small fraction (5%) of refuse shown in Table 3 as being converted to compost, discharged to the sewer, or salvaged is of special interest. As a method for producing a soil conditioner from the organic fraction of solid wastes, composting has been demonstrated to be technically feasible at all levels up to a full-scale plant. However, it has been demonstrated that large-scale agriculture is not interested in either compost or animal manures, and no other market is capable of utilizing the vast amount that could be produced from solid wastes.

The reasons for this lack of interest are that there is adequate organic matter in soil as a result of roots and the stubble of crops, and also that commercial fertilizers are much easier and cheaper to apply. Even in Europe, where municipal composting flourished for a generation or more, farmers are losing interest in its use as commercial fertilizers become more easily available. Therefore, composting does not appear to be a practical method of resource recovery. Moreover, as a way of protecting and enhancing the environment, it cannot compete with stockpiling in landfills.

Resource recovery

Much effort and imagination currently are being devoted to devising methods of extracting useful resources from the refuse pile and recycling them through the economy. These methods, to be economically feasible, require more than the simple direct reuse of waste products. Industry may have to develop new materials with the objective of disposability or reusability, and design its products so that their components can be isolated and reprocessed. There has been some work along these lines. For example, early in 1971 a plan to recover about 260 tons of material from a daily output of 570 tons of refuse was reported by Hercules, Inc., in Delaware. Like many earlier proposals, the project envisions subjecting the material to shredding, segregating, pyrolysis, and composting. Unfortunately, this series of processes ranges from a difficult but known technology (shredding) through two technologically and economically imperfected systems (segregating and pyrolysis), to a simple process of known technology for producing something that no one wants (composting). But in spite of the difficulties involved, the time is at hand for developing new technologies, a matter in which both industry and government are interested.

Unfortunately, the direct reuse of salvaged items has, in some cases, turned quickly from a profitable to a losing business. For example, the Sunset Scavenger Co. in San Francisco turned a profit for several years from the recycling of champagne bottles. The business collapsed with the introduction of the plastic stopper, which could not be used to seal the old cork-stoppered champagne bottles. Similarly, the substitution of synthetic fabrics for cotton cloth erased the profit in rag salvage. And a profitable return on recycling of paper in punched business-machine cards disappeared when plastic coating of the cards to increase their wear in computers was introduced.

As the first enthusiasm for resource recovery becomes sobered by the hard facts of reality, some puzzling problems will have to be resolved. Some of the most difficult arise from assumptions that overlook the equilibria of our social and economic system. They leave unanswered such questions as: When does recycling of residues achieve resource recovery? When does it simply lighten the waste load on the environment? And when does it merely transform one material that no one wants into another that no one wants? Examples of where these questions will arise abound in relation to the refuse materials most often mentioned for reprocessing—junked automobiles, paper, rubber tires, glass, and metal cans.

The junked automobile is an excellent example of a product from which the gain to society resulting from reclamation is almost all in the form of nonrenewable resources, being about 89% by weight reusable metal. There are, however, several problems with old automobiles: they are scattered widely throughout the country; burning them to remove combustibles generally is prohibited; and small operators often cannot afford the equipment necessary to crush the bodies. Giant shredders have been developed that are capable of reducing the metal of automobile bodies to recoverable components and of converting the paint and combustible fractions to filterable dust. They are suited especially to large cities such as New York, where 73,000 automobiles were abandoned in the streets in 1970. These shredders, however, themselves generate environmental problems of noise and the heating of a considerable amount of water. But they are effective, and if rail-mounted models capable of being moved about the country were developed, they could become a major factor in converting the junked automobile from an environmental insult to a valuable resource.

In contrast with the automobile, paper is an example of a product in which it is difficult to tell where recycling ceases to represent a recovery of resources and becomes merely a reduction in waste load on the environment. Although the salvageable fraction of the 50 million tons per year of paper reported in Table 1 has on occasion been equated to the yield of 91.5 million acres of forest land, it may not be assumed that a vast resource would be saved if that paper were recycled. Because pulpwood is planted and harvested like other crops on a continuing basis, it is far from clear how much resource might be saved by paper salvage. If recovery causes tree farms to be put out of business, there is little gain for the environment and a definite loss for the economy. In such a circumstance, who would inherit the once carefully managed forest environment and for what purposes? These are

Ron Sherman from Nancy Palmer Photo Agency

Tin is recovered from scrap metal by treating the scrap with chemicals. After being deposited in a drum (opposite page, top), the scrap is immersed in and then lifted from a chemical bath (opposite, bottom). The detinned metal (above) can be processed further to separate out its steel and copper components.

Margot Granitsas from Photo Researchers

Scrapped and compressed automobiles are a prime source of reusable metal. Giant shredding machines reduce the crushed metal bodies to recoverable sizes for further processing.

among the many problems that arise when man sets out with uninformed enthusiasm to save both the environment and natural resources through waste recovery.

Simplistic answers are heard most commonly in relation to the reuse of packaging wastes, especially glass bottles and metal cans. Most common is the idea that bottles should be returnable under a system of financial rewards or of penalties imposed by law. This practice generates new problems. Many retailers resist absorbing costs of floor space for storage and of manpower for handling returned bottles. In the system of glass use, between the man who quarries glass sand and the consumer who throws away the bottle, lawmakers have found it difficult to decide where the penalty for not returning a bottle should be imposed. The glass manufacturer, bottle maker, producer of bottled product, and seller of product have no clear system for passing the penalty along to the consumer who, after all, is the one who makes the bottle a waste. Then there are questions of racketeering in refilled liquor bottles, and dozens of technical problems related to the reuse of colored glass.

Plastic containers are even more vexatious. Recovery of used plastic is difficult, and its resistance to biodegradation gives it a distressing degree of permanence in the environment. Consequently, plastics are a prime candidate for the stockpile of resources man does not yet know how to use.

The problems associated with metal cans are no less complex than those related to glass and plastic containers, although they are somewhat different. Because the use of metal cans has been declining, rapid changes in packaging practice may be catastrophic to an enterprise that invests in reprocessing equipment for cans. Also, the economic incentive for returning cans, as well as bottles, is usually not sufficient in the present-day affluent society.

Thus, it seems evident that simple reuse of products is a naïve and unlikely answer to the problems of solid-waste disposal and resource recovery. Resource recovery must involve an intermediate reprocessing of wastes to produce new types of resource materials. In this way it can offer a new opportunity for industry to generate wealth from the residues of affluence.

Difficult as the technical problems may be, and however far the ripples created may extend into the economic and social structure of our culture, both environmental and conservation considerations require that resource materials be recycled. That is the task of reconquering the wilderness now confronting us. The alternative is for historians, recalling the demise of our culture, to record that we were buried in affluence.

See in ENCYCLOPÆDIA BRITANNICA (1971): REFUSE DISPOSAL.

FOR ADDITIONAL READING:

Bulrico, F. A., "Solid Wastes and Land Pollution?" *Current History* (July 1970, pp. 13–17).

Corey, R. C. (ed.), *Principles and Practices of Incineration* (Interscience, 1969).

Goldstein, J., *Garbage as You Like It* (Rodale, 1969).

Harrison, G., "Mess of Modern Man," *Natural History* (January 1970, pp. 68–69).

Schweighauser, C. A., "The Garbage Explosion," *Nation* (Sept. 22, 1969, pp. 282–284).

Sittig, M., *Water Pollution Control and Solid Wastes Disposal* (Noyes Development, 1969).

Small, William E., *Third Pollution: The National Problem of Solid Waste Disposal* (Praeger, 1971).

AUDIOVISUAL MATERIALS FROM ENCYCLOPÆDIA BRITANNICA EDUCATIONAL CORPORATION:

Film: *The Garbage Explosion.*

A lake 10,000 square miles in area and once rich with marine life is now virtually dead; a river is so filled with oil and chemicals that it bursts into flame. Lake Erie and the Cuyahoga River are only two of the dramatic chapters in a story that is being repeated constantly throughout the industrialized world—the fouling of our waters. How can we prevent further damage? Can we reverse the destruction that has already occurred? The author traces the historical development of the flowing-water treatment of sewage, one of the great engineering achievements of the 19th century. He explains how extreme pressures are being put on the system by growth of population and industry. Scientists are attempting to meet this new challenge by devising more thorough methods of sewage treatment. The author describes their progress to date and assesses the possibilities for future improvement.

Washing Our Dirty Water

by Abel Wolman

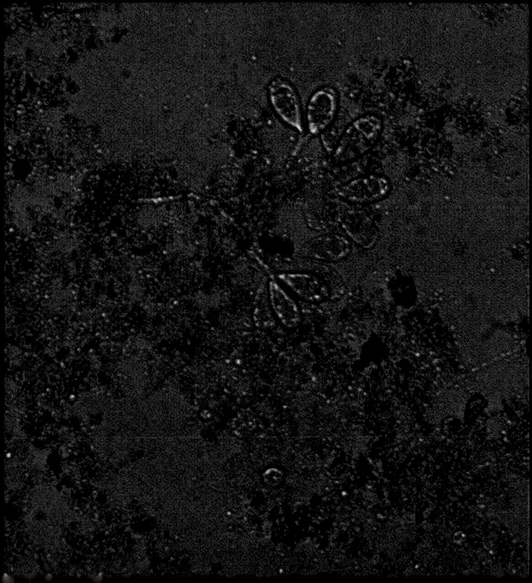

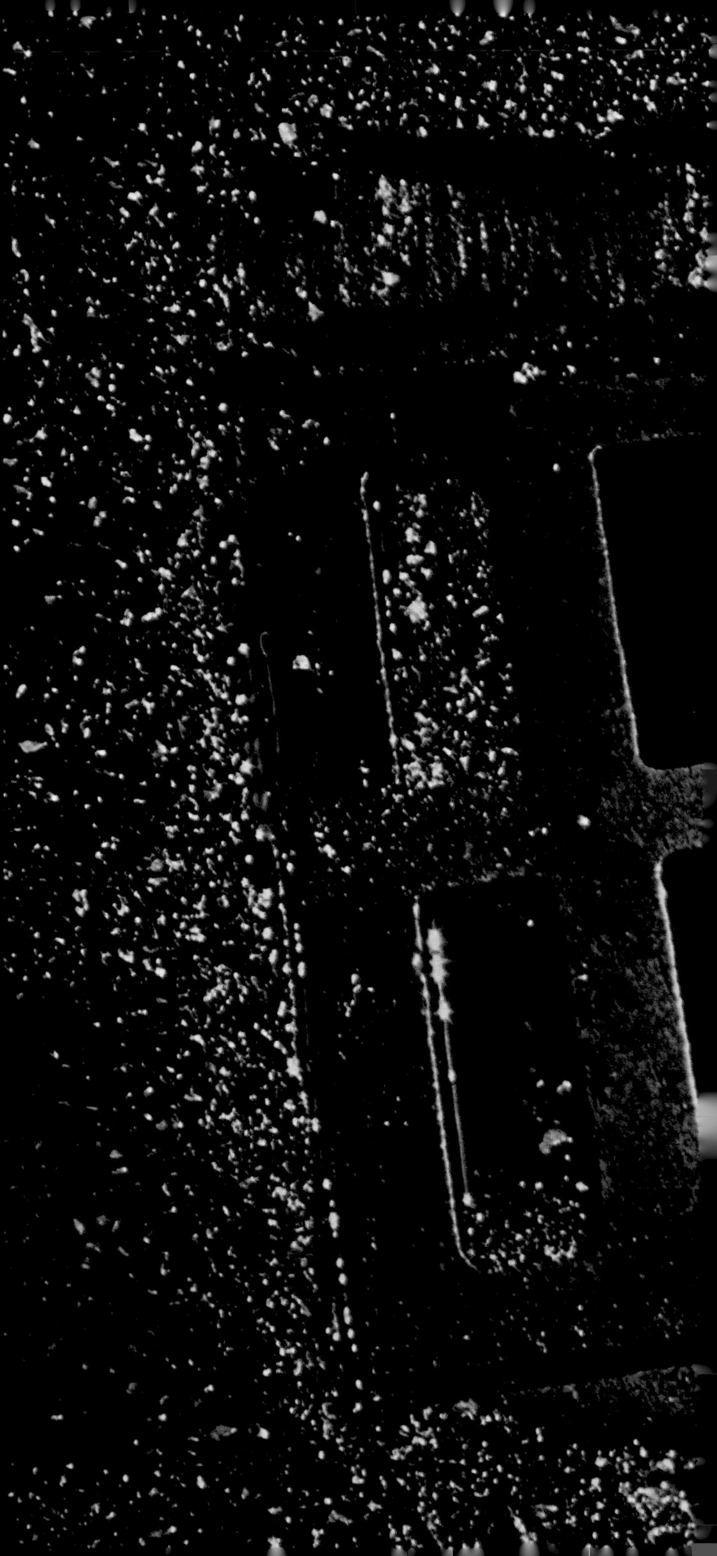

The Bettmann Archive

Sewage disposal in the Middle Ages generally was accomplished simply by dumping wastes onto the public streets and alleys (right). In the 20th century, openings on the streets (opposite) allow waste and refuse to flow into underground sewers.

Scientific sewage disposal is a comparatively recent development, even in the advanced countries of the world. While often profusely illustrated and lavishly described in history books, the great sewers of antiquity actually were storm-water surface drains rather than true sewers. Their function in the removal of wastes was purely incidental to the major purpose of providing for the rapid removal of rain runoff. In fact, until about 1815 the discharge of any wastes, other than kitchen slops, into the drains of London was prohibited by law. In Paris the same policy was continued until 1880.

The result of such prohibitions was the accumulation of extraordinary amounts of decomposing organic matter in the streets and alleys of all the cities of the Western world. Finally, however, an investigation of sanitary conditions in Great Britain in the mid-19th century provided the basis for the relaxation of these restrictions and the rapid introduction of human wastes into the storm drains. Even in the United States, where many storm-drainage systems existed as far back as the 17th century, their use for waste disposal did not become accepted on a broad scale until about 100 years ago.

ABEL WOLMAN, a member of both the U.S. National Academy of Sciences and the National Academy of Engineering, is professor emeritus of sanitary engineering at Johns Hopkins University.

With the advent of the water-carrying sewerage system, the community then assumed responsibility for the disposal of wastes of the individual. In this process, however, unsanitary conditions surrounding each dwelling were merely transferred to the outskirts of each city. There, the concentrated excretions from the entire population had to be disposed of. Since the water-carriage system diluted the wastes from the individual by almost one hundredfold, the new problem also involved a large volume of water that contained human excreta as well as kitchen and bathing wastes.

Even today, the disposal of vast quantities of sewage and the waste water that carries it continues to constitute a major challenge confronting many parts of the world, where sewage-disposal systems are being strained to their capacity by the rapid growth in population and industrialization. So far, however, the search for substitute systems, hopefully less costly and potentially less difficult to manage, has been elusive and frustrating.

Bill Arsenault

Magnitude of the waste-disposal problem

For each 100 gallons of sewage per capita per day delivered from the average U.S. community, less than 0.5% represents true waste ingredients. Although engineers concerned with sewage disposal are accustomed to thinking in terms of large volumes, few laymen appreciate the quantities of sewage produced by a modern city. Each day a typical U.S. city of one million people generates approximately 500,000 tons of waste water carrying about 100 tons of suspended solids.

The task of handling such large quantities of matter is obviously a major obligation of society. Although about 130 million Americans were served by sewers in 1970, the wastes of nearly 70 million others—those who live chiefly in small towns, suburbs, and in rural areas—still were being piped into backyard cesspools and septic tanks. When these devices are properly designed and the receiving soils are not overloaded, they create no particular sanitation hazard. Unfortunately, in many areas neither of these two criteria is met.

The principal pollution hazard arises where wastes collected by a sewerage system are discharged into a lake or river without adequate treatment or without any treatment whatsoever. As of 1970, the wastes of approximately 30 million Americans were discharged without being treated at all; the wastes of about 30 million more received treatment considered by federal agencies as inadequate. Generally adequate secondary treatment was applied to the wastes of 82 million people. (The term "secondary treatment" covers a variety of techniques, often used in combination. These techniques, which are discussed later, include extended aeration, activated sludge—an accelerated form of bacterial degradation—filtration through beds of various materials, and stabilization ponds.)

Cities, of course, account for only a part, and probably not the major part, of the pollution that affects the nation's waterways. Industrial pollutants are far more varied than those in ordinary sewage, and their removal often calls for specialized measures. Even in states with adequate pollution-control laws there are technological, economic, and practical obstacles to enforcing the laws. Several bills passed by Congress enlarged the role of those federal agencies concerned with the pollution of interstate waterways, and were sometimes helpful in strengthening the hand of local law-enforcement bodies.

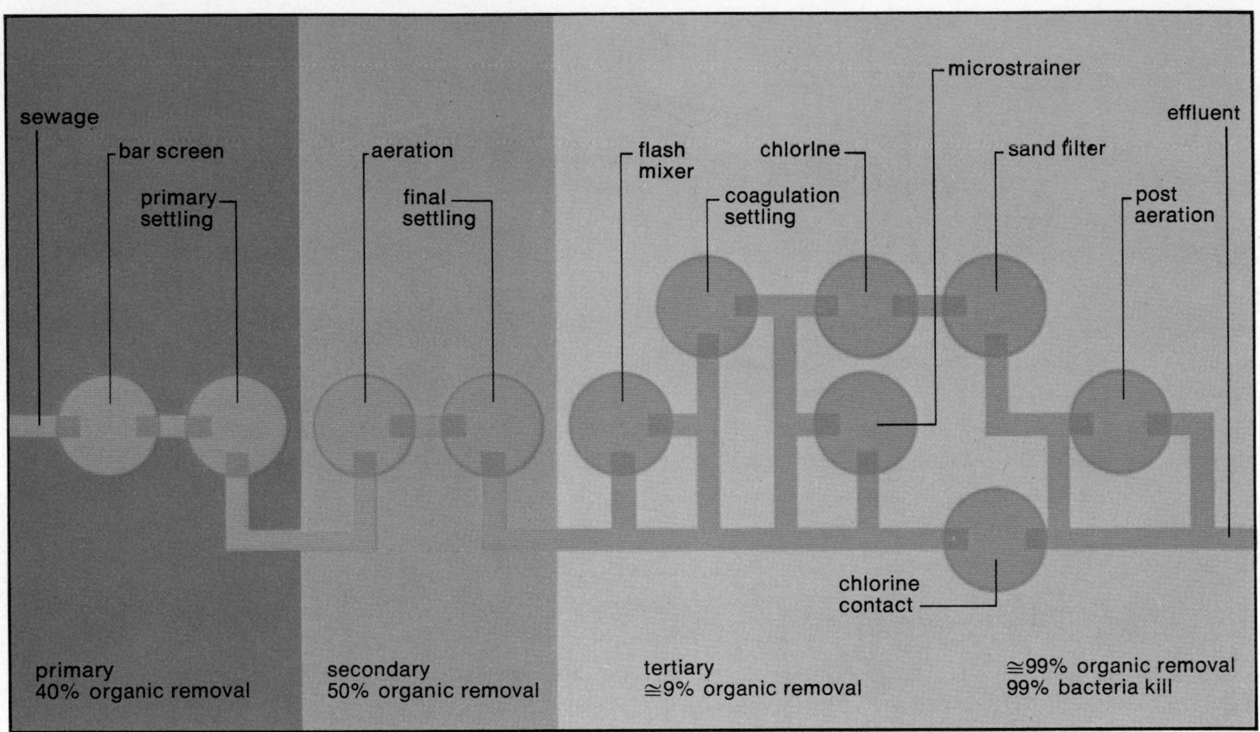

Flow line of a waste-water treatment plant shows the three main stages in the artificial processing of sewage. From the left edge to the first vertical dotted line is the primary stage, during which 40% of the organic matter is removed from the water. Secondary treatment is represented by the area between the two dotted lines; it removes an additional 50% of the organic matter. To the right is tertiary treatment, during which a final 9% of the organic wastes is removed.

Although local and federal laws have contributed to a significant improvement in sewage treatment in the United States between 1942 and 1970, much work remains to be done. Only during the last five years of this period did the rate of sewer installation begin to overtake population growth. Now, the present investment in sewers and sewage-treatment plants in the United States is well over $12 billion. Their replacement value, however, is much higher. It is estimated that replacing obsolete facilities, improving the standard of treatment, and providing for population growth will require an investment of more than $1 billion a year in treatment works for the rest of the decade. This figure does not include the cost of extending sewage-collection systems into new urban and suburban developments. This factor may add at least another $1 billion to the annual requirements, for an approximate total expenditure of more than $2 billion a year.

Waste disposal in less developed countries

In the world's less developed countries, where a large proportion of the people live in rural areas, the problem of waste disposal is both great and unresolved. Even most urban areas in these countries have only primitive water-carriage sewerage systems. The problem is further compounded by the cost or scarcity of water, which, in relation even to medium-sized aggregates of population, makes the installation of water-borne sewage almost impossible.

In the Middle East, for example, little has been done about sanitation. In a few places, where ministries of health have been especially active or internationally assisted, sanitation projects have been started; satisfactory latrines have been installed by individuals, and sometimes have been well maintained. By and large, however, these have been the exception rather than the rule, and the safe disposal of wastes remains a critical challenge. Discharges are promiscuously distributed in fields, on the banks of irrigation canals or drains, or in secluded corners of village streets. Latrines are rare, mainly because they are too expensive for most people. Use of human wastes for fertilizer is fairly widespread but relatively uncontrolled, and, as a consequence, generally ineffective.

In rural Brazil, several types of inexpensive waste-disposal facilities have been developed. These include the bored-hole latrine and two types of pit privies. But their continued maintenance remains a problem, and the use of their contents for any well-controlled agricultural purpose is minimal.

In the Philippines, surface disposal in rural areas has been found to be quite convenient—and quite dangerous. Bamboo groves in backyards provide privacy, wastes quickly dry up in many seasons of the year, and odor and fly problems have not been too severe. Under the pressure of the nation's Central Health Department, however, more effective means of waste disposal, such as the pit privy, gradually are being installed in village and farm areas. It is hoped that this will help control gastrointestinal disease, one of the leading causes of infant death in that country.

In East Pakistan, where population density is very high, many preventable diseases that originate in sewage prevail in great numbers. Because disposal of human wastes is extremely primitive, thousands of cases of cholera, typhoid, dysentery, and diarrhea occur each year.

More than 80% of India's population lives in small villages, all dependent for their existence on agriculture. Neither good water supply nor satisfactory waste disposal is to be found in most of them. The towns and smaller cities of India are not served much better. In these locations, where the dry-pail type of latrine is quite common, socially outcast scavengers remove and clean the latrines manually. They are washed in the street drains, and only a small quantity of the wastes ever finds its way to any central disposal ground.

Courtesy, The Metropolitan Sanitary District of Greater Chicago

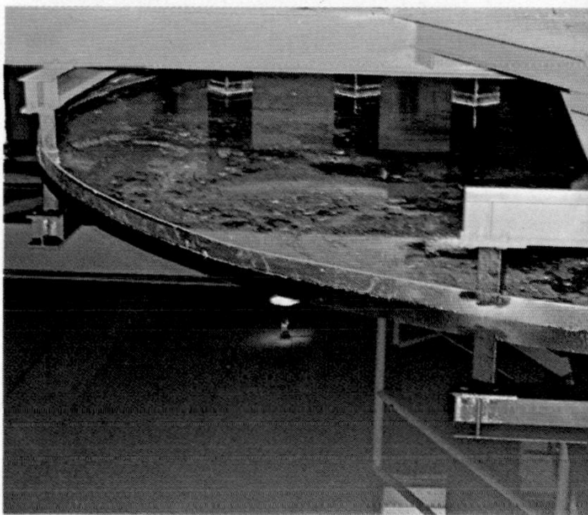

Sewage undergoing advanced processing is mixed with coagulation material in a flash mixer (top). It then goes to a coagulation tank (bottom), where all the solid particles suspended in the waste water coagulate and drop out.

The nature of sewage wastes

The undesirable constituents of sewage generally are classified as follows: living germs, dead organic matter, and chemical wastes. The first of these creates disease; the second produces nuisances; the third carries the potential of specific toxicity. The great epidemics of the past, such as those of typhoid fever and cholera, had their origins in human wastes that polluted private and public water supplies. Where these supplies were inadequately purified, which was often the case in the 19th and early 20th centuries, diseases of intestinal origin were extremely widespread.

The severe epidemics of Asiatic cholera in Hamburg in 1892–93 probably taught the lesson of the danger of water-supply contamina-

(Left) Stuart Finley Inc.; (right) The Bettmann Archive

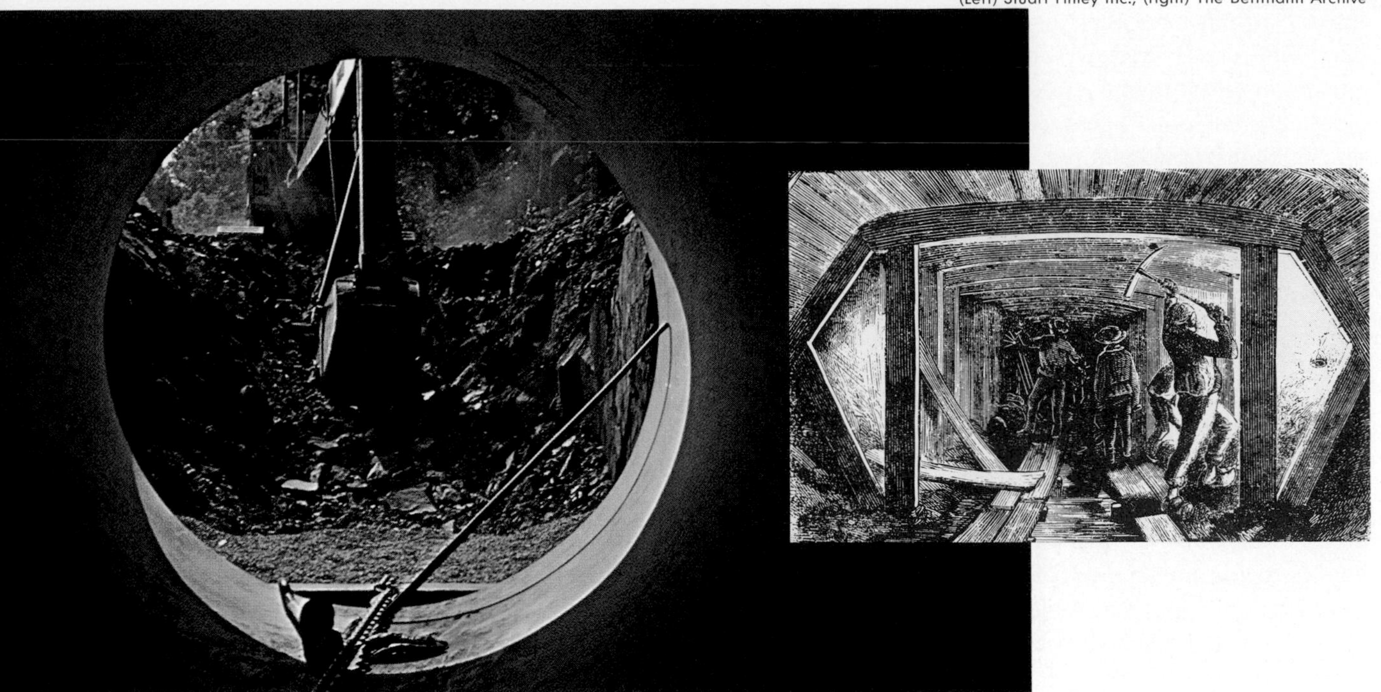

Building a sewer in 1858 (above, right) required the use of considerable manpower. In the mid-20th century, machines (above) have taken over much of the labor.

tion in a fashion unequaled by any other experience in sanitation. The acceptance of the germ theory of disease was at that time less than a year old. It is disconcerting, therefore, to record that in 1970–71 many parts of the world experienced a cholera pandemic (an outbreak of disease that covers a large geographic area and affects a high proportion of the population). Because it occurred in those areas that did not have effective sanitation, the probability of its spread into economically developed countries was low. Nevertheless, during the period 1961–70, the disease did spread to at least 46 countries. In 1970 alone, approximately 38,000 cases were reported in Africa, Asia, and Europe. Since many cases were neither detected nor reported, it was conservatively estimated by the World Health Organization that, in 1970, probably at least 400,000 persons had been affected by cholera.

Natural purification

Since sewage contains bacteria, cysts, viruses, and other biological forms capable of causing disease when ingested by man, the significance of its prompt removal from contact with materials to which man may be exposed is obvious. In the past, most of the large cities of the Western world disposed of their wastes by taking advantage of the natural purification capacities of nearby bodies of water. This method was practiced by New York City, London, Chicago, Cleveland, and other cities. It was and is a legitimate use of such resources, provided, however, that the amount and nature of the wastes to be assimilated are carefully and adequately related to the capacities of such waters. Because modern technology provides the tools by which to evaluate these balances, the discharge of regulated amounts of wastes with only primary treatment to remove suspended materials into such bodies of water is not only permissible but also of great economic value in reducing the investment in sewage treatment.

Unfortunately, such natural purification has been relied upon in many instances well beyond the point at which it was either successful or appropriate. With the growth of urban communities, in particular, natural purification procedures became increasingly less adequate. For this reason, major cities of the world have had to supplement natural purification in nearby waters with artificial processes to prevent those waters from producing such nuisances as odors, sludge deposits, oily surfaces, and objectionable physical appearances.

Artificial treatment processes

The development of artificial processes for treating sewage began approximately 100 years ago, because it was apparent even then that some rivers and harbors were becoming polluted. Early investigators turned promptly, therefore, to a search for artificial methods that would accomplish the same purposes as natural purification, would entail minimum costs, and might provide, if possible, for the recovery of important organic constituents.

The royal commissions on sewage disposal in Britain recognized the fact that the only way to prevent the increasing pollution of rivers was to purify community sewage. These commissions decided that the most appropriate way of doing so was to dispose of the sewage on land. This procedure also offered an economic incentive in that the nitrogen and phosphorus in the sewage could be reclaimed for use in fertilizer. However, the attempted disposal on land of vast quantities of sewage water, particularly in Britain, which has relatively few dry months in a year, was a dismal failure. It was also a failure in many

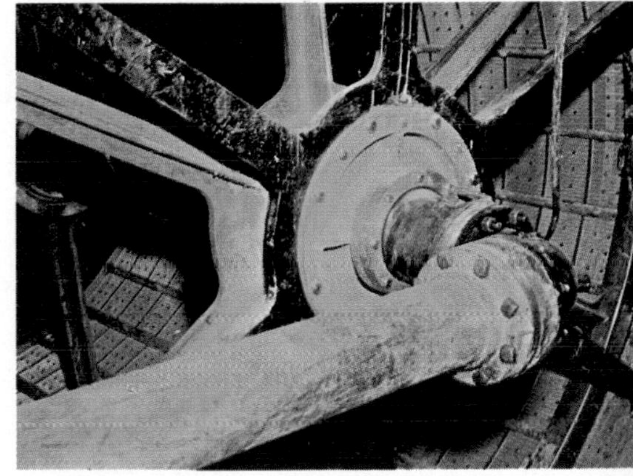

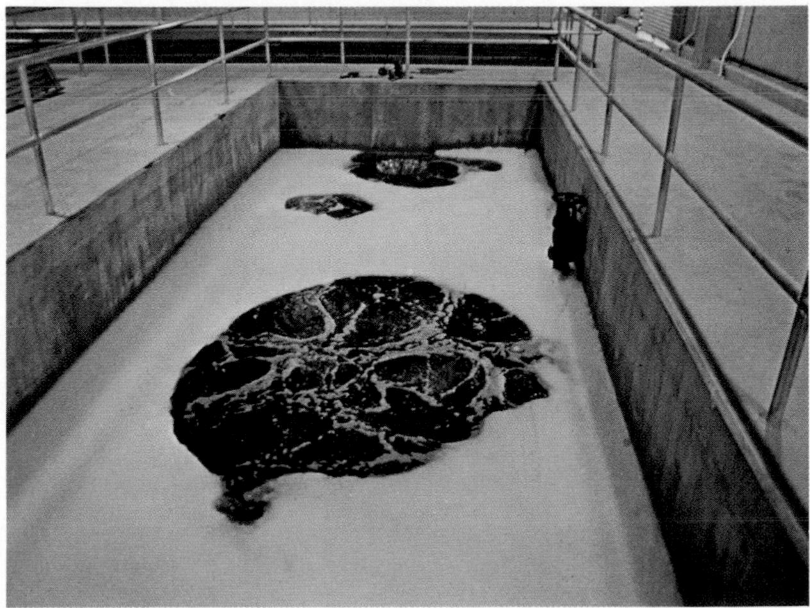

Courtesy, The Metropolitan Sanitary District of Greater Chicago

Two phases in the advanced processing of sewage include passing the wastes through the fine wire fabric (160,000/sq in.) of a microstrainer (above) and using a post-aeration tank (left). In the latter, pressurized air is added to treated sewage in order to cleanse it by introducing oxygen.

The Bettmann Archive

German soldiers (above) during World War I disinfect polluted water so that it can be used for drinking. In a chlorine tank (opposite), chlorine is added to treated waste water in order to combat bacteria.

other countries, in which equally inappropriate climatological conditions prevailed and where sufficient land was unavailable. In addition, great difficulties were encountered with soils too fine to permit continuous absorption of liquids and solids without clogging. Moreover, the emphasis on land disposal probably retarded the development of successful artificial procedures for treating sewage by at least a quarter of a century.

As scientists and technologists gradually realized that land disposal was not the solution, the development of artificial sewage-disposal procedures moved forward rapidly. Among the methods used were artificial sand filtration, contact beds filled with stone, trickling filters, and activated sludge and its many modifications. In virtually all of them the sewage was first settled in tanks to remove as much of the suspended material as possible so that the resulting liquids could be purified faster on smaller and smaller units of land. The effectiveness of these methods is illustrated by the fact that, whereas an acre of land in the original land-treatment process would provide for 10,000 gallons of sewage per day, the modern activated-sludge treatment plant and its appurtenances require less than one acre for more than one million gallons of sewage per day. In the developed countries of the world, therefore, current practices in the treatment of municipal sewage are predominantly by the activated-sludge process.

Generally, the activated-sludge method removes waste matter from the transporting water in the following steps, the first three of which are known as "primary" and the last three as "secondary" treatment:

1. Bulky floating and suspended matter is strained out by racks and screens.

Courtesy, The Metropolitan Sanitary District of Greater Chicago

2. Oil and grease are skimmed off.

3. Heavy and coarse suspended matter is settled out in tanks.

4. Nonsettleable and some dissolved solids are sometimes flocculated (aggregated into small lumps) and precipitated with chemicals.

5. With the aid of oxygen, organic matter is converted biologically into cellular substances that will settle. This is carried out either on beds of granular material, such as broken stone, over which the settled sewage trickles, or by returning the agitated and oxygen-rich organic matter continuously to the flowing sewage. The first method is appropriately called "trickling filters"; the second, "activated sludge."

6. Pathogenic (disease-producing) and other organisms are partly removed in these processes. More complete removal is accomplished by disinfection, usually by chlorination.

Advanced treatment methods

As requirements for still more advanced treatment have emerged, pilot plants at Lake Tahoe and near Washington, D.C., have been established to experiment with chemical precipitation of either raw sewage or of effluents treated with lime, alum, or iron salts. In addition, sand filtration, carbon purification, ozone treatment, chlorination, and mineral deionization are under detailed study, if their use becomes increasingly necessary. Many of these added treatments result from the desire to remove such nutrients as phosphorus and nitrogen in order to reduce or retard eutrophication (oxygen starvation caused by algae growth presumably stimulated by the richness of the dissolved nutrients in receiving bodies of water).

Although it is evident that tremendous strides have been made in the

Courtesy, The Metropolitan Sanitary District of Greater Chicago

Liquid fertilizer containing sludge (the solid matter removed from waste water) is sprayed on a farm field (above). Runoff from this process is captured by an automatic sampler (below), part of an experiment to determine whether this type of fertilizer is harmful to the soil.

treatment of municipal sewage, the search must be continued for a treatment process that is cheaper than any so far developed and that also provides for a more satisfactory disposal of sludge (the solids removed from the sewage water). Unfortunately, complete destruction of sludge not only is costly but also dissipates a material of some possible value for soil conditioning. Under existing conditions, however, this material generally is not worth the cost of its salvage.

Reclaiming waste water

The search for an economical recovery of waste products has been paralleled by efforts to obtain general acceptance of the concept of using the carrying water as a true water resource. For many years, waste water has been a potential resource for irrigating land, for the recharging of subterranean water sources, or for reuse by industries. Many examples of successful applications for each of these purposes can be found in the United States and, to some degree, in other countries of the world, particularly in arid and semiarid areas. In California, for example, some hundreds of thousands of acre-feet per year of sewage-plant effluents from private and public institutions and municipalities have been applied to the soil for agricultural purposes or for underground-water recharge. The extension of these uses will undoubtedly occur in the future because of the economic value of such applications and because of the increasing pressure for water conservation.

In the United States, industries now make use of waste water from municipal sewage-treatment plants, with important economic benefits. Perhaps the largest installation of this type is at the Bethlehem Steel Co. plant at Sparrows Point, Md., near Baltimore. This plant, which manufactures rod, wire, tin plate, rails, pig iron, nails, pipe, ships, and miscellaneous steel products, employs approximately 30,000 people and uses vast quantities of fresh, brackish, and salt water for various purposes, ranging from drinking to the quenching of coke. Since 1942, the plant has been using treated waste water from more than one million people in the Baltimore metropolitan area. In 1970 it used approximately 122 million gallons per day of sewage from this source after the waste water had passed through the municipal sewage-treatment plant and had been further processed for industrial purposes. This represents perhaps the largest amount of water-carried sewage reused in any one place in the world.

The processes used for the above purposes are neither complex nor hazardous. The activity has been successfully managed for about 30 years and has resulted in major economic and sanitary advantages to both the city of Baltimore and the Bethlehem Steel Co. It also offers a useful example of the advantages of recycling waste water, especially with regard to conservation and economic values. It requires, of course, a favorable geographic juxtaposition of a large user and a large producer of waste water, a situation not frequently at hand.

See also Feature Article: BURIED IN AFFLUENCE.

See in ENCYCLOPAEDIA BRITANNICA (1971): SEWAGE DISPOSAL.

FOR ADDITIONAL READING:

Fair, Gordon M., Geyer, John C., and Okun, Daniel A., *Water Supply and Wastewater Removal* (John Wiley and Sons, 1966).

Reynolds, Reginald, *Cleanliness and Godliness* (Doubleday, 1946).

Wolman, Abel, "The Metabolism of Cities," *Scientific American* (September 1965, pp. 178–190).

Wolman, Abel, "Disposal of Man's Wastes" in *Man's Role in Changing the Face of the Earth* (University of Chicago Press, 1955).

A Gateway to the Future

Challenging the Restless Atmosphere

by John W. Firor and William W. Kellogg

As the saying goes, "Everybody talks about the weather, but nobody does anything about it." This is certainly not true at the National Center for Atmospheric Research, where attempts are being made to further understand and control the world's weather.

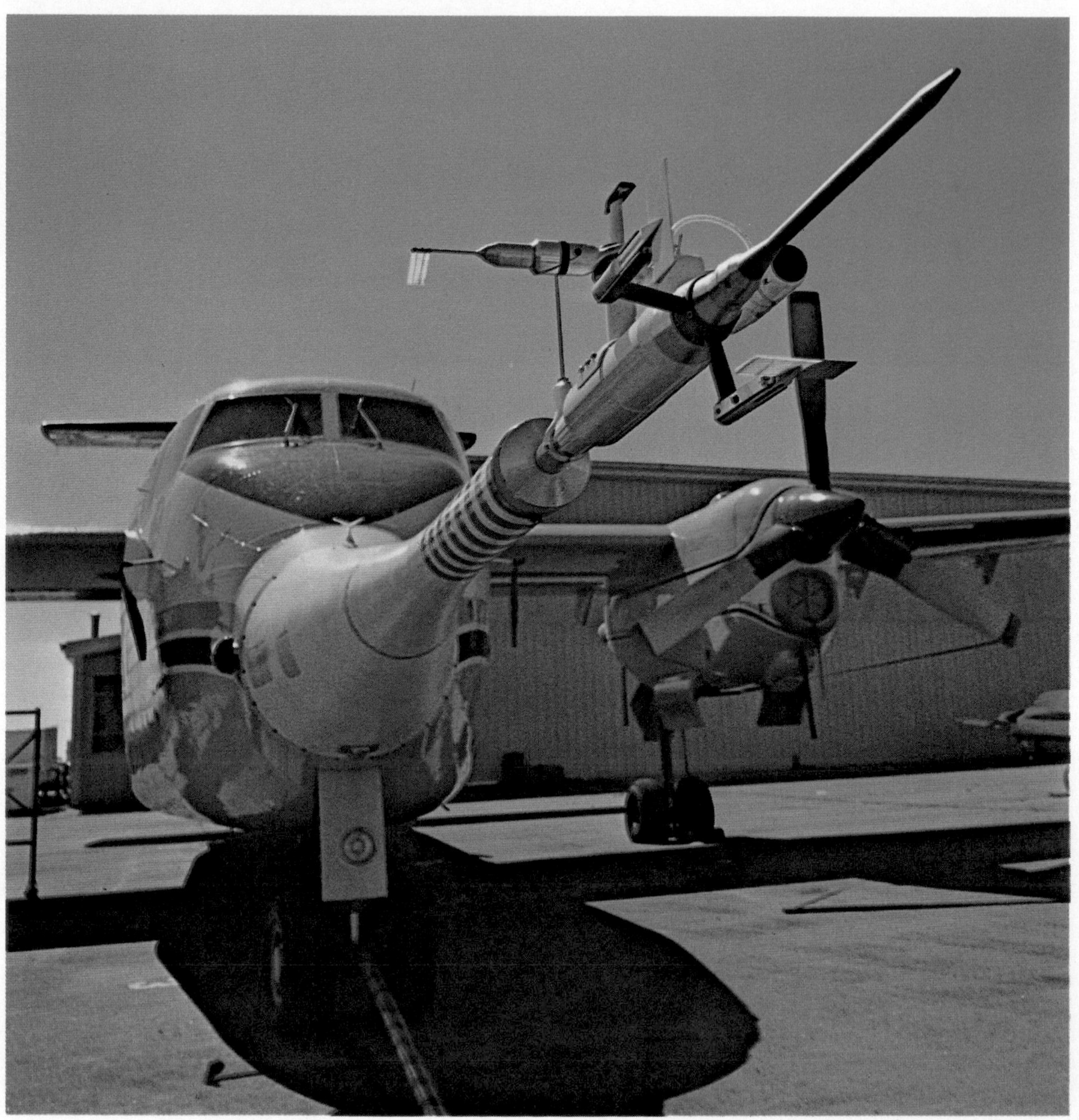

All photos, courtesy, National Center for Atmospheric Research

Twice a day, at more than 500 locations throughout the world, instrumented weather balloons called rawinsondes start their swift ascent into the upper atmosphere. Before bursting in the thin, cold air of the stratosphere, well above the highest-flying aircraft, they radio back to earth information about the temperature, pressure, humidity, and winds at different altitudes.

Every day, especially during the hurricane or typhoon season, long-range aircraft take off to reconnoiter the weather conditions over remote portions of the ocean and the polar regions. At the same time, weather satellites passing overhead in their tireless orbits send back pictures of the earth's clouds and measurements of air and surface temperatures below.

It is data from these various sources that the World Weather Centers in Washington, D.C., Moscow, and Melbourne, Australia, use to make the daily weather forecasts that are familiar to almost everyone within reach of radio, television, or a newspaper.

Less well known, however, are the gigantic balloons launched from Palestine, Tex., which carry hundreds of pounds of instruments to altitudes far above those attained by the rawinsonde balloons. Some stand over 800 feet high just before being released and, when fully inflated in the stratosphere, have volumes of up to one million cubic meters. Few people know about the war-surplus Buffalo airplane that now carries a 20-foot nose boom supporting sensitive air-motion and temperature sensors. Clamped to the boom's base is the most accurate inertial navigation system available. Nor are many people familiar with the U.S. Air Weather Service RB-57F jet aircraft that have oversized engines and wings to enable them to fly at altitudes unattainable by ordinary aircraft. These planes sample the rare upper atmosphere, where they can detect traces of volcanic particles and debris from man-made nuclear explosions. And little public recognition was given the crews of meteorologists and their assistants, who, in early 1967, set up camps on three small islands in the central Pacific, called the Line Islands, to make the first detailed measurements of the tropical atmosphere. These measurements have provided new insights about how that crucial part of our globe affects the world's weather.

NCAR's de Havilland Buffalo (left), a twin-turboprop aircraft, has been outfitted with a long boom that places wind and temperature sensors well ahead of the aircraft.

All of these unheralded adventures into the unknown have been motivated by man's desire to probe and understand the earth's atmosphere and its weather. They were sparked by a new organization whose sole mission is to make such studies: the National Center for Atmospheric Research (NCAR).

The history of NCAR

The weather that so greatly influences the affairs of mankind is determined by the complex atmosphere that covers the earth. During the Age of Discovery, as venturous sailors traveled about the earth, the general patterns of the atmosphere became better understood. These early explorers learned to use the steady winds of the tropical belts to push their square-riggers across the oceans; they learned something about the movement of hurricanes; and they developed simple ways of predicting the weather from the direction of the wind and the patterns of clouds in the sky. However, the real key to this complicated medium, each part of which interacts with every other part, came only when there were ways to observe it as a whole.

Recognizing the problem of studying the atmosphere in its entirety, meteorologists have for many years sought a cooperative approach. As early as the 1850s, under the leadership of Matthew F. Maury of the U.S. Navy, sea captains of several countries began, on a regular basis, to exchange weather observations from all parts of the globe.

This informal organization eventually resulted in the present World Meteorological Organization, in which the weather services of most of the countries of the world cooperate in exchanging information on a twice-daily basis. This exchange allows meteorologists to draw complete weather maps twice a day and to trace changing patterns as they develop.

But exchanging observations is not enough. More than 50 years ago Cleveland Abbe, an American meteorologist wrote:

> What I most long to see, and what I believe is of fundamental importance in atmospherics—the want of which is a real obstacle—is the existence of a laboratory building specifically adapted to atmospheric experiments and the association therewith of able students trained in mathematics, physics, and mechanics. When all this is realized, the intellectual work that will be done there will gradually remove all obstacles to the perfection of our knowledge of the atmosphere. Does this seem like a long look ahead? Not so. The time is ripe for the institute.

It was not until about 50 years later, however, when, on the recommendation of the U.S. National Academy of Sciences, the U.S. National Science Foundation (NSF) appropriated funds for the creation of an institution along the lines envisioned by Abbe. A group of universities, each with its own teaching and research program in atmospheric science, worked out the details of the new organization and provided the management structure. It was the responsibility of this group to convert plans and hopes into reality.

These universities saw the institution as an important extension of their own capabilities, and as a valuable mechanism for progress in understanding the atmosphere. Many tasks and activities were planned for this new creation, but the highest hopes of the founders were distilled into a single sentence that appeared in a planning document: "The institution . . . is to be thought of as a center at which high scientific competence and consummate technological skill can be combined in a free and natural alliance to master our atmospheric environment."

Under the leadership of Walter Orr Roberts, its first director, the new organization was formally established as the National Center for Atmospheric Research, in Boulder, Colo., in 1960. With support from the NSF, Roberts began the job of assembling the expert staff and powerful facilities needed to foster the combined attack on atmospheric problems by scientists from many disciplines and many countries.

NCAR is situated on a magnificent site on the eastern slopes of the Rocky Mountains. Most of the NCAR staff of approximately 550 people work in Boulder, but there are several contingents elsewhere. In Palestine, Tex., a group launches large weather balloons; a smaller group in Christchurch, New Zealand, flies globe-circling, constant-density balloons; and scientists and technicians at Climax, Colo., and Mauna Loa, Hawaii, observe the corona and activity of the sun.

Modeling the atmosphere

Man always has been subject to the vagaries of the weather. His dream has been to predict what the weather will be a few hours or days ahead, perhaps to control it, and, more recently, to preserve the quality of the air in which he lives and breathes.

The basic driving force of the weather is easy to understand. The atmosphere is a restless fluid covering the entire planet. Near the Equator this fluid receives much more heat from the sun than it can radiate back into space, while near the poles the opposite situation exists—the air loses more heat to space than it receives from the sun. Somehow, heat must be transported from the Equator to the poles. It is in the process of producing this

JOHN W. FIROR is the director and WILLIAM W. KELLOGG is an associate director of the National Center for Atmospheric Research in Boulder, Colorado.

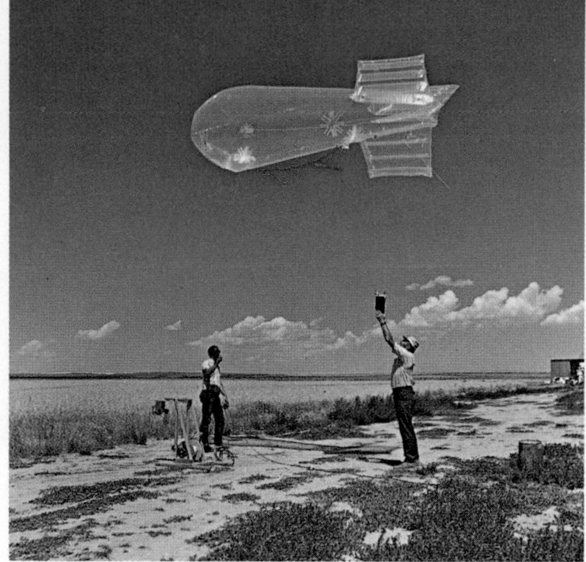

By raising and lowering a balloon with a winch on the ground (left), researchers can monitor atmospheric conditions with sensors in a package beneath the balloon. Convective thunderstorms (opposite page, left), huge cumulus clouds that often penetrate into the stratosphere, are of general concern because of their dangerous winds, lightning, and hail. The dropsondes (opposite page, right) being loaded into a pod on the belly of an NCAR Sabreliner research aircraft will be dropped through a storm as the plane flies over it. The sondes transmit signals to ground stations, providing a record of their movement through the air and the winds and updrafts through which they fall.

transport that motions of the atmosphere are generated.

Even a casual look at a weather map indicates that these motions are very complicated. Not only are there large-scale, relatively stable motions, such as the trade winds in the tropics and the higher-latitude westerlies, but there are also motions of lesser size—sea breezes and land breezes, cyclones and anticyclones, hurricanes and squall lines—that range in intensity from gentle to ferocious and in size from a few miles to many hundreds of miles. While it is true that the atmosphere as a whole must be observed before one can hope to understand its variety and complexity, observation alone is not sufficient. Ways also must be developed to digest the vast amount of information collected. To refine our understanding of atmospheric processes, one of the more difficult tasks confronting meteorologists is to perform calculations on these data within a framework of theory.

The main tool being used to relate theory and weather data is a "model" of the atmosphere. This is not a model in the laboratory that one can see or feel; it is, instead, a mathematical model stored in the memory of a large computer and manipulated by its computational ability. While theoreticians continue to isolate problems that can be studied analytically, and field studies measure such processes as motions, budgets of energy, water, and momentum in limited portions of the atmosphere, it is in the numerical model in the computer where all this information is brought together. Only then can scientists determine the impact of an isolated process on the system as a whole, or attempt to predict an accurately measured atmospheric quantity as a check on the degree of their understanding.

The working of a mathematical model, like the real atmosphere it hopes eventually to simulate, is rather simple in concept but immensely complex in execution. The model obeys well-established laws of physics that are written in the form of less than a dozen equations. These equations relate, for example, pressure forces and the rate of change of momentum of air masses, the addition of heat to the air and the rate of temperature change, and the water evaporated from the oceans and the heat released when this water falls as rain or snow. The application of these general laws must be made, however, while also taking into account the *actual* shape of the oceans and continents, the *actual* temperature of the ocean surface, and the effects of *actual* mountain chains. Hence, the computer must be supplied with a great deal of factual information.

To complicate matters further, not only must *all* the laws be applied simultaneously, it also must be understood how each event in the calculations affects the future course of the model atmosphere. For example, if the model (or the real atmosphere) produces snowfall at some location, the color of the ground there changes from brown or green to white. This, in turn, changes the amount of heat absorbed by the ground from the sun, which then affects the heat radiated into the air by the ground, and so on through a lengthy chain of interactions. The computer must therefore keep up with the results of its calculations and adjust all subsequent steps accordingly.

Finally, because no computer yet built can deal with each individual particle of air, one point in the model must be used to represent a fairly large volume of air. The most sophisticated model now in use at NCAR divides the air into volumes, each one about 250 kilometers square and 3 kilometers deep, for purposes of calculation. Each volume is considered to have a constant temperature throughout, with clouds the same throughout the volume, and so on.

Predicting weather and climate

A model of the atmosphere can be used to forecast the weather, assuming that it is sufficiently realistic, that its calculations are based on accurate observa-

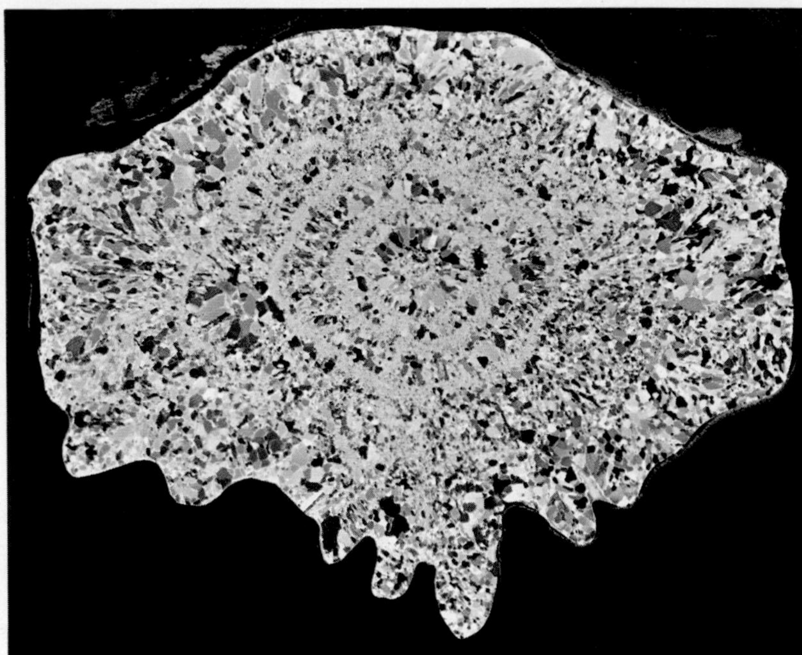

tions of the real weather, and that the computer can calculate ahead faster than the weather advances. Numerical models are being used by the National Weather Service to make predictions of the weather from one to four days in advance. The NCAR model and those used by other research institutes, however, are somewhat more elaborate than the ones used by the National Weather Service because the research version must be changed constantly to test new approaches. Eventually, the research models will serve as the basis for greatly improved operational forecasting.

Forecasting long-term changes in climate involves very different considerations: the time scale for such predictions must be tens or hundreds of years instead of days. While scientists believe that many factors influence the climate, there is as yet no suitable way of showing how these factors operate over very long periods of time. The question of predicting climate has become particularly urgent, now that some of man's activities are large enough to have some influence on an entire region of the earth, and possibly on the planet as a whole. These influences include the accumulation of particulate matter and of carbon dioxide that are injected into the atmosphere by burning vast amounts of fuel, and the large amount of heat that is released from sources such as power generators, automobiles, and the heating of buildings in cities.

In order to understand the long-term changes in climate, it is necessary first to understand the relationship between the major source and the major reservoir of all the earth's energy—the sun and the oceans, respectively. Astronomers have not been able to establish to an accuracy of better than about 1% the changes in the amount of heat the sun sends to the earth, but even a change of only 1% would probably cause an appreciable change in global weather. It is known that the sun's output of X-ray and ultraviolet radiation does change, but these variations are small compared to the total amount of heat radiated by the sun. Furthermore, they are clearly noticeable only in the upper atmosphere. Such changes occur in an 11-year cycle, which is sometimes called "the sunspot cycle." There is quite a bit of evidence, such as the changing sizes of tree rings in certain parts of the world, to indicate that perhaps even this subtle 11-year change in the sun can affect the climate in a small way.

The National Hail Research Experiment

Among the most obvious features of the weather are the local storms that pass across the face of the earth, bringing with them abrupt changes of temperature, along with winds and rain or snow. The more intense ones can also produce damaging hail. NCAR is well located for the study of such severe storms because northeast Colorado has one of the highest hail frequencies in the United States. Since 1967 the study of thunderstorms in northeastern Colorado has been a cooperative project between NCAR, the Environmental Science Services Administration (now called the National Oceanic and Atmospheric Administration, or

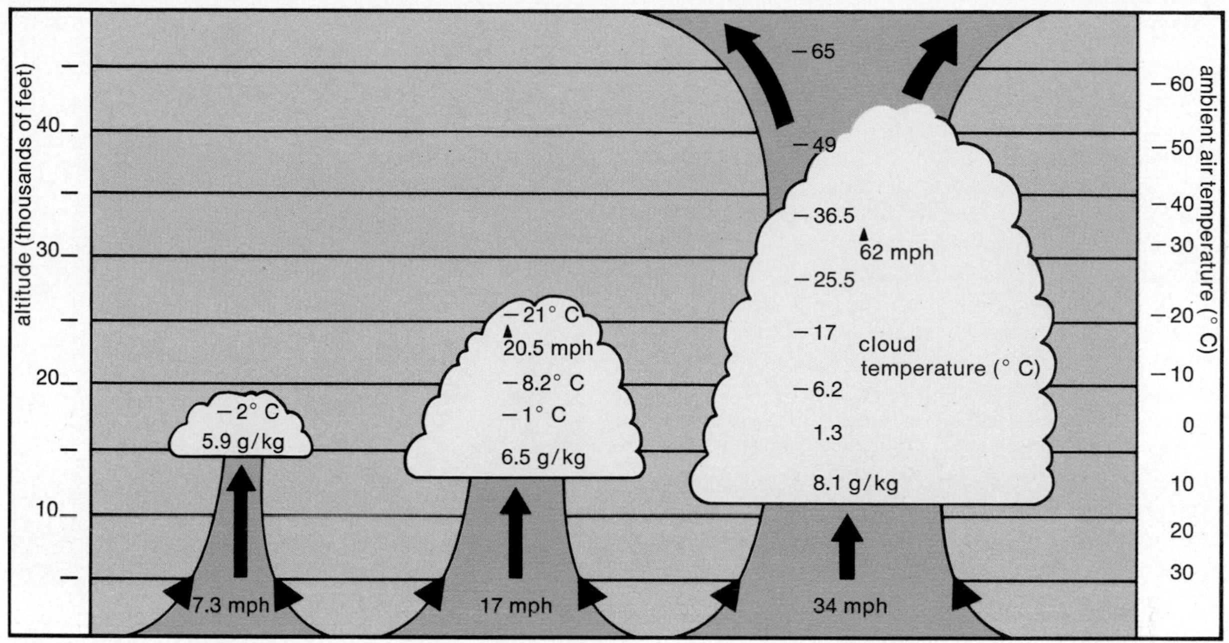

Using a computer model and a predicted temperature profile (above), the probability of the formation of large clouds at hail-producing levels can be predicted. The model, developed at NCAR, predicts cloud-base and cloud-top heights, cloud temperature, maximum updraft velocity, and cloud moisture content. The cross section of a huge hailstone, about 17 in. in diameter, is photographed with polarized light to show the patterns of ice crystal growth within the stone (opposite page, right). A radar antenna in Colorado searches the lowering clouds for signs of hail, while a weather balloon studies the atmosphere below the developing storm (opposite page, left).

NOAA), and Colorado State University. This work has involved the combined use of radars to detect the presence of storms and to observe their size, shape, and changes; aircraft to measure conditions around the storm; instrumented balloons to measure and radio back conditions aloft; and ground networks to observe conditions under the storms.

NCAR is now expanding this activity into a program called the National Hail Research Experiment. The goal of this program is to make major advances in the understanding of severe convective storms. One of the purposes of the experiment will be to determine whether a technique used in the Soviet Union for "seeding" thunderstorms will reduce hail in Colorado thunderstorms. This technique involves shooting into the heart of a hailstorm small rockets that contain lead iodide or silver iodide crystals, which promote the freezing of the supercooled water found in a hailstorm.

Meanwhile, as NCAR scientists and their colleagues study thunderstorms in nature, a companion effort is under way to learn how to make mathematical models of thunderstorms. Just as a group of equations describing various physical processes can be combined in a large computer to simulate the global atmosphere, in the same general way it is possible to simulate a cumulus cloud as it grows to a giant thunderstorm. Such cloud models must take into account the way the air is drawn into the cloud at the bottom, the growth in the air updraft of the small cloud droplets into raindrops, and, as they reach the colder upper levels, the freezing of the raindrops to snow and ice and their continued growth.

Investigating gravitational waves

To describe how energy moves from one part of the atmosphere to another is a problem of major importance to NCAR. It has become clear in recent years that, in addition to the usual methods of convection and radiation, sizable quantities of energy may be transported by what are called gravitational waves.

To study these waves, once again NCAR finds itself not only with the diverse talents needed but also in a fortunate geographical location. When the wind blows strongly across the Continental Divide, just west of NCAR, it sets up waves in the atmosphere, just as the motion of water over a large boulder in a rapids produces a disturbance downstream. These mountain waves can extend for

100 kilometers (1 kilometer = 0.621 mile) or more downwind over the Great Plains and also far upward into the stratosphere. In fact, they probably can extend to the very fringe of the atmosphere, influencing the propagation of radio signals that bounce off the ionosphere at heights of 100 kilometers or more above the surface.

To study these gravitational waves, NCAR has organized cooperative experiments with the Air Weather Service, NOAA, the U.S. Army, and a number of universities, in which many aircraft fly in a coordinated pattern to observe the air motions above the mountains. One early result of the study was of immediate relevance to the future of aviation. Wave effects of large amplitude were found as high as 20 kilometers (65,600 feet), altitudes at which man had expected to fly in commercial aircraft in the "calm serenity of the stratosphere." Man undoubtedly will fly there, but he must prepare for interruptions of his serenity.

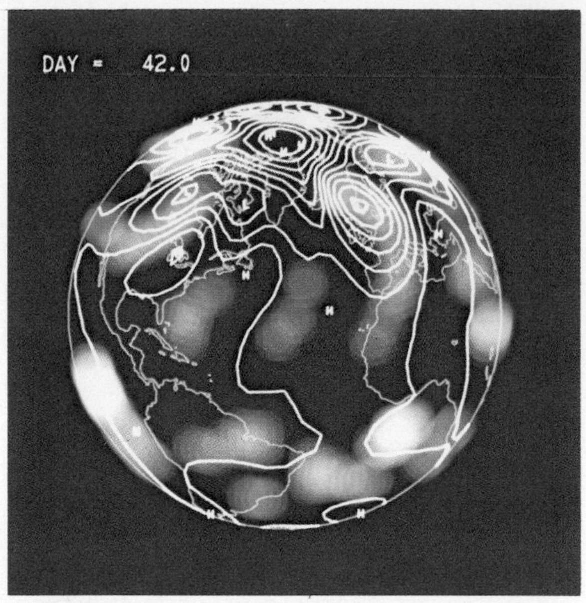

Global chemical reconnaissance and air pollution

In the area of air chemistry, NCAR has for many years studied the mysteries of the natural sources and sinks of trace gases and particles in the atmosphere. Some of the trace gases, such as carbon dioxide and ozone, have a very important effect on the global heat balance because they can absorb both solar and infrared radiation and thereby influence the air temperature, and it is becoming increasingly obvious that dust and smoke particles also affect the air temperature. It is in this realm that industrial activity plays an important role.

Since the beginning of the Industrial Revolution, enough fossil fuel has been burned to cause an appreciable increase in carbon dioxide in the atmosphere, and it is now predicted that there will be about a 25% increase from the present level of 320 parts per million by mass to 400 parts per million by the year 2000. This will slightly raise the surface temperature and will therefore have a small but possibly significant effect on the climate of the world unless there is some counteracting influence. A possible counteracting influence is the increase in the production of particulate matter from industry and from agricultural practices that involve the burning of large quantities of plants or trees, since these particles may cause a net cooling of the lower atmosphere by reducing the sunlight reaching the ground.

To see what trends there are in the atmosphere as a whole, there is a need for observations in remote parts of the world, far from the direct influence of mankind, though there is now virtually no part of the globe where it can be said that man has no influence at all. NCAR scientists have made measurements of trace gases and dust particles in Antarctica, the Arctic, the jungles of Panama and Brazil, and on the peaks of volcanoes in the Hawaiian Islands. Their purpose is to determine the natural background of such materials against which man's activities can be measured and thereby try to determine whether man can influence the atmosphere as a whole.

The question of preserving the atmospheric environment is increasingly urgent, the air in our large cities having become dangerously polluted. Nevertheless, there are many mechanisms, particularly precipitation of rain and snow, which continually cleanse the air in and around cities. Scientists at NCAR, along with their colleagues in universities and government laboratories, are attempting to define the chemical processes that produce smog and haze particles, as well as the more irritating byproducts of air pollution. Perhaps they also will discover ways of either altering the chemistry of the atmosphere or reducing the sources of pollutants.

Studying the sun and upper atmosphere

Finally, the weather cannot be considered without taking into account the sun—the ultimate source of heat energy for the earth as well as the determining factor of conditions in the earth's upper atmosphere and the space between the earth and the sun. The sun has a fascinating and complex atmosphere of its own, and scientists at NCAR are seek-

initial

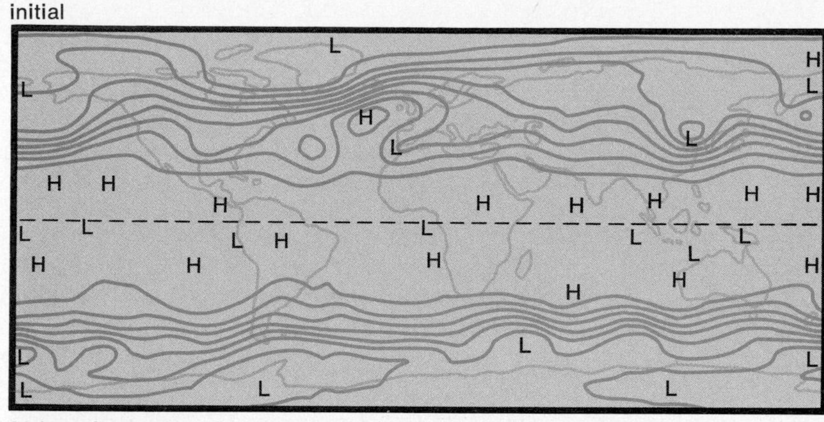

96-hour forecast

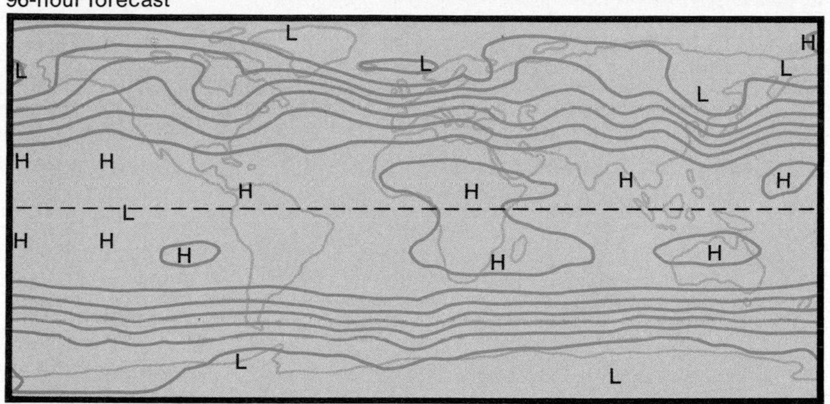

96-hour verification

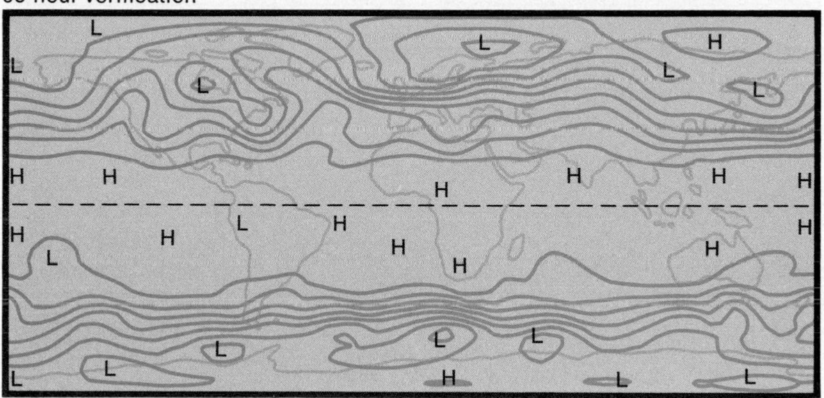

Produced by running the NCAR general circulation model on a computer, this weather map shows simulated cloud cover and surface-level pressure patterns, or highs and lows. It also shows graphically how computer simulation produces features that closely resemble those of the real atmosphere. To test the forecasting effectiveness of NCAR's mathematical model of the global atmosphere, experiments are conducted using real weather data as a starting point; the result is compared with the conditions that developed in the real atmosphere. These surface-level pressure patterns, recorded on microfilm by NCAR's computer, show the results of one forecasting experiment (right). The top frame shows the real weather conditions that served as the starting point. The second frame shows the computer-produced forecast, and the bottom frame the weather conditions that actually existed in the real atmosphere at the end of 96 hours. A weather map of weather that never happened, at least not in the real atmosphere is shown (opposite page).

ing to unravel the secret of the 11-year cycle that seems to govern solar activity.

For many years, scientists studying the upper atmosphere have been aware that this region responds to changes in solar activity. For example, from time to time there are major emissions of energetic particles, X rays, and ultraviolet radiation from the sun that cause storms at levels above 80 to 300 kilometers, the region called the ionosphere, and, still farther up, in the region called the magnetosphere. During such an event the magnetic field of the earth changes quite rapidly. Moreover, these magnetic storms often are accompanied by intense displays of aurorae in high latitudes and by difficulties in transmitting high-frequency radio waves over long distances.

Tools of the atmospheric scientist

The atmosphere must be studied in depth and on a global basis to understand and predict the larger scales of atmospheric phenomena. For smaller-scale studies, such as gravitational waves and thunderstorms, the phenomena must be investigated in three dimensions and in more detail. In order to do this, scientists have had to develop an array of vehicles and instruments with which they can probe and measure this complex medium.

The most obvious way to probe the atmosphere is to carry an instrument directly into it. Because aircraft have proved extremely useful in this respect, NCAR owns and operates two twin-engine Beech aircraft, a twin-jet Sabreliner, and a de Havilland Buffalo; it also has been furnished access to a Schweizer sailplane owned by the National Oceanic and Atmospheric Administration. All of these are instrumented to do special kinds of research, such as measuring winds and temperatures with great accuracy, sampling the air for trace gases and particles, and measuring the radiation from the sun and earth.

The Climax, Colo., observing station of the High Altitude Observatory of NCAR (bottom); the solar corona, photographed during the total solar eclipse of March 7, 1970 (below).

Long before aircraft were employed, balloons were used to probe the atmosphere. About 1904, the first successful observations of the stratosphere by balloons were made by a French scientist, Leon Teisserenc de Bort. Balloons have come a long way since the days of Teisserenc de Bort. They now have volumes of nearly one million cubic meters (30 million cubic feet) and can carry loads of more than 500 kilograms (1,100 pounds) far into the stratosphere, where they can float for many hours or even days. NCAR usually launches these from a site at Palestine, Tex.

NCAR has also developed smaller, constant-density balloons that can float at the same level for more than a year, circling around the South Pole. In addition, balloons of various sizes can be used to carry instruments for special purposes, such as measuring the distribution of ozone or the vertical flux of radiation that determines the air temperature.

The dream of the meteorologist always has been to be able to sit on the ground and to observe the behavior of the air above him. He still cannot do this in every respect, but NCAR's radars can probe the heart of a thunderstorm and see the distribution of rain and hail inside. The pulsed laser, called a "lidar," can measure the distribution of airborne particles, and a new sensitive ultraviolet spectrometer is being used to determine ozone distribution high in the atmosphere from the scattered sunlight. Many of these efforts still are experimental, but the potential of being able to observe from a remote location spurs NCAR's scientists to improve such techniques.

As already mentioned, the computer is another important tool of the atmospheric scientist. NCAR operates the world's largest computer facility devoted exclusively to atmospheric research. These computers are used for the large numerical models of the general circulation of the atmosphere and also for the many other studies that need to be made, including those of convective clouds, gravitational waves, interactions between the atmosphere and the surface of the land and oceans, and circulations in the oceans themselves. This computer complex is used by many scientists outside of NCAR as well as by the NCAR staff.

Aiding global atmospheric research

Because atmospheric research deals with a global system, it is a logical candidate for important international undertakings. One that emerged in the 1960s is called the Global Atmospheric Research Program (GARP). As a key U.S. participant in

View from the east of the NCAR laboratories, built on the edge of the Rockies in Colorado.

GARP, NCAR has joined with scientists in NOAA, NASA, and many universities to undertake observations of the circulation in the lower atmosphere on a global basis.

Combinations of balloon and satellite systems will aid in measuring wind velocities and in obtaining pressure and temperature data to supplement observations from satellites alone and from the ground-based global meteorological network. At the same time, NCAR has also undertaken a number of experiments with its numerical models to simulate and compare various combinations of observing systems.

A large and significant part of the atmosphere that has been sorely neglected until recently is the tropical belt. This is partly because of the lack of interest in this region on the part of the more-developed countries, most of which are in the temperate zones of the Northern and Southern Hemispheres, and also because of the mistaken belief that it was not necessary to understand the tropics in order to understand and predict the weather in the temperate latitudes. It is known now that this is not true; hence, an important part of NCAR's work with GARP is in tropical meteorology. In 1967, NCAR organized the Line Islands Experiment, an ambitious expedition to three small islands located south of Hawaii in the Pacific Ocean, almost on the Equator. This experiment consisted of intensive observations of the tropical atmosphere by means of rawinsonde balloons, aircraft, radar, and surface weather stations, combined with pictures from a U.S. synchronous satellite directly overhead. This combination permitted a view of tropical weather systems that had not been obtained before. Since that time NCAR has cooperated in a number of tropical studies that are generally centered in the area of Barbados. It is now expected that another, even more ambitious, tropical experiment, one involving research teams from many countries, will take place in the Atlantic Ocean in 1974. NCAR scientists are helping in the planning for this experiment, which is also an important part of GARP.

The culmination of GARP will probably be an intensive observation program in the late 1970s to obtain data that can, for the first time, adequately define the conditions in the atmosphere on a global basis. This program will involve the combined use of all the tools available to the meteorologist, including meteorological satellites, constant-density balloons, radiosondes, aircraft, and drifting buoys.

In a very real sense, this culmination of GARP symbolizes the kind of enterprise for which NCAR was created. Because there are problems too large and too complex for individual atmospheric scientists to undertake, NCAR provides a mechanism by which scientists in the United States, in cooperation with those from other countries, can approach the atmosphere of the planet as a single system and piece together the riddle of the gaseous envelope that covers the earth.

See also Feature Article: KILLER STORMS.

A Gateway to the Future

A Freedom to Excel

by Howard J. Lewis

Founded in 1911 as the Kaiser Wilhelm Society, the Max Planck Society for the Advancement of Science carries on a proud tradition of fundamental scientific research in the Federal Republic of Germany.

Courtesy, Max Planck Society for the Advancement of Science

The world's largest steerable radio telescope is operated by the Max Planck Society at Effelsberg/Eifel in West Germany. Opened in May 1971, it has a reflector 100 meters in diameter.

One day in the spring of 1971, at the Wallops Island, Va., launching site of the National Aeronautics and Space Administration, a Scout rocket was fired to an altitude of 20,000 miles. Its purpose was to provide a means of analyzing the electric and magnetic fields in the magnetosphere, the outermost layer of the atmosphere, by releasing an immense, glowing cloud of barium ions whose behavior could be observed by instruments on the ground.

Several thousand miles away, in Seewiesen in the Federal Republic of Germany, a small group of scientists was studying the reactions of an adult male Cichlid fish to repeated appearances of a painted model of another adult male Cichlid. In Munich, a social psychologist was making a sound motion picture of people conversing. Later he studied the films frame-by-frame, noting eye contact as well as apparently unnecessary movements of arms, legs, head, and trunk. He was determining how such nonverbal modes of communication reinforced or contradicted what the subjects actually were saying to each other.

About 25 miles southwest of Bonn, the mammoth bowl of the world's largest movable dish-type radiotelescope was pointed slowly toward a distant galaxy. This instrument was designed to reach farther into space—and consequently farther backward in time—than any other telescope in existence. Its builders hoped it would provide a conclusive answer to the question of whether the cosmos started in an instantaneous explosion, whose remnants are still flying outward, or whether the cosmos is constantly replenishing itself.

In the Munich suburb of Garching, a group of physicists and engineers was working on a project whose goal was to achieve an almost limitless supply of pollution-free energy. To do so, they were attempting to design a vessel that could contain a fire five times hotter than the interior of the sun.

Each of these diverse scientific undertakings had one common feature. Each was being conducted by a branch of a single organization, the Max Planck Society for the Advancement of Science.

A framework for research

The Max Planck Society had its origins in the Kaiser Wilhelm Society, which was founded in 1911 under the patronage of Kaiser Wilhelm II to promote a unified German scientific research effort. After World War II, the organization was renamed for Max Planck, the German physicist known for his work regarding the quantum theory of energy.

In its 60-year history as both the Kaiser Wilhelm Society and the Max Planck Society, the organization has achieved a worldwide reputation for excellence in scientific achievement. It has provided the research base for a vast number of Nobel laureates, including such illustrious names as Max Planck, Albert Einstein, Walther Bothe, James Frank, Werner Heisenberg, and Max von Laue in the field of physics; Otto Hahn, Richard Kuhn, Fritz Haber, Otto Meyerhof, Karl Ziegler, and Manfred Eigen in chemistry; and Otto Warburg, Adolf Butenandt, and Feodor Lynen in biology and biochemistry. In addition, Rudolph Mössbauer did much of his brilliant work as a fellow at the Max Planck Institute for Nuclear Physics.

Today, the Max Planck Society, which is funded by the West German federal and state governments, shares with the German Research Association (GRA) the responsibility for maintaining that nation's scientific research program. The budget of the GRA is not only somewhat smaller than that of the society but is derived from private as well as government funds. The GRA mainly supports research in German universities and other independent institutes. The Max Planck Society defines its tasks in these terms:

—To support new trends as well as new methods of research, in particular where these are developing on the borderlines between traditional disciplines and have not yet found a place in the universities because of the institutional ties between research and teaching;

—To develop new types of institutes and to take charge of research projects that demand equipment of such size and specialization that the universities do not take them on for fear of disturbing their internal balance;

—To provide scientists of exceptional ability, who wish to devote themselves to pure research, with working facilities adapted to their specific requirements so that they can bring their entire energy toward achieving their scientific aims.

To carry out these tasks, the society operates some 53 institutes, nine of which are devoted to the humanities. About 7,500 individuals are permanently employed by the society; of these, 1,800 are scientists. In addition, there are normally some 1,200 visiting German scholars and scientists and some 500 from other countries also working at the institutes.

To finance its activities, the society's operating budget for 1970 was about DM. 400,000,000 (approximately $109 million). On this basis, the society ranks among the top 150 West German corporations. About one half of its total budget is funded by the federal government and one half by the state governments, but the allocation of these funds to individual institutes is made by the society itself. Both funds and the freedom in deciding how to spend them have been granted to the society to a degree that would be astonishing to a U.S. scientist. Even as late as 1971, the budget of the society has been climbing at a dizzying rate of 15% a year, which is twice the current rate in the United States.

The missions of the institutes

The society's freedom to pursue scientific research is nowhere more apparent than in the tremendous flexibility of its organizational structure and program. The work under the society's direction is carried out by the diverse variety of its institutes, each of which has a specific scientific mission. The mission may be broadly generic, embracing an entire field of research, or it may reflect the research interests of a single investigator.

The basic freedom of the society, and the one in which the judgment of its scientific members is crucial, is the independent authority to open new institutes, to change their leadership and direction, to merge one into another, and, when it seems advisable, to dissolve them or turn them over to universities. During the 60-year history of the organization, more than 30 out of 80 institutes were dissolved or handed over to universities.

An institute may be decommissioned for a number of reasons—because the field is no longer productive, because the subject matter has been entirely absorbed by a suddenly developing new line of research, or simply because the institute has succeeded in its mission. In the view of the society leadership, when a research program has progressed from fundamental research to applied, it is time for the program to seek other auspices, such as direct support by the government, industrial sponsorship, or as part of a university department. A small number of institutes, however, conduct applied research as a natural part of their programs.

An institute can also undergo metamorphosis, usually because one of several concurrent lines of research suddenly develops exciting potentials. One such example is the institute directed by Manfred Eigen, who received the Nobel Prize in 1967 for his work on exceedingly fast chemical reactions. Even though Eigen himself is a physical chemist, the work of his institute began to concentrate on the physical and chemical reactions taking place in living things, consequently attracting new specialists interested in these fields. In the summer of 1970, Eigen declared that he was hot on the trail of the most exciting scientific adventure of his career, the investigation by physicochemical methods of the Darwinian theory of evolution. It is this capacity of the institutes to direct a concerted multidisciplinary attack on a single research problem that the society feels is one of its most valuable contributions.

The Max Planck's "middle way"

A unique characteristic—its relationship with academic centers—is what distinguishes the Max Planck Society from research institutions in the two leading scientific countries, the United States and the Soviet Union. In the United States, basic research frequently is conducted within a university environment as an integral part of the production of new scientists. In the Soviet Union, both basic and applied research is conducted in a formally established institute whose research programs are more or less fixed according to designated fields of inquiry and whose existence is quite separate from academic regimes. In contrast to these cases, the institutes of the Max Planck Society are located not within but conveniently near leading universities. Scientific staff members are encouraged to participate on these university faculties, according to their own inclination, but they have neither formal teaching assignments nor burdensome academic administrative responsibilities. However, many of the institutes do find places for doctoral candidates. In the words of one society spokesman, "We think we have found a middle way."

HOWARD J. LEWIS, director of the Office of Information at the U.S. National Academy of Sciences-National Research Council-National Academy of Engineering, is also editorial consultant for the Britannica Yearbook of Science and the Future.

Courtesy, Max Planck Institute for Plasma Physics

A Wendelstein II stellarator is used by the society's Institute for Plasma Physics to help determine how to achieve controlled thermonuclear fusion. The machine measures the rate of particle loss in a variety of closed magnetic confinements.

Another unique characteristic of the Max Planck Society is that directors of many of the institutes (or of departments in a single institute) have almost total authority in determining the course of the entire scientific program within their own organization. And when a director is replaced, the entire thrust of the institute may well change with his successor, sometimes to the discomfiture of junior staff members. A U.S. visitor to the Max Planck Institute for Biophysics in Frankfurt noted recently that a new director there had redirected the entire research program of the institute from radiation effects to transport through membranes. Spectroscopes only recently acquired and set up for operation were replaced by apparatus for measuring nuclear-spin echo resonance. Lately, however, such absolute freedom has been restricted by the fact that many of the institutes have simply grown too big to become the creature of any one man.

Officials of the society readily admit that the broad authority delegated to the head of an institute sometimes works hardship on junior scientists and results in costly ruptures in continuity of research. But that freedom is defended on two grounds. One is that it is an integral part of the philosophy of the society, as expressed by its current president, Nobel laureate Adolf Butenandt: "The scientists of the Max Planck Society conduct their research in a broad domain that lies between the ideal of freedom postulated and guaranteed in principle by the Society and the need for their cooperation with many other scientists in order to reach a distant goal. It is the task of the Society to allow its members utmost latitude in their spontaneity while assuring that the overall program permits systematic advances over finite periods of time."

The second defense is in the caliber of the society's scientific membership. As a society official explained to a recent visitor, "It is our belief that freedom to pursue interesting lines of research is an absolute necessity. It is for that reason that the selection of academic members of the society is so strict. Because of that, it is easier to give them greater freedom."

Currently, the institutes of the society are engaged in a wide range of research programs. An examination of the ongoing work of two of the institutes, the Max Planck Institute for Plasma Physics, in Garching, and the Max Planck Institute for the Physiology of Behavior, in Seewiesen, provides some idea of the scope of research and the methods of operation of the society.

Harnessing thermonuclear power

One of the most exciting areas of research within the society is that being undertaken by the Institute for Plasma Physics. This institute is unique in that it has been committed to a prescribed course of research by a charter from the Federal Republic. The charter specifies that the mission of the institute will be to determine the feasibility of controlling thermonuclear fusion to produce energy.

The intellectual source of the institute dates back to World War II, when work taking place within the Max Planck Institute for Physics and Astrophysics (under the direction of the world-famed Werner Heisenberg), in Munich, yielded a theoretical basis for thermonuclear fusion. This work eventually led in the United States and then in the Soviet Union and, later, in France and China, to the development of the hydrogen bomb. However, the processes of the thermonuclear reaction also bear the promise of solving three of man's most pressing problems: (1) how to obtain sufficient energy to meet his growing needs without (2) using up all his fuel resources or (3) polluting the globe beyond resuscitation with either the residue of fossil fuels or the radioactive byproducts of atomic fission.

The fusion reaction requires a set of circumstances which man so far has not been able to reproduce. These requirements, which are measured in terms of heat, density, and time, are no problem for the sun, whose energy transformation is thermonuclear. In the laboratory, however, the situation is quite different. The chief problem is that in order to maintain the very high temperatures required by the reaction, it must be contained in a vessel whose walls cannot touch the contents. Theory indicates, however, that the three criteria of heat, density, and time are interdependent. Hence, if the duration of the reaction can be increased, for example, less heat will be required. The same holds true for the density of the material in the reaction. Therefore, one possible set of circumstances that might produce a controlled thermonuclear reaction calls for a temperature of 100,000,000° C, a density of 10^{14} protons per cubic centimeter, and a duration of one second.

A number of parallel approaches to this problem are being made by laboratories all over the world. Normally, when national interests involved are so high, nations are moved to conduct their research with a certain amount of secrecy. But in the case of fusion, the global stakes, in a sense, are so great and man's knowledge is so limited that there has been a tacit understanding that—for the time being at least—the study of the peaceful uses of thermonuclear energy will proceed rather openly. It is for this reason that the current work at the Institute for Plasma Physics reflects not only the work begun years before by Werner Heisenberg at the Institute for Physics and Astrophysics, but also work taking place in the United States, the United Kingdom, and the Soviet Union.

Scientists at Experimental Division I of the institute operate Isar 1, a pinch device which enables the researcher to clamp a colossal magnetic field around a stream of ionized particles with such suddenness that particle temperatures of 60,000,000° C and a density of 10^{16} particles per cubic centimeter are obtained without great difficulty. Indeed, says research scientist Arndt Eberhagen, with larger condensers, it is likely that the goal of 200,000,000° C can be reached.

Not that the present bank of condensers in the Institute's Isar I is a weakling; release of the electrical energy stored in its 2,500 capacitors sends 20,000,000 amperes surging through a coil about 18 feet long and generates a magnetic field of 49 kilogauss. So violent is the surge of energy in Isar I that its designers had to lay some 20 tons of lead on top of the copper buss bars carrying the current to keep them from buckling.

But for all the heat and density of flux produced by Isar I, it has one disabling flaw. Its magnetic field is cylindrical and its configuration cannot prevent the plasma from streaming out of either end in a matter of microseconds. A hollow torus the shape of a doughnut would contain the plasma satisfactorily, but this configuration produces irregularities in the wraparound magnetic field that sends the hot plasma skittering against the cooler internal walls of the torus, thereby reducing the temperature of the plasma far below its effective range.

One of the most promising efforts at the Institute for Plasma Physics toward containing the fusion reaction is based on a very successful project being carried out at Kurchatov in the Soviet Union. There, Lev Artsimovich has induced an electric current to flow through the magnetically compressed plasma in a toroidal shape with the apparent effect of both heating the plasma further and reinforcing the magnetic field. Indeed, the results of the so-called Tokamak project were too good for Western scientists to accept until visits to Kurchatov verified the astonishing data. German scientists at Garching are now constructing their own version of the Soviet Tokamak, which they hope to have ready by the end of 1971. (Similar Tokamaks are also being built in the United Kingdom and the United States.) The scientists working on the project hope that the success of this device will permit them, in the not-too-distant future, to build a much larger and more powerful machine, thereby enabling them to make use of greater energy sources that would then be available.

Although the practicality of the mission of the Max Planck Institute for Plasma Physics makes it unlike most of the other Max Planck institutes, its practical role is limited only to proving that the production of thermonuclear power is possible.

Harry Redl from Black Star

Konrad Lorenz, a controversial scientist in the field of animal behavior, heads one of the four divisions within the Max Planck Institute for the Physiology of Behavior.

Once this has been done to the satisfaction of the government, the responsibility for carrying out the development of the technology will more than likely be transferred to another organization outside the society, possibly the government-operated nuclear research center in Karlsruhe.

Studies of animal behavior

In sharp contrast to the high-technology equipment at the Institute for Plasma Physics, the research program at the Max Planck Institute for the Physiology of Behavior, in Seewiesen, a wooded lakeside area about 20 miles south of Munich, proceeds in almost bucolic serenity. This institute, which is devoted almost entirely to the study of the biological basis for animal behavior, has become popularly known as the site of the laboratory of Konrad Lorenz, although Lorenz directs only one of the four divisions within the institute. Very much in the traditional pattern of the society, each director in the institute charts the course his entire division will pursue. In Lorenz's division, the work is concerned primarily with the observation of animals in nature and how their behavior is affected by instinct and learning.

The philosophical differences that exist in the study of ethology, the field of animal behavior, are well known. For many, Lorenz represents the archetypal view that there are highly significant parallels between the behavior of man and that of lower animals. In his own words, "ethology permits the study and observation of man without philosophical, religious, or ideological spectacles that presuppose that man is a supranatural being who does not obey the laws of nature; if you know animals well, you know yourself reasonably well."

Because of this argument for the existence of a phylogenetic program in man that gives "instinct" a strong role in determining behavior, Lorenz has been accused by other students of behavior as justifying violence and aggression as a normal characteristic of human behavior and thereby ineradicable from contemporary society. Lorenz counters with the argument that he is merely trying to "show the existence of internal forces that man must know in order to master." The intensity of this criticism, especially in the United States but also from such a noted British ethologist as Robert Hinde and even from some of his own colleagues, can also be attributed in part to the attraction that Lorenz holds for such dramaturgical ethologists as the amateur Robert Ardrey, who has applied Lorenz's ideas on territorialism in birds to human aggression. (For a criticism of Lorenz's ideas by Hinde, see *1971 Britannica Yearbook of Science and the Future:* AGGRESSION AND VIOLENCE, pages 64–77.)

Evolution under observation

Although Lorenz is the best known of the scientists at the Institute for the Physiology of Behavior, there are other impressive researchers, among them Wolfgang Wickler, a lively generator and expositor of some important ideas based on his own observations of animal behavior. Wickler describes his field of interest as that of the adaptive value of social systems in animals, particularly in the evolutionary "misuse" of brood care and sexual behavior. He has selected the word "misuse" to describe the situation in which a specific behavioral pattern has a positive value to the species other than the apparent reason it was favored by

Dr. Irven DeVore

Baboons are among the animals studied intensively at the Institute for the Physiology of Behavior. The adult male above, in a typical fierce gape, displays incisors that are more than twice as long as those of an adult female.

the selective processes of evolution. In the animals he is studying, the behavioral patterns selected out by evolution to favor brood care and sexual behavior have the vitally important side effects of stabilizing social groups and neutralizing aggression.

Wickler finds in the Arabian or sacred baboon (*Papio hamadryas*), for example, a most unusual demonstration of this relationship. In general, male baboons are hostile to other males when their paths cross. They are not hostile to females, however, because the female, when she is ready for mating, displays a red posterior which abruptly turns off the hostility mechanisms in the male and, with the stimulus of other sex attractants from the female, allows the primarily sexual mechanisms to come into play. Male Arabian baboons, however, naturally have red hindquarters. This permanent coloration acts to turn off automatically any hostility mechanisms among the males because of its similarity to the female mating signal. Thus, Arabian baboons are not aggressive toward each other. Wickler further noted that this lack of natural intraspecific aggression among the Arabian males is particularly beneficial in that it allows them to work together to defend the group against a common enemy.

More recently, Wickler has begun to study vocal stimulus and response in paired animals. Although these vocal exchanges are neither "duets" nor "conversations" in the common meaning of those words, it is Wickler's view that, quite possibly, there is some kind of communication taking place, very likely to permit better synchronization of action by the pair. He has observed this phenomenon not only in songbirds but also in other birds and in large monkeys, such as the siamang gibbon. In some species, pairs will engage in such vocal exchanges every time they meet.

The computer as observer

It is interesting to note, however, that there is also a countercurrent to the observational, narrative tradition at the Institute for the Physiology of Behavior. It is best exemplified by the work being done there under the direction of Horst Mittelstaedt. Mittelstaedt endeavors to filter out the bias of the observer by training the animal under study to act as its own reporter. For example, one of his current areas of research concerns the behavioral patterns of the migrating bird: how such a bird finds its way over long distances in fair weather and foul, and how it reorients itself in direction when it is moved from one location to another by human intervention.

In earlier attempts to explain this phenomenon, behavioral scientists studied the wing and head movements of captive birds under shifting projections of night skies and other techniques in order to duplicate the circumstances by which the migrating birds may determine their course of flight. In evaluating such movements, however, the observer can deceive himself. To eliminate all possible influences on the judgment of the observer resulting from his own preconceptions of avian response mechanisms, Mittelstaedt's group has conditioned its feathered subjects by giving them slight shocks whenever they face north relative to a projected night sky. After a period of training, the shock is discontinued, but the birds continue to register an increase in heartbeat when they sense a northern orientation. The bird's heartbeat is monitored by a computer so that its *involuntary* responses register quantitatively its sensing of north. In this manner, the bird's effort to align itself to north will be registered whether or not the bird is conscious of its efforts in some part of its primitive brain.

"We know about two cybernetic systems [in these birds]," says Mittelstaedt, "a closed [communication-information] system employing feedback and an open system that is superior [to the first] but requires far more precise measurements. Perhaps there is a third. We must learn more about the animal's ability to stabilize his internal environment despite the disturbing influence of external stimuli. This can be done only by observing these responses under controlled conditions in a precise

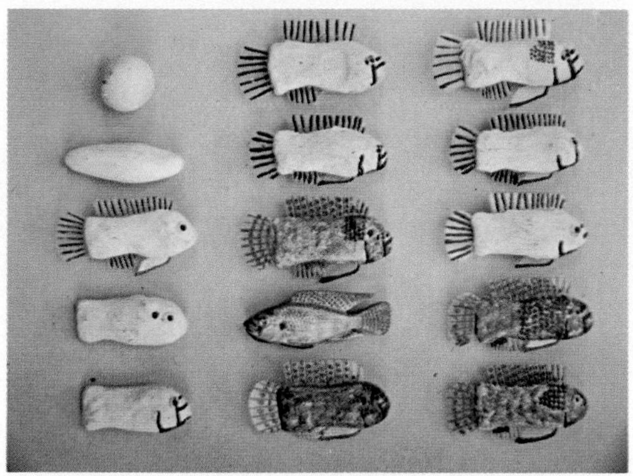

Dummies of the adult male fish Haplochromis *resembling the real fish to varying degrees were created to discover the specific attributes of one male that stimulate aggressiveness in another. A vertical marking under the eye was found to be the stimulus.*

and rigorous manner through the use of advanced instrumentation and computer technology."

The uncanny ability of insects to detect minute amounts of airborne sex attractants is one of several interests of Dietrich Schneider, another research scientist at the Institute for the Physiology of Behavior. He said:

We have the ability here to organize research teams drawing on a wide variety of disciplines. In my own field of insect olfaction, for instance, we have gathered together people from biochemistry, biophysics, cybernetics, experimental psychology, as well as insect physiology. The scientific leadership in the United States talks about such arrangements, but it is difficult to find examples."

Toward the future

Other Max Planck institutes have been engaged in equally significant work. Among the recent achievements of the Max Planck Institute for Cell Physiology, for example, is the development of techniques that made possible the first quantitative measurements of photosynthesis. The Max Planck Institute for Cell Chemistry has recently succeeded in tracing the biochemical processes involved in the formation of acetoacetic acid and the biosynthesis of fatty acids, rubber, and squalene (the most important precursor of cholesterol and numerous other natural substances). Research efforts at the Institute for Physics and Astrophysics have yielded important results in connection with the mathematical structure of quantum field theories and have also succeeded in providing the principles for a fundamental theory of elementary particles. Currently, this institute is conducting studies in magnetohydrodynamics and its application to the understanding of cosmic magnetic fields; it is also engaged in producing models to explain the development of stars and to elucidate their mechanisms of energy production. The Max Planck Institute for Nuclear Physics is presently conducting experimental research on elementary particles and atomic nuclei in order to gain insight into nuclear forces and the laws governing the structure of nuclei.

It is clear that the techniques that have been developed in West Germany for the support of fundamental research through the Max Planck Society have been highly successful with regard to the quality of its leadership and the work produced. The society also seems to have been able to sustain the independence, flexibility, and sensitivity to developing potentials that have been the principal arguments for its continued existence as a keystone of the German research endeavor.

This has been done at a time when support of its efforts has increased at a rate larger than any other item in the federal budget, an increase that will be impossible to sustain for any extended period. It has also been done during a period when the ability of the German university system to continue to produce talented young people has been seriously questioned. And it has been done at a time when, in almost every other country of the world, the stress has been on applied rather than on fundamental research—the quick fix and the fast payoff. It has yet to be seen whether the Max Planck Society, which has demonstrated its capacity to generate really first-class research in two different incarnations and to survive two disastrous world wars, can survive these latest threats to its integrity and independence.

However, the president of the society, Adolf Butenandt, is not depressed by any bleakness of outlook for basic scientific research in other countries, particularly the United States. In fact, he looks to the United States as the spur to even greater efforts by European science. Recently, he addressed these remarks to some fellow Europeans:

The Kaiser Wilhelm Society was founded 60 years ago in order to overcome political fragmentation within Germany by creating *one* nation of science. . . . Why shouldn't we seriously consider bringing about a similar condition in our now small Europe? Attempts in single areas of study —for example, the European Molecular Biology Organization—already exist. We know that political institutions resist consolidation, but that basic research needs a broad base of freedom, especially from political influence. Organizational forms in which the scientist can isolate himself from local political pressures have proved themselves to be scientifically creative on a national scale. It is to be expected, then, that the same would be true of the international domain. We owe our assistance and participation in the furtherance of any organization that might bring about these ends. . . .

Materials from the Test Tube

by Gene Bylinsky

Glass that bends without breaking, metals that stretch, plastics as tough as steel—these are among the products being created in laboratories throughout the world as man learns to reshape the basic structures of his raw materials.

Although a decade of widespread economic depression, the 1930s were also the years when polyethylene was discovered and when men first understood fully what a metal was. Today, we reap the benefits of these and similar discoveries in the form of a steady stream of new materials that sometimes seem almost magical—glasses that darken in sunlight and brighten indoors; polymers that endure heat better than some metals; stretchable metals; and plastic substitutes for familiar materials, even for the stainless steel of a surgeon's scalpel.

In truth, there has been a revolution, not only of materials but of outlook. Until well into the 20th century men took what nature had given them and did the best they could: clays were hardened, wool spun, and metal ores smelted. Now, man has taken the initiative and fashions materials to suit his needs.

The future is even brighter. Fibers of carbon and boron embedded in metals and plastic have already been produced and someday will meet the engineer's dream of the ideal material: high strength, little weight, and low cost. Materials will be made in zero-gravity space, where liquids can be processed like solids. The growing practice of ion implantation—bombarding pure crystals with selected impurities—not only will feed the insatiable demands of the electronics industry for new materials but also will revolutionize lens design.

Coils used in "flyback" transformers in television sets show the effects of exposure to intense heat. The newly developed coil at left, impregnated with a silicone resin, remains intact, while the coil at the right begins to melt. Such melting has caused detuning of sets and has created fire hazards.

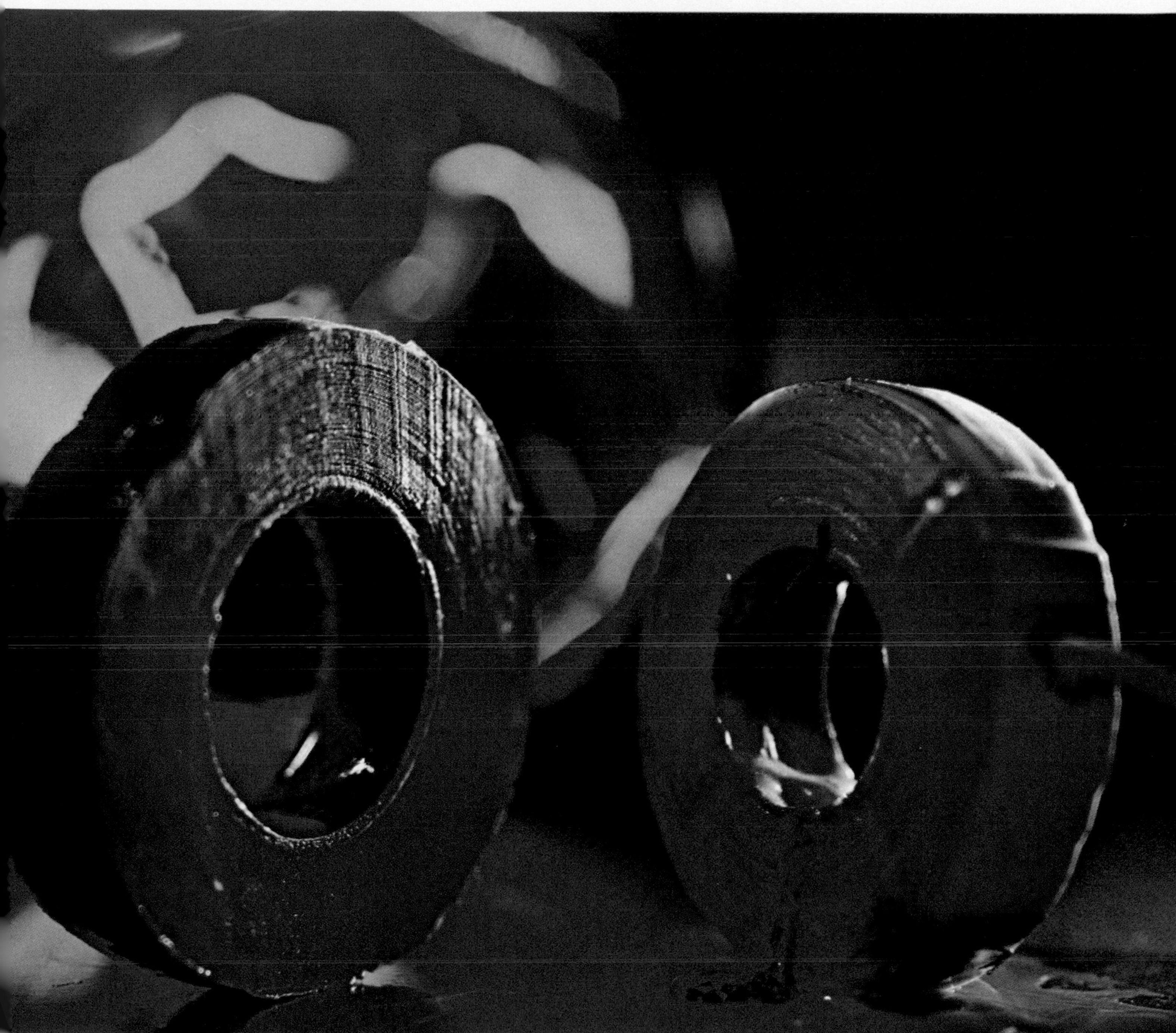
Courtesy, Dow Corning MATERIAL NEWS

A boron filament, shown in the photomicrograph at right, can be embedded in metals and plastics to add strength to the material without adding much weight. The fire-resistant suit below is made of an organic textile fiber called Kynol. It allows the wearer to walk for three minutes through flames of 2,500° F.

Courtesy, United Aircraft Corporation

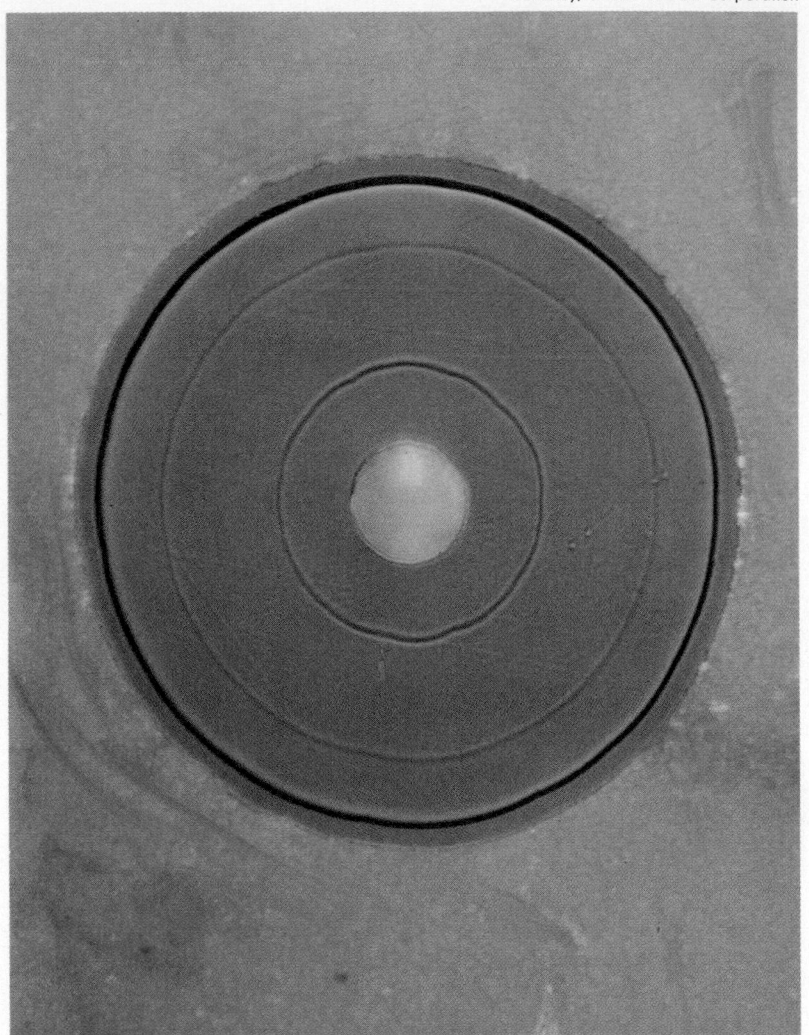

Courtesy, The Carborundum Company

Courtesy, General Electric Research and Development Center

Gross structure of a silicone polymer is seen in this photomicrograph to be partly crystalline and partly amorphous. The radially symmetrical arrays fanning out from the bright centers are spherulites; other parts of the polymer are amorphous.

Order and disorder

The science of materials is called a "battlefield between order and disorder." Early in this century, X rays were used to probe the crystal structure in salts, minerals, and metals, finding beauty in the regularity of their atomic arrangements. In the 1930s metallurgists learned that there was also beauty in imperfection. They began to realize that aberrations in the regular alignment of atoms gave materials their special nature, such as chemical reactivity, electrical conductance, and resistance to stress. New theories evolved, specifically, quantum and dislocation theory. The quantum theory of solids, which is based on the concept of the subdivision of electronic energy into finite parcels, or quanta, provided a clearer picture of what held atoms together in a solid. It also revealed the nature of metals and nonmetals, the crucial difference between the two being the freedom of movement of electrons within energy levels. Dislocation theory, which postulated that growing crystals were vulnerable to mistakes, produced a new vocabulary to describe those errors, such as point defects and dislocations.

GENE BYLINSKY, an associate editor of Fortune *magazine, has written numerous articles on scientific subjects for many publications and was the recipient of the 1970 Albert Lasker Medical Journalism Award.*

Courtesy, Corning Glass Works

Courtesy, General Electric Research and Development Center

As metallurgists became familiar with the quantum theory, they developed a new awareness of the potentialities of their materials. This new understanding bore fruit in the 1950s, when it was used in conjunction with computers to guide the development of semiconductors, materials that made possible the transistor age. Dislocation theory, on the other hand, led to the creation of new metal alloys, often bizarre combinations whose recipes were devised not by ancient metallurgical skills but by complex equations. A participant at a recent materials science conference assigned $10 billion to $30 billion as the economic value of the quantum theory of solids, and $1 billion to dislocation theory.

Glass that bends without breaking (opposite page, top) is the result of chemical strengthening achieved by an ion exchange process. The glass-crystal mixture shown in the photomicrograph (bottom) was produced by melting ordinary window glass in a crucible and holding it at a temperature of a few hundred degrees for several days. During this period the crystal needles, seen in the picture as predominantly yellow and blue against the magenta glass matrix, grew inward from the slightly cooler edges of the crucible.

Manipulating giant molecules

In order to understand how chemists manipulate such giant molecules as polymers and other compounds, one must first know how atoms unite to form molecules. The mechanism for such unions is the chemical bond, in which two or more atoms join together by sharing electrons that orbit their nuclei. When atoms share a pair of electrons (one from each atom), they form a single bond; those that share two or three pairs of electrons form double or triple bonds, respectively. A union of single bonds is weaker than one of double bonds, while one with triple bonds is the strongest of the three.

Polyethylene, the first of the synthetic giant molecules, was discovered accidentally in 1933, when chemists at Imperial Chemical Industries Ltd. (ICI) in Britain were testing the effects of high pressure on the chemical behavior of the compound ethylene, an unsaturated hydrocarbon. (A hydrocarbon is a compound that consists solely of hydrogen and carbon atoms. In an unsaturated hydrocarbon, the carbon atoms are joined together by two chemical bonds, which is one more than is absolutely necessary. It is this double bond that makes ethylene a highly reactive compound.) All the ICI men got for their trouble was a small bit of white powder at the bottom of the reaction vessel. It was polyethylene. Under the inducement of high pressure, the ethylene molecules had joined together, forming a chain of ethylene molecules, or, more accurately, one very large molecule—a polymer. Plastics are one class of polymers; others include proteins, carbohydrates, and rubber.

For some time following the discovery of polyethylene nothing happened. There were years of trial and error, but not until July 1939, six years after its discovery, was polyethylene first sold commercially, as a cable covering. The material came into its own, however, when it enabled radar to take to the sea and air. With its need for high voltages and low electrical losses, radar was originally limited to land, where it could be protected from dampness and corrosive sea water. Polyethylene solved the problem: it repelled water, was an electrical insulator, and could be molded to fit the intricate design of radar units. Polyethylene plants soon went into production in 1942 in Britain and in 1943 in the United States.

The largest piece of glass ever cast in the noncommunist world, a 25-ton mirror for a telescope in Chile, cools during a stage in its production. The mirror is made of a new glass material called Cer-Vit, developed by Owens-Illinois, Inc. It was poured at a temperature of about 3,000° F, and the heat shield in the center of the photograph was lowered over the mold to control the cooling.

Courtesy, Owens-Illinois

Then, in the mid-1950s, chemists in West Germany and the United States discovered how to use low pressures to make "high-density" polyethylene, a plastic that is stiffer and less pliable than the low-density material discovered by ICI. The difference between high- and low-density polyethylene is somewhat like the difference between straws bunched together and those linked end to end: one arrangement is stiffer than the other. The ethylene chains of the high-density polymer are not only linked together end to end, they are also bunched, giving them their stiffness. The world still makes more low- than high-density polyethylene, although production of the latter is climbing faster. Low-density polyethylene is used for films and electrical insulation; high-density for bottles, containers, and crates.

During the same period that high-density polyethylene was created, Italian chemists not only polymerized propylene (also an unsaturated hydrocarbon but one step above ethylene in complexity), they also were able to produce in polypropylene the same arrangement of atoms as in a crystal. Thus, just as many metals attain their strength and stability from their crystalline structure, so crystalline (really semicrystalline) polypropylene was stiffer, less prone to heat-softening, and more stable than noncrystalline polypropylene. Polypropylene soon became a major commercial polymer, used for transparent packaging, piping, and textiles among other things. The application dictates the type of polypropylene used—semicrystalline and stiff, or noncrystalline and flexible.

Other polymers then were created, all offering a spectrum of properties to architects, builders, car makers, and others to ease their particular problems of design, production, and cost. By 1971 a new car contained about 110 pounds of plastics; a company expected to market an all-plastic refrigerator in five years; and a city in Michigan built its city hall entirely of reinforced plastic, a combination of plastic and concrete. Some of the plastics are gradually supplanting natural materials—polyurethane and fiber-reinforced vinyls for leather; acrylics and polystyrene for glass. Specialized plastics to withstand high and low temperatures as well as corrosive atmospheres are being created in laboratories as the chemistry and physics of polymers become better understood.

A wider spectrum of high-temperature plastics, including ones that can be easily processed, will be created when researchers learn how to dampen vibrations produced by high temperatures. These vibrations snap chemical bonds and destroy polymers. In ladder polymers, which are one kind of high-temperature plastic, two chains are linked periodically in the way that two sides of a ladder are joined by rungs. Such polymers are particularly advantageous in high-temperature use for two reasons: (1) if one "rung" is broken, the ladder stays intact; and (2) it is statistically unlikely that the two vertical joints between two rungs will snap at the same time and break the polymer. If one joint is broken, it can probably be repaired before the opposite joint is snapped.

Polymers thin as film but tough as steel may emerge as more is learned of "extended-chain crystallization." Crystalline polymers crystallize in tightly folded positions, held in place by secondary but fairly weak chemical bonds formed within the same chain. If polymers can be crystallized when the chains are fully extended rather than folded, their strength would be determined solely by the primary carbon-to-carbon bond, a notoriously strong one in chemistry. The end product would be a very thin but very tough plastic.

When first put on the market, plastics got off to a bad start because they yellowed in sunlight and broke apart, some literally crumbling to a powder. Chemists solved that problem, but now—in a society newly concerned with waste disposal—they have had to revise their course by trying to find ways to make sure plastics do not outlive their usefulness. One approach has been to attach ultraviolet-sensitive chemical groups to polymer molecules. As the ultraviolet rays in sunlight act upon these groups, they snap polymer bonds, reducing the plastic to a form that is susceptible to degradation by weathering and microbes.

Courtesy, Monsanto Research Corporation

Cellulose acetate hollow fibers, seen here in cross section, have proved valuable in desalinating water. Salt water under high pressure is passed through the fibers, which separate the salt and the water. The walls of the fibers exclude the salt, while the pure water flows through the hollow centers to a reservoir. Each fiber is about 300 microns in diameter.

Ceramics and glass

Without ceramics, archaeology would be a thin subject—no broken pottery, no puzzling statuary, no Portland vase. Earthenware, stoneware, and porcelain are ceramics, combinations of various materials, usually including silicates, that are hardened in a fire. Put another way, ceramics are a mixture of order and disorder: crystalline and amorphous materials, or crystals suspended in a glassy network.

While ceramics are ancient materials, their value to such fields as optics, electronics, and architecture has increased only recently as men learned to manipulate their properties. Normally opaque ceramics can be made transparent with new firing methods. Although brittle in comparison to metals, ceramics can be toughened by fashioning them into fibers. Such ceramic fibers as silicon carbide could, according to some authorities, lead to materials "capable of use at higher temperatures than currently available metals."

Glass has also become more widely useful as scientists have learned how to cope with its amorphous structure, the exact opposite of the fine atomic arrangements of crystals. There is now clear glass that bends without breaking, glass that can stop bullets, and glass that acts as a semiconductor, passing or blocking electrical current in response to exacting cues. Fiber optics is the use of fine glass wires that can transmit light. Its probable applications include image intensifiers that can bring dim starlight up to working surveillance levels, waveguides for "bug-proof" communications systems, and transmitters of laser energy into inaccessible parts of a living body.

Glass owes its versatility to its accommodating chemistry; it is an effective medium for a great variety of chemical reactions, and almost all elements can be dissolved in it. Glasses that darken or lighten in response to light usually are produced by incorporating into the glass

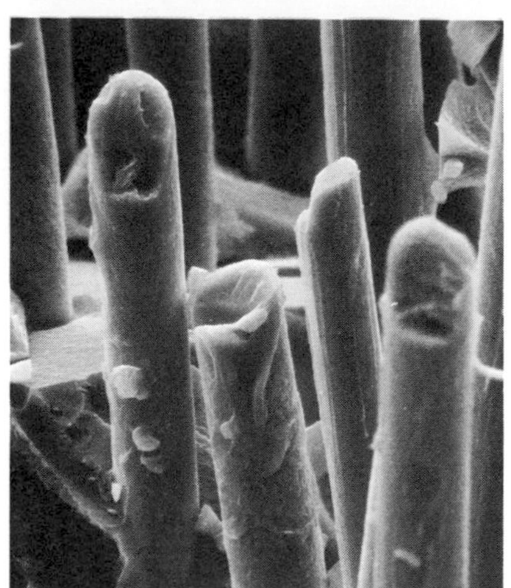

microcrystals of silver halide, each so small that one cubic centimeter of glass contains about a quadrillion (5×10^{15}) crystals. The microcrystals are so small that they do not scatter visible light nor are they affected by it. Hence, in such light, the glass remains clear. However, when exposed to the shorter wavelengths of ultraviolet light, such as from the sun, the microcrystals undergo a change, causing the glass to darken.

Actually, the amorphous character of glass is deceptive. True, overall, the atoms are scattered in random fashion, but there are short-range effects between neighboring atoms and ions that create momentary, highly local centers of order within amorphous solids. This short-range order is the basis for amorphous semiconductors, which are glassy electrical devices whose switching and memory capabilities may make them useful in electronics, particularly in systems exposed to radiation or temperatures that can wreck the regularity of crystalline semiconductors. In addition, their relative ease of fabrication—crystalline perfection is not needed—may make amorphous semiconductors the most desirable materials in some situations.

Superconductors and alloys

A truism in materials science is that the material is often at hand for years before it is developed for commercial use. This was true of fiber optics, polyethylene, and even with some metals, such as titanium. But in some cases the development awaits the material. Witness superconductivity. A superconductor has no electrical resistance and theoretically can pass an electrical current in perpetuity. A number of superconductors have been developed, the most successful being a niobium-germanium-aluminum alloy. None, however, works at practical operating temperatures, which are usually around −320° F.

The search for new metal combinations that can raise superconducting temperatures even a fraction of a degree is intense. A jump of several degrees above current levels would nurture a whole new technology of power transmission. Superconducting cables could completely eliminate the costly 20% loss in energy that occurs when electricity is passed through overland wires from generating station to user. Using the superconducting materials now available instead of waiting for better ones, Great Britain and Switzerland have already installed experimental superconducting power transmission cables. (See *1971 Britannica Yearbook of Science and the Future*, CRYOGENICS: THE WORLD OF SUPERCOLD, pages 318–323.)

Some dream of superconductors that work at room temperatures. At Stanford University, physicists and chemists in a collaborative effort are attempting to design and create organic molecules—materials with a carbon chain bearing a chemical resemblance to certain dyes—that in theory should superconduct at room temperature. Another approach, based largely on Soviet theory, is to create layers one to two atoms thick of known superconductors. By late 1971, however, neither these nor similar approaches had proved fruitful.

Still another source of superconducting alloys could be a new metallurgical technique that many view as one of the most exciting developments in the field of materials science in recent decades. This is fast quenching, or cooling, of a liquid metal by splashing it against a copper or silver base. Because the technique allows precise control of the metal's structure through manipulation of the cooling rate and the specific alloys used, research metallurgists have been able to produce materials that vary in structure from the amorphous to the microcrystalline, as well as alloys with an extremely fine grain size. The last property is basic to improved toughness, strength, hot plasticity, and cold ductility.

A related process, dispersion-strengthening, is used to produce alloys that can withstand extremely high temperatures. If refractory (heat-resistant) metals with a melting point above the present limit of 4,000° C could be developed, the life of materials as diverse as light bulbs, rocket parts, and turbines could be dramatically lengthened. In dispersion-strengthening, a refractory metal oxide, usually a ceramic, is dispersed throughout a metal matrix (the principal constituent of an alloy). Just how the dispersed particles affect the atomic structure of the alloy to make it stronger is still not certain. Presumably, the dispersion reduces the slippage along tiny cracks, or planar defects, of the alloy grains. Although alloys strengthened in this way are very expensive, often brittle, and sometimes crack under local stress, scientists expect that they will be available commercially well before 1980.

A great deal of effort is also going into hardening pure but soft metals, such as aluminum and copper. In particular, two methods are being explored to help these metals keep their shape under stress: work hardening and precipitation hardening. Both methods, while differing in technique, are designed to create a maze of dislocations in the crystalline structures of these metals so that new cracks, like cars stuck in a traffic jam, cannot move far.

Superplastics and fiber composites

Stretchable, or superplastic, metals represent another facet of contemporary metallurgy. Titanium alloys, brasses, and aluminum alloys are among the important stretchable metals. Basically, superplastic alloys can be stretched up to 20 times their original length without breaking. Because they can be worked almost as if they were liquid rubber or plastic, superplastic metals could open up new methods of manufacturing large and small metal articles, ranging from cups to computer frames. Many of the same techniques used in plastic manufacture could be adapted to superplastic metals.

Although superplastic metals were discovered by a British scientist in the early 1930s, they are not yet fully understood. Scientists are uncertain as to why some metal combinations can be stretched without breaking or snapping. One guess is that certain crystal defects,

Courtesy, Phillips Petroleum Company

Glass fibers (opposite page) and glass cloth (above) are mixed with Ryton® PPS (polyphenylene sulfide), a resin that allows the glass to be molded and makes it resistant to high temperatures and corrosion. The scanning electron photomicrographs show the ends of tensile test specimens; the upright material is the glass fiber and cloth, while the adhering substance is the Ryton®. Magnifications of the photographs are: opposite top, 200 times; opposite bottom, 1,000 times; above top, 100 times; above bottom, 20 times.

Courtesy, Stanford University News Service

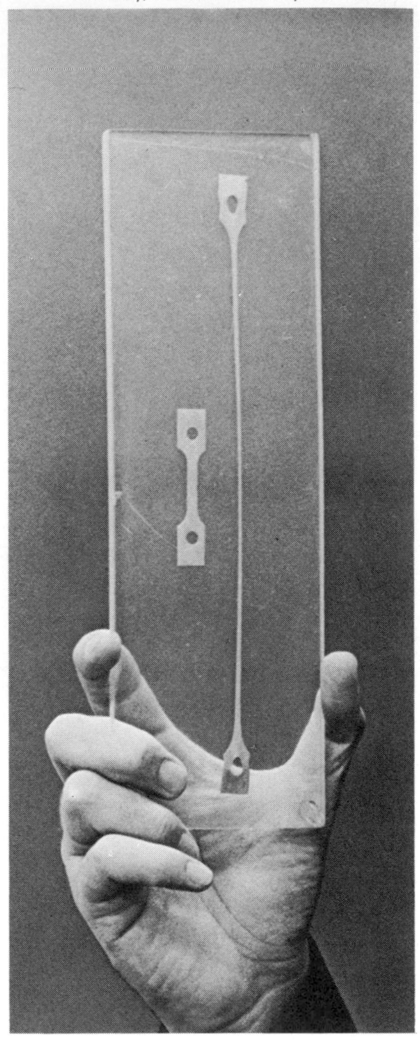

Superplastic metal, encased in protective plastic, is displayed before and after being stretched. The process was developed by Charles Packer of Stanford University.

called line defects, enable crystalline planes to slip over each other. Put another way, superplasticity probably owes its existence to dislocation motion, which is the parallel movement of defective crystal planes.

While the superplastics are only starting to shed their image as laboratory curiosities, the fiber composites have almost arrived. They promise the engineer useful mechanical strength along with low weight and—someday—low cost. Composites are products of a mixture between strong but often brittle fibers of various materials and a weak but ductile metal or plastic matrix. For example, by embedding almost invisible strands of sapphire into aluminum, the aluminum becomes six times stronger and about twice as stiff.

Because the price of advanced fibers is still high—anywhere from $100 for a pound of carbon fiber to a spectacular $5,000 for a pound of ultrathin beryllium wire (10 times the price of gold)—composite materials are in limited use, principally in the aircraft industry. But just as the development of plastics was spurred by wartime and the special needs of radar, so the development of composite fibers is being spurred by government money for spacecraft and new airplane design.

Some serious obstacles to the large-scale manufacture of composite materials remain. Tiny whiskers of carbon, boron, and similar materials are difficult to handle mechanically; inserting long filaments into a metal matrix is another difficult problem. In one approach, filaments are wrapped around a spindle and sprayed or electroplated with metals; in another, the metal-clad filaments are pressed together into a compact bundle. Large-scale manufacturers, however, will need simpler, more efficient methods of merging fiber and matrix. Moreover, manufacturers may also find that the matrix softens when fiber composites are used in high temperatures. This problem may be resolved, however, by using high-temperature plastics for the matrix.

Materials for electronics

Advances in electronics are inexorably linked to advances in materials. "It used to be that materials could be selected almost incidentally in the development, say, of a new electron tube," said one specialist. "Now we find that materials and processes are an inseparable part of the design process and indeed mainly determine our progress."

The rise of solid-state electronics has emerged from a marriage of the theories of modern physics, such as quantum mechanics, with metallurgy. The offspring has been a great variety of devices for guiding the movement of electrons through solids. In order to produce the semiconductor materials used in these devices, very pure crystals are contaminated with certain impurities or dopants. Basically, that is accomplished by heating the pure crystal enough to allow the dopant to diffuse into the material. The heating, however, may damage the semiconductor lattice (the array of atoms that must be kept intact to assure the proper movement of electrons).

It has been discovered that damage to the semiconductor lattice can be reduced by shooting rather than diffusing ions into the crystal with ion accelerators. This technique—ion implantation, as it is called—is already in use, particularly with such a difficult-to-soften material as silicon carbide. It may be especially valuable in preparing semiconductors that can withstand high temperatures and intense radiation.

Ion implantation may, however, have an even wider future, based largely on the fact that the number of ions implanted in a material can be varied with depth. For example, by bombarding optical glasses with ions at different depths, the refractive indexes of the glasses can be varied at will, making possible the design of new lenses. Ion implantation can also be used to insert ions of one metal into another. Thus, lead ions can be inserted into aluminum, a metal in which lead normally is virtually insoluble. The value of such new products is still uncertain, but the technique will give metallurgists yet another tool in their continuing quest for new materials.

Ion implantation may be the opening wedge to the ultimate goal in materials design: materials whose properties vary throughout and are locally responsive to different requirements. To cite a simple example, the ends of a piece of metal and its middle may vary in elasticity. Thus, whereas the structure of materials once dictated their use, in the future their structure will be determined by the properties that man wants his materials to have.

See in ENCYCLOPÆDIA BRITANNICA (1971): LOW-TEMPERATURE PHYSICS, *The Phenomenon of Superconductivity; PLASTICS.*

FOR ADDITIONAL READING:

"Alloys That Stretch, Bounce and Bend," *The Sciences* (April 1969, pp. 12–15).

Frados, J., "Plastics in the 1980s: A 15-Year Outlook," *Modern Plastics* (Nov. 1968, pp. 120–138).

Industrial Research, Special issue on composites (October 1969).

Scientific American, Special issue on materials (September 1967).

Courtesy, United Aircraft Corporation

Continuous boron filament is wound on a mandrel and then sprayed with molten aluminum to form high-strength tapes for use in the aerospace industry.

Instant Intimacy

by Irvin D. Yalom and Samuel Moffat

In an ever more impersonal society, man wants to belong, to be understood and accepted. The rapidly growing encounter group movement, which seeks to meet these needs, has many dangers but also offers much promise.

Not since Sigmund Freud's discoveries about the human mind became widely known after the turn of the century has there been such ferment as there is today among the sciences concerned with the changing of human behavior. One major phenomenon that has stimulated great professional and public interest is the development of the encounter group.

An encounter group consists of a small number of people who meet with a leader and try to explore their feelings about themselves and others in the group. It has been estimated that as many as six million Americans have participated in such groups. Psychologist Carl Rogers wrote in 1968 that the intensive group experience movement was "one of the most rapidly growing social phenomena in the United States . . . perhaps the most significant social invention of this century." The movement also has spread to Europe, especially to England, West Germany, and the Netherlands.

Groups go by so many different names today that it is often difficult to be sure just what each one does. "Encounter" is the best word to characterize them all, however, for they all depend fundamentally upon a face-to-face meeting between human beings. The term "encounter group" is used arbitrarily to cover sensitivity training groups, human relations training groups, T-groups (T meaning training), human awareness groups, human enrichment groups, personal growth groups, sensory awareness groups, Synanon games, marathon groups, and so on. Any single group may use one or several techniques, including verbal confrontation and self-disclosure, Gestalt (based on the unity of mind and body), sensory awakening, massage, hypnosis, body movement, Yoga, or nudity.

A rapidly growing movement

Groups are particularly numerous in college communities, where bulletin boards and underground newspapers are crowded with advertisements for them. An informal survey indicates that in many large universities more than 50% of the students may have participated in one or more groups. But experience is not limited to those of college age. Many participants are, in fact, middle-aged. Groups are attended widely by employees of large corporations, teachers at all levels, and members of the helping professions, such as ministers, social workers, and mental health professionals.

One index of the rapid expansion of encounter groups has been the proliferation of "growth centers" that use various group methods designed to help individuals enhance their creativity, self-knowledge, and ability to work with others. The best known of these centers is Esalen Institute at Big Sur, Calif., but since Esalen's founding in 1962 approximately 100 similar centers have sprung up around the country, many of them modeled on the Esalen design.

IRVIN D. YALOM is associate professor of psychiatry at the Stanford University School of Medicine. SAMUEL MOFFAT, a science writer, is also a contributing editor for the Britannica Yearbook of Science and the Future.

Growth centers often are isolated geographically and may be located in a setting of great natural beauty; hot baths, saunas, swimming pools, and similar facilities frequently are included. Encounter sessions at these centers rarely last more than a few days, and fees may range from nothing to several hundred dollars for one- or two-week workshops.

In addition to groups at growth centers, there are countless informal groups run by individuals who were "turned on" by an encounter experience and then decided to lead a group. At present, anyone who can enlist enough members can start an encounter session, for there are no laws regulating groups and no professional standards for their leaders.

This rapid growth, together with the lack of a regulatory body, has allowed irresponsible practitioners to enter the field. There are groups whose avowed aims are to encourage personal change and growth, but whose true purpose is to permit unrestrained sexual indulgence, or to satisfy the leader's own need for recognition, money, sex, or power over others.

These groups' excesses are more publicized than actually practiced, but they still represent a real danger for potential participants. They have been rightly condemned by professionals in the mental health fields. These few groups should not be allowed, however, to detract from the positive contributions to behavioral science made by responsible practitioners in the encounter group movement.

A social need

In light of this sudden proliferation, it is natural to wonder whether encounter groups may turn out to be just another American fad—more significant than the Hula-Hoop but not much more enduring than the jogging craze. Some observers feel this is so, but we disagree. It is probable that the basic format of the encounter group will be with us for a long time.

The principal reason that encounter groups are likely to persist is that they meet an urgent need. They offer one means by which men and women can accommodate, if not overcome, powerful dehumanizing trends in our technocratic society. Man is naturally gregarious; he has deep-seated needs to associate with and be accepted by others. But our society does not meet these needs. The family is breaking down; religious and social institutions are losing their importance; and individuals move about so much in their lifetimes that they rarely have real roots in the communities where they live. Furthermore, man's creative urges are not satisfied by the monotonous, often meaningless work required in many jobs. Faced with isolation and alienation, many people seek the instant intimacy an encounter group provides. In a sense they say, "Touch me," emotionally if not physically.

This is not to suggest that short-lived, artificially formed small groups perform ideally. In a curious way the small-group movement is at the same time an antidote for a disease and a symptom of the same disease. Growth centers can be just as bureaucratic as other organizations, and many employ slick Madison Avenue advertising styles. Moreover, despite their apparently intense intimacy, encounter groups can have a disquieting impersonality. The members depart after a few hours or a few days together, and with rare exceptions they will play no active role in each other's future lives.

Despite their shortcomings, however, encounter techniques offer much promise. Some research studies have shown that the more educational version, the sensitivity training or human relations training group, has had lasting effects on the behavior of individuals and the performance of organizations. Participants have demonstrated increased sensitivity to others' needs, better understanding of their own and others' behavior, and increased ability to work effectively with others. For some corporations sensitivity training has increased profits—one study showed that a team of pharmaceutical salesmen made more sales after sensitivity training. Another study showed it improved the job performance of elementary school principals, as measured by objective observers. Some organizations have been less satisfied with such training, however, and have discontinued it.

Some encounter techniques are now widely used in group psychotherapy for disturbed patients. In addition, encounter groups provide an informal auxiliary mental health system in some parts of the country. It is perhaps not exaggerating to estimate that in California as many people attend encounter groups because of their individual psychological problems as seek out professional therapy.

What happens in an encounter group?

With the multiplicity of names applied to encounter groups, one might expect totally different experiences in different groups. Actually, however, many features are common to all. The participants are usually strangers, and groups are small, averaging about 12–16 members plus a leader. The leader usually provides few guidelines; he or she may say simply that they are there to talk about themselves and their feelings about the other members.

The group begins gingerly. The members may talk about the strangeness of the setting, everyone's silence, or their initial reactions to each other. Gradually, they share more and more feelings. Some members may admit how shy they feel when facing strangers, or will acknowledge they fear others in the group who seem intimidating.

The leader usually tries to draw silent members into the conversation and discourages abstract or intellectual discussions. If, for example, someone talks about "man's propensity to respond defensively when faced with danger," the leader might try to bring the conversation around to "here and now" by asking, "What is dangerous *to you now in this room*?" Drawing conclusions about others' motives is frowned upon, but "owning" your reactions is encouraged. If someone said, "I think you're trying to make me mad!" he would probably be urged to say what he feels instead: "When you talk that way I feel angry." The leader will try to get members to reveal themselves, loosen up, and express their emotions more vigorously.

The leader or more experienced members may introduce "exercises" to stimulate interaction. The participants may "go around" and describe their feelings about everyone else in the group. Or, one at a time, they may describe their impressions of a single member, so he can get a better idea how he affects others. Nonverbal techniques are widely used, and members may be asked to communicate by touching or gesturing. In some groups the barriers to free expression are broken down by prescribed techniques—special physical exercises, physical contact, meditation, or new sensory experiences, such as taking a fantasy trip through your own body.

A "social microcosm" with feedback

One of the first tasks of an encounter group is to establish conditions that encourage members to behave the way they do in real life. Thus the group becomes a "social microcosm," a miniature version of the real world. Shy people will display their shyness and aggressive people their aggressiveness. Vain, exploitative, selfish, or self-denying persons will reveal themselves, too. Then the group members feed back information to each other about this "presentation of self." Feedback enables the participants to see blind spots in their own understanding of themselves.

The group need not stop here, however. Once you get a better picture of how you really behave, by seeing how you appear in others' eyes, you may be able to change your behavior. In some groups behavior change is, in fact, a major goal. If someone is guarded and secretive, he may be urged to take risks by saying things he has never before dared to share with others. If he is too timid, he can try to be assertive or express anger; emotions may even be displayed physically by pounding a pillow, arm wrestling, or, occasionally, having a fist fight. A distant, aloof, or hostile individual may be encouraged to express close, affectionate feelings either in words or by holding or caressing another group member.

The other principal feature of the encounter group is that it offers a chance to be part of a cohesive, functioning unit for the life of the group. The members experience at first hand the power of a group and the pressure it can exert on human affairs; and they also learn what pleasure one can gain from being a valuable and accepted member of a tightly knit group.

The leader's style

What happens in an encounter group depends more on the personal approach of each individual group leader than on the "school" with which he is associated. In general, there are four features by which any leader may be characterized:

1. *Group or individual focus*. Some leaders focus more on the group than on the individual. Those who were trained in the older T-group or sensitivity training style of the National Training Laboratories Institute for Applied Behavioral Science, located in Washington, D.C., which essentially founded the encounter group movement, tend to concentrate on group dynamics, or what is happening in the entire group. They try to help members learn how groups operate so the participants can function more successfully with associates at work or in any of the boards or committees widely found in business, education, and other endeavors.

Group focus is not as popular now as focus on the individual members. Most leaders today view the group experience as a form of therapy or growth for normal people. They try to help each person explore his own personality and his relationships with others. Some leaders carry this approach to even greater degrees and neglect the group entirely by focusing on each of the individual members in turn.

2. *Leader activity*. Some leaders are relatively passive. They participate only enough to make the group become an effective agent for change. In the beginning they bring in silent members, model openness by being open themselves, and guide the group in a variety of other ways. But their ultimate aim is to bring the group to such a level of

autonomy that it can function essentially without a leader. Other leaders feel that it is their responsibility to guide the group personally throughout its lifetime. Often they are so active that the group consists principally of interactions between the leader and individual members.

3. *Confrontation*. Some leaders are gentle and supportive. If a member is threatened or upset, the leader will let him withdraw temporarily or take part at his own pace. Other leaders stimulate direct confrontation. They are outspoken, express their own anger and other strong emotions, and push others to do the same at an accelerated pace. There is nowhere to hide in these groups.

4. *Structure*. Some leaders leave their groups almost completely unstructured. There is no agenda, and there are no prescribed approaches for getting people involved. Other groups follow a series of exercises laid down by the leader or an experienced member. The group might break up in twos or threes to talk, to communicate other than in words, or to massage each other. Or they might take part in a psychodrama, acting out various roles to get insight into their own or others' behavior. An extreme version of the structured group is the completely programmed group that consists of a series of tape recordings. There is no leader, but participants follow the transcribed instructions.

The four features just described can be used to classify the various sorts of encounter groups. There is a spectrum of types. At one end is the T-group—concerned with group dynamics, having a leader who facilitates the group rather than actively directing it, and characteristically nonconfrontive and unstructured. At the opposite end of the spectrum is the Gestalt therapy group—oriented toward individual exploration and with an active leader, much personal confrontation, and a structured program. Other groups can be classified in terms of these same four variables.

Evolution: from education to therapy

The two major types of encounter groups—T-groups at one end of the spectrum and individually oriented, confrontive groups at the other—reflect the history of the encounter group movement. It began with a workshop, organized by the psychologist Kurt Lewin in 1946 for the Connecticut Interracial Commission, to train leaders who could deal effectively with racial tensions and who could change racial attitudes.

The workshop's discussion groups were the first encounter-like groups. Their potential for human relations education was discovered almost by accident. Lewin believed strongly in research and had social psychologists observe each of the discussion groups. Afterward the observers described what they had seen, and the participants were allowed to listen. The participants discovered that the objective review of their behavior gave them unexpected insights about themselves, their effects on others, and on group behavior in general.

Lewin died a few months after the Connecticut experience, but his three group leaders held larger sessions the following three summers in Maine. Then, in 1950, their sponsoring organization, the National Training Laboratory in Group Development (since renamed NTL Institute), became a year-round institution within the National Education Association. Its first director was Leland Bradford, one of Lewin's original group leaders. The NTL Institute now has a network of 600 trained leaders and has held laboratory sessions for more than 2,500 participants in one year. The professional acceptance and success of early encounter groups, indicated by the demand for groups and the outcome of research showing their effectiveness, was due in large part to the NTL Institute.

T-groups of the 1950s placed increased emphasis on feedback, honesty, self-disclosure, and "unfreezing" (challenging participants' values and beliefs). Groups proved most effective when they focused on the "here and now" (behavior occurring at that moment in the group) rather than on past history or "back home" problems of group members. Despite members' great emotional involvement, however, T-groups were not considered therapy but a sophisticated technique of participatory education in group relationships.

In 1962 three NTL Institute trainers from California published an influential article that signaled a major change in emphasis. They broadened the concept of education. No longer did it mean just acquisition of skills useful in dealing with others; now it meant total enhancement of all of an individual's potential. These writers argued that the demands of a fast-moving, competitive society created a neurosis shared by a majority of the population, and that the T-group could help overcome it. In other words, the T-group could provide therapy for normal individuals. Since that time the encounter group movement has evolved toward stressing individual dynamics more than group dynamics, and enhancing personal growth rather than teaching interpersonal and group skills.

continued on page 420

"I don't know whether it helped . . ."

What do people do in an encounter group? How do they feel after confronting an angry group member, or having their own hang-ups exposed for everyone to see?

There are too many kinds of groups and too many individual reactions for anyone to answer these questions simply. Verbatim transcriptions of several sessions would be needed to reveal the diversity of approaches used, and the variety of individual response. Even then, the picture might be misleading, for what seems a great success to one participant might have been a disaster for another.

A better way to capture, in a few words, the feeling of what happens is to let some participants tell their reactions in their own words. These comments were obtained as part of the research project described in the accompanying article.

. . . There was this guy in the group that I absolutely hated. There was something about him that made me flex up from the first time I saw him. And every time he'd say something, some nasty thought would enter my mind. Finally I started talking about this and we really went at each other. At one point the leader asked me to take this guy by the shoulders and push him down to the ground as hard as I could, and then when he was on the floor I had to pick him up again. Then I had to talk about the feeling I had when I pushed him down and the feeling I had when I picked him up. He had some similar feelings for me and we both talked about why we hated each other so much. Then we talked some more about the kinds of persons we were—especially the kinds of things we were afraid of, the kinds of things we wanted in life, and how we got to be the way we are. I began to see that he was a lot like me, and I was surprised to find out that he was really scared of me. In the third meeting he told about a dream of his in which we were playing tennis and I had this huge racquet and he only had a little match stick. I think it took a lot of guts for him to tell me that dream, and we ended up really close buddies. This was all very important to me. The fact that I could work through these feelings of real hate toward him means that I could probably do it with anybody if I really made the effort . . .

. . . The person that got hurt in that group was me. Afterwards I got really depressed and afraid. I went to the first meeting, which lasted 10 hours, and never came back. I was really scared about what I went through. The leader told me I was stupid and I didn't know how to participate. I felt really hurt and alienated from everybody in the group. The other members, especially the girls, attacked me. The leader attacked me but at the same time I felt warmth from him too, and that made him even more scary. One minute he was supporting me, the next he was attacking. I was an outsider, I was treated as the lowest thing on earth, my opinions were not valid, I was a toy being played with. For months, I'd been trying to build up some kind of identity, and I lost it all. [This student was identified as a severe casualty. He went into a long-term, chronic depression as a result of the group experience.]

. . . The group was really important for me. I said a lot of things there that I've never been able to say to anybody else before. I talked about a lot of things that I was ashamed about, and found that other people had the same kinds of feelings. I talked about my feeling that I was a fake, that I was really stupid, pretending to be smart, competent, and that whole adequate game. I found out that almost everybody in the group felt the same way . . .

. . . The thing that bugged me most about the group was the big hypocritical game that everybody played. All the guys were on the make and it looked like a lot of the girls were there for the same thing—and yet nobody really talked about that . . .

. . . I guess the thing that really helped me was the feedback. I really thought I was ugly and unattractive, and all the little successes I'd had before were never enough. This time when I spilled my guts, everybody went around and told me what they thought of me, and most of it was positive. I really felt on cloud nine . . .

. . . I was just totally stupefied . . . because the whole focus is on negative orientation. You don't get any reinforcement at all, and for a lot of the kids and for me, it was just a totally unreal experience. I was totally off balance for a couple or three weeks. I didn't know how to react to it . . .

. . . One guy in our group said he'd been a sissy all his life—he never had guts enough to fight anyone or approach a girl or anything. So the leader made him fight with a couple of guys there and he even got a bloody nose. Then the guy had to go up to every girl in the room and ask her whether he turned her on, and ask her why or why not, and what he should do differently. The whole thing was kind of scary and I sat there thanking God it wasn't me who had to do it. I don't know whether it helped him or not. He says it did . . .

continued from page 417

Research—but not enough

The T-group has been studied extensively because its founders were behavioral scientists who early established a tradition of research. The current "anything goes" groups, however, are for the most part unresearched. There are several reasons why: many groups are informal, records are few, and long-term follow-up has been almost impossible. Perhaps more basic, however, is the attitude of leaders and participants. They have opposed systematic investigation because they believe it reduces emotions to numbers and therefore is contrary to their concern for human needs. Most of the information about the newer groups, therefore, has come from anecdotal accounts by single participants instead of systematic investigation of entire groups. (*See* "I don't know whether it helped . . . ," pages 418–419.)

There is only one research study to date that offers reliable data about the effects of the newer individual-oriented groups. In this study, 210 university undergraduates took part in a group for 2–10 sessions totaling 30 hours; about 15% dropped out before the last session. The style of the leaders ranged from a nondirective and group-oriented approach to a very aggressive, confrontive, and individual-oriented approach.

At the end of the study the overwhelming majority of participants found their experience pleasant and constructive. Three-quarters felt the group had changed them for the better, and 75% of these expected the changes to last. Six months later, however, the participants viewed the encounter differently: far fewer rated it positively and the number rating it as poor increased significantly.

The outcome of each group depended a great deal upon the leader and his approach. Nevertheless, some effects were common to all groups. Self-esteem rose significantly compared with that in a matched sample of nonparticipants (the control population), and in most groups it continued high six months later. The students felt they understood their inner feelings better, were less work-oriented and more permissive, valued new experiences more, valued the expression of anger more, and reported that they had more friends and spent more time with them. On the other hand, life values, the ability to make major decisions, and mental health scales were not influenced. And yet these are changes that are often listed as goals of encounter groups. Friends perceived fewer positive changes in the participants than did the participants themselves or, for that matter, friends of the control population.

Danger for some

Some critics have attacked encounter groups as dangerous, but they have based their charges on anecdotal information because good numerical data were lacking. Participants have become psychotic or committed suicide after group sessions, but so do people who never

attend encounter groups. Advocates of the movement, however, have often tended to ignore possible adverse effects, or have not been aware of them.

The only systematic data come from the university study just cited. At least 16 of the 210 students had real emotional difficulties afterward that were a result of their group experience, as documented by extensive personal interviews. For a student to be considered a casualty, he had to have a severe, long-lasting reaction that could be traced to the group encounter. Some participants were severely disturbed and suffered nervous breakdowns. Others were depressed, lost a great deal of self-esteem, became discouraged about ever changing, and were more isolated from other people than before being in a group. We should also stress that there may be other problems besides frank psychiatric disturbance. Some participants go away from a group so enthusiastic that they use encounter techniques at the wrong time or place, and thereby sometimes jeopardize their business careers or social relationships.

As yet there is no completely reliable way of predicting whether someone will have a "bad trip" in an encounter group. So far it appears that the type of group leader is the best advance indicator. Those groups in the university study with aggressive, intrusive, charismatic leaders who confronted individuals directly and relentlessly had casualty rates significantly higher than groups whose leaders were more supportive and focused on group dynamics. Another indicator of possible trouble is an individual's motivation for joining a group. Some of those with adverse reactions had serious personality difficulties that really justified psychotherapy.

By now it should be clear that one needs to choose an encounter group leader with great care. At the least a leader should have some basic clinical training; some work in the behavioral sciences, such as psychology, organizational behavior, psychiatry, or sociology; and should also have had supervised experience in leading groups. He or she need not be a behavioral scientist, however. Many good leaders have had careers in education, religion, or personnel work, and have taken specialized training to learn about encounter methods.

What a leader requires is appreciation of the forces that work in groups; sensitivity to others' thoughts, feelings, and needs, even when they are incorrectly expressed; the ability to help others use feedback constructively; knowledge of how to resolve conflicts in groups and between individuals; and awareness of his own psychological needs and blind spots, so he does not exploit the group to serve his personal needs. If one is concerned that a certain group might be hazardous, the best thing to do is confer with a professional in the field of mental health.

It is of the utmost importance that participants in encounter groups give informed consent to their participation. They should be fully aware of what will be expected of them, and they should attend on a voluntary basis. Some organizations have made the mistake of ordering employees to attend groups, a practice that may lead to a highly undesirable invasion of privacy.

Encounter groups are like other human institutions—schools, churches, business, or whatever. One can overestimate them and be disappointed or hurt by them. But if one understands their limitations, their weaknesses as well as their strengths, one can gain a great deal from them. An honest encounter between human beings can be a profound, enriching experience. An encounter group can liberate and exhilarate healthy individuals capable of facing the stresses of self-exposure and criticism. It can offer a meaningful education in group or individual dynamics—under the proper circumstances. Abused or uncontrolled, however, it can be, like any other human endeavor, devastating.

FOR ADDITIONAL READING:

American Psychiatric Association, "Encounter Groups and Psychiatry," Task Force Report 1 (Washington, D.C., April 1970).

Blank, Leonard, Gottsegen, Gloria B., and Gottsegen, Monroe G. (eds.), *Encounter: Confrontations in Self and Interpersonal Awareness* (Macmillan, 1971).

Bradford, Leland P., Gibb, J. R., and Benne, Kenneth D. (eds.), *T-Group Theory and Laboratory Method: Innovation in Re-education* (John Wiley and Sons, 1964).

"Human Potential: The Revolution in Feeling," *Time* (Nov. 9, 1970, pp. 54–58).

Lieberman, Morton A., Yalom, Irvin D., and Miles, Matthew, *Encounter Groups: First Facts* (in press).

Luft, Joseph, *Group Processes: An Introduction to Group Dynamics* (National Press, 1963).

Rogers, Carl, *Carl Rogers on Encounter Groups* (Harper & Row, 1970).

Schein, Edgar H., and Bennis, Warren G., *Personal and Organizational Change Through Group Methods: The Laboratory Approach* (John Wiley and Sons, 1965).

Yalom, Irvin D., *The Theory and Practice of Group Psychotherapy*, ch. 14, pp. 340–373 (Basic Books, 1970).

An Issue Essay

The UN: Its Science Mission

by Walter M. Kotschnig

The UN was created to help prevent war and to promote economic and social gains for all mankind. Increasingly, it looks to science to achieve these peaceful goals.

To most people, the United Nations is intimately connected with the world of high politics. To "go to the UN" is to visit the huge glass building on the East River in New York, to hear matters of peace and war being debated in the Security Council or the General Assembly. The organization's usefulness is judged by its relevance to the latest world crisis.

But, though peacekeeping was a prime object of the UN's founders and though the organization's most visible activities are directed toward that end, there is another central objective enshrined in the UN Charter: "to promote social progress and better standards of life in larger freedom," and "to employ international machinery for the promotion of the economic and social advancement of all peoples." This objective was written into the Charter, on the initiative of the United States, "with a view to the creation of conditions of stability and well-being which are necessary for peaceful and friendly relations among nations. . . ."

Thus, the UN is charged not only with keeping the peace, but with building it, and for this purpose the UN in New York is only the center of a network of a dozen international agencies, organically related to it by special agreements, with headquarters in Geneva, Paris, Rome, London, Vienna, Montreal, and Washington, D.C. The very names of several of these organizations denote their technical character and their concern with scientific progress and the use of advanced technologies. The Food and Agriculture Organization (FAO), for example, seeks to improve nutrition, increase agricultural production, and better the condition of rural populations by using the latest scientific findings in these fields. The World Health Organization (WHO) is dedicated to "the attainment by all peoples of the highest possible level of health." The World Meteorological Organization (WMO) aims to enlarge man's understanding of the weather through international cooperation in the atmospheric sciences. The promotion of scientific cooperation is one of the prime objectives of the United Nations Educational, Scientific and Cultural Organization (UNESCO). Also included among these organizations are the International Civil Aviation Organization (ICAO) and the International Telecommunication Union (ITU). In recent years the UN General Assembly has created, in addition, the United Nations Conference on Trade and Development (UNCTAD) and the UN Industrial Development Organization (UNIDO).

The ramifications of these agencies are so far-reaching that it is impossible to do justice to the subject within the compass of a short article. If, however, there are threads that tie many of their activities together, they lie in the recognition of the essential role of science and technology in achieving worldwide economic and social progress and in a concern with the unfortunate side effects of technology. To understand these emphases, it is necessary to consider briefly the changes that have taken place in the UN during its 26 years of existence.

Development: a new priority

The UN of today is radically different from the institution created in 1945. From an original membership of 51 mostly Western-oriented countries, it has grown to a total of 127, almost 90 of which are less developed countries, characterized in many instances by prescientific methods of production, low levels of living, rapid growth of population, massive illiteracy, disease, and premature death. This membership explosion made the UN and its specialized agencies worldwide in scope and has brought about a reordering of their priorities.

Under the Charter, it is the task of the General Assembly and the Economic and Social Council to coordinate, by way of review and recommendations, the policies and activities of the specialized agencies. In the early days of the UN, this meant primarily the avoidance of overlaps and duplication of effort, the exchange of information, and the establishment of statistical and administrative arrangements beneficial to all the agencies concerned. Some of the agencies stressed their "autonomy" and tended to go it alone. They achieved remarkable results within their areas of competence, but the lack of cooperation between them tended to bring about imbalances of socioeconomic development in the countries where they were active.

As basic changes took place in the priorities and modes of operation of the UN, the emphasis shifted toward concerted planning and common action. A degree of cooperation and interdisciplinary action has now been reached that warrants the present practice of speaking of the United Nations as a system of organizations. When Pres. John F. Kennedy, in 1961, invited the General Assembly to designate the 1960s as the UN Development Decade, he spoke to the condition of this new UN in the making. The acceptance of his proposal not only reflected the needs and aspirations of the less developed countries but also gave a clear focus to the economic and social activities of the UN and established development as the overriding objective. High on the list of priorities were surveys of natural resources with a view to their exploitation; modernization of agriculture; power development and industrialization; better transport and communications; literacy and

WALTER M. KOTSCHNIG was formerly deputy assistant secretary of state in the Bureau of International Organization Affairs, U.S. Department of State. In this capacity he was closely associated with the UN and many of its agencies.

Courtesy, United Nations

Experiments in raising irrigated crops during the dry season are being conducted at the Battambang Experimental and Demonstration Farm on the Battambang River in Cambodia. After a crop has been harvested, field workers burn the waste and prepare the ground for the next crop. The UNDP and the FAO assist the Cambodian government in the farm's operation.

trained manpower; improved management—anything, in a word, that would help bridge the economic and social gap between the less developed countries of the South and the industrially advanced countries of the North.

The response of the UN organizations, while far below the expectations of the less developed countries, was spectacular. Earlier technical assistance programs, initiated on a modest scale in the 1950s and before, assumed new dimensions. Many thousands of projects and programs were undertaken in Asia, Africa, and Latin America, including some in countries that were not and are not beneficiaries of bilateral aid. Tens of thousands of experts spread to all corners of the less developed world. Contributions to earlier technical assistance funds, which in 1966 were combined in the United Nations Development Program (UNDP), rose from $72.4 million in 1960 to $226.1 million in 1970. The World Bank and its affiliates, which are also part of the UN system, increased their loans to the less developed countries from $659 million in 1960 to $2,286,000,000 in 1970.

Transfer of the requisite knowledge and technology to the less developed lands was the central issue at the United Nations Conference on the Application of Science and Technology for the Benefit of the Less Developed Areas, held in 1963. It was followed by the establishment of an Advisory Committee on the Application of Science and Technology to Development (ACAST), composed of leading scientists and development experts. ACAST is about to complete a detailed World Plan of Action that focuses on scientific and technological advances most likely to be applicable to economic and social development. The plan also is concerned with building up in less developed countries the institutional structure needed to enable them to make the best use of science and technology. This is an area in which UNESCO, in cooperation with other UN organizations, the International Council of Scientific Unions, and other bodies is doing pioneering work, including the promotion of science teaching at all levels and of national science councils and associations, information centers, and translations.

On the road to achievement

The results of all these development efforts are difficult if not impossible to measure. Although the UN system, at this time, is the major multilateral source of technical assistance to the less developed countries, it shares due credit with bilateral aid programs and the signal contributions made by national foundations and the private sector. In general terms, it is probably safe to say that the UN has been the single most important factor in awakening even the least developed countries to the role that science and technology can play in their future.

Progress can be more clearly attributed to the UN system of organizations in some specific sectors of development. For example, FAO, supported by UNDP, played an important role in the Green Revolution by encouraging use of the high-yielding varieties of rice, wheat, and corn (maize) that had been developed under the auspices of private foundations at the International Rice Research Institute in the Philippines and at the International Maize and Wheat Improvement Center in Mexico. These "miracle" grains, with yields two or three times as high as the older varieties, require irrigation and careful water control, fertilizers, pesticides, insecticides, and farmers able to use them. All these are areas in which UN organizations are actively cooperating with national governments and the private sector. In Asia alone, the estimated acreage planted with the high-yielding varieties rose from 200 in 1964–65 to 20 million in 1969. The Philippines and Pakistan are becoming self-sufficient in grains, while other countries such as Malaysia and Indonesia expect to achieve self-sufficiency in the early 1970s. ACAST's World Plan of Action envisages continued highly organized, multidisciplinary research, aimed toward achieving similar breakthroughs in winter and durum wheats, barley, millet, and sorghum, and the adaptation of new varieties to different soils and climates.

A related problem concerns protein deficiencies, which affect one-third of the population of the less developed countries. Among the victims are children, who may suffer serious impairment of health and growth and, in many cases, permanent mental retardation. Not only is this a human tragedy, but the resulting loss of productivity may seriously interfere with development. Taking advantage of advanced research, FAO, WHO, and the United Nations Children's Fund (UNICEF) have cooperated in an effort to encourage the cultivation of protein-rich food, the increase of available animal protein, the enrichment of food with synthetic amino acids, increased use of the food resources of the sea, the development of nonconventional sources of protein, and tests of their acceptability.

Another example of the application of science by a UN agency is WHO's antimalaria campaign. As a result of this effort, 78% of the world's population was protected against malaria in 1968, as compared with 65% in 1959. Further, by the end of 1969, three years after WHO began its smallpox eradication campaign, the worldwide incidence of that disease had declined almost 69%. It is expected that by 1971 endemic foci of smallpox will be restricted to East Africa and Southeast Asia, and complete eradication may be achieved by 1976. The fight against trypanosomiasis, schistosomiasis, and tropical virus diseases, as well as the general question of vector control (including the problems raised by the development of resistance to residual insecticides), are high on the agenda of WHO, in terms of both research and application.

The specter of overpopulation

Unfortunately, population growth in the less developed countries threatens to defeat the best efforts to achieve higher levels of living. To quote a recent report of the World Bank: "It is clear that there can be no serious social and economic planning unless the ominous implications of uncontrolled population growth are understood and acted upon." Thus, in spite of the remarkable absolute increases in food

M. Jacot, courtesy, WHO

A WHO nurse in Niger (above) explains how to make a balanced diet from a millet gruel, while a child in Guatemala (below) waits for milk donated by UNICEF.

Courtesy, United Nations

Courtesy, United Nations

A medical assistant in Cameroon (above) tests blood for malaria, while a worker in Ghana (below) sprays a dwelling to destroy mosquitoes. The WHO and UNICEF assist both antimalarial programs.

production, per capita food production in Africa and the Far East declined between 1961 and 1967. In countries with annual per capita incomes of less than $150, annual growth figures of up to 5% of gross national product frequently have meant only an additional two or three dollars a year in the pockets of hundreds of millions of people. Sharp rises in population absorbed the rest.

The population explosion was triggered by falling death rates. As we have seen, medical science and modern technology have drastically lowered death rates in the less developed countries. On the other hand, birthrates in these areas, which are more than double those in the developed countries, have remained more or less stable, and little has been accomplished to reduce them. For example, in 1946 WHO inaugurated a malaria-eradication program in Ceylon, along with other health measures. This campaign brought about an increase in life expectancy from 43 years in 1946 to 54 years in 1948 and to 62 years in 1960. Yet when Ceylon, plagued by food shortages and unemployment, asked WHO in 1949 to consider a family-planning program on a worldwide scale, no action was taken.

Throughout most of the 1960s neither WHO nor the UN found it possible to initiate an effective program of population control. In both organizations strong opposition was voiced, for religious and other reasons, by most of the Latin-American countries and by Catholic countries in Europe and Africa. The Soviet Union and some of the Eastern European countries expounded the traditional Communist position that Western discussions of the population problem were based on "neo-Malthusian fallacies." Some of the less developed countries argued that family planning was a form of genocide designed to prevent the poorer countries from achieving power and wealth.

This situation is beginning to change. In the last three years, a growing number of requests for aid in setting up family-planning programs have come from less developed countries, and the U.S.-proposed UN Fund for Population Activities, under the administration of UNDP, has helped to finance such assistance. WHO has substantially extended its research on human fertility and reproduction and, together with the UN, is now committed to action programs in the field, including the provision of contraceptive devices. In 1971 six UN organizations (UN, UNICEF, FAO, UNESCO, and WHO, and the International Labour Organization) published a comprehensive report on "Human Fertility and National Development: A Challenge to Science and Technology." It lays out a comprehensive program for research and action in the 1970s.

Degradation of the environment

Another area of scientific interest that has emerged recently in the UN concerns the problems of the environment. As with population, the subject contains seeds of disagreement between advanced and less developed countries. In fact, environmental deterioration, to a considerable extent, is a result of rampant population growth. Beyond this, however, attempts to save the environment may well come into conflict with the drive toward economic development. Historically, at least, industrialization has been marked by a single-minded pursuit of growth and material satisfaction, ruthless exploitation of natural resources to the neglect of ecological consequences, and the use of technologies irrespective of their adverse by-products. To continue these patterns may well place such a burden on the world's ecology as to threaten human survival.

Sweden and the United States, supported by other industrialized countries, took the initiative in the General Assembly to involve the UN system in environmental research, planning, and action on a global basis. This is as it should be, since the highly developed countries are the major polluters, the prime victims of pollution, and, thanks to their scientific and technological capacity, the best able to cope with it. The less developed countries, however, tended to feel that pollution was not "their" problem, and some of them hotly contended that this new initiative was merely another attempt on the part of the rich countries to escape their obligation to assist the poorer ones in their development. Smokestacks were more important to them than pure water and air, and they needed DDT to control the locust pest, even if it affected the health of some of their people. The Soviet Union remained indifferent until the sturgeon in the lower Volga River and the Caspian Sea began to die from the discharges of hot water and the chemical effluents of power stations and industrial plants that had been built along the Volga and its tributaries in disregard of any ecological considerations.

Despite these clashes of concept and interest, the General Assembly, in December 1969, resolved that a worldwide UN Conference on the Human Environment should be convened in 1972. Preparations for this conference, to be held in Stockholm, are now under way. Consideration is to be given to all forms of environmental deterioration and to their man-made causes. Emphasis will be placed on plans for remedial and preventive action and on the initiation of international agreements required to cope with regional and global problems. Special attention will be paid to problems of defining, establishing, and applying appropriate standards, and the establishment of monitoring systems.

In Tunis, young eucalyptus trees are irrigated by the Reforestation Institute, which was established by the government and the UN Development Program. The trees are grown for use as windbreaks in low-rainfall areas.

F. Botts, courtesy, FAO

About the weather

Just as pollution knows no national boundaries, so weather—its prediction and possible control—is a subject of worldwide concern. Indeed, the two are closely tied together in the activities of the World Meteorological Organization. WMO's work is more deeply rooted in scientific research and the application of sophisticated technology than that of any other UN agency.

WMO's central program is the World Weather Watch (WWW), to which its 130 member states committed themselves in 1967. The purpose of the program is to improve weather services, including the provision of longer-range forecasts to all countries. It seeks to accom-

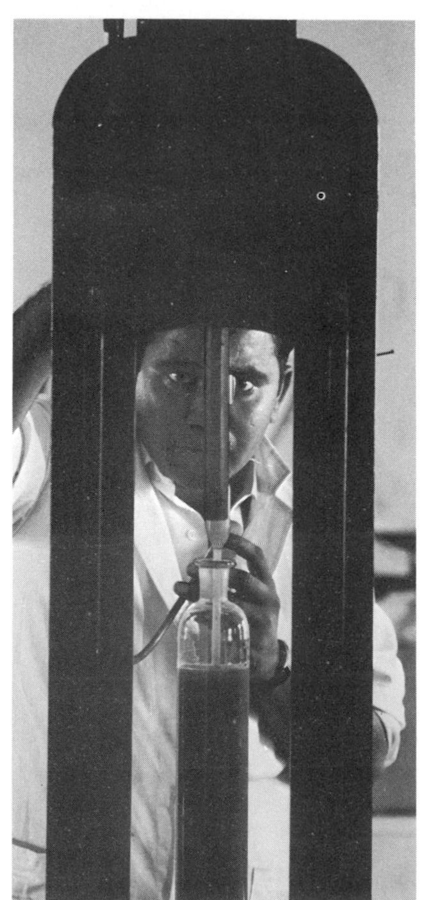

plish this goal by increasing the accuracy of weather observations through a better understanding of the properties of the atmosphere, of atmosphere and hydrosphere interaction, and of many other elements that influence the weather. All this calls for application of the latest scientific findings and use of the most sophisticated tools of the space age, including satellites for automatic picture transmission and instant communication. The basic research involved in the WWW is centered in the Global Atmospheric Research Program, which is jointly planned by WMO and the International Council of Scientific Unions. Most of the work is accomplished through the voluntary effort of members. (*See* A Gateway to the Future: CHALLENGING THE RESTLESS ATMOSPHERE.)

Closely related to the WWW are other WMO activities carried out in cooperation with other organizations, including studies of urban climate and an agricultural meteorology program to monitor and prepare reports on plant injury and reduction of yield by nonradioactive air pollutants. WMO has set up a Technical Commission for Atmospheric Sciences with a Working Group on Atmospheric Pollution and Atmospheric Chemistry. The latter has proposed the establishment of a global network to monitor low-concentration (background) pollution. It is also concerned with reviewing the techniques available to analyze and forecast dispersion of high-concentration pollution in cities and industrial areas.

Atoms for peace and development

Since the founding of the UN, its political debates have taken place in the shadow of nuclear weaponry, and much of its effort has been directed toward attempting to restrain the development and use of such devices. But the UN has also played a major role in promoting peaceful uses of atomic energy. Following Pres. Dwight D. Eisenhower's Atoms for Peace initiative in the General Assembly in 1953, the UN created the International Atomic Energy Agency (IAEA) in 1956 and convened three major conferences on the peaceful uses of atomic energy (1955, 1958, and 1962). These conferences made a signal contribution by fostering closer cooperation among the scientists of the United States, the Soviet Union, and other scientifically advanced countries. They helped to achieve a focus on peaceful uses most likely to benefit mankind.

According to its statute, IAEA seeks to accelerate and enlarge the contribution of atomic energy to peace, health, and prosperity throughout the world, while attempting to ensure, under a system of mutually agreed "safeguards," that its assistance is not used to further any military purpose. It cooperates with such other UN organizations as FAO, UNESCO, WHO, and the UNESCO-sponsored European Organization for Nuclear Research in promoting peaceful uses of nuclear power. It also assists individual countries in the development of nuclear reactors, other applications of nuclear science and technology, and in training needed personnel. In both industrially advanced and less developed countries, IAEA has taken the lead or participated in such projects as the establishment of "dual purpose" nuclear power-desalinization plants; the use of radioactive tracers to determine the optimum methods of fertilizer use in rice and corn production; the application of radioisotopes to medical purposes and the protection of food against spoilage; and the "sterile male" technique for the eradication of such insect pests as the Mediterranean fruit fly and the tsetse fly, the carrier of African sleeping sickness.

Wider applications of technology

Space technology is another area in which military and quasi-military applications by the major powers have tended to obscure the possible benefits to mankind in general and to the less developed nations in particular. We have already noted the use of artificial satellites in weather reporting. Advanced space technology also promises to revolutionize surveys of the natural resources of the earth. Multispectral photographic and remote-sensing instruments can be installed in satellites or aircraft. Enough is known for us to foresee the use of such tools for national, regional, and global surveys of geologic formations and mineral resources, groundwater, streams and river basins, vegetation, agricultural and forestry resources, areas of plant diseases, and the migration of fish. Data thus obtained are, of course, of the utmost importance to development planning and management.

In 1969, Pres. Richard M. Nixon, in his first appearance before the General Assembly, committed the United States to sharing, through the UN, the benefits from the earth resources satellites being developed in the United States, the first of which is to be launched in 1972. Considering its preoccupation with development, the UN is almost certain to play a major role in facilitating worldwide earth resources surveys and in the establishment of comprehensive inventories and the evaluation of natural resources of entire regions and continents. In addition, as a global organization of governments, the UN, over time, should be able to modify the traditional concept of unrestrained sovereignty, which conflicts with the most effective utilization of the new space technology.

Problems of national sovereignty also loom large in plans and programs for the exploration and exploitation of the seas and oceans. The 1960s witnessed a great increase in intergovernmental and scientific interest in the oceans and in the search for additional food and mineral resources from the sea. Responding to this, the Intergovernmental Oceanographic Commission, under the aegis of UNESCO, formulated on request of the General Assembly a long-term, expanded program of oceanic research providing for the use of new instruments, borrowed in part from space technology. However, conflicting claims to sovereignty on the part of coastal states threatened to limit the exploration of vast areas of the oceans, impede their effective exploitation, and interfere with scientific research. It was not until 1970 that the General Assembly reached agreement on a Declaration of Principles, solemnly

Courtesy, United Nations

To help solve social problems in Pakistan, a teacher (above) conducts a reading class with materials supplied by UNICEF. At the Bhabha Atomic Research Center (opposite page, top) in Trombay, India, the IAEA is assisting in various scientific studies, as is the UNDP in soil studies being conducted at the Central Water and Power Research Station near Poona, India (opposite page, bottom).

Courtesy, United Nations

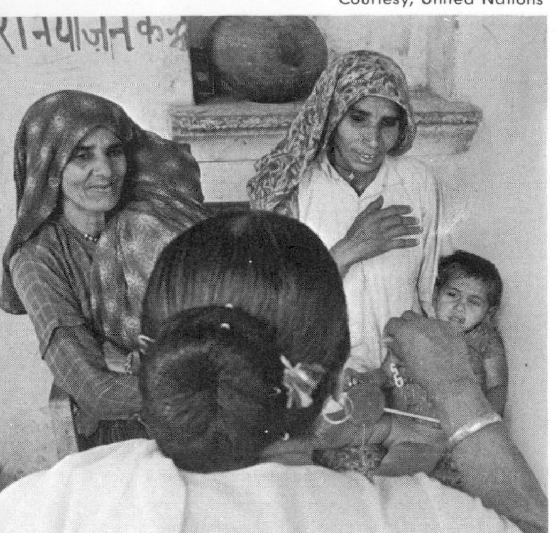

(Above) In a family-planning program recommended by UN experts, a fieldworker in Deoli, India, explains the use of contraceptive devices. (Below) The FAO is helping direct the construction of an open percolation well in Bihar, India.

Courtesy, FAO

affirming, among other things, that the seabed and ocean floor beyond the limits of national jurisdiction "are the common heritage of mankind."

A comprehensive Conference on the Law of the Sea has been called for 1973 to define this area and to establish an "equitable" international regime for the area and its resources. It will deal with such matters as the regimes of the high seas, the continental shelf, the territorial sea and contiguous zone, fishing and the conservation of the living resources of the high seas, the preservation of the marine environment (including the prevention of pollution), and scientific research. This conference will put the UN to a supreme test. If successful, it will vastly strengthen the fabric of international cooperation and assure the international community, for the first time in history, of a substantial, independent source of income to be used in the cause of development and peace.

This record is by no means complete. Lack of space prohibits more than brief mention of the projected use of satellites for educational purposes in India and several Latin-American countries (UNESCO); the studies and regulatory activities of the International Civil Aviation Organization related to the abatement of noise produced by aircraft, and its panel on the sonic boom problem; or the many-faceted efforts to build defenses against natural disasters through the establishment of early warning systems, which, in turn, depend on continuing studies and observation of such hazards as hurricanes, typhoons, and tsunamis (seismic sea waves). The International Telecommunication Union, at its 1971 World Administrative Radio Conference, considered the allocation of new frequency bands to satellite service. One of the least known achievements of the UN is a small laboratory in Geneva that has developed the capacity to determine the origin of raw opium and cannabis wherever grown, thus becoming an important link in the fight against drug abuse.

Future promise

Clearly, there is a broad recognition within the UN system of the present and potential contributions of science and science-based technology to the achievement of UN objectives. There is less awareness of the fact that advanced technology is extending the scope of these objectives, creating new objectives, and providing the means for their attainment. Attention is shifting to worldwide programs that call for global action and new approaches, and the UN, as the only intergovernmental organization of worldwide character, is best suited for these purposes.

As has been shown, we are moving toward the establishment of global monitoring systems to collect continuously data needed for meteorological purposes; to ascertain the incidence of worldwide pollution; to provide a global inventory of earth resources; and to obtain information in many other areas of common concern. What is needed now is international agreement on these and other aspects of the contemporary world that deserve synoptic observation and, as monitoring programs are developed, the means to carry them out systematically and to communicate the results in a form useful to the international community. By strengthening its working relationships with the scientific community, the UN is in a good position to achieve these objectives.

The UN is also well qualified for action in such related areas as standard-setting and the establishment of legally binding instruments having to do with scientific matters. These are needed for effective monitoring as well as for remedial or innovative action. Most multilateral legal instruments of the last 25 years have been formulated and are administered by UN organizations.

Even more important is another development, now only dimly perceived. As the world recognizes that life on this earth may be saved only by concerted international action, basic changes in political concepts and attitudes may reasonably be expected. There are limited but definite signs that in this new era of "unlimited opportunities" provided by science and technological progress, politics may also move into the modern age. The proposed Law of the Sea conference may be a portent of a new UN capable of resolving basic political and juridical issues.

In these instances, the opportunities—and the perils—of science and technology to some degree shape the UN. But the possibility also exists for the UN to guide science and technology. There are those who insist that science needs no guidance since it has only positive social value, but this notion is challenged by many who cannot be accused of antiintellectualism. Nor is it universally supported by the scientific community itself, as is clear from the series of articles published in earlier issues of the *Britannica Yearbook of Science and the Future* by such contributors as René Dubos, Philip Handler, Christopher Wright, and Detlev Bronk. While these writers urge freedom of inquiry as essential to science, several of them also suggest a need for "social mechanisms" or "institutions" to provide guidance and controls if the potentials of modern science are to be used for the benefit of mankind.

The UN clearly has the potential for becoming such an institution, and an important one. It includes most governments and their peoples, irrespective of race, religion, level of development, socioeconomic system, or ideology. It is moving pragmatically toward a worldwide consensus on desirable objectives and the best ways of achieving them, and it is beginning to reach agreement on the optimum use of science and technology and on avoidance of their perils.

The UN is far from perfect. Its normative functions and its potential for action continue to be limited by political disagreements, antiquated notions of sovereignty, or simply human cussedness. It is marked by future promise rather than present achievement. That promise will not be fulfilled without better knowledge and understanding in depth of the scope and role of the UN and its relevance to contemporary man and to generations yet unborn.

Per Gunvall

UN technical assistance programs for less developed countries include a UNESCO-sponsored course in handicraft skills conducted in a small village in the Rufiji district of Tanzania.

Index

Index entries to feature and review articles in this and previous editions of the *Britannica Yearbook of Science and the Future* are set in boldface type, *e.g.*, **Astronomy.** Entries to other subjects are set in lightface type, *e.g.*, Radiation. Additional information on any of these subjects is identified with a subheading and indented under the entry heading. The numbers following headings and subheadings indicate the year (boldface) of the edition and the page number (lightface) on which the information appears.

Astronomy 72–187; **71**–133; **70**–119
  Colonizing the Moon **72**–12
  detection of gravity waves **70**–382
  honors **72**–260; **71**–205; **70**–260
  Mount Wilson and Palomar Observatories **70**–356
  photography
    stars' trajectory **71**–367
  Soviet Union **70**–418
  spacecraft navigation **70**–184
  space probe research **71**–129

All entry headings, whether consisting of a single word or more, are treated for the purpose of alphabetization as single complete headings and are alphabetized letter by letter up to the punctuation. The abbreviation "il." indicates an illustration.

A

B

F

G

H

I

J

K

L

M

N

O

P

Q

R

T

Acknowledgments

4	Illustration by (top, left) Ben Kozak; photographs by (top, right) Dan Morrill; (center) Richard Keane; (bottom) courtesy, National Center of Atmospheric Research
10–11	Illustration by Bill Chambers
12–33	Illustrations by Ben Kozak
34–35	Photographed by Bill Arsenault
50–63	Illustrations by Jan Wills
64–75	Illustrations by George Suyeoka
76–87	Illustrations by John Everds
100–113	Photographed by Bill Arsenault
128–129	Photograph by Schaefer & Seawell from Black Star
128–141	Illustrations by George Suyeoka; photographs by Bill Arsenault
159	Photographs by (top) courtesy, Dr. Wilfred Roth, University of Vermont; (center, left) courtesy, Dr. J. W. Butcher, Michigan State University; (center, right) courtesy, Los Alamos Scientific Laboratory; (bottom) David I. Owen, with permission of Giuseppi Foti
175, 189, 221, 236, 269, 327, 334	Illustrations by Victor F. Seper, Jr.
344–345	Photographs by (left) Dan Morrill; (right) R. Blair from Pix
352–353	Photographs by (left) courtesy, Adolph Coors Company; (right) Ron Sherman from Nancy Palmer Photo Agency
366–367	Photographs by (left) courtesy, The Metropolitan Sanitary District of Greater Chicago; (right) Bill Arsenault